I0056223

Power Distribution Network Design Methodologies

Executive Editor
István Novák, PhD
Distinguished Engineer, Signal and Power Integrity

Power Distribution Network Design Methodologies

Also by István Novák (with Jason R. Miller)
Frequency-Domain Characterization of Power Distribution Networks

ISBN 978-1-949267-67-9

Cover Design by Guy D. Corp, www.GrafixCorp.com

Faraday Press—An Imprint of Stairway Press
1000 West Apache Trail—Suite 126
Apache Junction, AZ 85120 USA

Foreword

MOST OF WE practitioners in the field of Power Delivery Networks consider István Novák as the father of power integrity and the leading expert of our time. For me, István is a mentor, and he is also a friend, and thus, I am honored to write this foreword for the republication this book.

Power Distribution Network Design Methodologies presents a good balance of theoretical and experimental knowledge, encompassing all stages of a typical power distribution network, including the VRM and the bulk/decoupling capacitors. From this point of view, this book is unique. István served as the Executive Editor, while incorporating his own work and the knowledge of dozens of leading industry experts, each contributing key perspectives to the effort.

The underlying physics of Power Integrity has not changed in the 13 years since this book was first published, though many of the implementation details have. The book remains relevant, though it is important to look at what progressed in this period, in part because of this work.

One thing that changed is that power levels have increased significantly with ASIC's now operating at peak power of up to about 1.5kW with power rail voltages now at 0.7V or lower. There have been other changes as well. Edge speeds have increased—as have signal path frequencies. Solution sizes have decreased with increasing power densities. These updates magnify Power Integrity challenges but also brought much more awareness to the topic of Power Integrity. In part, this is due to the awareness that many Signal Integrity and EMI issues result from poor Power Integrity, so the system really needs to be evaluated holistically, end-to-end.

Another change is that many of the measurement techniques described in this book are now well understood, if not mainstream. Companies like Picotest, founded a few years after the original publication date of this book, focused on many of these Power Integrity measurement challenges. A multitude of 50 Ohm 1-port and 2-port probes, ground shield isolators and specialized cables for PDN impedance measurement are now readily available from Picotest and other companies. There have been many papers, application notes and webinars on the details of performing measurements below 100μOhms, so engineers are more knowledgeable on the subject. One result of this increased awareness is that the questions are now more specific. For power engineers, lower impedance magnitudes create more difficult challenges to overcome.

Not all changes since this book was originally published have been for the better—some were for the worse while other challenges remain unchanged.

Many fewer ESR controlled ceramic capacitor options exist, while ESR being too low is still a common issue. Digital controllers and higher-frequency GaN devices are more prevalent to shrink the size of the voltage regulator and/or to reduce switching related noise. More semiconductor manufacturers offer simulation models, though many are as inaccurate as ever and many of the manufacturer's recommendations are still poor—or in many cases worse.

This all means this book remains an important body of work. Rather than focusing on specific details, it presents the significant elements and their sensitivities in relation to Power Integrity. The difficulty of intelligent decision making and design trade-offs remain. The measurement challenges, differences between self-impedance and transfer impedance, management of Q and flat impedance, proper selection and location of capacitors and the stability of the voltage regulator are as relevant today as they were when this book was originally published in 2008.

I encourage every SI, PI and EMI engineer to read this book. It will improve their awareness of the ramifications of a poor power distribution network design. It will also provide insight into the sensitivities of the elements, allowing a path to troubleshooting deficiencies and optimization of the power distribution network.

For those that have the ability, learn the suggested measurement techniques and try them. You will learn where the manufacturers data often falls short. Not to say that their data is always incorrect, but the details of the measurement, where the reference planes were set and how the calibrations were performed are often missing. One example I often write about is the ESL of the ceramic capacitors. This is predominantly a function of the ship size much more than the manufacturer. Yet routinely, different manufacturers present very different values for the same size chip. Independent measurements confirm these devices, from different manufacturers, present quite similar inductances when measured and calibrated in the same way. An added complication is that often that—independent of the method of measurement—none of the data is correct for use with the simulator. SPICE simulation requires one model while EM simulators require a different model.

In the years since this book was originally published, we focused on these details. While we know more now than we did in 2008 and many of these issues are understood and test equipment is readily available, the bar has moved. Welcome to the wonderful, ever-changing world of Power Integrity.

Thank you to István and all the contributors to this work. It was a valuable reference in 2008, it is a valuable reference now and it will be a valuable reference for years to come.

—Steve Sandler

Publisher's Note

THIS BOOK IS an updated reprint of the historic IEC publication originally released in 2008. We are proud to bring this book back to print and are confident in its value and utility.

In the years since its original publication, technology has marched on, but the enclosed content still has value for the working engineer.

First of all, the principles of physics do not go obsolete or out of fashion, so all coverage of time-and frequency-domain trade-offs and RLC component/parasitic interaction will be valid forever.

Secondly, some of the technology covered in this book—like DDR, CPU and server platform generations and certain component families—are obsolete at this point, but it never hurts to understand the history of the state of our rapidly and continually advancing technology.

This book pays tribute to the researchers, engineers and technologists who discovered and developed extraordinary ideas and practical solutions to complex problems—many of whom are still practicing today.

To all practitioners in the fields of power and signal integrity—often laboring in obscurity—who immeasurably contribute to the comfort, safety and joy of modern life, we see you and we appreciate you. Please keep doing your great work.

As always, published reviews and commentary are welcome. If you find an error or have additional information to share, we would love to hear about it.

—Ken Coffman, Ken@StairwayPress.com

Preface

A BOOK ON power distribution design methodologies was long overdue. At the time of this book's original publication, one might have asked the question: with all the great books on signal integrity, why have power-integrity books not yet appeared on the market? The answer was multi-fold.

To understand the reasons why power-integrity books are just now becoming available, we have to put the topic into perspective and look at related fields, which need to be addressed in a book.

Power integrity (PI), as a principle, was relatively new, and follows the footsteps of the electromagnetic compatibility (EMC) and signal integrity (SI) principles. The first of these three principles to address the quality of electronic designs was EMC, which also encompasses electromagnetic interference (EMI). These topics emerged following the need to avoid unnecessary and harmful interaction between and among electronic circuits.

With the rapid progress of wireless transmissions during the first half of the twentieth century, EMC rules were created to primarily protect radio-broadcasting and radio-communication services. Legal limits were created for radio waves to keep interference under control. Soon, it was realized that interference may come not only from intentional radiations, but also from electronic circuits, which otherwise do not need to radiate to perform their intended functions. To address these issues, the generation and reception of electromagnetic waves, as well as the wave propagation through various media and boundaries, had to be studied. Since the proper understanding requires a good knowledge of both close-field and far-field waves and three-dimensional propagations as well as wave scattering, the EMC topic is the most demanding and most complex to analyze of these three principles.

Traditionally, EMC covers the interaction of circuits belonging to different systems, and except for especially demanding systems, there has been less need, until recently, to deal with the interference within the boundaries of a given computing system. There are many good books on EMC; one of my early favorites was [1]. Signal integrity, as a principle, gradually emerged from EMC during the last quarter of the twentieth century. With increasing circuit speeds, the rapidly falling rise and fall times of signals quickly became comparable to or shorter than the time-of-flight delay on passive interconnects, raising the need to study reflections, matching, and distributed interconnect behavior. The ever-increasing densities called for the understanding of cross-talk.

Signal-integrity simulations started to become common but, for some time, the simulations and analyses assumed that the noise on the supply rails was negligible. Over time a lot of good books emerged on SI as well. Two of my early favorites are [2] and [3], although [2] did not yet mention signal integrity in its title.

Each SI book devotes a chapter or two to power distribution since, even if we stay strictly with signal quality, it has been both understood and accepted that clean power is eventually the key to maintain clean signals.

From an analytical perspective, SI is a relatively simple principle, since we deal mostly with conducted signals, where—as opposed to the three-dimensional propagation of EMC waves—the propagation path is dictated by the known geometries of our interconnects. One major "complication" is that in planar printed circuit board (PCB) interconnects, the return signal on the planes can spread out and follow the path of least impedance at any given frequency. Since planes in PCBs usually serve multiple purposes such as return path for the signals and power distribution interconnects; this aspect of SI already brings us closer to PI.

As circuit densities and speeds continued to rise, it became obvious that the most challenging designs required detailed analysis and a conscious design flow of the power distribution networks. The additional driving forces were the dropping noise tolerance on the supply rail and the proliferation of independent supply domains.

In the 1990s, there were many publications on the different aspects of power distribution, including plane resonances, target impedance calculations and synthesis, and simulation options of power grids. The increasing customer demand resulted in a large number of application notes from silicon vendors with various suggestions to provide acceptable power distribution for their chips. Other papers offered design suggestions to placement and connection geometry of bypass capacitors. In separate conferences and papers, the power sources (alternating current [AC]–direct current [DC] and DC–DC converters) were treated and analyzed.

While a lot of excellent papers emerged on these particular subtopics, there has been no visible activity to summarize the board-level PI principles in books.

In terms of complexity, PI problems lie somewhere between the 3D EMC and 1D SI problems: most PI calculations require the proper understanding and analysis of 2D current distribution on PCB planes and the components connecting to them. Except for some extreme cases, today, the vertical interconnects in PI analysis can mostly be modeled with lumped equivalent circuits, leading to 2D problems to solve.

In addition to just being recently recognized as a separate principle, PI is very broad and wide. While SI deals primarily with intentional signals—and therefore we should have a good knowledge about the signal shapes and levels—PI deals with the result of noise currents, which are very hard to measure, simulate, or get any data on. In addition to not always knowing the source of the noise accurately enough, more often than not the tolerable noise level is equally undetermined. As a result, we have to deal with a very large solution space, vague input data, and usually having multiple solution options for each problem.

In addition, some circuits or systems may be sensitive to instantaneous peak noise, and instantaneous glitches exceeding some levels may lead to catastrophic errors, while others may sense the "average" or root-mean square (RMS) noise and can tolerate occasional errors. Central processing unit (CPU), application-specific integrated-circuit (ASIC), and field-programmable gate array (FPGA) cores usually fall into the first category; the optimization for maximum computing throughput allows little or no resilience against instantaneous supply-rail glitches. High-speed serial interconnects where—due to the inherent delay of interconnects, latency matters less—error checking and correction as well as handshaking make these interconnects tolerant of a small level of statistical noise. In such applications, the average noise may be a better measure of power distribution network (PDN) performance.

By studying the numerous publications on various aspects of PI, one can find many different, sometimes contradictory, presentations and conclusions on the same topic. Placement and selection of bypass capacitors is probably the most notorious in this respect. Some papers just present different solutions to the same problem, which can be explained by the fact that in PI there are many independent knobs to turn to get the same result. Some other papers seem to offer outright contradictory advice, including where to place bypass capacitors. Assuming that all of the papers are correct, the reason behind the differences usually stem from the different assumptions (sometimes stated, sometimes implicit) or from using different metric.

A further coloring of the picture comes from the impact and interaction of PI with other design constraints, including geometry restrictions, overall space available, efficiency and cooling requirements, and, most important, cost. One could argue that, strictly speaking, none of these constraints are PI electrical requirements, and still they can have a large influence on how the PDN network is built up, resulting in seemingly contradictory solutions.

All of these mentioned reasons contributed to the lack of PI books. As a pioneer attempt, [4] was published a couple of years ago with limited distribution; an edited version of one of its chapters is included here as Chapter 5.

During recent years, sub-topic coverage of PI has started to emerge. There is at least one book currently available on PDN characterization and modeling—[5] from Artech House—and another one appeared late 2007 from Prentice Hall [6]. This book is an attempt to fill the void with a compendium of selected publications on PDN design procedures. The compendium format allowed me to bring together various, and sometimes conflicting, views in an organized way so the reader can put each contribution into perspective.

The contributions are not necessarily in chronological order, although the goal was to give an overview about the evolution of various approaches.

The selection of papers is necessarily subjective, and it reflects my own opinion. The goal was to represent each sub-topic with at least one of the experts in the particular area and to cover all major trends. Papers from various sources have been selected, including papers and conference articles from IEEE, papers from DesignCon, application notes, and book excerpts.

In the selection process, many good papers had to be necessarily left out; some of them are listed as references at the end of this chapter.

In the editing process, the guiding principles were to retain the original messages of the articles with only the necessary corrections of occasional minor errors in the original publications, sometimes shortening the paper, and in case of excerpts from longer works, by adding and/or deleting material to make the stand-alone article understandable. However, "smoothing" the individual contributions together was intentionally avoided. This also means that some minor overlaps and duplications were not removed either; they were, however, minimized by the selection of contributions.

A large amount of editorial work has been devoted to the figures, that is, since the contributions came from various sources with the constraints and requirements of the original publications, almost all of the graphs and figures have been at least partially redone so that they are readable in black and white (BW) printing and in the overall size that fits this book.

The book is organized around five parts.

Part 1 has four chapters about PDN components: two about DC-DC converters, one on bulk capacitors, and one on ceramic capacitors. Part 2 contains five chapters on various aspects of systematic synthesis of PDN impedance. The four chapters in Part 3 compare the PDN design methodologies along different requirements. Part 4 has four chapters on PDN filters, jitter, and simultaneous switching noise (SSN). Finally Part 5 shows a few selected aspects of PDN characterization.

The first chapter in Part 1 shows a case study around a common issue with switching DC sources: improperly selected output capacitors result in an un-damped feedback loop and ringing in the transient response.

Chapter 2 points toward the underlying reason and elaborates on additional potential traps and pitfalls around DC–DC converters: unspecified high-frequency ringing, too narrow loop bandwidth resulting in overly large bulk-capacitor requirement, and the challenge of designing and validating with non-linear digital loops.

The third and fourth chapters describe novel bulk and ceramic bypass-capacitor constructions with controlled ESR values, enabling the synthesis of smooth impedance profiles with guaranteed performance, as shown later in the fifth chapter in Part 2.

Part 2 starts with an excerpt from [4]. The article walks the reader through the main steps of creating a core power distribution system for given specifications. The main message of this piece is that the quality factor (Q) of each consecutive PDN stage has to be carefully selected and should be low. After describing the overall design steps, the other four articles describe in chronological order powerful and yet simple design methodologies to minimize PDN noise, starting at low frequencies and moving toward high frequencies.

Chapter 6 in Part 2 shows how to match a DC–DC converter output to bulk capacitors to minimize transient noise. Although not necessarily described in the frequency domain, the adaptive voltage positioning matches the DC output resistance of the converter to the mid-frequency impedance target, thus cutting the transient noise by half [7].

Chapter 7 in Part 2 introduces a systematic design approach to synthesize the mid-frequency target impedance by placing sequentially along the frequency axis the series resonances of different-valued bypass capacitors [8]. As a direct continuation of the adaptive voltage positioning method,

Chapter 8 in Part 2 shows how to minimize the transient noise not only at the interface of DC–DC converter and bulk capacitor, but also in the entire mid-frequency range. The extended adaptive voltage positioning matches the capacitor-resistor-inductor (C-R-L) values of bypass capacitors to create flat impedance profile [9].

Chapter 9 in Part 2 completes the frequency range by showing how to combine the previously introduced methods and to continue the impedance matching all the way to the PCB planes [10]. It also introduces the concept of bypass resistor, which is commercially realized by the components described in the last two chapters in Part 1.

To the careful reader, it will be obvious that each paper presents a given state of the art, that have tailored the examples and details to the state of technology at the time of writing. Although not included in this book, PCB laminates (regular or thin, as well as embedded capacitance) can also be pulled into the systematic PDN design process. Chapter 9 does that from a characteristic impedance point of view. To read further details on PCB laminates for PDN application, the reader is referred to, for instance, [11].

The chapters in Part 3 compare some of the widely used design methodologies. The works included here show the variety of opinions and also the differing assumptions and metrics.

Chapter 10 gives an overview of how PDN design requirements have changed over the years at a major computer company and what analysis techniques can be utilized to best support the design process. It also focuses on the return-path role of the PDN planes by elaborating on appropriate simulation models for this purpose.

Chapter 11 compares three popular synthesis methods for PDN target impedance: multi-pole, capacitor-by-decade, and big-V. By showing sensitivity, performance, and stress data, the paper gives the needed tool set for the designers to make the proper choice among the three listed methods. One should note that the paper remains mostly in the frequency domain in terms of performance comparison.

Chapter 12 in Part 3 compares the multi-pole, Big-V, and distributed matched bypassing (based on the fifth chapter in Part 2) by first establishing a simple metric. It ranks the design methodologies based on their worst-case transient noise requirements. It is shown that flat impedance response, in fact, minimizes the worst-case transient noise. Based on the impedance profile, a simple rule is deduced of when and how the locations of bypass capacitors may be important.

Chapter 13 compares the basic design methodologies from two standpoints: SI and EMI. The article is part of a series; the other articles cover the location of bypass capacitors [12], PCB laminates and materials

[13], and the various sources of PDN noise [14]. Papers from the author of [15] also provide a cross-discipline approach to PI.

There are a couple of common themes among the widely varying opinions: all of these works agree that inductance is a very important parameter for all bypass capacitors, not just for high-frequency multiplayer chip capacitors (MLCCs), and that flat impedance target should be the overall design goal.

Part 4 starts with an article on simultaneous-switching analysis of a double data rate (DDR2) memory system. The paper derives models for the power distribution network and the switching profiles alike [16], calculates the self-induced and transfer noise components, and shows correlation between simulated and measured noise waveforms.

Chapter 15 extends the analysis to obtaining supply-noise-induced jitter on multi-gigabit seralizer/deserializer (SerDes) interconnects. Multi-gigabit embedded-clock interconnects are characterized by bit-error rate (BER) rather than instantaneous transmission errors, and as such, instead of the instantaneous worst-case noise values, the impact of noise is better described by its spectral contents and RMS value [17].

Chapter 16 explains the design procedure for SerDes phase-locked loop (PLL) supply filters, takes a vendor-recommended supply filter network and compares it with an integrated interposer-type power-distribution filter. Finally, Chapter 17 is the power distribution design guide for a recent FPGA family. The article provides a full design overview of the PDN section, including component selection and placement, component hook-up layout recommendation, and simulation and measurement suggestions [18].

Part 5 addresses the characterization of PDNs. Since [5] gives a detailed account of frequency-domain PDN characterization, here some of the major articles dealing with only time-domain characterization are included at the beginning.

Chapter 18 provides the key solution to obtain the worst-case transient noise in point-of-load PDNs, assuming linear and time-invariant PDN circuit and known limits to current-step magnitude and slew-rate. The method is based on the step response of the PDN, and thus it provides the bridge between frequency-domain and time-domain solutions, allowing the user to calculate worst-case noise based on the complex impedance profile of PDN. The inverse link is provided, for instance, by [19], which describes a measurement solution to obtain the impedance profile by using controlled load-current excitation.

Chapter 19 in Part 5 describes modeling techniques to obtain the noise on PDN planes. This chapter, similar to the second chapter in Part 4, uses the RMS noise value as a metric, so it is well-suited to assess the effective impact of noise on system performance.

Chapter 20 grew out from a panel discussion at DesignCon 2004 [20], and it addresses the need for standardization of the board-package interface, which could facilitate the development of better software tools for the purposes of simulation board-package or board-package-silicon PDNs.

Chapter 21 in Part 5, which is a revised and updated DesignCon paper, gives a brief overview of the frequency-domain measurement methodologies, based on the two-port VNA setup from [21]. Chapter 22 gives a summary of some of the capacitor modeling options and introduces a simple causal model to capture secondary resonances in MLCCs. It also shows that ultimately neither ESL nor ESR is unique properties of the capacitors; they are being influenced by the external connection geometry through the shaping of the current distribution inside the capacitors.

Readers who are interested in more details of power distribution design and analysis are encouraged to browse the comprehensive collection of recent power-distribution publications listed in [22].

—*István Novák*
Distinguished Engineer, Signal and Power Integrity

References

[1] Henry W. Ott, *Noise Reduction Techniques in Electronic Systems*, John Wiley & Sons, 1976.

[2] Clayton R. Paul, *Introduction to Electromagnetic Compatibility*, John Wiley & Sons, 1992.

[3] Howard Johnson, Martin Graham, *High-Speed Digital Design: A Handbook of Black Magic*, Prentice Hall, 1994.

[4] K. Barry A. Williams, *Designing Power Distribution Systems for Electronic Circuits*, Aikman Engineering LLC, 2005.

[5] István Novák, Jason R. Miller, *Frequency-Domain Characterization of Power Distribution Networks*, Artech House, 2007.

[6] Madhavan Swaminathan, Ege Engin, *Power Integrity Modeling and Design for Semiconductors and Systems*, Prentice Hall, 2007.

[7] Richard Redl, Brian P. Erisman, Zoltan Zansky, *Optimizing the Load Transient Response of the Buck Converter*, Applied Power Electronics Conference, Anaheim, California, February 15-19, 1998.

[8] L. D. Smith, R. E. Anderson, D. W. Forehand, T. J. Pelc, T. Roy, *Power Distribution System Design Methodology and Capacitor Selection for Modern CMOS Technology*, IEEE Transactions on Advanced Packaging, Vol. 22, No. 3, August 1999, p. 284.

[9] Waizman, Chung, *Extended Adaptive Voltage Positioning (EAVP)*, Proceedings of EPEP Conference, October 23-25, 2000, Scottsdale, Arizona.

[10] István Novák, *Reducing Simultaneous Switching Noise and EMI on Ground/Power Planes by Dissipative Edge Termination*, IEEE Transactions of CPMT, ITAPFZ, Vol. 22, No. 3, August 1999, pp. 274-283.

[11] István Novák et al., *Lossy Power Distribution Networks with Thin Dielectric Layers and/or Thin Conductive Layers*, IEEE Tr. CPMT, Vol. 23, No. 3, August 2000.

[12] James L. Knighten, Bruce Archambeault, Jun Fan, Giuseppe Selli, Liang Xue, James L. Drewniak, *PDN Design Strategies: II. Ceramic SMT Decoupling Capacitors—Does Location Matter?,* IEEE EMC Society Newsletter, Winter 2005, pp. 56-67.

[13] James L. Knighten, Bruce Archambeault, Jun Fan, Giuseppe Selli, James L. Drewniak, *PDN Design Strategies: III. Planes and Materials—Are They Important Factors in Power Bus Design?,* the IEEE EMC Society Newsletter, Summer 2006, pp. 58-69.

[14] James L. Knighten, Bruce Archambeault, Jun Fan, Giuseppe Selli, Abhilash Rajagopal, Samuel Connor, James L. Drewniak, *PDN Design Strategies: IV. Sources of PDN Noise*, IEEE EMC Society Newsletter, Winter 2007, pp. 54-64.

[15] Todd Hubing, *Effective Strategies for Choosing and Locating Printed Circuit Board Decoupling Capacitors*, Proceedings of IEEE EMC Symposium 2005.
[16] R. Schmitt, X. Huang, L. Yang, C. Yuan, *System-Level Power Integrity Analysis and Correlation for Multi-Gigabit Designs*, DesignCon 2004, Santa Clara, California, February 2004.
[17] R. Schmitt, C. Yuan, W. Kim, *Modeling and Correlation of Supply Noise for a 3.2GHz Bidirectional Differential Memory Bus*, DesignCon 2005, Santa Clara, California, February 2005.
[18] Virtex-5 PCB Designer's Guide UG203 (V1.0) December 15, 2006, available at www.xilinx.com under Virtex-5 User Guides.
[19] I. Kantorovich et al., *Measurement of Milliohms of Impedance at Hundred MHz on Chip Power Supply Loop*, Proceedings of the 11th Topical Meeting on Electrical Performance of Electronic Packaging, October 27-29, 2002, Monterey, California.
[20] Board and Package Level PDN Simulations. Panel Discussions at DesignCon2004, High-Performance Systems Design Conference, Santa Clara, California February 2-5, 2004.
[21] István Novák, *Measuring Milliohms and Picohenrys in Power Distribution Networks*, DesignCon 2000, High-Performance Systems Design Conference, Santa Clara, California, February 1-4, 2000.
[22] John Barnes, *Power Distribution on Printed Circuit Boards*, Bibliography, www.dbicorporation.com/pwrbib.htm.

About the Executive Editor

DR. ISTVÁN NOVÁK IS a distinguished engineer in the field of signal and power integrity. He is an IEEE fellow recognized for his contributions to signal-integrity modeling, measurements, and simulations. Dr. Novák worked on high-speed signaling and power distributing designs of Sun's V880, V480, V890, V490, T1000, T2000, T5120, and T5220 mid-range server families. His new technology developments with laminate suppliers, printed-circuit fabricators, and component vendors resulted in the introduction of the first sub 2-mil laminates and controlled-ESR bypass capacitors for Sun servers.

Dr. Novák developed a new methodology for the measurement of a wide range of power-distribution components, such as DC-DC converters, bypass capacitors, and printed-circuit-board power-ground laminates. The methodology has been presented in several conference papers, two of which won best paper awards. Dr. Novák carries many years of international consulting, design and instruction experience in the field of high-speed and high-frequency circuits and systems. He is an international consultant serving high-technology companies in the fields of high-speed and high-frequency circuits and systems. Dr. Novák holds 25 patents in power distribution, signal integrity, and digital signal processing, is the co-author of *Frequency-Domain Characterization of Power Distribution Networks* and has published more than 100 technical papers.

Dr. Novák has worked and consulted for several companies in the computer and telecommunications industry, to do clock- and power-distribution networks, switch-mode converters as well as various high-speed backplanes, and copper and optical interconnects in the Gb/s range.

Dr. Novák obtained his Ph.D. degree from the Hungarian Academy of Sciences, and received his technical education from the Technical University of Budapest.

Acknowledgements

MOST OF THE thanks should be directed to the contributors of the articles in this book, almost 50 of them altogether. I have had the privilege to have known and worked with several of them, and I am utterly grateful for their time to support the adaptation and adjustments of their articles. As a special thanks, their LinkedIn profile URLs are presented in a separate chapter.

The titles and affiliations listed in the header of each chapter were in effect when the particular work was written.

From early on, I received tremendous support from the DesignCon conference organizers to expand on these technical papers and to turn them into multicompany TecForums and panel discussions. Over the years, there were joint publications with the best industry experts, among others, on thin laminates, laminate characterization, PDN measurement methodologies, component characterizations, simulation tools, PDN problem definitions and DC–DC converter characteristics. I am deeply indebted to the contributors of these joint appearances. Some of the most important of these articles found their way into this compendium.

I am especially thankful to the IEC leadership, in particular Barry Sullivan, who first conveyed the invitation to put together a compendium on power distribution design methodologies. As the manuscript later emerged, I enjoyed the professional support of André Sulluchuco, who tirelessly helped the publishing process.

It turned out that at the time of the original invitation, about four years ago, the material available in the publications was not enough to cover all major aspects of power distribution design. Previously published papers mostly focused on analyzing various segments of the problem, and apart from a few isolated design papers, they did not cover the design methodologies and choices in a comprehensive way. Over the past few years, fortunately, the situation has changed. The ongoing power-distribution series of TecForums and panels at DesignCon, as well as IEEE papers and independent publications, have made it possible to offer a more complete collection of articles to cover all major aspects.

Last but not least, I acknowledge the contribution of a large number of individuals, who in the past aroused my curiosity and knowledge about power distribution. It started about 50 years ago, when our late father, a pharmacist and medical doctor by profession, showed me and my elder brother how to make AC-main transformers with handmade bobbins and hand-winding, to light up small incandescent bulbs. In the 1970s, while working in the Space Research Group of the Technical University of Budapest in Budapest, Hungary, I had the privilege to work with a number of talented experts. Among them was Richard Redl, one of the contributors to this book, who taught me the theory and analysis of power-conversion circuits, and

Antal Banfalvy, from whom I learned important technology details of power circuits. Later, in the 1980s, I was honored to be invited to work at Design Automation, Inc., where the president and founder of the company, Nathan Sokal, deepened my knowledge about switching-mode power converters.

Beginning in the fall of 1997 at Sun Microsystems in Burlington, Massachusetts, I worked with outstanding talents of the power-distribution field, among them Larry Smith and Raymond Anderson. I am also grateful to the management, Jeff Elsmore, Wayne Bolivar, Steve Klosterman, Nick Laplaca, and Kenneth Weiss, who encouraged and supported the work to create robust, cost-effective, and reliable power-distribution networks for Sun products.

—*István Novák*
Bedford, Massachusetts

Contents

Chapter 1—Power Supply Compensation for Capacitive Loads

Jonathan L. Fasig, Principal Engineer, Mayo Clinic
Barry K. Gilbert, Director, Mayo Clinic
Erik S. Daniel, Deputy Director, Mayo Clinic

This paper was presented at DesignCon 2007, Santa Clara, California, January 29-31, 2007.

AS APPLICATION-SPECIFIC INTEGRATED circuit (ASIC) supply voltages approach one volt, the source-impedance goals for power distribution networks are driven even lower as well. One approach to achieve these goals is to add decoupling capacitors of various values until the desired impedance profile is obtained. This approach can unintentionally result in a reduced power supply stability and even oscillation.

In this chapter, we present a case study of a system design where these problems were encountered, and we describe how these problems were resolved. Time-domain and frequency-domain analysis techniques are discussed and measured data is presented.

Introduction

Much research has been published on strategies and methods to design power distribution networks that minimize disturbances on the system direct current (DC) supply voltages, and it has become common to invest considerable time and effort in the optimization of circuit-board constructions and decoupling capacitor selections to ensure that the system power nets are disturbance-free [1, 2, 3]. It can indeed be a surprise when an oscilloscope is connected to a poorly behaved prototype and significant "noise" is observed on a supply voltage.

This chapter describes the diagnosis of such a system and the steps we took to remedy it.

System Overview

Our main focus here was a system comprised of one large motherboard and two optical-interface daughtercards. This motherboard incorporated several very large field-programmable gate arrays (FPGAs) along with various clock generation and support circuits. DC power for the system was supplied from a single alternating current (AC)-powered unit capable of generating multiple constant-voltage outputs between +12 V and -5.2 V.

Early in the design of the motherboard, the designers became aware that significant current surges could be produced when the FPGAs transitioned from program mode to functional mode. To mitigate supply voltage droop during these transient events, the motherboard designers incorporated several large-capacitance low equivalent series resistance (ESR) tantalum capacitors into the decoupling networks for the FPGA supplies.

In the course of prototype testing, it was discovered that the system exhibited various functional and performance limitations including difficulty programming the FPGAs and highly elevated bit-error rates (BERs). Further investigation revealed that all these problems abated when the motherboard was powered from lab-grade bench supplies instead of the internal multiple-output power supply. No issues were observed when the outputs of the internal supply were monitored driving just purely resistive loads, but when the internal supply was connected to the motherboard, then significant ringing was observed on the FPGA supplies coincident with steps in the load current as shown in Figure 1.1.

The salient difference between the purely resistive load (provided using an Agilent N3306A programmable load) and the system motherboard was the 4,700 microfarads of bulk storage capacitance added to the motherboard power distribution networks. Clearly, our internal supply had compatibility issues with our motherboard bulk capacitors. If these capacitors were removed, then the ringing abated as well, but the supply voltage at the FPGAs fell excessively during the current surge. Consequently, outright elimination of the bulk capacitors was not a viable solution.

Another solution was required.

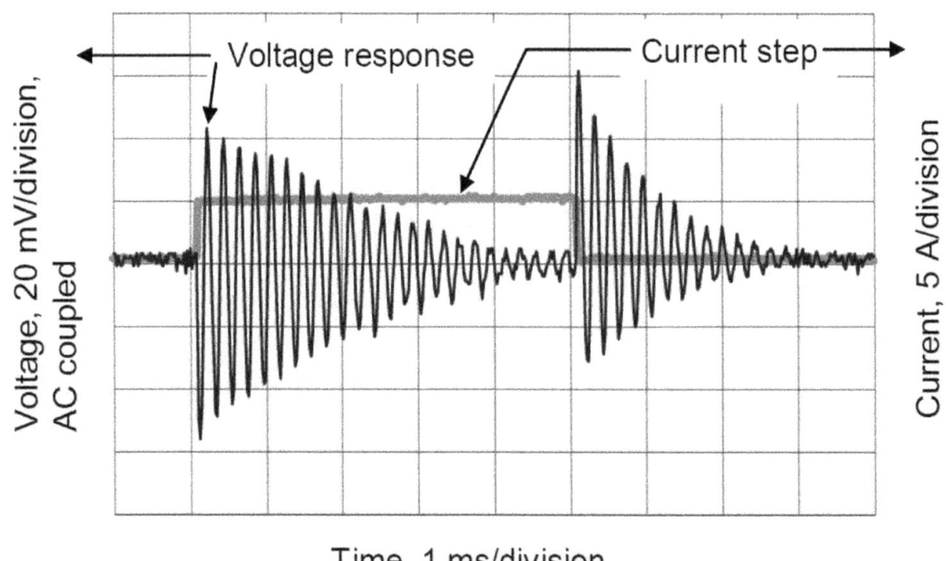

Figure 1.1: Transient Response of the Internal Uncompensated Power Supply at 3.3V, Driving a System Motherboard with 10 470μF Tantalum Capacitors Installed and a Static Load of 40A

Power Supply Stability Primer

A regulated power supply is actually a complete control system.

While a thorough discussion of control theory is beyond the scope of this paper, it is necessary to introduce a few concepts. Control systems employ negative feedback techniques whereby the output of the system (V_{OUT}) is sampled (V_{fb}) and compared to a stable reference (V_{REF}) as shown in Figure 1.2. The difference (V_{ERROR}) stimulates the control system to correct itself. When the feed-forward path includes high gain, the magnitude of the error signal is driven toward zero and the output of the supply is made to track the reference. This tracking is largely independent from variations in the load current.

It is possible to show [4] that the frequency-dependent closed-loop transfer function of the power supply is then given by:

$$\frac{V_{OUT}(s)}{V_{REF}(s)} = \frac{G(s)}{1 + G(s)H(s)}$$

...where the feed-forward response [G(s)] and feedback response [H(s)] are functions of frequency represented by the complex variable "s."

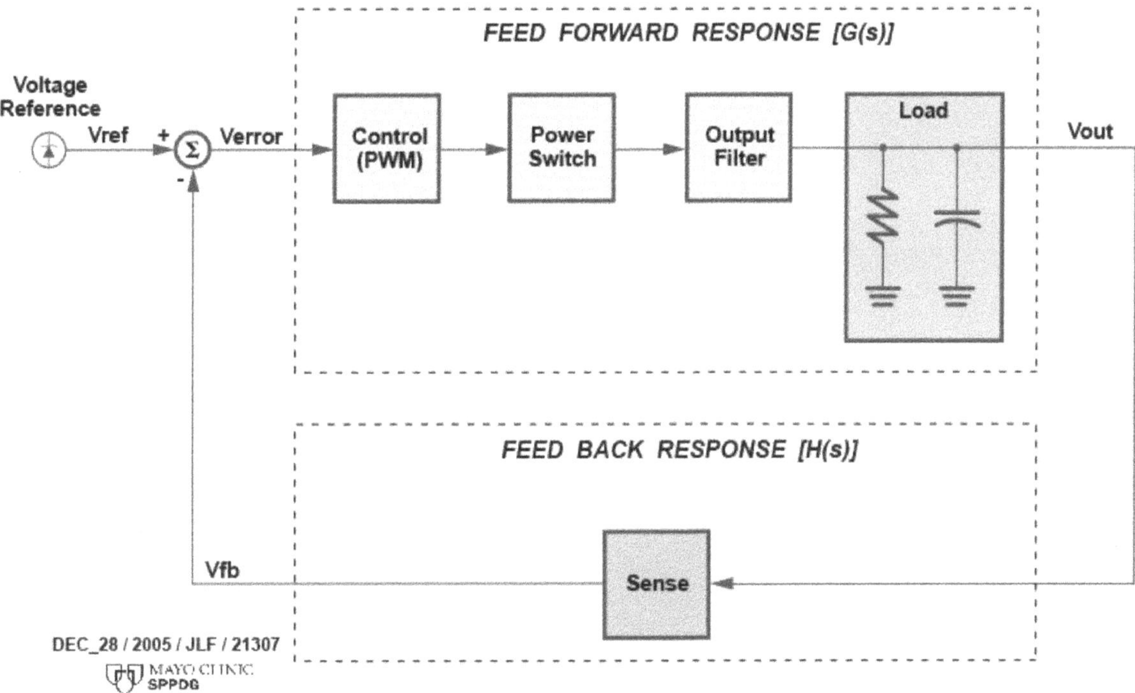

Figure 1.2: Block Diagram of an Inverter-Based Voltage Regulator and its Load

The V_{OUT}/V_{REF} equation shows that the quantity G(s)H(s) has a profound impact on the operation of the control system, and we can see in Figure 1.2 that this corresponds to the net gain around the open loop from V_{ERROR} to V_{fb}. This open loop gain or "loop gain" characterizes the dynamic performance of the supply.

Since the control system feeds its output back into its input, certain values of loop gain may possibly cause the output to become unstable. In particular, when the magnitude of the loop gain is exactly equal to one (that is, 0 dB) and the phase shift of the loop gain is exactly 180 degrees, then the additional 180-degree phase shift from the inverting input of the summing node places the feedback signal in phase with the error signal and a sine wave oscillator is created. Since self-oscillation is not desirable in a constant-voltage power supply, two metrics are defined to gauge how close a system may be to oscillation.

As seen in Figure 1.3, "gain margin" refers to the magnitude of the loop gain when the phase crosses the critical 180-degree threshold, while "phase margin" refers to the phase delay of the loop gain when the magnitude crosses the critical 0 dB threshold. Well-designed systems typically have at least 10 dB of gain margin and at least 45 degrees of phase margin.

Loop gain can be measured by adding a low-magnitude AC test signal to the feedback path [5], typically by inserting a transformer secondary winding in series between the load and sense circuit, as shown in Figure 1.4. This transformer is necessary because the feedback signal node (V_{fb}) is not at ground potential.

Our analysis was conducted using a Tamura part number MET-60 wideband audio transformer[1] as described in [11]. When applied to a "well-behaved" power supply, this measurement technique produces a magnitude and phase plot as shown in Figure 1.5.

Using the definitions from Figure 1.3, we see that this supply has a gain margin of 28.6 dB and a phase margin of 90.2 degrees.

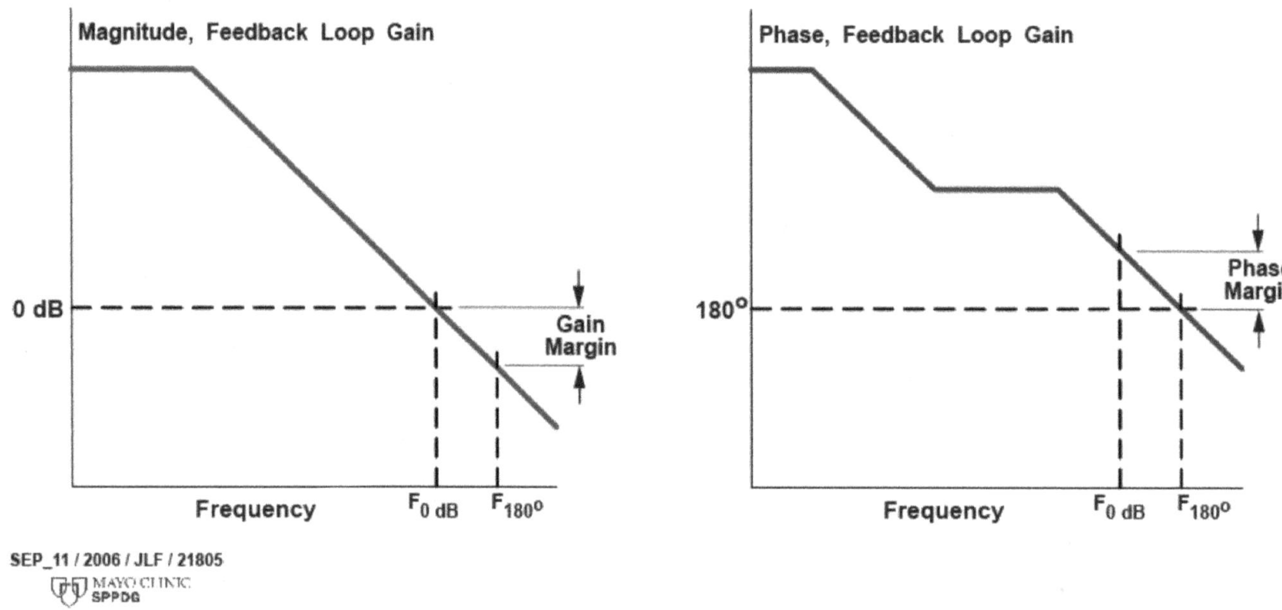

Figure 1.3: Definition of Stability Metrics Used to Characterize Feedback Control Systems

[1] Details about the transformer are absent from [5], perhaps because the requirements depend upon the bandwidth of the control loop being investigated. Those interested in performing loop-gain measurements are encouraged to request documents [6] through [10] from their Agilent (Keysight) representative.

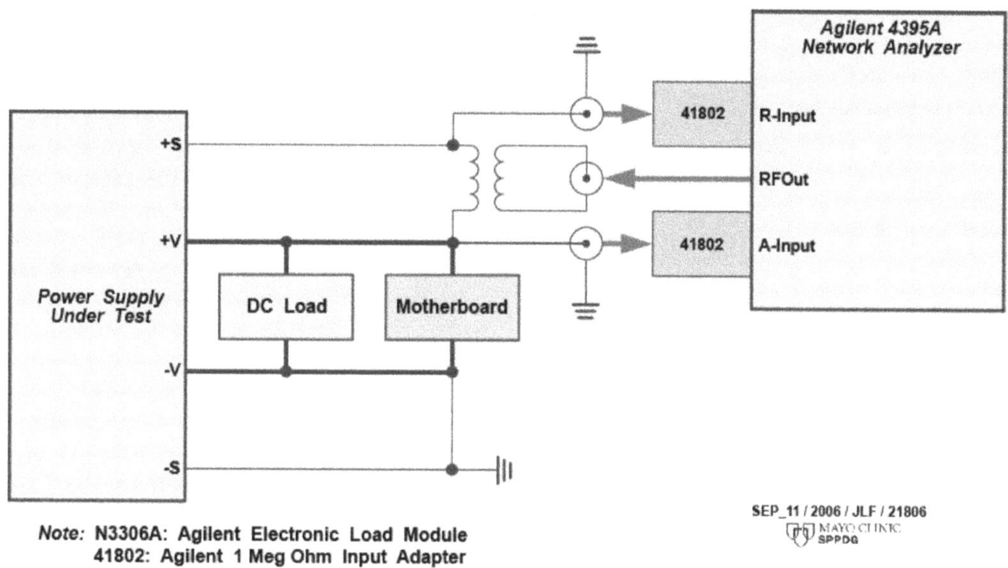

Figure 1.4: Test Setup for Power Supply Loop-Gain Measurement

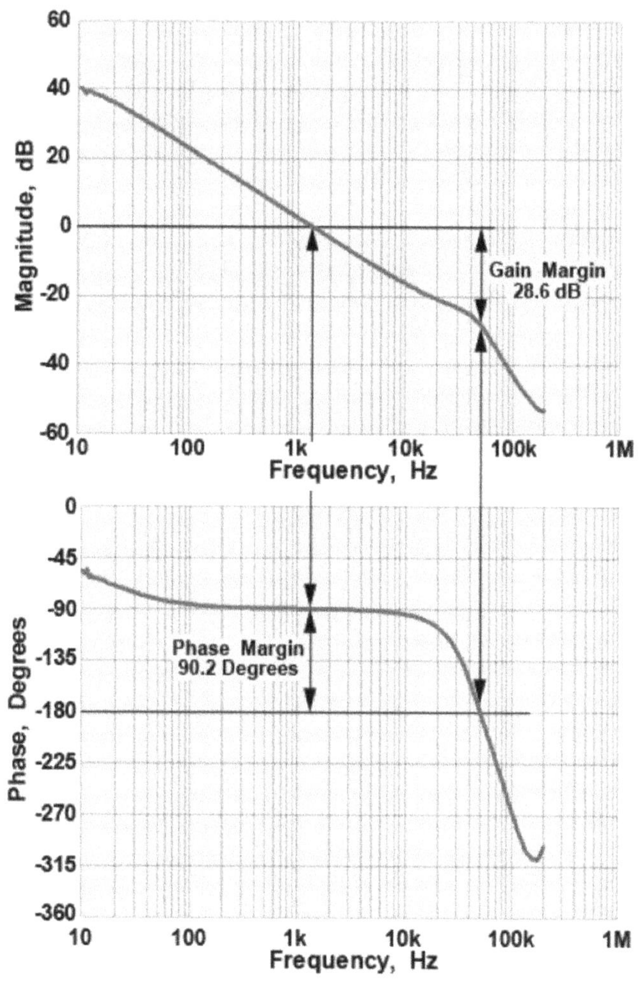

Figure 1.5: Measured Loop Gain of a Well-Behaved Laboratory Power Supply

Analysis

Insight into the internal supply stability issue may be obtained by referring to the magnitude and phase plots shown in Figure 1.6.

From these plots, one can see that the magnitude curve crosses 0 dB at 5.04 kHz, yielding only 13 degrees of phase margin, while the phase curve crosses -180 degrees at 6.05 kHz, yielding only 3.2 dB of gain margin.

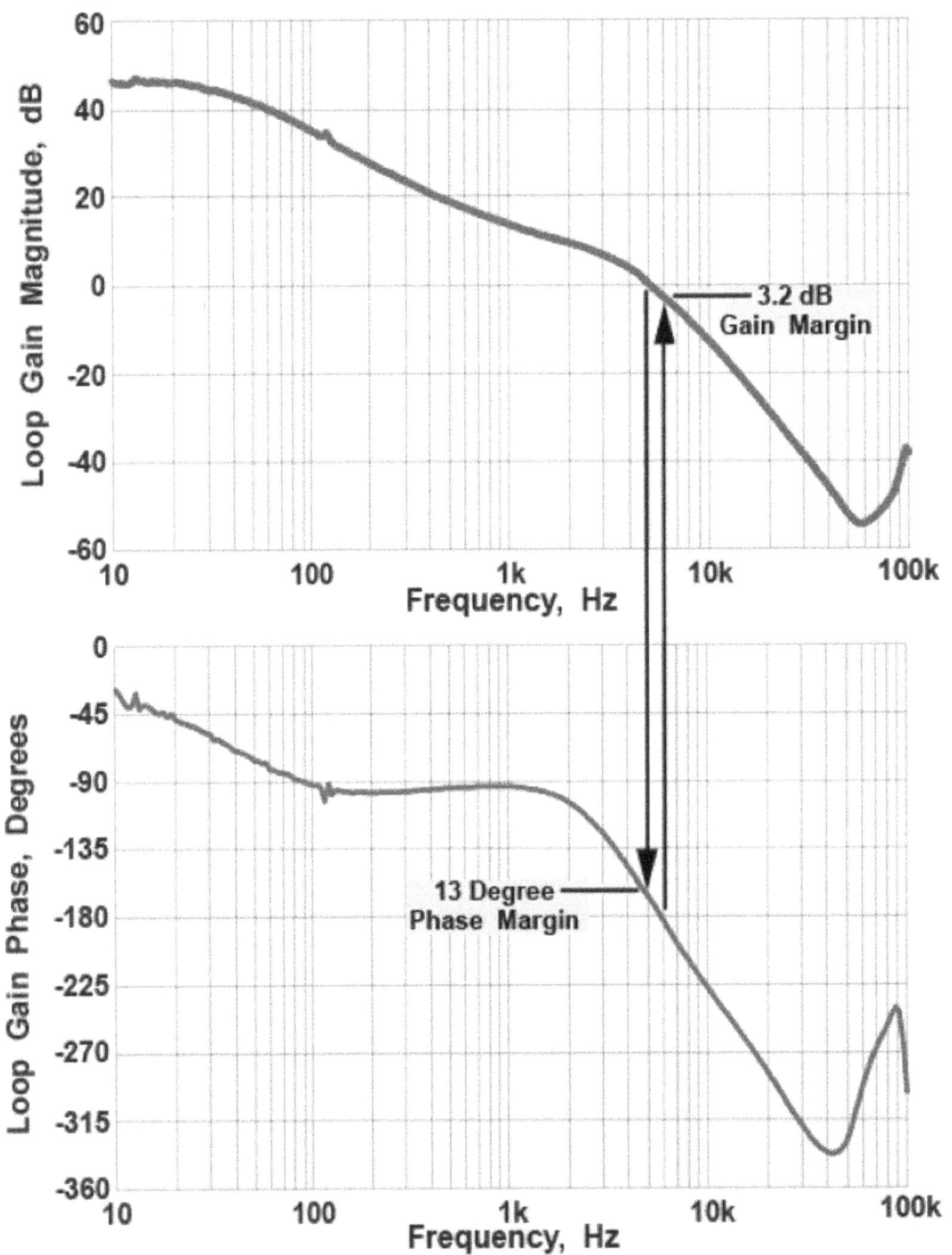

Figure 1.6: Measured Loop Gain of Internal Supply at 3.3V with 20A Load and Driving a System Motherboard with 10 470µF Tantalum Capacitors

The remedy chosen for this situation is the addition of a phase-lead network to the feedback path. A phase-lead network is essentially a high-pass filter that provides a low-impedance path from the supply output back to the sense (feedback) input without encountering the load. The effect of such a network is to postpone or delay the accumulation of phase until the loop-gain magnitude has passed safely through the 0 dB threshold. From the magnitude plot of the uncompensated loop we see that the curve increases its downward slope between 3 kHz and 4 kHz. For maximum effectiveness, the natural frequency of our phase-lead network should be at or below the 3 kHz value. For simplicity, we chose 2 kHz as our starting point. Recalling that the natural frequency of a simple resistor capacitor (RC) network is given by:

$$f_o = \frac{1}{2\pi RC}$$

...and accounting for the fact that the power supply already has an internal 100 Ohm resistor (Rint), which parallels Rcomp in Figure 1.7, then it can be shown that:

$$R_{comp} = \frac{R_{int}}{(2\pi\, f_o\, C_{comp}\, R_{int} - 1)}$$

Furthermore, we see that R_{comp} is 20.38 Ohms when fo is 2 kHz and C_{comp} is 4.7 µF. Close inspection of Figure 1.7 also reveals that we implemented the compensation network on the "-S" side of the feedback network so that the resistance of our injection transformer did not enter into the calculation of the compensation resistor value, and the frequency response of the supply did not change when the transformer was removed.

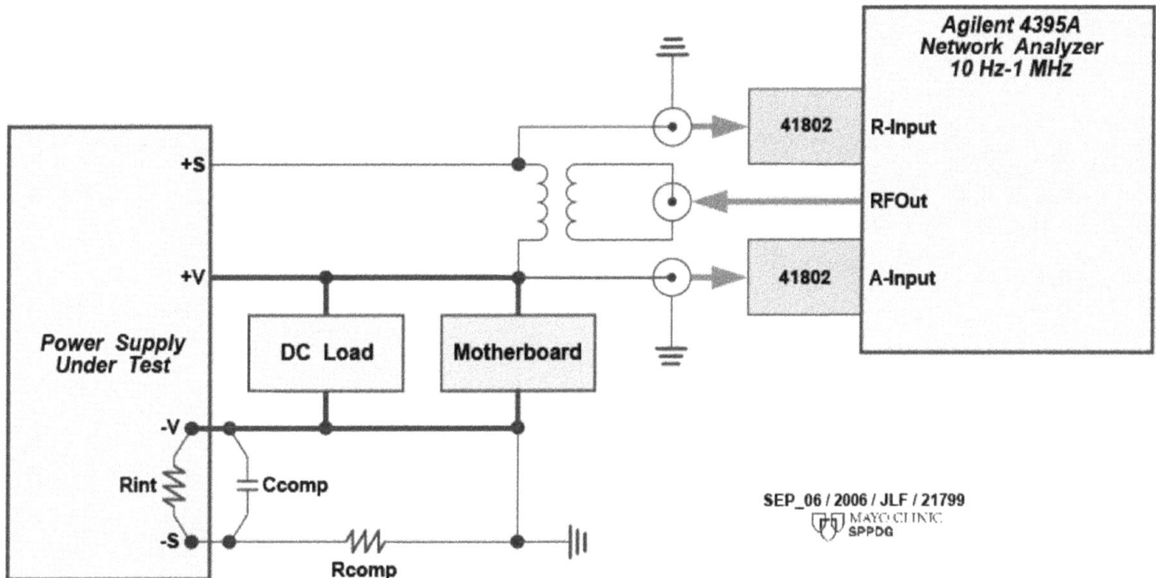

Figure 1.7: Test Setup for Compensated Loop Gain Measurement

The effect of this lead network is observed in the compensated loop-gain measurements shown in Figure 1.8. While the magnitude plot is not substantially affected, the -180 degree crossing of the phase plot is pushed out to 18.2 kHz (which is three times the uncompensated 0 dB crossover frequency), and the gain margin and phase margin are increased to 17 dB and 51 degrees, respectively.

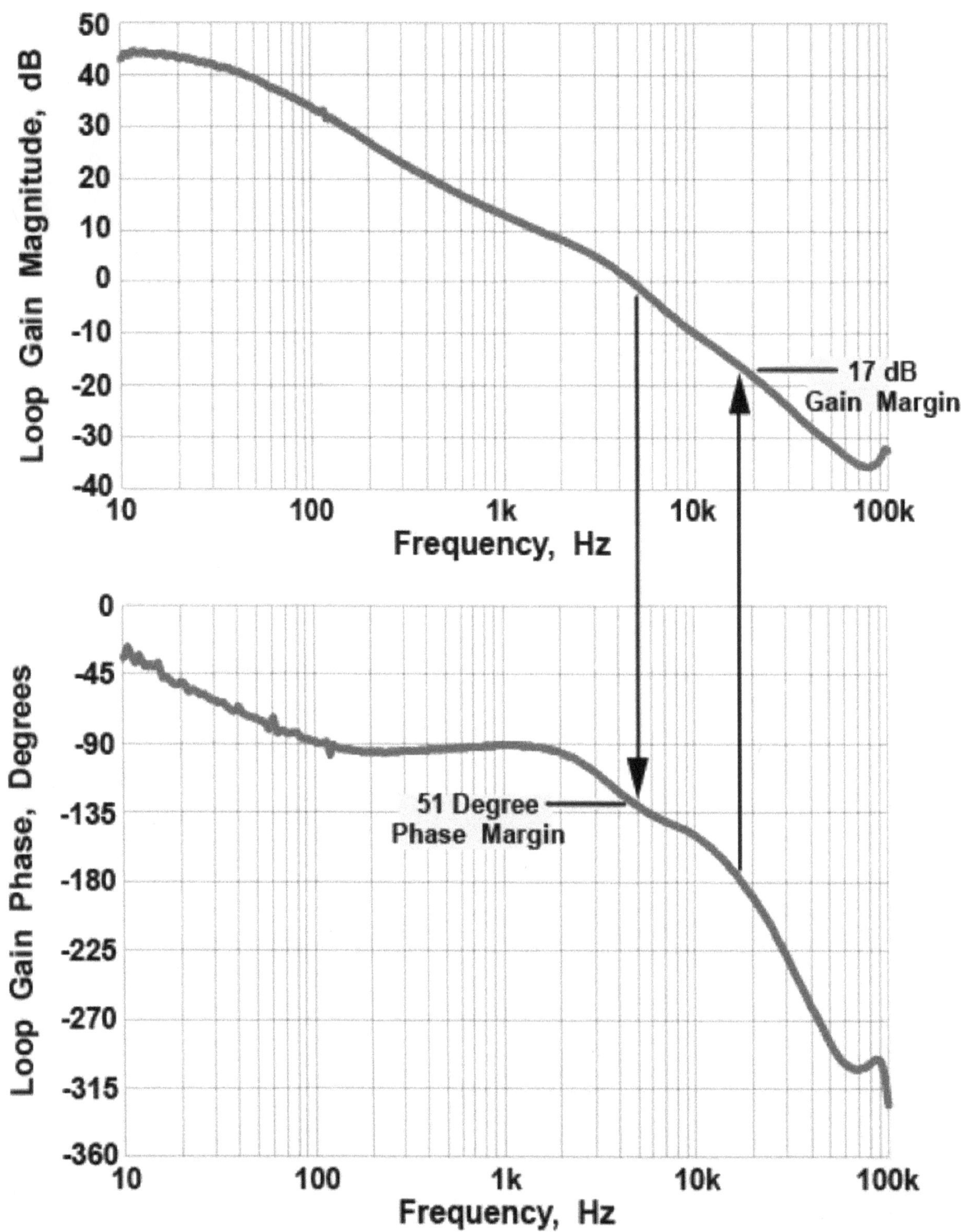

Figure 1.8: Measured Loop Gain of Compensated Internal Supply at 3.3V with 20A and Driving a System Motherboard with 10 470μF Tantalum Capacitor. The Regulator is Compensated by a Phase-Lead Network of 4.7μF and 20 ohms.

The effects of compensation are dramatically shown in the transient response in Figure 1.9. The same 5 Ampere current step now produces a well-damped 60 mV peak (30 mV steady-state) droop in the supply voltage, while the persistent ringing of the supply voltage observed with the uncompensated supply has been eliminated.

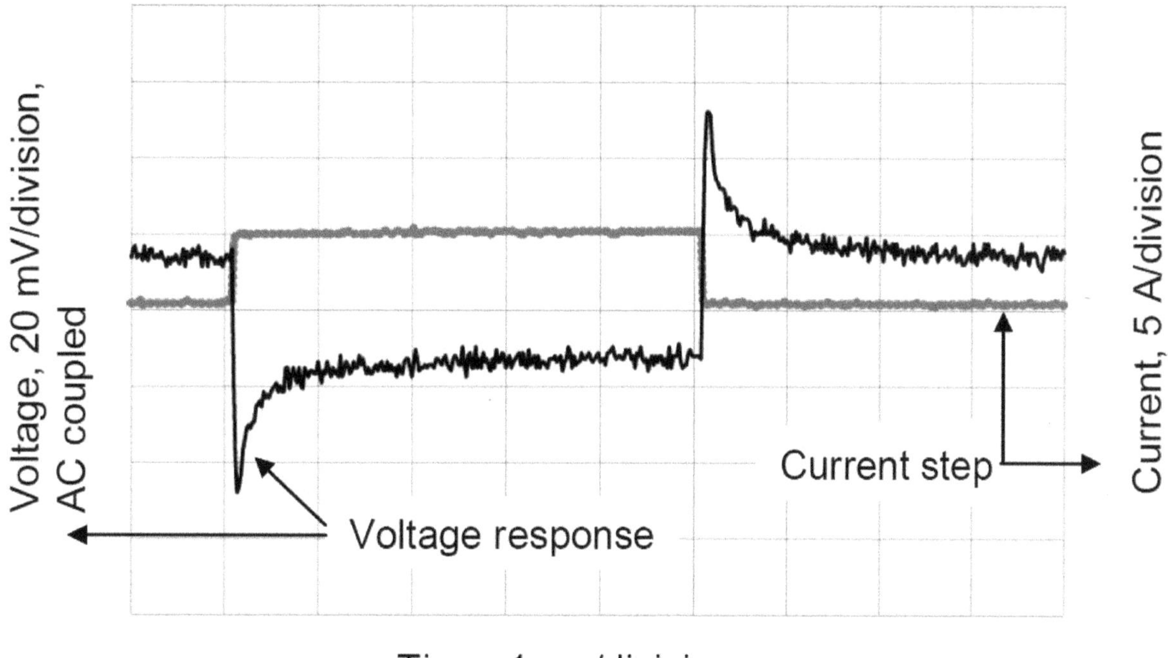

Figure 1.9: Transient Response of the Internal Compensated Power Supply at 3.3V, Driving a System Mother Board with 10 470μF Tantalum Capacitors and a Static Load of 40A

Conclusion

The primary lesson of this effort is that, in reality, decoupling design can no longer be done independently from power supply design or selection. Traditional decoupling design strategy typically asserts that more capacitance is better, and from a narrow decoupling perspective that assertion is still valid. However, the development of very-low-ESR capacitors and fast-response regulators with wide loop-bandwidth means that it is now possible to accumulate loop phase so rapidly that regulator stability can be compromised.

In future systems, the power distribution network and the power source must be concurrently developed and power system stability must be characterized. Effective time-domain and frequency-domain techniques to facilitate this effort were presented. In addition, a simple technique for power supply compensation was demonstrated.

Acknowledgments

The authors would like to thank Elaine Doherty, Theresa Funk, Deanne Jensen, and Steve Richardson for their assistance in the creation of images for this report, along with Jeffrey Bradley, Kevin Buchs, and Wendy Wilkins for their swift reviews of the manuscript.

References

[1] Larry Smith, Raymond Anderson, Doug Forehand, Tom Pelc, and Tanmoy Roy, *Power Distribution System Design Methodology and Capacitor Selection for Modern CMOS Technology*, IEEE Transactions on Advanced Packaging, Vol. 22, No. 3, p. 284, August 1999.

[2] Larry Smith, Dale Becker, Steve Weir, and István Novák, *Comparison of Power Distribution Network Design Methods*, DesignCon 2006.

[3] Andrey Mezhiba and Eby Friedman, *Power Distribution Networks in High Speed Integrated Circuits*, Kluwer Academic Publishers, 2004.

[4] Robert Saucedo and Earl Schiring, *Introduction to Continuous and Digital Control Systems*, Macmillan & Co., 1968. [5] Agilent Technologies, Product Note 4395-2: Switching Power Supply Evaluation, 1999.

[6] Agilent Technologies, *Switching Power Supply Evaluation with Agilent 4395A: Transformer Selection for loop gain measurement, Measurement Examples with Tips*, document # SWP-1, 2002.

[7] Agilent Technologies, *Switching Power Supply Evaluation with Agilent 4395A: RF Output level of the 4395A, Measurement Examples with Tips*, document number SWP-2, 2002.

[8] Agilent Technologies, *Switching Power Supply Evaluation with Agilent 4395A: Results of Loop Gain Measurement, Measurement Examples with Tips*, document # SWP-3, 2002.

[9] Agilent Technologies, *Switching Power Supply Evaluation with Agilent 4395A: Output Impedance Measurement, Measurement Examples with Tips,* document # SWP-4, 2003.

[10] Agilent Technologies, *Switching Power Supply Evaluation with Agilent 4395A: Loop gain measurement with inappropriate transformer, Measurement Examples with Tips,* document # SWP-5, 2003.

[11] http://www.tamuracorp.com/clientuploads/pdfs/engineeringdocs/MET-60.pdf

Chapter 2—DC–DC Converters: What Is Wrong with Them?

István Novák, Senior Signal Integrity Staff Engineer

This chapter is a revised version of the contribution to Emerging Challenges of DC-DC Converters at DesignCon 2007, January 29-February 1, 2007, Santa Clara, California.

DIRECT CURRENT—DIRECT CURRENT (DC-DC) converters are widely used in power distribution networks (PDNs). While they can conveniently generate any regulated voltage, off-the-shelf converters come with several inevitable compromises. These compromises stem from the fact that the converter designer does not know the exact nature of the load: the bypass capacitors and their interconnects are different depending on the application. This paper illustrates four major concern areas associated with off-the-shelf converters:

- Limited control-loop bandwidth may require large output capacitance.
- Non-flat and/or non-stable output impedance profile results in increased transient noise.
- Unspecified high-frequency ringing may risk sensitive clock signals.
- Non-linear control loops may make validation more tedious.

Introduction

DC-DC converters are the staples in today's PDNs. The reasons for the popularity of DC-DC converters are multifold:

- Various kinds of semiconductor pieces (e.g., memory, input/output [I/O], core) continue to follow their own supply-voltage scaling and therefore the number of independent supply rails in the systems is on the rise.
- The continued push for higher efficiency in both high-power and low-power systems calls for generating the supply voltages near their destination.
- Distribution losses are reduced with higher voltage of raw power distribution.

DC-DC converters are switching circuits that chop the incoming DC voltage, and by controlling the duty cycle of the pulse stream, the output voltage can be translated and tightly controlled [1].

There are numerous topologies and circuit solutions; with the proper control and output circuit, one can create isolated or non-isolated converters, and output voltages can be in any relation with respect to the input voltage. The same-polarity step-down converter is commonly referred to as the buck converter. The same-polarity step-up converter is called the boost converter.

Finally, the converter, which changes the polarity between input and output voltages, is called a buck-boost converter.

Figure 2.1 shows the simplified switching scheme of a synchronous buck converter. The upper and lower field effect transistors (FETs) switch in opposite phase, creating a pulse stream with steady levels approximately equaling zero and the input voltage. The inductor-capacitor (LC) low-pass filter on the output suppresses the main switching frequency and its harmonics and produces a DC average voltage equaling the input voltage times the duty cycle. This results in the output-input voltage relationship approximately following the duty cycle. Static and dynamic switching losses slightly modify this ratio. By controlling the duty cycle in closed loop, one can keep the average output voltage constant in spite of (slowly) varying input voltage, load current, and losses. By assuming ideal lossless components, no power is dissipated in the converter circuit, the entire input power is transported to the output, and the conversion efficiency is 100 percent.

In real circuits, there are always finite losses and the efficiency with close-to-nominal output power is usually in the 75 to 95 percent range. As output power decreases, efficiency inevitably goes down due to the current necessary to feed the standby and control circuits.

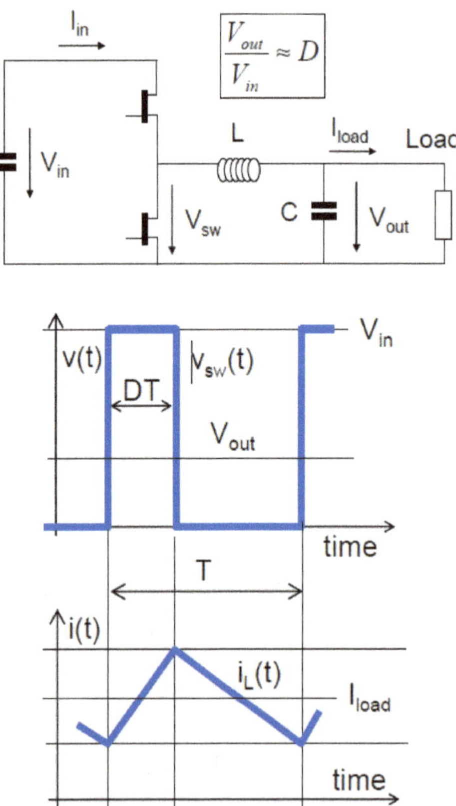

Figure 2.1: Switching Scheme and Characteristic Waveforms of a Non-Isolated Buck Converter

DC-DC Converter Parameters Related to Signal Integrity

The switching circuit theoretically can work with a large range of frequencies. As frequency increases, the circuit requires less inductance and capacitance in the output filter circuit, which results in smaller and lighter reactive components in the output filter circuits. However, dynamic losses increase and this balance creates a sweet spot of switching frequency with any given available technology, which is a balance between losses and size/weight of reactive components. Today, the main switching frequency (per phase) typically falls between 200 kHz and 2 MHz. The equivalent switching frequency can be increased, and thus the required inductance and capacitance can be reduced, by using multiple interleaved phases, running at the same frequency.

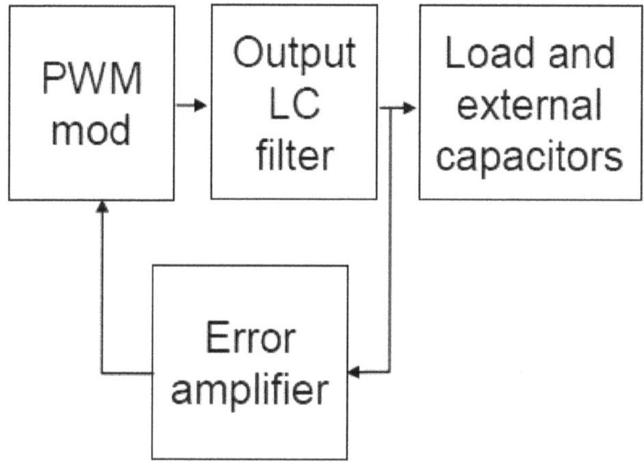

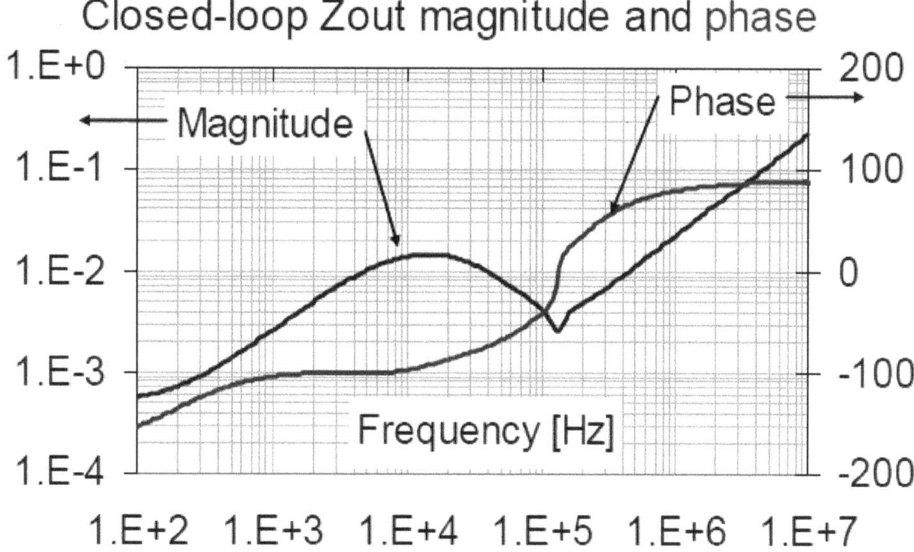

Figure 2.2: Block Schematics of a DC-DC Converter (Top) and Simulated Small-Signal Output Impedance (Bottom)

The simulated small-signal output impedance plot on the bottom of Figure 2.2 shows that this control loop maintains constant output voltage in spite of load-current changes, which manifests itself in a low output impedance, in this case, less than a milliohm at 100 Hz.

However, as frequency increases, the output impedance increases due to the finite bandwidth of the control loop. Above a few hundred kHz, the output impedance follows an inductive slope, dictated by the parasitics of the output capacitor network. Beyond the switching frequency, the control loop has a diminishing effect on the loop parameters.

Important quality parameters of the DC-DC converter are the transient noise due to load-current and input-voltage changes as well as due to its own internal switching. Figure 2.3 illustrates two major signatures of the stead-state switching noise: switching ripple (top chart) and high-frequency switching ringing (bottom graph).

The switching ripple is an inevitable consequence of how the converter works. The peak-to-peak ripple can be reduced by increasing the inductance and/or the capacitance in the output stage. The tradeoff is that larger values will result in slower transient response. The high-frequency switching ringing, on the other hand, is a side effect, generated by the main switching signal through the layout and component parasitics. As the main switching edge gets faster, larger and potentially higher-frequency ringing is induced. When minimizing the high-frequency ringing, the tradeoff is between ringing magnitude and efficiency versus size and cost.

The load transient noise behavior of the converter is determined by the switching elements, control-loop circuit, and output capacitors. We need to realize that the user load is inside the control loop, determining its dynamic performance. If the load comes with a set of its own bypass capacitors, those will become part of the control loop and thus will have an impact on the transient noise. The converter designer has no way of knowing what bypass capacitors the end user will attach to the converters. On the flipside, off-the-shelf converter users usually have very little or no information about the control loop parameters, so it is hard to determine how the externally attached bypass capacitors impact the dynamic performance.

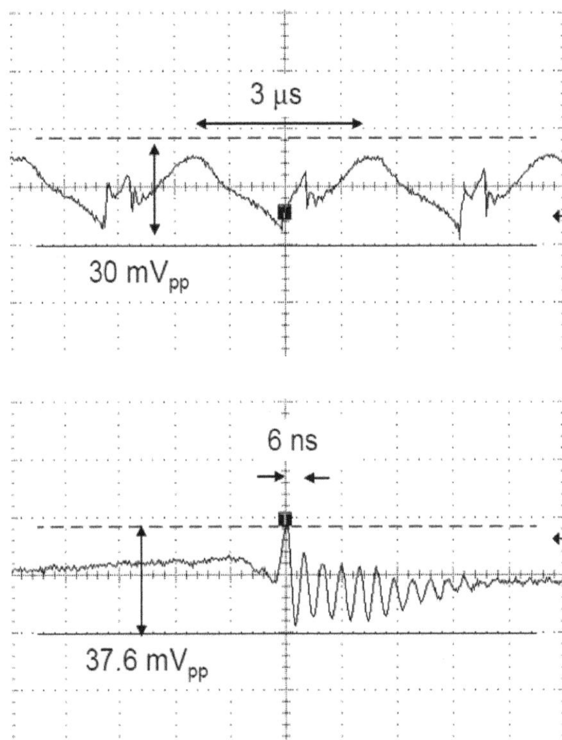

Figure 2.3: Steady-State Transient Noise Signatures of a 5-A Point-of-Load Buck Converter: Switching Ripple Measured with 20 MHz Bandwidth on the Top, High-Frequency Ringing Measured with 500 MHz Bandwidth on the Bottom.

Since the converters are based on switched waveforms, they are inherently nonlinear. However, well-designed analog control loops result in a stable averaged response, and when the output has the proper set of capacitors, the resulting network can be well approximated as a linear circuit. For linear circuits, determining the worst-case transient noise created by step-like excitations is a straightforward process based on the impedance profile [2].

The reverse pulse technique takes the step response of the PDN and composes the worst-case transient response from the time stamps and values of subsequent maxima and minima.

It is easy to show that the worst-case transient noise is minimized if the frequency-domain response, namely the output impedance, is flat [3].

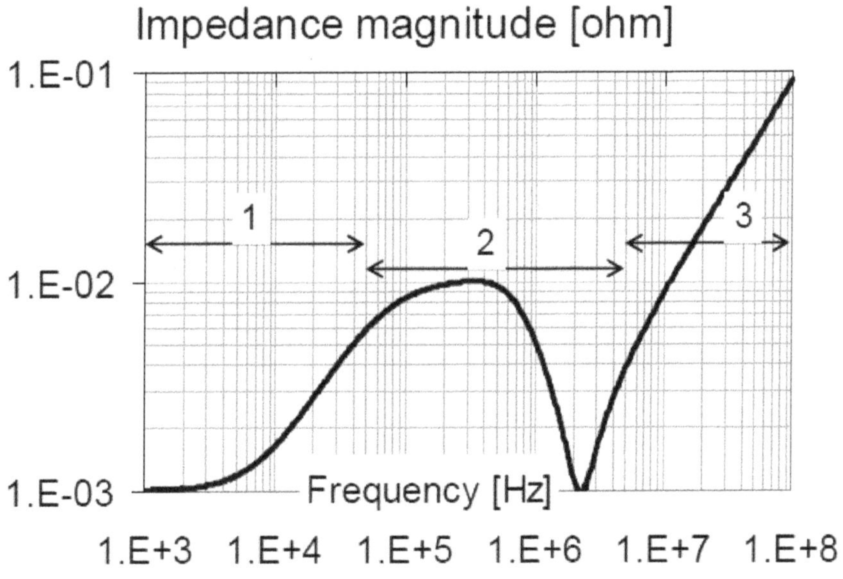

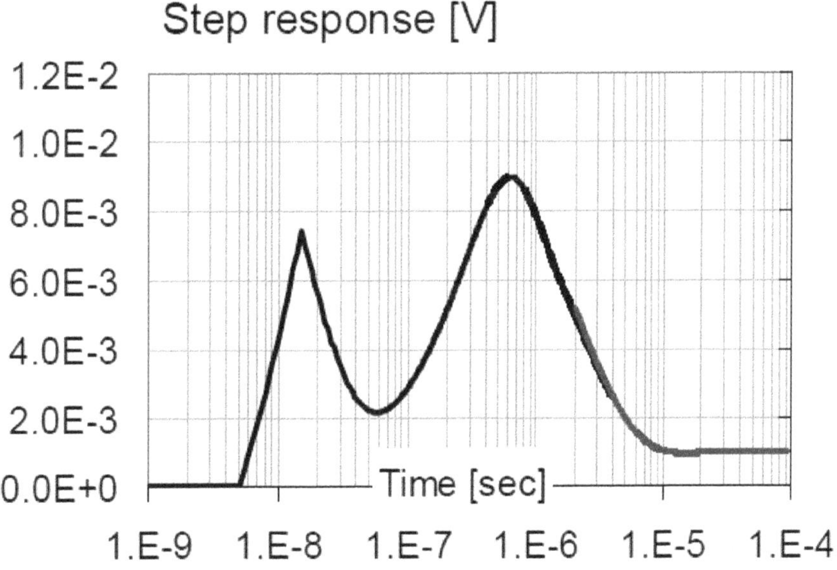

Figure 2.4: Simulated Output Impedance and Step Response of a DC-DC Converter with External Capacitors. One Should Note That Both Horizontal Scales Are Logarithmic.

Figure 2.4 illustrates a pair of output impedances and its corresponding step response. In the impedance profile, we can identify three frequency ranges. Range 1 extends from DC to a few times 10 kHz. In this frequency range, the components and the control loop of the DC-DC converter determine the impedance; at such low frequencies, we can hardly influence the impedance with external capacitors. At the high end, Range 3, which starts a few times above the switching frequency, behaves the opposite way: high above the switching frequency, the control loop has a diminishing influence on the output impedance and therefore, in this frequency range, the performance is primarily determined by the passive components. Finally, Range 2 is the critical interim gray zone in question, where the output impedance is determined by the interaction of the open-loop output impedance of the power stage, the control loop, and the output capacitors.

Potential SI Problems from the User's Perspective

Off-the-shelf converters are generic products that aim for a wide range of applications. Some converters have a wide input-voltage range; some also have a wide adjustable output voltage range. The load current may also widely vary in range, namely, between zero and the full rated current. The wide parameter ranges, together with a huge set of potential input and output capacitors, make the converter design a real challenge. It is equally challenging, however, to take an off-the-shelf converter and design a PDN network around it for a given set of requirements. The difficulties mostly come from the middle frequency range, Range 2, in Figure 2.4, where two blocks of components, the DC-DC converter and the external PDN, interact, but both are designed with minimal, if any, information about the other.

Low Loop Bandwidth

Figure 2.5 shows small-signal output impedance profiles, measured on a point-of-load buck converter. The converter had a wide input voltage range (3.0V to 13.2V) and adjustable output voltage range (0.7V to 3.6V). The maximum rated current was 17A. The impedance profiles were measured in small increments along all three parameters: input voltage, output voltage, and DC load current. The figure shows two representative summary plots at 1.0V output voltage. The top plot shows the minimum, average, and maximum impedance with 3V input voltage, as DC load current steps through 0 to 17A. The bottom plot shows the same with 10V input voltage. The min/max range is fairly tight, indicating a stable loop, and raising the possibility that the small-signal impedance profile can be used to get the load-step transient response through the inverse fast Fourier transform (IFFT).

Instead of carrying out IFFT on the measured impedance profile, we can make simple approximations to evaluate the usability of the converter. Let us assume that the load current can have 5-A steps, which for a 17-A converter is a modest assumption. If the 5-A steps happen to occur with a periodicity aligning with the 0.1-ohm impedance peak at 16 kHz, the resulting transient noise with 3V input is 5App*0.1ohm = 0.5Vpp (Actually, since the impedance profile will significantly attenuate the harmonics of the assumed 16 kHz square-wave load current, the PDN will pick out the fundamental waveform of the square wave, which is $4/\pi = 1.27$ times the fundamental). On a 1.0V supply rail, we can very seldom afford 0.5Vpp transient noise, and most probably the control loop would also become nonlinear for this large excursion. On a 1.0V rail a more realistic noise target could be 50 mVpp. To achieve 50 mVpp noise with 5-A load step changes, the impedance profile should stay below 10 milliohms, and it must be preferably flat. The output impedance can be reduced by adding low-impedance capacitors. With 3V input voltage, the average impedance curve of the converter crosses the 10-milliohm line at 3 kHz. To get 10 milliohm capacitive reactance at 3 kHz, we need approximately 5,000 μF capacitance. To provide a peak-free transition in the impedance profile, we would need to use at least twice of this capacitance, or at least 10,000 μF. This simple illustration shows that often times the converter's advertised rating cannot be fully utilized in demanding applications for low-noise rails without a large amount of external capacitance.

The large output capacitance requires extra board space and increases the total cost of ownership. In addition, dependent on the converter design, some loops may become unstable if the output is loaded with capacitors below a certain impedance (ESR) value.

Moreover, too large external output capacitance may prevent the converter to start up properly.

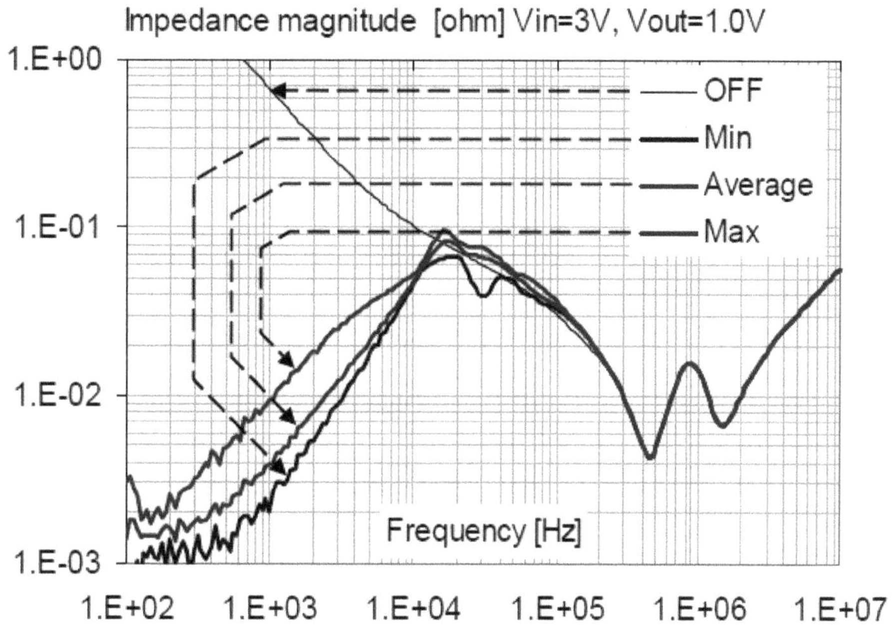

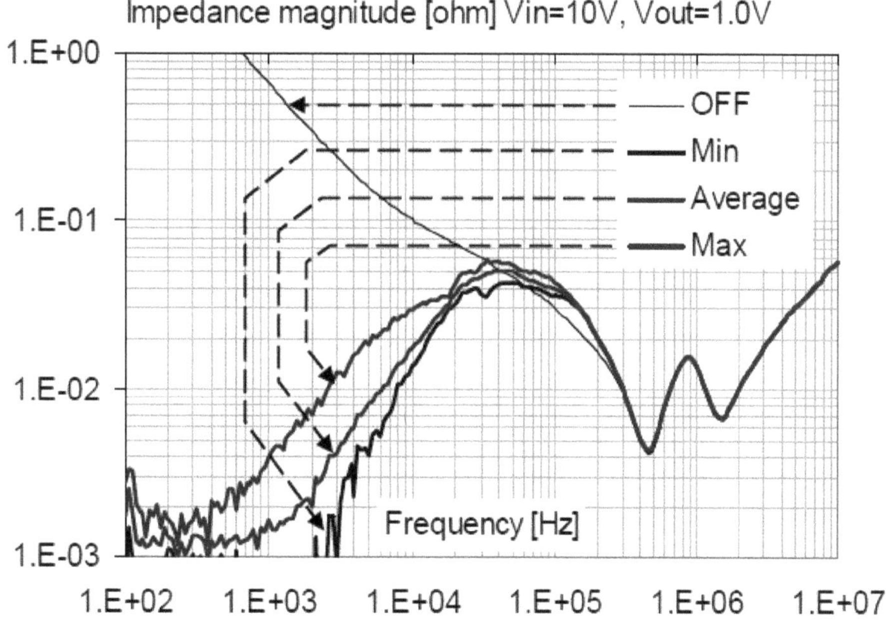

Figure 2.5: *Measured Small-Signal Impedance-Profile Summaries of a 17-A Point-of-Load Converter. Top: 3V Input Voltage. Bottom: 10V Input Voltage.*

Stability and Impedance Profile

Traditionally, the converter stability is described with the phase and magnitude margins of the transfer function of the control loop.

Usually, a 45-degree phase margin is required. While this ensures that the control loop will not exhibit a large class of self-oscillations, we need to consider that for the user, the ultimate performance parameter is the worst-case transient noise. As long as the small-signal output impedance does not change much with the DC load current, the transient noise can be minimized with a flat impedance profile. So, for practical purposes, the small-signal impedance surface is a good indicator of quality. Figure 2.6 illustrates the interrelation of phase margin and impedance-profile flatness.

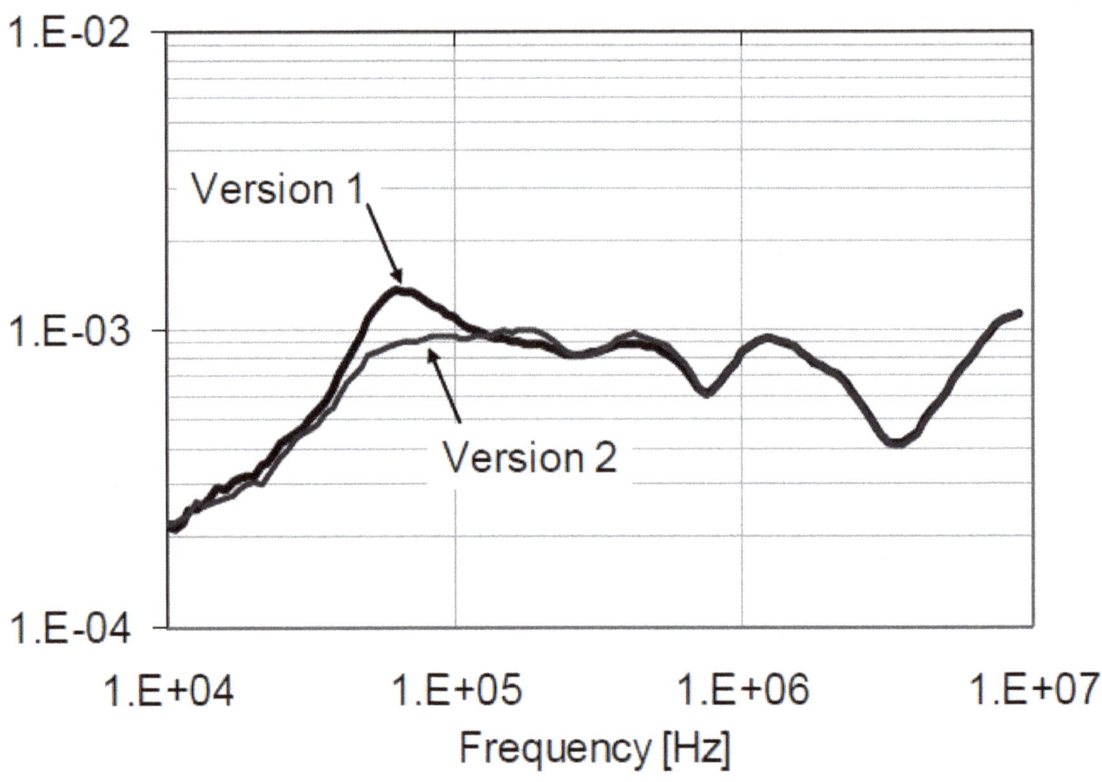

Figure 2.6: Impedance Profiles of a 100-A Multiphase Converter with 45- Degree Phase Margin (Version 1) and 70-Degree Phase Margin (Version 2)

This example uses the measured impedance profiles of a multiphase non-isolated 100-A converter with two different loop compensations. Both converters had the same external capacitors.

Version 1 of the converter had a 45-degree phase margin, whereas Version 2 had 70 degrees.

Let us note that the impedance response of Version 1 has a peak at 67 kHz, with a peak value 40 percent higher than the broadband flat plateau exhibited of the Version 2 design.

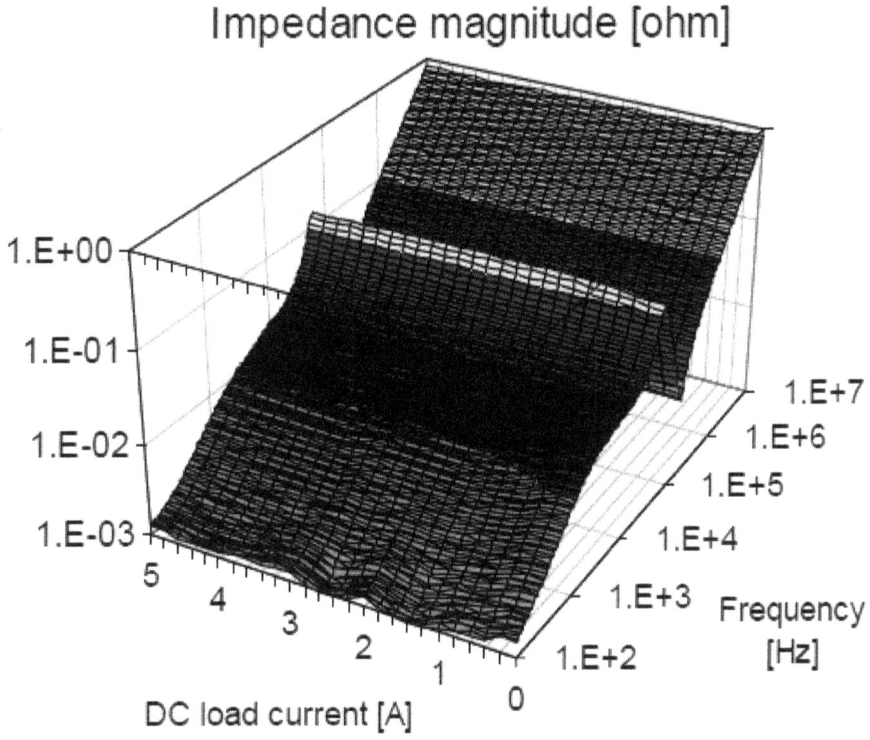

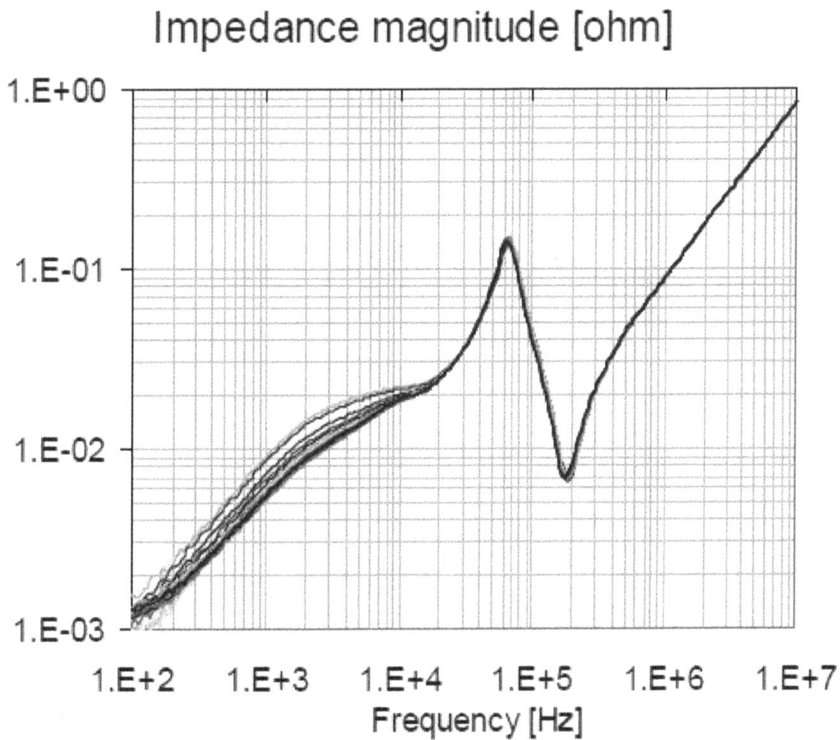

Figure 2.7: Impedance Surface (Top Graph) and Cumulative Traces (Bottom Graph) of a Point-of-Load Buck Converter with Stable Impedance Profile

The converters in Figure 2.6 had very stable impedance response with the given external capacitors; within the specified ranges of input voltage and DC load current, the impedance profile changed very little.

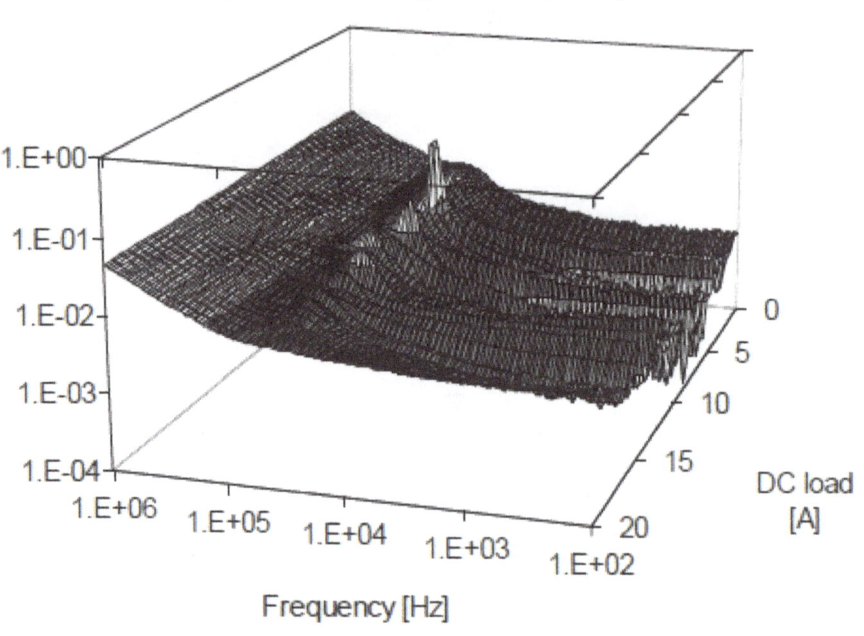

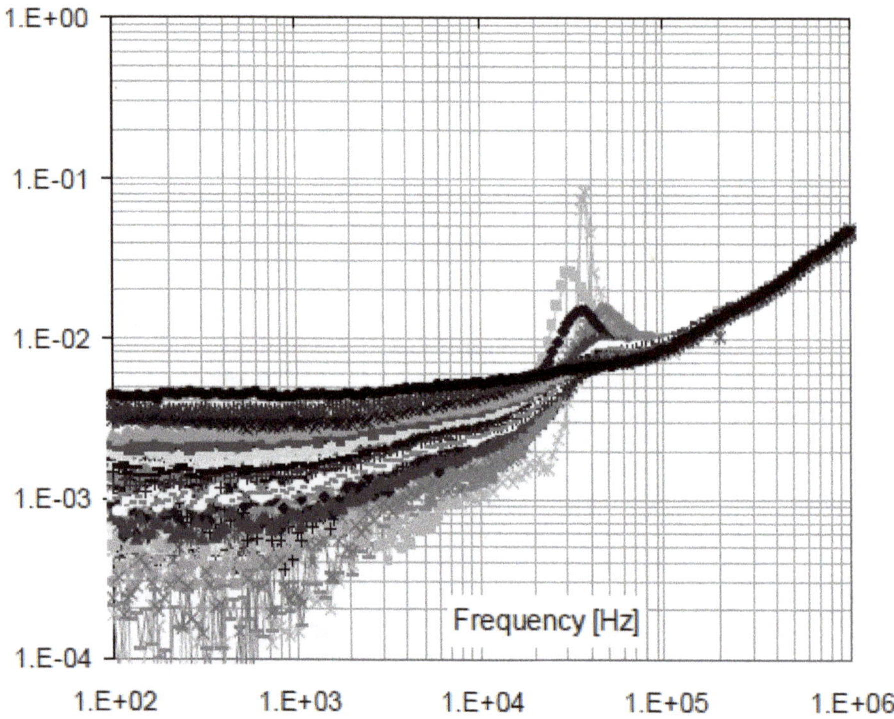

Figure 2.8: Impedance Surface (Top Graph) and Cumulative Traces (Bottom Graph) of a Point-of-Load Buck Converter with Unstable Impedance Profile

If the control loop is not properly designed, and/or if the converter layout allows the various circuit loops of the converter to interact, the impedance profile may show large variations.

Figure 2.7 shows the measured impedance plots of a converter with stable impedance profile, whereas Figure 2.8 illustrates an unstable impedance profile. Both converters were point-of-load 20-A buck converters measured with no external output capacitor (a legitimate operating condition according to the converters' data sheets).

While one could argue that the impedance profile of Figure 2.8 is on average very flat, and that peaks in the impedance profile occur only in narrow ranges of frequencies and load-current parameters, the real challenge is how to validate the design and how the performance around the sharp peaks may further degrade with time or due to part-to-part tolerances. Experience shows that converters exhibiting such anomalies in their impedance profile also exhibit larger unit-to-unit variations, raising questions about the quality and stability of design.

High-Frequency Ringing

Figure 2.3 shows the high-frequency ringing on the output of a 5-A converter. The ringing is 37.6 mVpp in amplitude and 150 MHz in frequency. The first issue is that most off-the-shelf converters specify the steady-state noise, unrelated to load-step changes (e.g., periodic and random deviation [PARD]), with 20 MHz test bandwidth.

Clearly, with 20 MHz measurement bandwidth, the 150 MHz ringing is completely missed. A second issue is that while the ringing signal is originated in the converter, the ringing magnitude that appears on the user's PDN is strongly dependent on its high-frequency impedance, which may also be highly dependent on the frequency and location. The ringing magnitude then will be dependent on where it is measured, and the noise may very likely become bigger at hot spots further away from the converter.

Many times, the argument to dismiss this concern is that the board should have sufficient high-frequency bypassing and it will suppress this ringing. While this is a valid argument, the high-frequency ringing current leaving the converter is not specified, so the user does not have numbers to design with.

How the ringing magnitude depends on the input voltage and load current appears to be dependent on circuit and layout details of the converter. Several converters were found where the high-frequency ringing was almost independent of DC load current but varied linearly with input voltage. The high-frequency ringing in some other designs showed very weak dependence on input voltage, but the DC load current had a strong influence. Figure 2.9 shows data from such a converter.

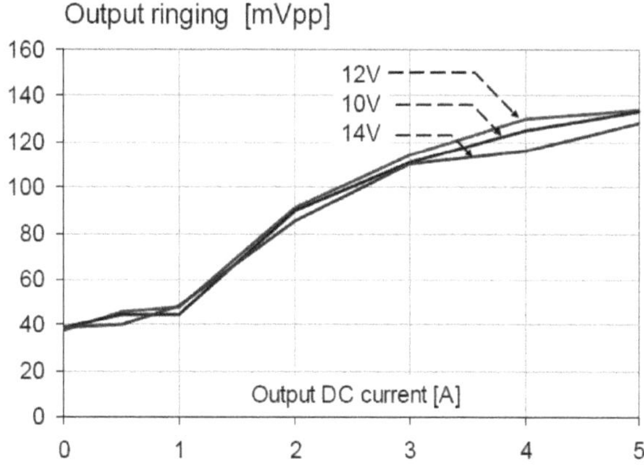

Figure 2.9: High-Frequency Ringing Magnitude in a Point-of-Load DC-DC Down-Converter Measured with 250 MHz Bandwidth as a Function of Input Voltage and Load Current

In today's converters, the ringing frequency is in the 100-500 MHz frequency range, which overlaps with several reference clock frequencies of high-performance buses. The clock signals represent a few mA current in each line, whereas the ringing is associated with the switching edge of the converters, which may represent tens of amperes.

Having an unspecified noise component related to these high-current edges that may accidentally line up with clock frequencies represents a design risk in high-performance systems.

Challenges with Digital Power

Digital power is a promising new trend in power conversion. It enables not only a versatile on-the-fly control and monitoring of the usual parameters such as output voltage and turn-on and turnoff delays and ramps, but also the implementation of flexible nonlinear control loops.

Nonlinear control loops make it possible to significantly improve the transient behavior. Although this seems to be all and only good, there is a hidden catch: the worstcase design and validation of a system design becomes more challenging.

The expectation is that an adaptive nonlinear control loop eventually reduces the worst-case peak-to-peak transient noise, or alternately, it enables the user to achieve the same worst-case noise with fewer or cheaper external capacitors. In the design phase, the challenge is the lack of a systematic design approach, which could quantify the improvements and benefits.

Validation of a high-performance system design should never be skipped, especially when off-the-shelf converters are used. Since the dynamic performance of the converter depends on the external capacitors we connect to the converter, and chances are that the converter during manufacturing is tested with a different set of capacitors, we have to validate our design in situ.

As illustrated earlier, a well-behaved analog control loop with a properly designed set of external capacitors will show an approximately constant impedance profile, regardless of the static load current and input voltage. This means that the validation is reduced to checking the small-signal output impedance profile over the entire input parameter ranges, such as input voltage, output voltage, and DC load current. The small-signal output impedance can be obtained in a straightforward manner [4], and because it does not require injecting a test signal in series to the control loop, in general it is easier to perform than measuring the Bode plots for the classical loop stability.

When the control loop is made purposely nonlinear, the validation can not be done any more with small-signal output impedance measurements. Nonlinear loop control may tighten the loop for larger output deviations and loosen for small deviations. This means that any measured small-signal output impedance may show unrealistically large impedance values. An illustration of this is shown in Figure 2.10. One should note the low-frequency impedance magnitude of about 18 milliohms. If this was linear impedance, a 1-A load-current step would produce 18 mV output transient, whereas a 10-A load step would produce 180 mV transients.

A quick time-domain scan showed that this was not the case; in fact, as expected, larger load-step currents did not produce proportionally larger transients, which is the whole purpose of the nonlinear loop.

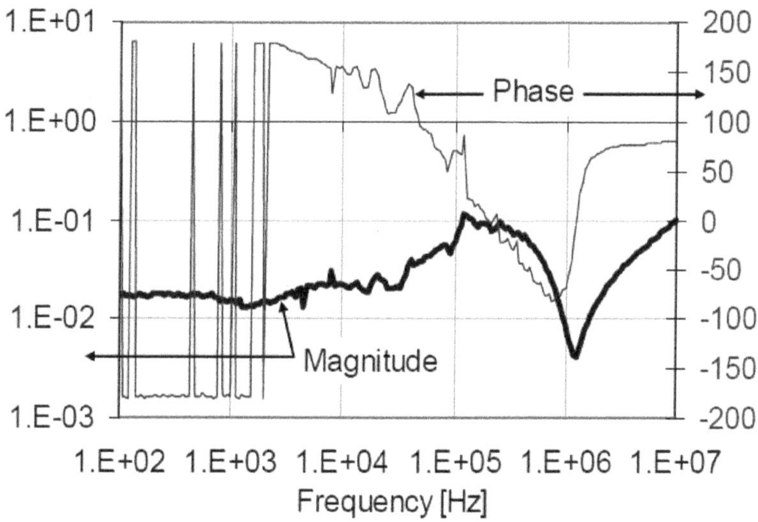

Figure 2.10: Measured Small-Signal Output Impedance of a DC-DC Converter with Nonlinear Digital Loop

While the nonlinear control loop helps to improve the transient behavior, now we have two more variables to scan against: DC load current and load-step magnitude. The frequency parameter of the impedance plots is replaced by the slew rate or rise time of the excitation current step. For a full validation the testing has to be done in the time domain, which, due to its wide-band nature, is more prone to external noise contaminating the data.

Similar to the additional challenge of validation, the thorough design procedure also requires additional effort. With the understanding that the nonlinear control loop reduces the worst case transient noise, the actual design procedures to provide the improved noise level are still to be worked out.

Conclusion

The critical missing link in the proper PDN design is the middle frequency range, where the PDN performance is determined by the interaction of the DC-DC converter and external bypass capacitors. Bypass capacitors and DC-DC converters with well-behaved analog control loops can be modeled as linear circuits; hence, their worst-case transient noise can be estimated from the inverse Fourier transform of their combined impedance profile.

The worst-case transient noise can be minimized by flattening out the impedance profile. It was also illustrated that traditional phase margin requirements do not necessarily result in optimum impedance profile.

To help proper system design with off-the-shelf DC-DC converters, the following is necessary:

- A simulation model of the converter's output stage and control loop should be available, which allows the user to properly design the external capacitors to meet user requirements.
- The PARD specification should be changed so that the high-frequency ringing is included.

It is usually known that a detailed simulation model may reveal proprietary information about the converter design. A usable compromise is to have a sufficiently detailed behavioral model, which would first require the definition of a standard interface for such simulations.

It was also shown that nonlinear control loops, though they can improve the noise performance significantly, lack systematic design procedures and increase the validation challenges.

References

[1] A. S. Kislovski, R. Redl, and N. O. Sokal, *Dynamic Analysis of Switching-Mode DC/DC Converters*, Van Nostrand Reinhold, 1991.

[2] V. Drabkin et al., *Aperiodic Resonant Excitation of Microprocessor Power Distribution Systems and the Reverse Pulse Technique*, Proceedings of EPEP 2002, p. 175.

[3] István Novák, *Comparison of Power Distribution Network Methods: Bypass Capacitor Selection Based on Time Domain and Frequency Domain Performances*, in TecForum TF-MP3, DesignCon2006, Santa Clara, California, February 6-9, 2006.

[4] István Novák, *Frequency Domain Power Distribution Measurements—An Overview*, in HP-TF1: Measurement of Power-Distribution Networks and Their Elements, at DesignCon2003 East, June 23-25, 2003, Boston, Massachusetts.

Chapter 3—The Advantage of Controlled-ESR Polymer Capacitors

Hideki Ishida, Design and Application Section Manager, Sanyo Electric Co.

This chapter is a revised version of the paper presented as part of TF-MP3, *Controlled ESR Capacitors Have Arrived*, DesignCon 2007, Santa Clara, California, January 29-31, 2007.

LOW—EQUIVALENT SERIES resistance (ESR) capacitors have become very popular in the market, but a too-low ESR value capacitor sometimes causes problems. Inappropriate ESR value may cause oscillation or ringing in the circuit; therefore, it is important to control both the minimum and maximum ESR value of capacitors.

Being aware of these problems, Sanyo has started shipping some of the POSCAPs with specifications on both minimum and maximum ESR value.

Controlling ESR Value of Tantalum Polymer Capacitor

In order to supply controlled ESR value capacitors, the capacitors' ESR value must be measured. When screening is performed to select the capacitors within certain range of the ESR, the out-of-range capacitors significantly impact the productivity. The only way to gain the productivity is to precisely control the ESR distribution so the center value of the distribution curve falls between specified minimum and maximum ESR.

This paper explains two key factors to control the ESR.

Structure

The dominant factors to determine ESR value are terminal frame, welding resistance of tantalum lead wire, interface resistance of each layer, and conductivity of cathode material. For example, the resistance of copper is lower than that of 42 alloy, therefore, low-ESR POSCAP series such as TPE/TPF/TPL/TPLF use copper as the terminal material.

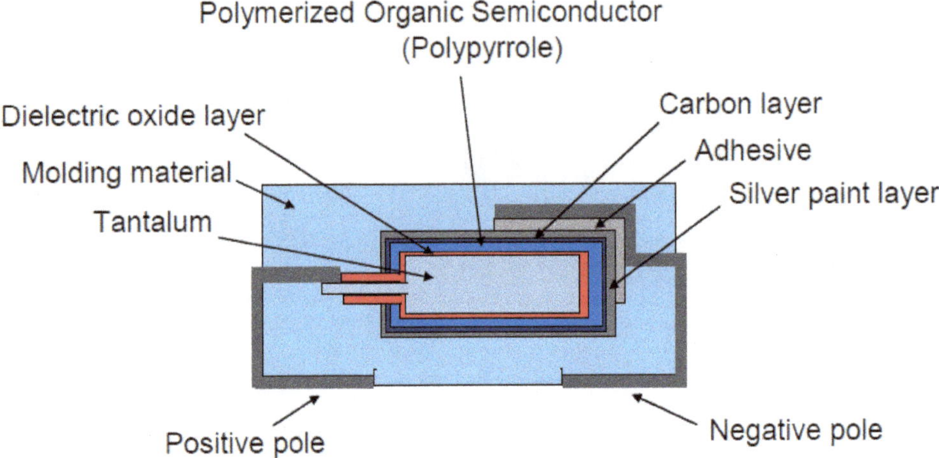

Figure 3.1: Cross-Section of POSCAP, TP Series with Gull Wing Terminals

Cathode Material

While a tantalum polymer capacitor shares a similar structure with a conventional standard tantalum capacitor, one of the significant differences is the conductivity of the cathode material.

POSCAP uses polypyrrole as a polymer and the conductivity is 1,000 times higher than that of manganese dioxide. Since the conductivity contributes the most within the above-mentioned four factors to determine ESR, polymerization process plays an important role in the distribution of ESR.

Over many years of fine-tuning the process, Sanyo achieved the lower ESR value and is now able to control the ESR distribution of the particular POSCAP models within the 5 mOhm range in the low ESR (less than 12 mOhm) products.

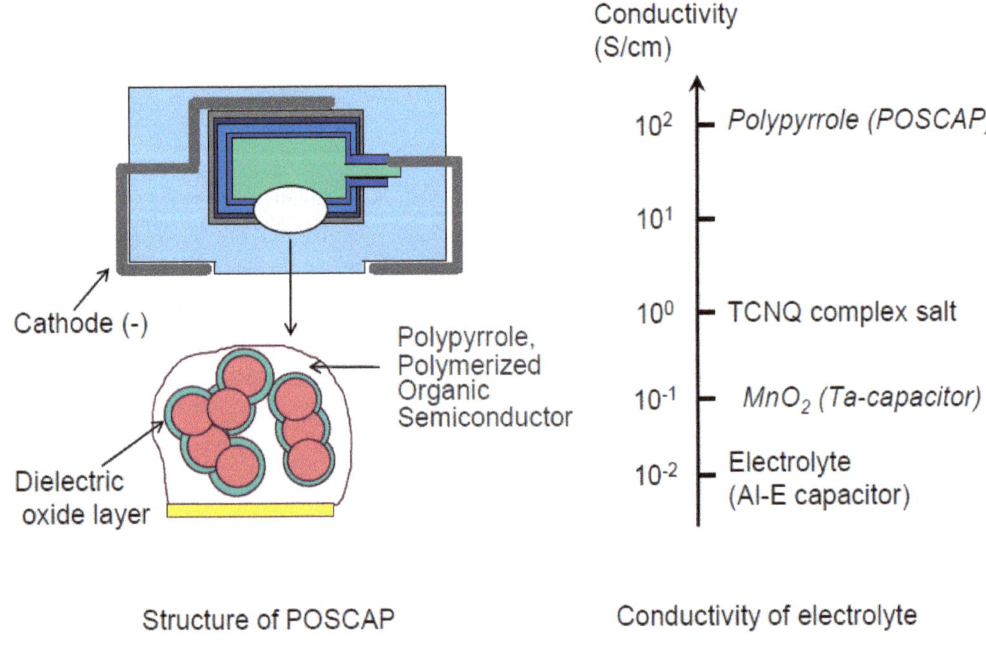

Figure 3.2: Close-Up View of Dielectric and Comparison of Electrolyte

Importance of the Controlled ESR Value Capacitor

The load current of central processing unit (CPU) and the slew rate are increasing every time new CPUs are introduced. V-core circuit requires a combination of several pieces of multiplayer chip capacitors (MLCCs) and a few bulk capacitors. MLCC mainly functions in a high-frequency range due to its low ESL value. Bulk capacitors operate in the low to middle frequency range. The total impedance is calculated by the parallel sum of impedances of the previously mentioned capacitors. A sudden change in the impedance curve implies that some kind of mismatch took place. A two-step solution needs be taken to prevent the mismatch from taking place. A first step is to use capacitors with a well-balanced ESR, ESL, and capacitance. A second step is to use the proper combination of capacitors in parallel.

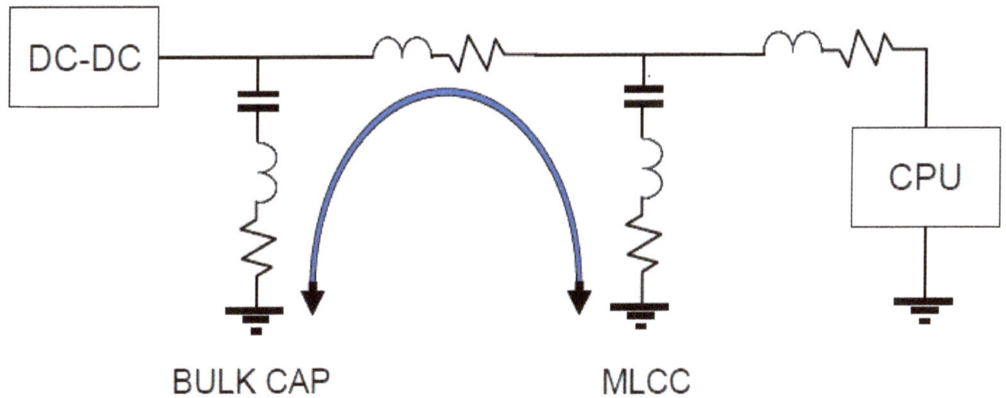

Figure 3.3: Standard CPU Power Distribution Circuit

Well-Balanced Capacitors

A three-element model is used to express the basic equivalent circuit of a capacitor.

The damping condition is as follows:

$$D = ESR^2 - 4(\frac{ESL}{C})$$

$D > 0$	Stable	ESR is big and, therefore, not likely to cause ringing noise
$D < 0$	Ringing	ESR is small and, therefore, likely to cause ringing noise

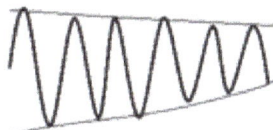

Case 1: Capacitance = 330 µF; ESR = 4 mOhm; ESL = 1.7 nH

$$D = .004^2 - 4(\frac{1.7n}{330\mu}) = -4.6 \times 10^{-6}$$

Case 2: Capacitance = 330 µF; ESR = 8 mohm; ESL = 1.9 nH

$$D = .008^2 - 4(\frac{1.9n}{330\mu}) = 41 \times 10^{-6}$$

It is a common mistake to believe that the lower ESR is always better. As it is proven by Case 1 and Case 2, although the actual ESR value is half in Case 1 compared with Case 2, the damping condition formula turns negative and it is likely that Case 2 results in ringing.

Case 3: Capacitance 330 µF; ESR = 3 mohm; ESL = 0.5 nH

$$D = .003^2 - 4(\frac{0.5n}{330\mu}) = 2.9 \times 10^{-6}$$

In Case 3, ESR value is as low as 3 mOhm, but when ESL value can be improved to 0.5 nH, the damping condition formula shows a positive result, and it is unlikely to cause ringing.

The above two cases explain the importance of balanced capacitors and the necessity for the capacitor manufacturers to specify both maximum and minimum ESR values for some case.

From a manufacturer point of view, specifying the minimum ESR may cause a severe impact on the yield rate in a manufacturing line, and thus it is essential to use a technique to produce controlled ESR capacitors.

Good Combination of Capacitors

As mentioned earlier, using a well-balanced capacitor is only the first step, and without a good combination of capacitors in parallel, ringing may still however occur.

An application case study from a game console is provided in Figure 3.4. One piece of aluminum polymer capacitor with two kinds of MLCCs was originally used in this application. Since the through-hole type aluminum polymer capacitor was used in this case, a big bump can be observed around 400 kHz due the nature of the capacitor (high ESL). The positions of capacitors and the pattern layout have direct impact on transient response; therefore, it is difficult to design the best pattern layout. The test card to measure the transient response of each solution is shown in Figure 3.5.

Three values of parallel bypass caps	C1	tol. [%]	C2	tol. [%]	C3	tol. [%]	
Capacitance C [microF]:	2060	20 -20	2.35	20 -20	0.05	20 -20	Fmin[Hz] 1.00E+03
Ser. resistance ESR [ohms]:	0.008	20 -20	0.014	20 -20	0.289	20 -20	Fmax[Hz] 1.00E+08
Ser. inductance ESL [nH]:	7.83 ACTUAL	20 20	0.317	20 -20	0.564	20 -20	Total:
Number of caps in each value:	8		72		225		305

2012MLCC 1005 MLCC

SANYO RECOMMEND MODEL the number of MLCCs around the CPU is estmate only

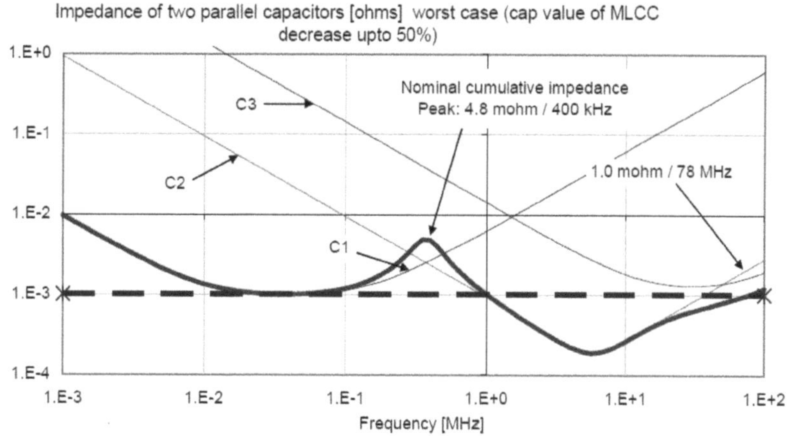

Figure 3.4: Parallel Impedance of Three Groups of Capacitors

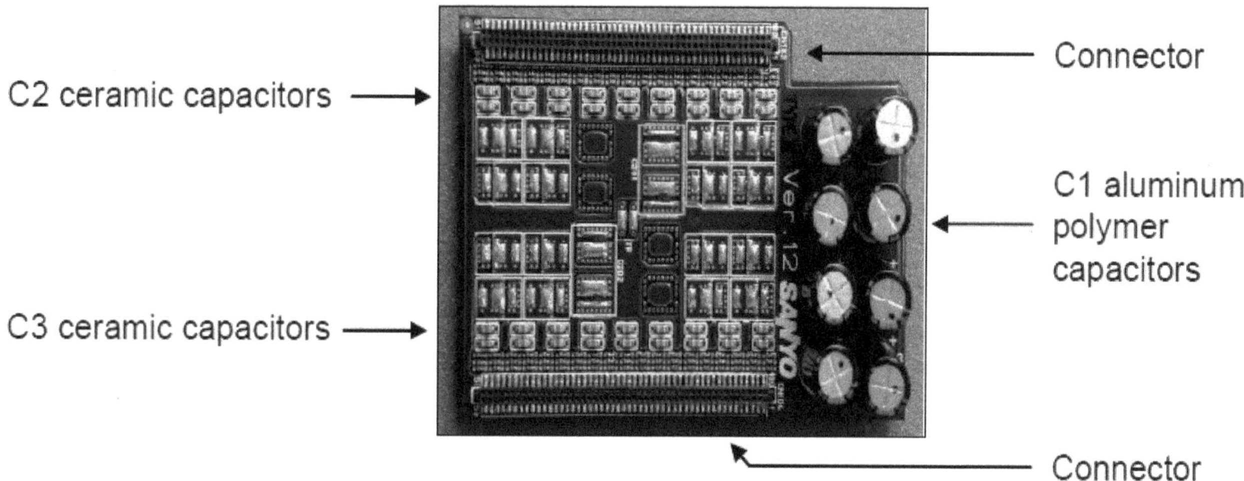

Figure 3.5: Capacitor Test Card

D-case size SMD capacitors can be populated in this test card as well.

The current goes through the connector on each side.

Transient Response

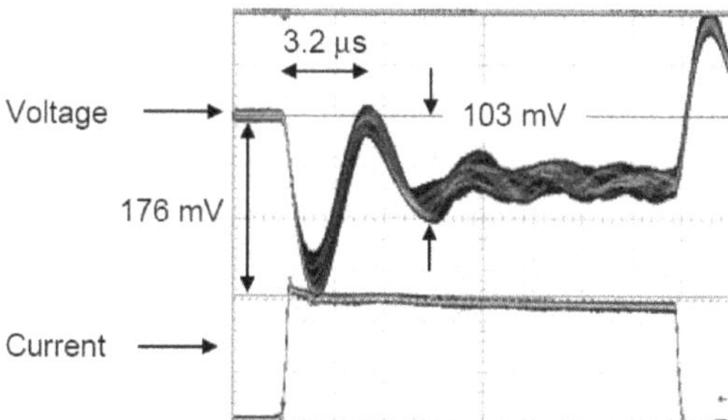

Figure 3.6: Transient Response of Original Design with 8 x 220 μF 6.3V 8 mOhm (typ) Aluminum Polymer, 72 x 4.7 μF 14 mOhm MLCC, and 225 x 0.1 μF 0.285 ohm MLCC. Note the Large Ringing.

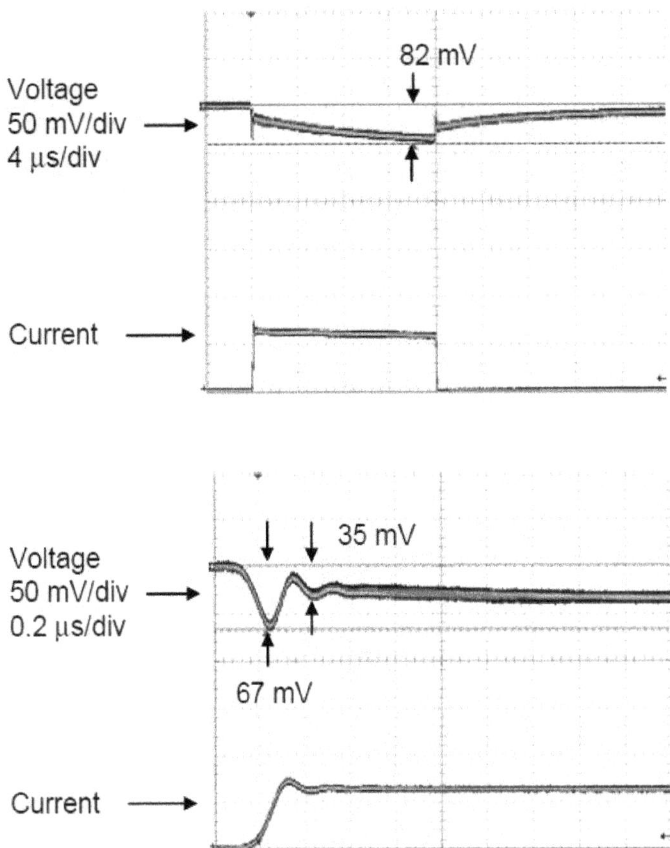

Figure 3.7: Transient Response of Modified Design with 10 x2R5TPLF470M7. Note the Stable Response.

We determined the test condition that imitates the high load current of a game console or server:

- Load current: change from 10A to 80A
- Slew rate: 480A/µs
- Applied voltage: 1.5V
- Voltage regulator module (VRM): 250 kHz x 2 phase

We would like to emphasize that this test card is made solely for simulation purposes and that the design does not represent the best solution. However, it does prove that the good combination capacitors prevent the ringing issues even on the non-optimum design test card.

Tantalum Polymer Capacitor with DC-DC Converter Switching in the MHz Range

MHz switching direct current-direct current (DC-DC) converter has become popular in order to supply more than 100A to the CPU or to reduce the capacitance value of bulk capacitors.

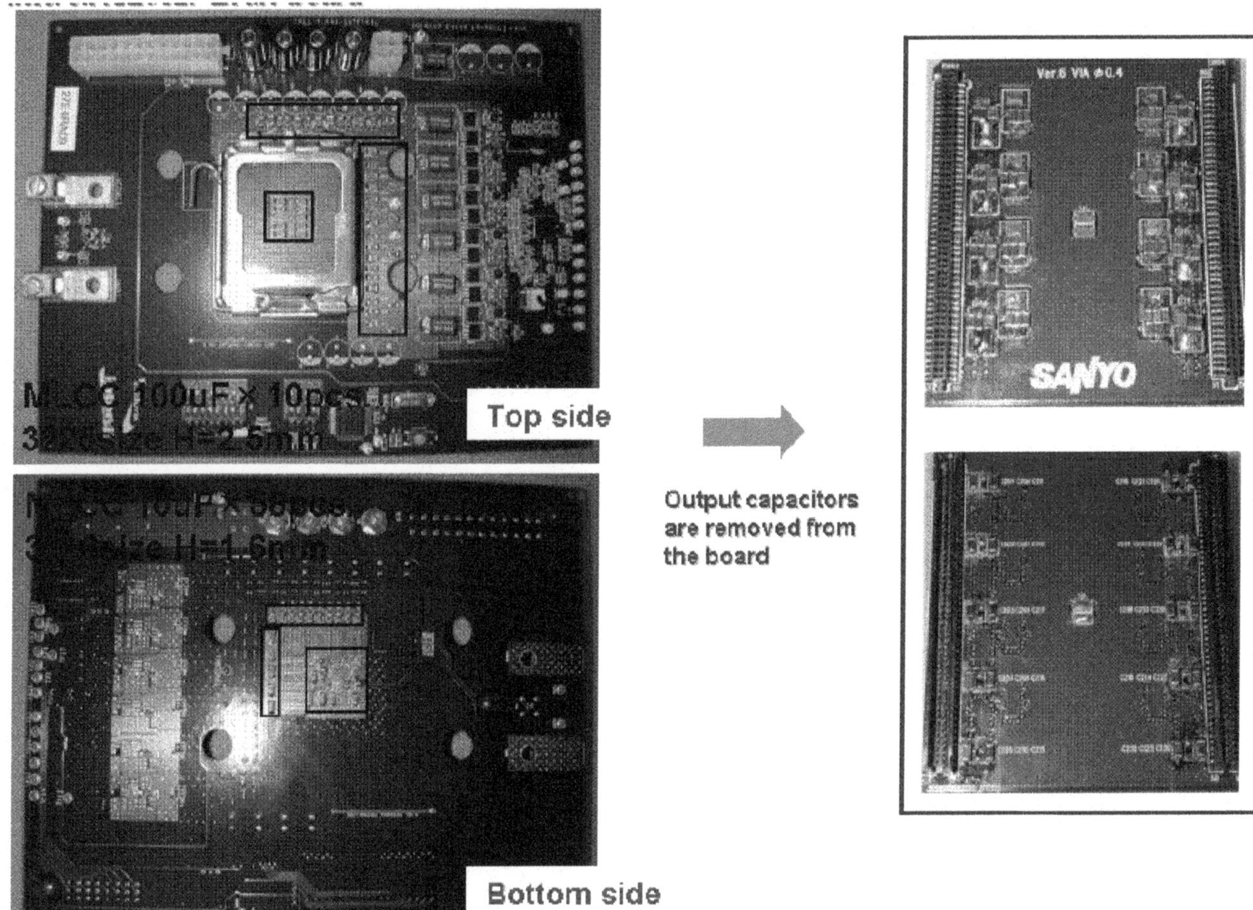

Figure 3.8: Intersil ISL6327 Test Board (Left), Sanyo Test Board (Right)

In this solution, ceramic capacitors are the only players and no bulk capacitors can be seen on board. However, by using well-balanced bulk capacitors with the good combination, a tantalum polymer capacitor can still play an important role.

The test condition is as follows:

- Load current: 10A to 120A
- Applied voltage: 1.2V
- Slew rate: 960A/µs
- Switching frequency: 3.2 MHz (533 kHz x 6phase)
- Evaluation board: Intersil ISL60327

Original Design

All ceramic capacitors for output side 100 µF x 10 pieces + 10 µF x 58 pieces.

In order to populate different kinds of capacitors on the test board, the board shown on Figure 3.8 (on the right) is used.

Modified Design

Polymer tantalum capacitor (2TPLF330M6) + 10μF x 20 pieces.

Test Result

The test result of transient response for each design is shown in Figure 3.9.

In the original design, the total ESL value is very low; therefore, transient response is very fast and voltage does not drop easily. In the new design, tantalum polymer capacitor supplies energy after 400 nsec. A good balance of impedance does not support the creation of big ringing and the transient response becomes stable soon.

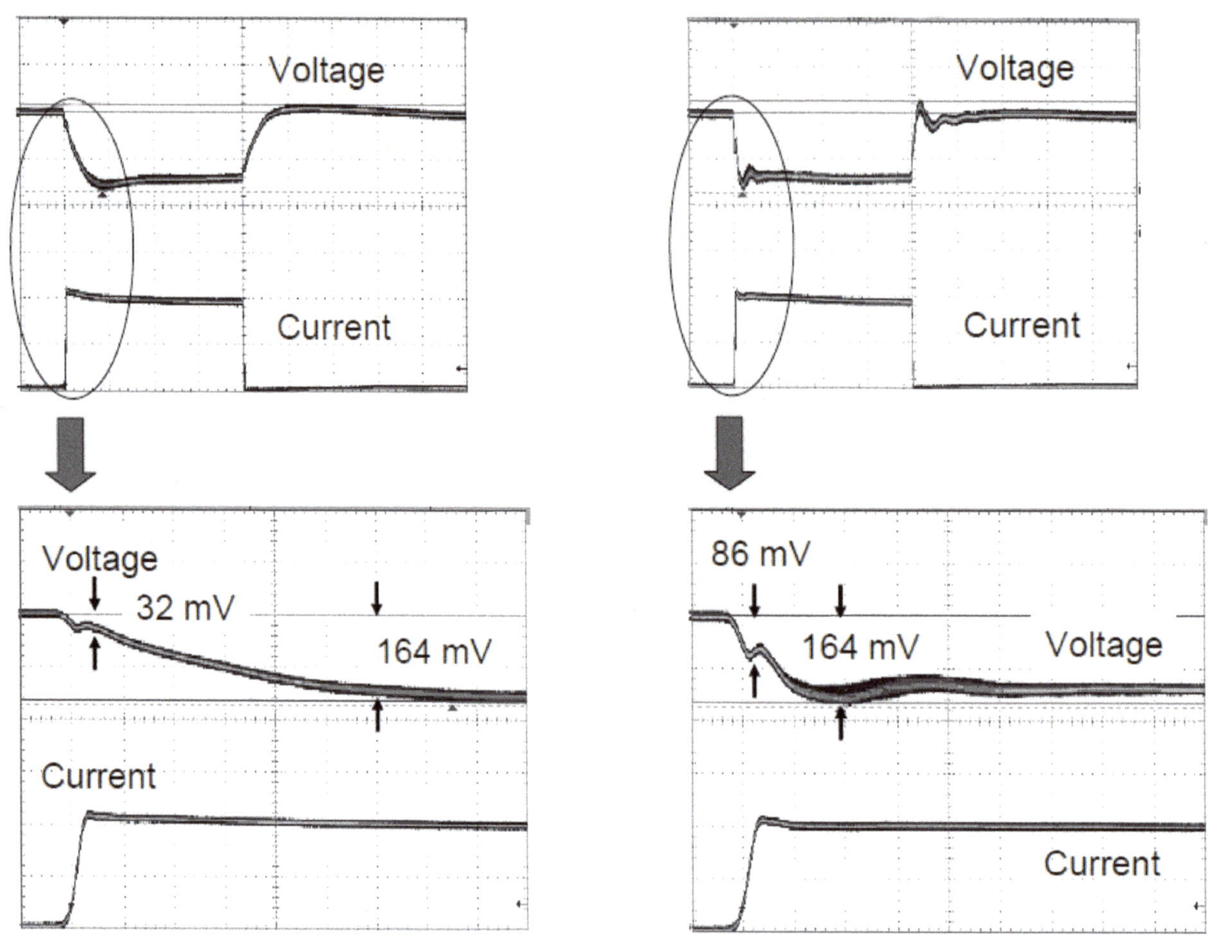

Figure 3.9: Transient Response of Original Solution (on the Left) with 10 x 100 μF 3225 MLCC and 58 x 10 μF 3216 MLCC; and Transient Response with a Better Combination of Capacitors (on the Right) with 3 x 2TPL33M6 POSCAP and 20 x 10 μF 2012 MLCC. Both Bottom Graphs have 400 ns/div Horizontal and 50 mV/div Vertical Scale.

Total impedance is calculated as follows and is shown in Figure 3.10.

Parallel bypass capacitors	C1	tol. [%]	C2	tol. [%]	
Capacitance C [microF]:	100	20	10	20	Fmin[Hz]
		-20		-20	1.0E+03
Ser. resistance ESR [ohms]:	0.005	20	0.005	20	Fmax[Hz]
		-20		-20	1.0E+08
Ser. inductance ESL [nH]:	0.6	20	0.6	20	
	ACTUAL	20		-20	Total:
Number of caps in each value:	10		58		68

SANYO RECOMMENDED MODEL	MLCC 100uF 3.2X2.5mm	MLCC 10uF 3.2X1.6mm

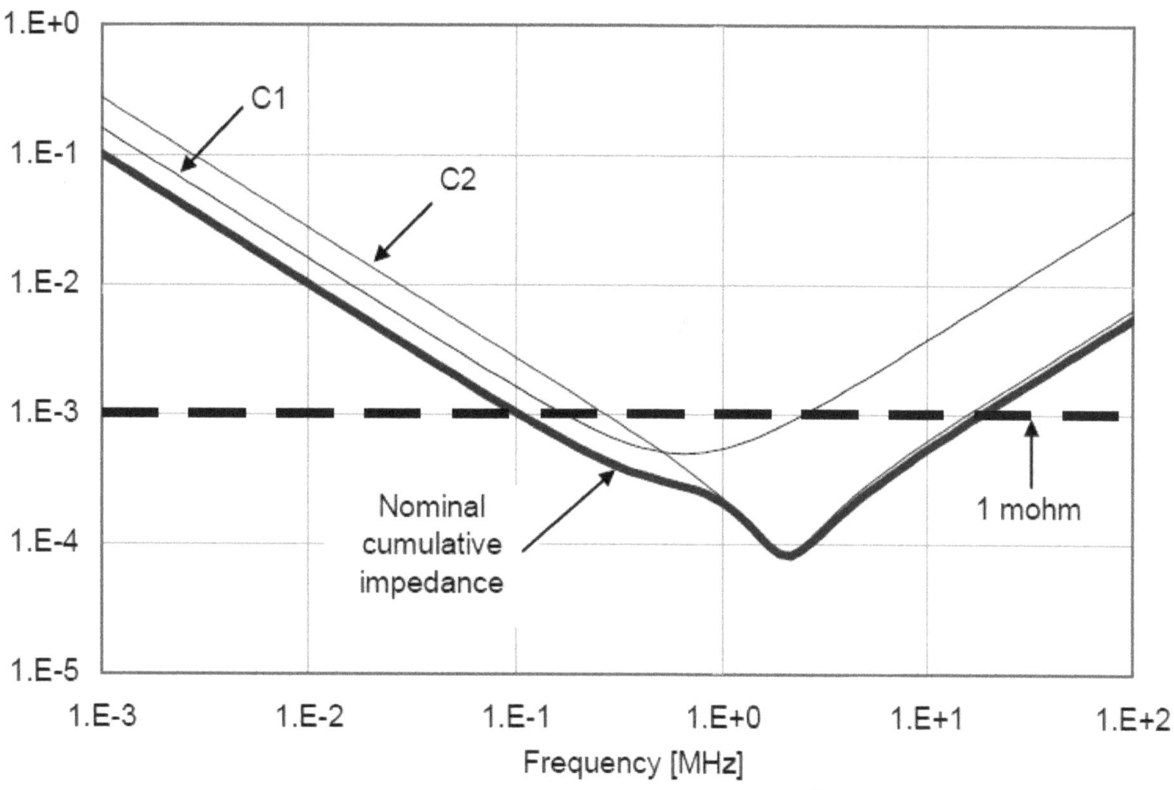

Figure 3.10: Total Impedance of Original Design

Due to the bias and temperature characteristics of MLCCs, the capacitance value will be lower under working conditions. Voltage regulator module (VRM) ver.11 specifies the target of total ESR value as 1 mOhm. It also specifies to use 3.2 MHz multi-phase solutions and to keep the target of impedance (= 1 mOhm) from the frequency of 1/6 of 3.2 MHz (= 533 kHz).

Therefore, it is not necessary to keep the target impedance (= 1 mOhm) from 100 kHz, at which ESR value is specified.

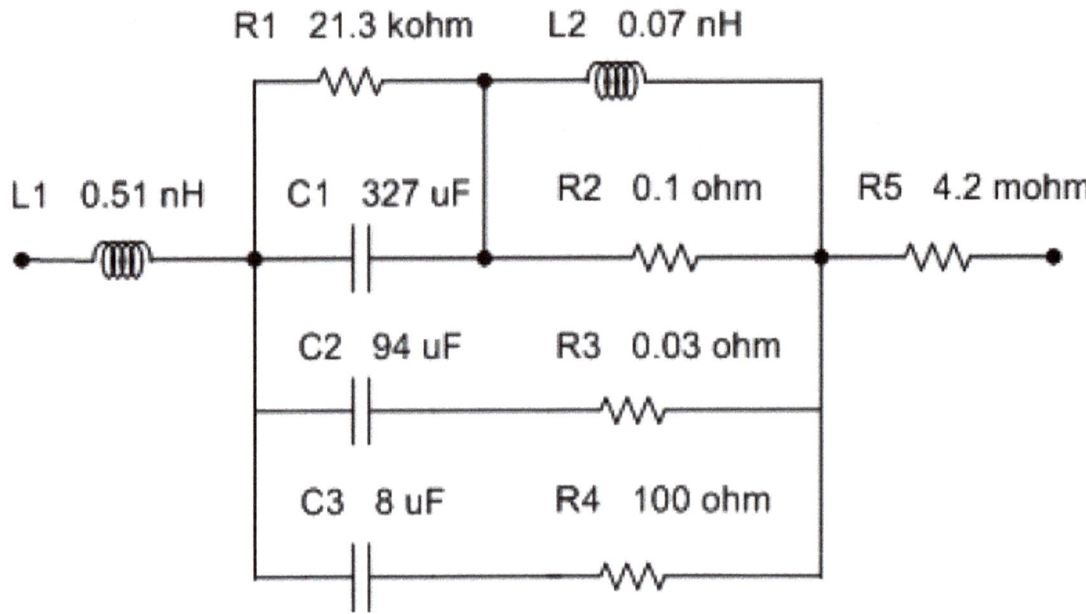

Figure 3.12: Equivalent Circuit of Tantalum Polymer Capacitor

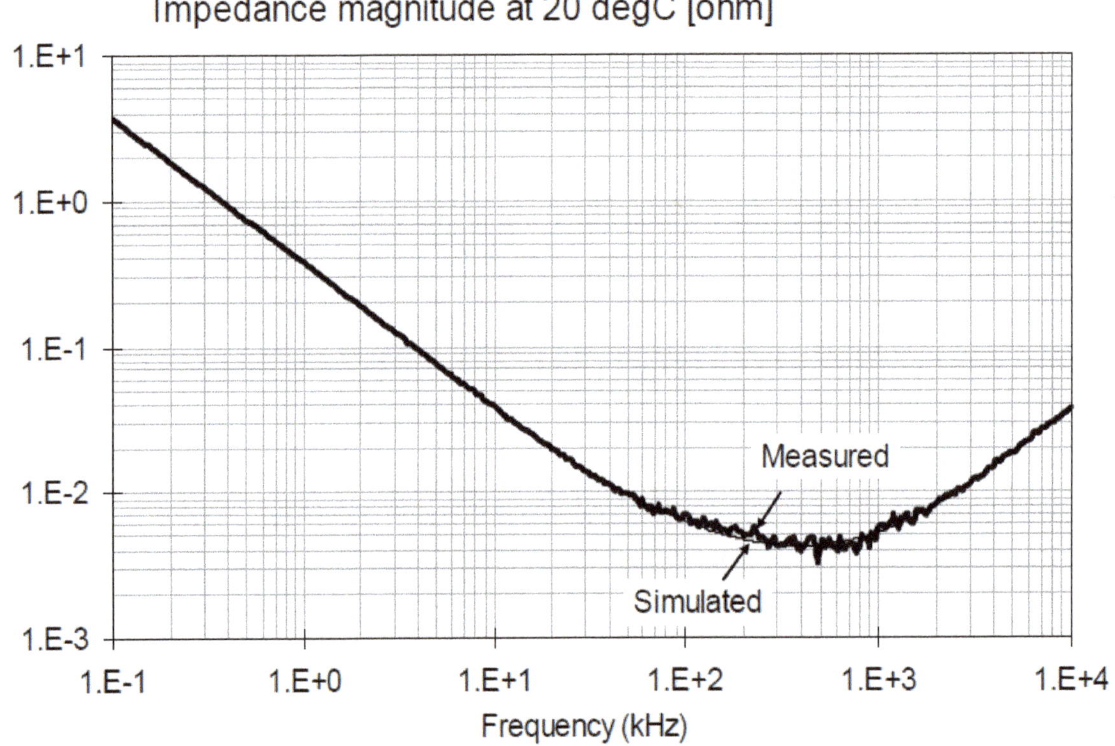

Figure 3.13: Comparison of Simulated and Measured Impedance Magnitude of 2TPLF470M7 (2V-470 μF): D2T SIZE (7.3x4.3x1.8mm) POSCAP

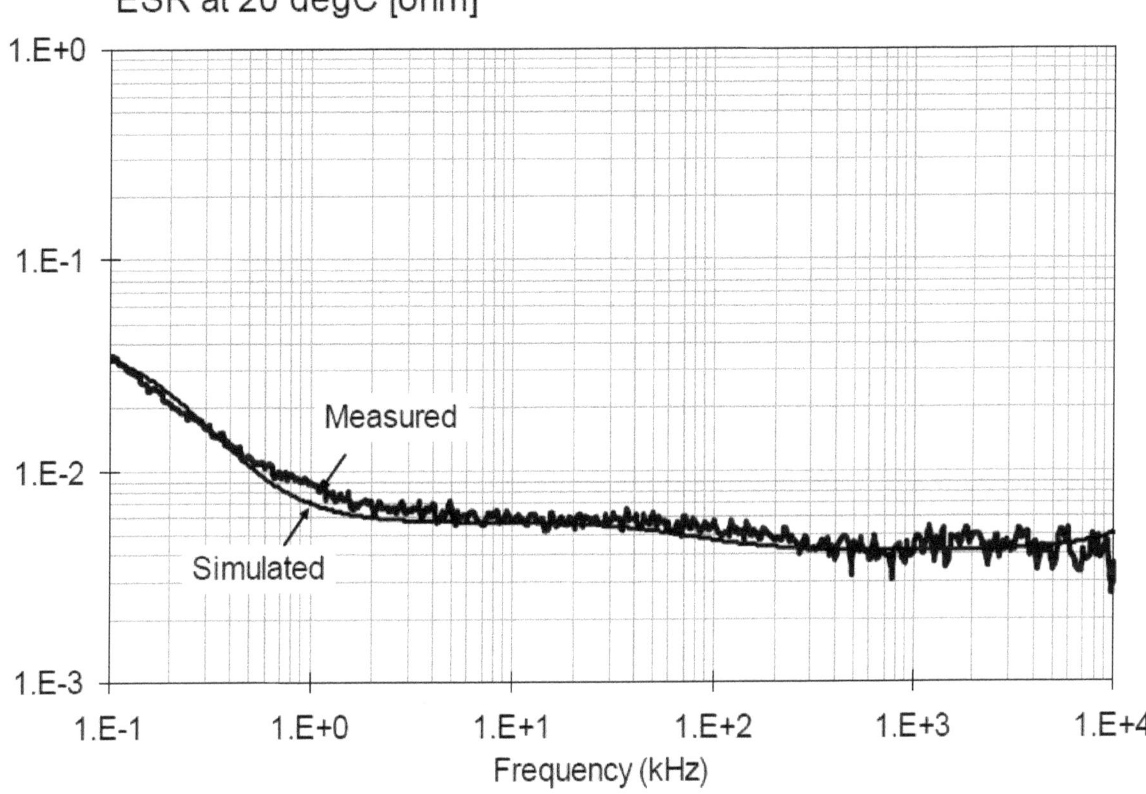

Figure 3.14: Comparison of Simulated and Measured ESR of 2TPLF470M7 (2V-470 μF): D2T SIZE (7.3x4.3x1.8mm) POSCAP

An ESL measurement method was specified in Japan in March 2006. To this day, Sanyo continues to work with almost all of the capacitor manufacturers at the JEITA Association based in Japan.

The working group in JEITA has been contacting the Electronics Industries Alliance (EIA) for more than three years now, so the International Electrotechnical Commission (IEC) will adopt this specification as the world standard.

We believe that a universal standard for the ESL measurement will benefit all engineers, since it is critical in comparing the various capacitors' true characteristics. Knowing the true ESL value is essential in creating the accurate equivalent circuit of capacitors.

As a result, engineers will realize the advantage of using the controlled ESR value capacitors, and thus it will result in the peace of mind without ringing issues.

Chapter 4—ESR-Controlled MLCCs and Decoupling Capacitor Network Design

Masaaki Togashi, Senior Development Engineer, TDK Corp.
Chris Burket, Senior Applications Engineer, TDK Corp.

This chapter comes from a revised paper presented in TecForum TF-MP3, *Controlled ESR Capacitors Have Arrived*, DesignCon 2007, Santa Clara, California, January 29-31, 2007.

WIRELESS NETWORK DESIGN engineers have capitalized on the low equivalent series resistance (ESR) and high-Q characteristic of multilayer ceramic capacitors (MLCCs) for impedance matching in radio frequency (RF) and filter circuits. In addition, MLCCs used in electromagnetic compatibility (EMC) applications to bypass and/or decouple have traditionally viewed this characteristic as beneficial.

However, recent increases in microprocessor operating speeds, coupled with a desire for more tightly controlled voltage variation, are challenging this point of view. The low ESR and high-Q MLCC characteristic is not always an advantage and, in fact, it may be considered an obstacle in some designs and applications.

Selecting a capacitor with specific ESR characteristics and applying it correctly can form a more stable, low-impedance floor across a wider frequency range. Such performance greatly improves the efficiency of low-impedance power distribution network (PDN) designs. By utilizing a unique MLCC design, TDK can provide controlled-ESR capacitors for these market needs.

Introduction

The higher processing speeds of microprocessors have caused engineers to be more focused on the signal integrity of circuitry design. Smaller internal integrated circuit (IC) interconnects and lower energy consumption have reduced the power supply voltage and, as a result, the degree of voltage variation has been tightened and has become more important. Design engineers need to reduce the impedance of the PDN as much as possible for a given frequency range, which makes verifying the power integrity difficult. The PDN consists of a voltage regulator (VR), printed circuit board (PCB) power plane, main processor unit (MPU) socket, MPU package, and decoupling capacitors. See Figure 4.1.

The PDN is provided through a current demand by the MPU and an impedance level target is mandated by the transient current and voltage allowance of the power supply, which yields the following:

$$V_{TOL} = I_{TRANS} \text{ x } |Z_{PDN}|$$

The total impedance of the PDN is expressed as a combination of impedances that are equivalent to VR response, PCB and MPU packaging, and decoupling capacitors. Today's designs require the PDN to exhibit lower-impedance profiles due to the MPU's high current and lower operation voltage. Due to its large influence on the PDN, the selection of the decoupling capacitors in this design is extremely important.

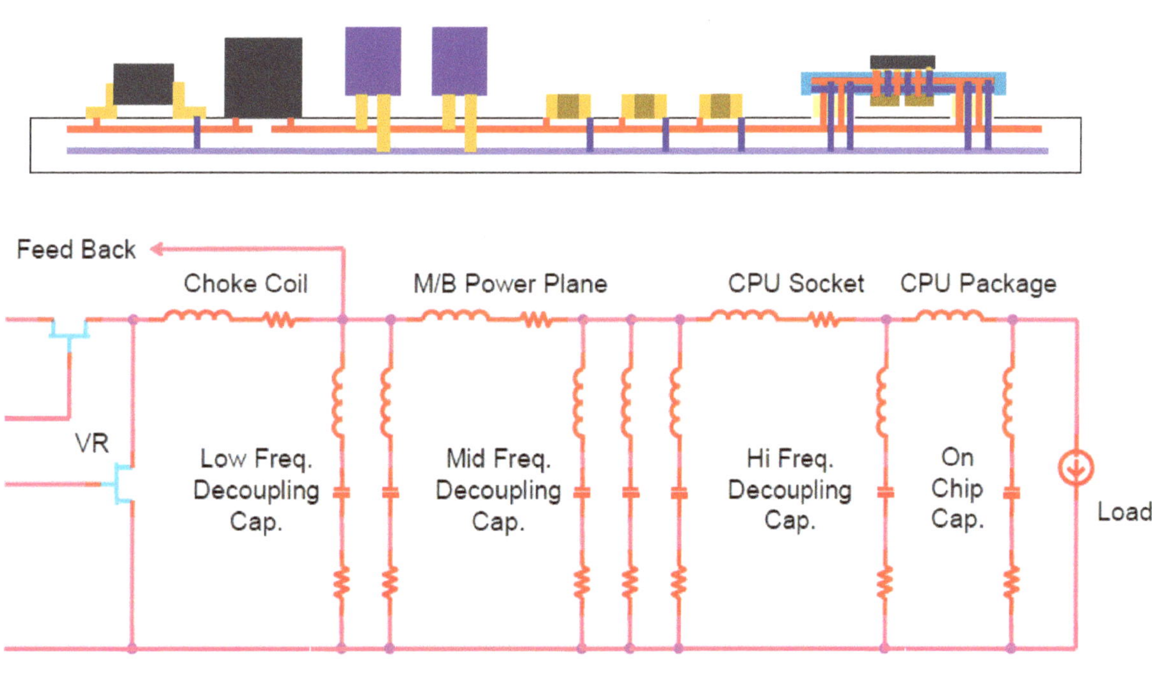

Figure 4.1: Power Distribution Network and Decoupling Capacitor Network

Decoupling Capacitor Network

Currently, many physical sizes and values of MLCCs are used as decoupling capacitors. The various MLCC self-resonant frequencies (SRFs) must be considered in the decoupling capacitor network (DCN).

By dispersing the various SRFs across a wide frequency range, the subsequent PDN impedance could be lowered, forming a low and flat impedance floor. Three to five discrete MLCCs of various capacitance values are commonly used for this result.

See the left plot of Figure 4.2.

However, this design method is prone to some problems. For instance, when parallel resonance causes the resultant impedance to rise, the impedance peak ("anti-resonance") may exceed the targeted impedance.

See the right plot in Figure 4.2.

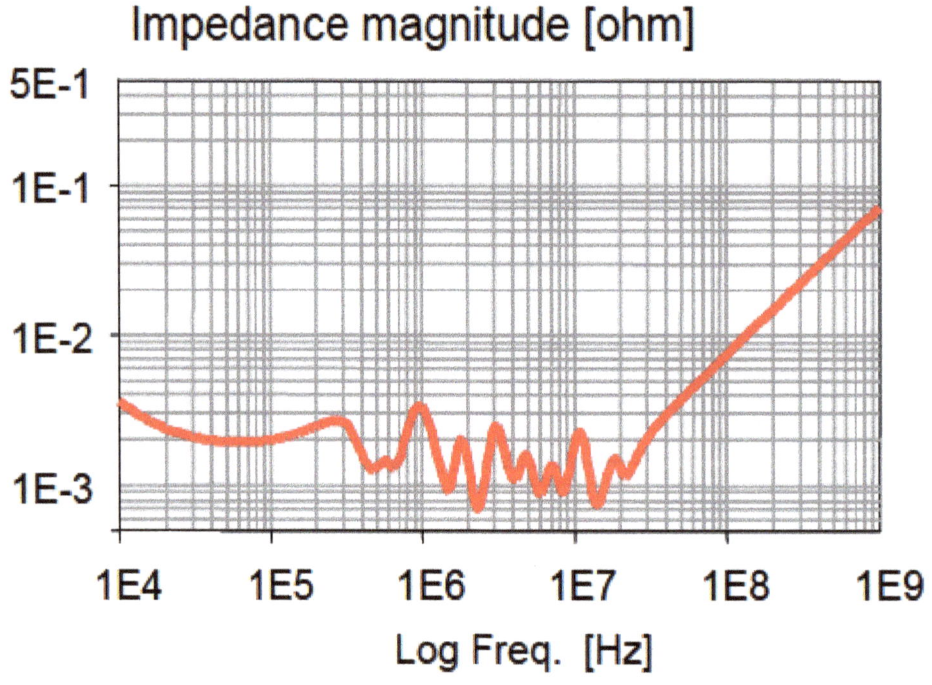

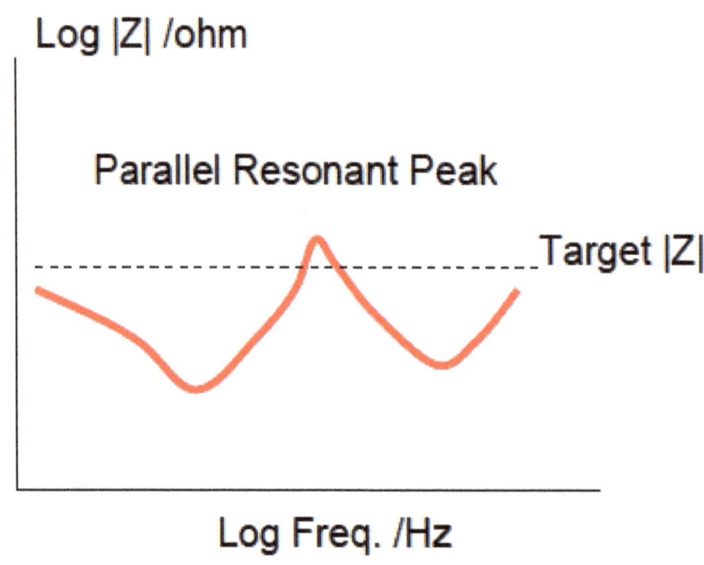

Figure 4.2: Impedance of PDN (Top) and Parallel Resonance (Bottom)

This variation in peak impedance could permit the permissible voltage to be exceeded, which may lead to error or insulation failure within the IC itself.

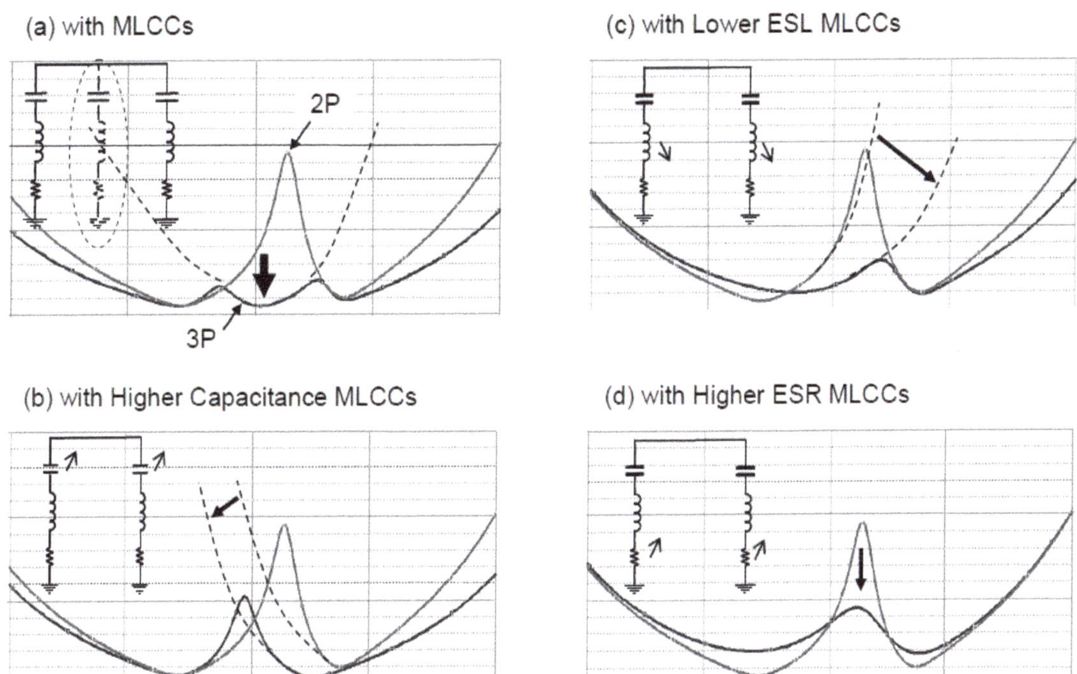

Figure 4.3: Impedance Peak Reduction with Parallel Resonance

Here are actually two ways to reduce the peak impedance:

1. Use an additional capacitor value to negate the parallel resonance peak
2. Optimize the MLCC's capacitance, equivalent series inductance (ESL), and ESR

For (1), the peak impedance can be reduced by adding one more MLCC with a SRF equivalent to the two-piece MLCC parallel resonance. See Figure 4.3(a). For (2), it is possible to use a single MLCC to offset the parallel resonance by utilizing higher capacitance, lower ESL, and/or higher ESR.

See Figures 4.3(b), 4.3(c) and 4.3(d).

With advancements in ceramic dielectric material, coupled with thinner layer processing technology, higher capacitances in smaller case-sized MLCCs are now available. See Figure 4.4. In addition, lower ESR values can be found with reverse geometry type and multi-terminal type capacitors. See Figure 4.5.

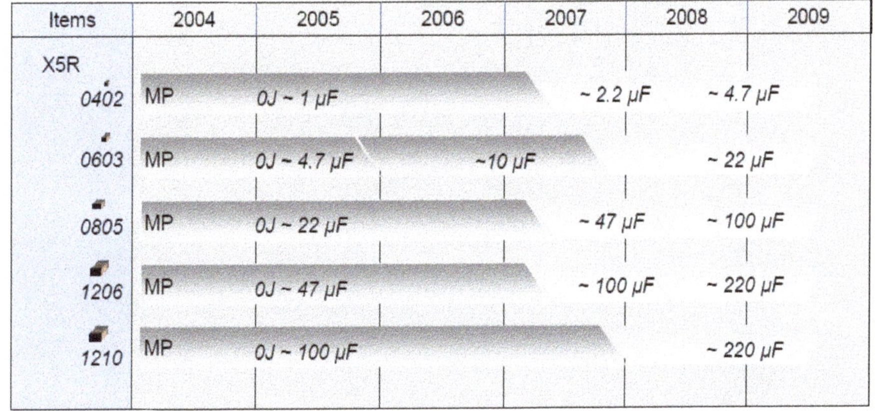

Items
2004
2005
2006
2007
2008
2009
X5R
0402
MP
0J ~ 1 μF
~ 2.2 μF
~ 4.7 μF
0603
MP
0J ~ 4.7 μF
~10 μF
~ 22 μF
0805
MP
0J ~ 22 μF
~ 47 μF
~ 100 μF
1206
MP
0J ~ 47 μF
~ 100 μF
~ 220 μF
1210
MP
0J ~ 100 μF
~ 220 μF

Figure 4.4: High-Capacitance MLCC Road Map

MLCCs with low ESR (high-Q) commonly exist in RF and microwave applications.

However, intentionally higher ESR MLCCs rarely exist today. In low-impedance PDN designs, it is widely known that MLCCs, with adequate impedance, can be a very effective solution. Currently, there are no high-ESR MLCCs available for PDN designs. For the past year, TDK has been developing MLCCs with high capacitance and controlled ESR for these specific decoupling applications.

Reverse Geometry MLCCs

Case Size	Capacitance.(uF)	ESL(pH)typ.	TDK Type
0204(0510)	~0.1	100	C0510
0306(0816)	~2.2	110	C0816
0612(1632)	~10	150	C1632

Multi-Terminal MLCCs

Case Size / Terminal#	Capacitance.(uF)	ESL(pH)typ.	TDK Type
0306(0816) / 8T	~2.2	50	CLLE
0508(1220) / 8T	~4.7	42	CLLC

Figure 4.5: TDK's Low–ESL MLCCs

Figure 4.6: Simplified Equivalent Circuit Model

ESR and ESL

Historically, the equivalent circuit of a capacitor has been expressed in terms of a capacitive element as well as a parasitic resistive and inductive element. See Figure 4.6. The ESR of a MLCC is inversely proportional to the number of inner electrodes. Capacitors with a higher number of inner electrodes exhibit lower ESR.

See Figure 4.7.

ESL is expressed by the self-inductance of the inner electrodes and mutual inductance between the inner electrodes and the power plane. See Figure 4.8. Since the ESL of an MLCC depends on its case size and length/width (L/W) aspect ratio and ESR depends on its case size, L/W aspect ratio, and the number of inner electrodes, it is quite difficult to select an optimum combination of Cap/ESR/ESL from currently available products. Until now, PDN designers would spend an extraordinary amount of time with simulation models searching for the optimum combination of performance and cost. Now, thanks to the availability of ESR-controlled MLCCs, the designer is provided another alternative.

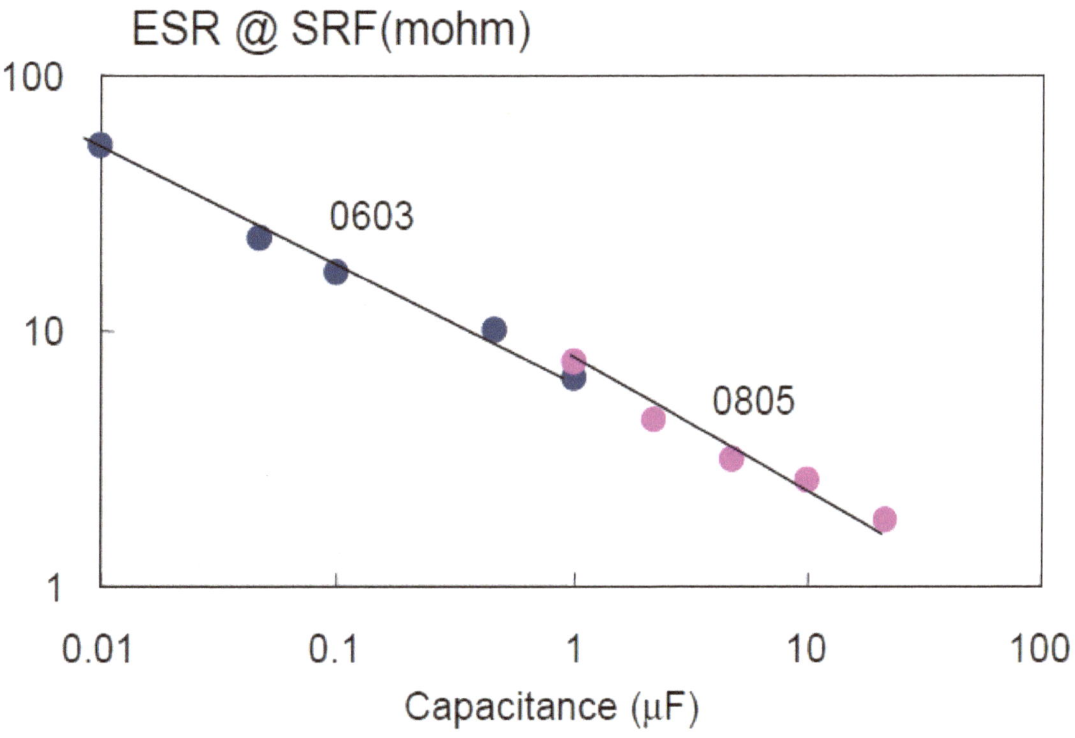

Figure 4.7: Capacitor Package—ESR Dependence

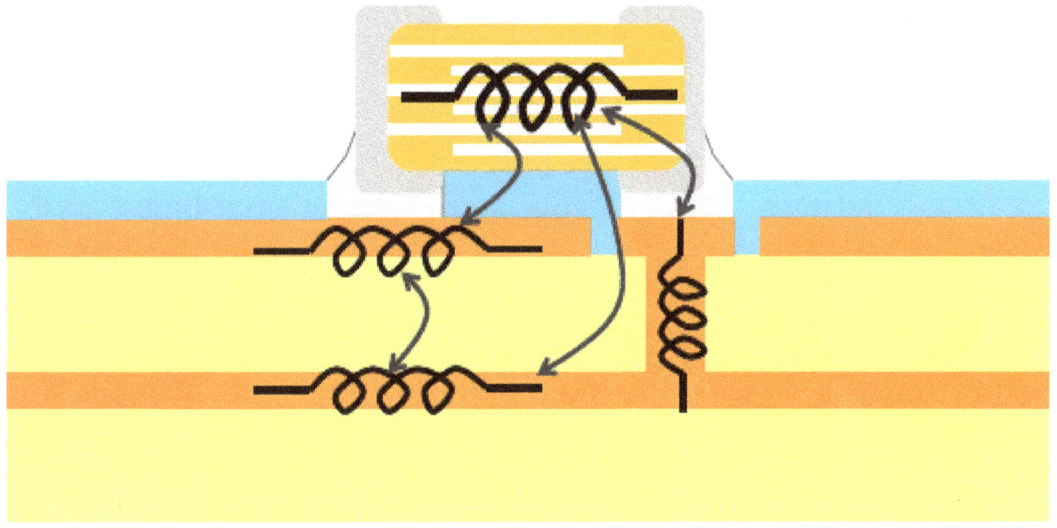

Figure 4.8: Mutual Inductance between MLCCs and PCB

ESR Control Method

In traditional MLCC designs, the ESR decreases with increasing capacitance for a given geometry. Controlling the ESR in MLCCs has traditionally utilized material methods such as semi-conductive terminal electrodes or construction methods such as altered inner and outer electrode shapes. TDK will introduce the latter method, a revised inner electrode approach, coupled with a new outer electrode construction.

This construction can be achieved with current materials, processes, production methods, and equipment. The ESR of an MLCC is determined by the number of inner electrodes connected to the outer termination. With this new MLCC design, the inner electrodes that are not connected to outer termination will be common through the NC terminal. See Figure 4.9.

The NC terminal will not be connected electrically to the circuit on the PCB. Although the number of inner electrodes connecting to the outer termination is reduced, the capacitance value will still depend on the total number of inner electrodes. With this construction, it is possible to change the ESR with the same capacitance value without changing the case size.

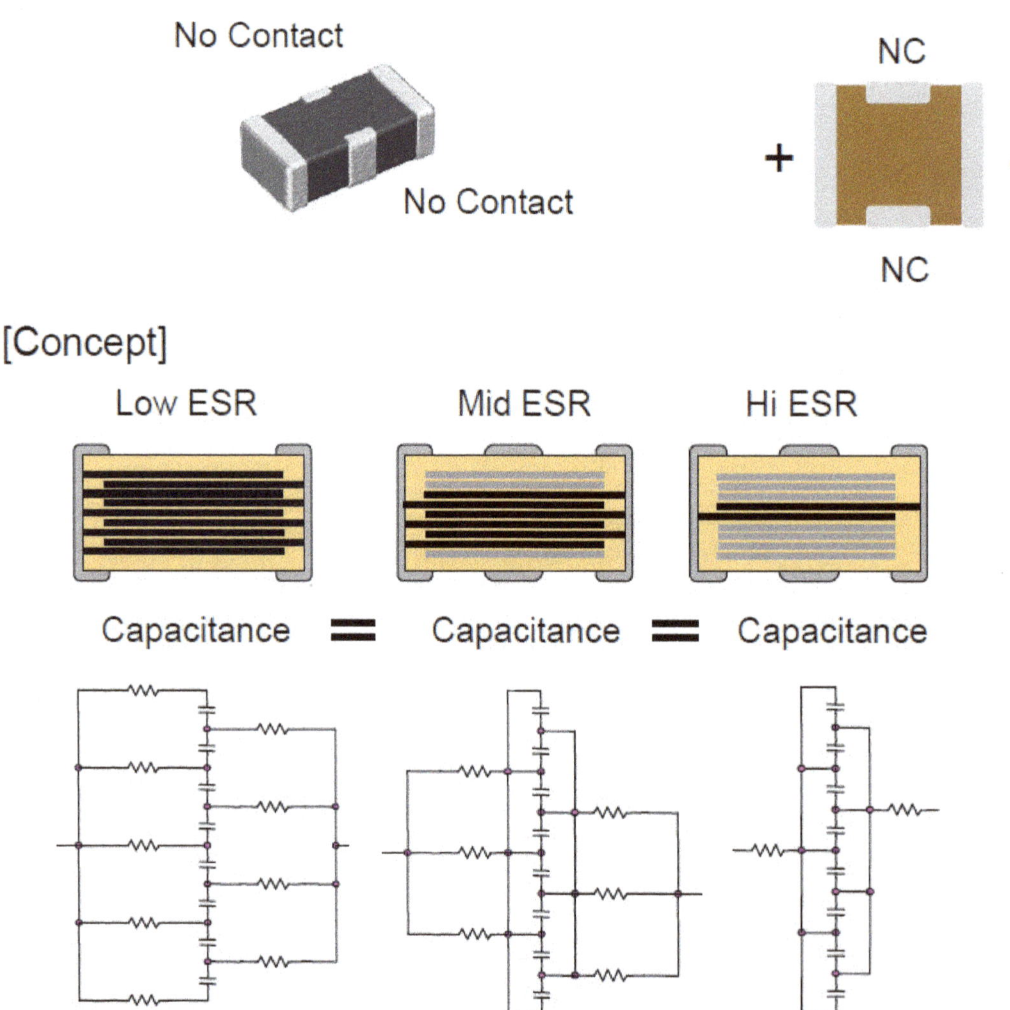

Figure 4.9: TDK-Designed Inner Electrodes for ESR Control

ESR/ESL Measurement of MLCCs

With such low-ESR and -ESL MLCCs, the impedance contribution of the test board must be removed to accurately report these parasitic values. The real characteristics of the device under test (DUT) cannot be obtained without even removing the test board impedance.

In this paper, we discuss the measurement method of ESR and ESL and the calculation method used to compensate for the equivalent circuit model. The test board used is a dual-layer board. Each layer is equivalent to power and ground (GND).

See Figure 4.11.

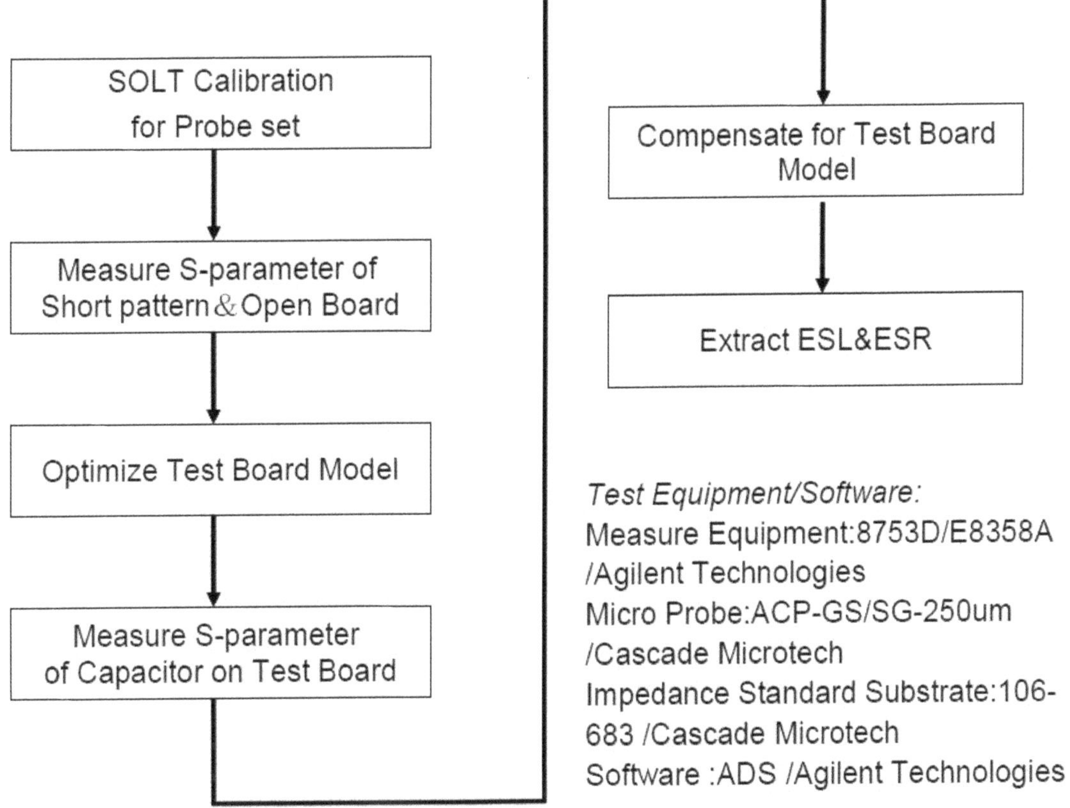

Figure 4.10: Measurement Procedure

Probe pads are located at the test board corners.

In order to measure ESR and ESL, it is necessary to make a test board with an inductance as low as possible, e.g., with large via size and thin layer thickness. For measuring purposes, a vector network analyzer (VNA) and calibrated microprobes were used. The microprobes were calibrated by using a standard impedance board and the VNA calibration guide (open/short/load/thru).

After the calibration, the first step is to form an equivalent circuit model of the test board characteristics. The test board model is expressed with inductance L_b, resistance R_b and capacitance C_b. L_b and R_b are the inductance and resistance of the test board when connected with capacitance in series.

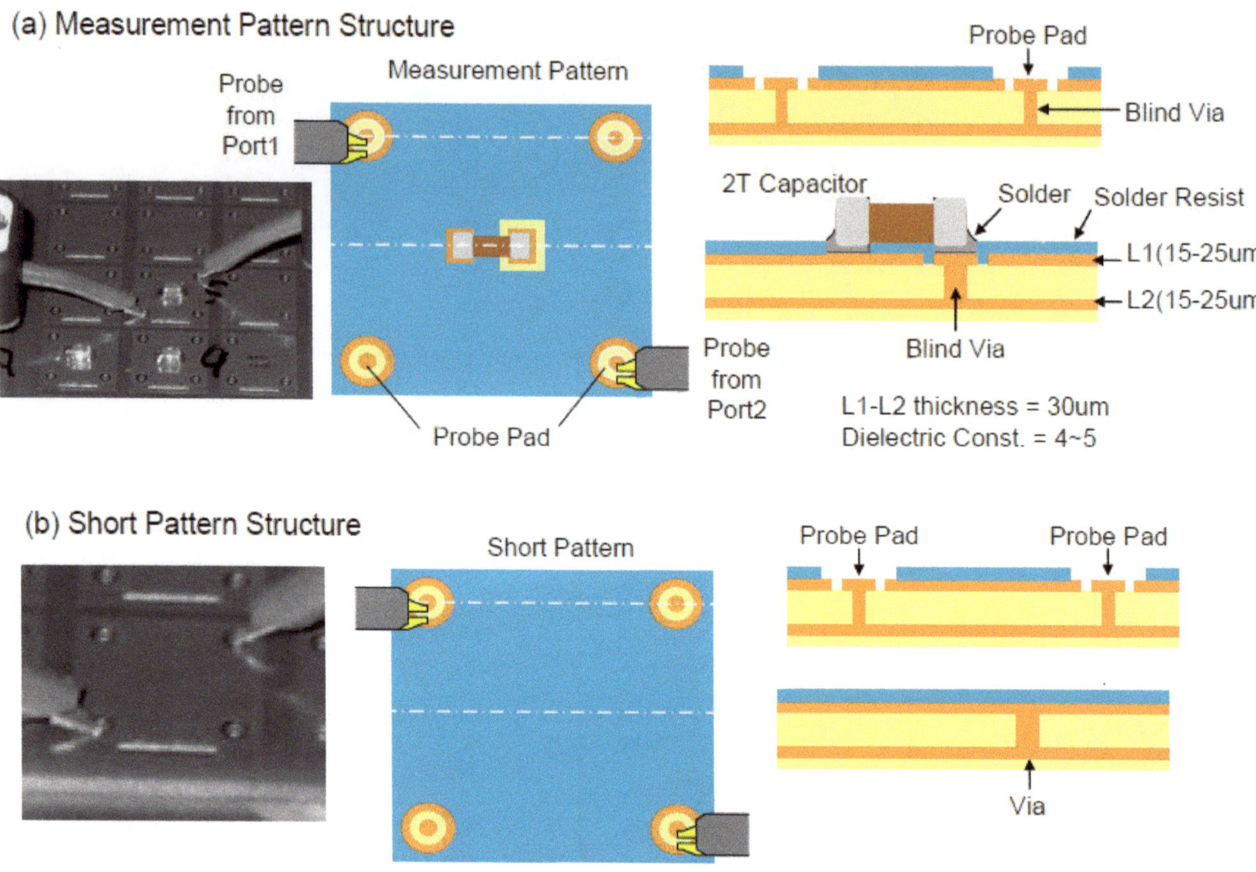

Figure 4.11: Test Board Structure

C_b is the capacitance formed between layer 1 (L1) and L2. In order to calculate L_b and R_b, measure the short pattern of test board. The short pattern is the short construction of the capacitor pad and L1.

See Figure 4.11(b) and convert the obtained S parameter to a Z parameter using the following formula:

$$Z = 25\,x\,\frac{S_{21}}{1 - S_{21}}$$

Let us calculate L_b and R_b so that the measured data and test board model impedance match.

See Figure 4.12(a).

Next, we measure the open board to calculate C_b. Open board is the measurement pattern of no capacitors, and the C_b is obtained to match with measured data.

See Figure 4.12(b).

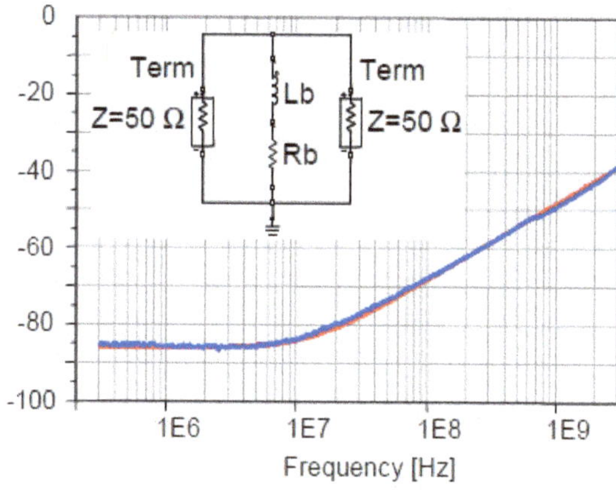

(a) Match Short Pattern Model (Rb and Lb) to Measurement Data

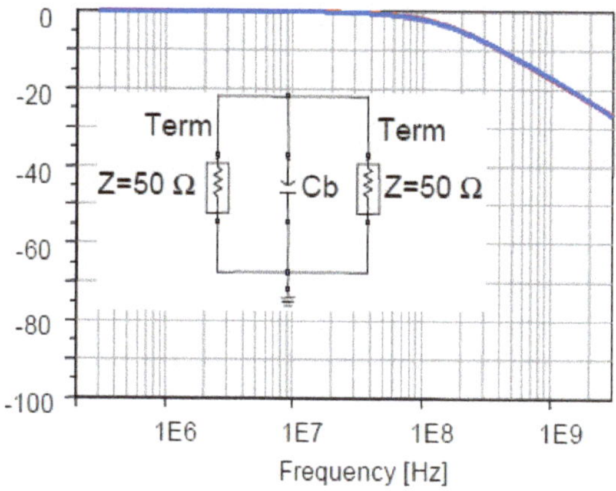

(b) Match Open Board Model (Cb) to Measurement Data

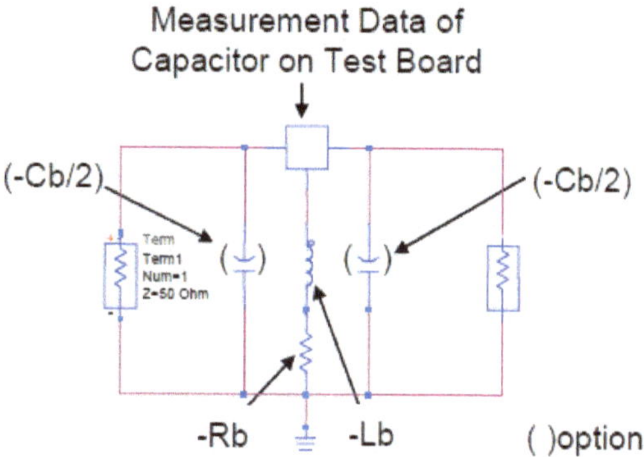

(c) Schematics added Test Board Parameter (Lb/Rb/Cb) to obtain Capacitor Characteristics

Figure 4.12: Extract Test Board Model and Compensation Schematics. (a) Match Short Pattern Model (R_b and L_b) to Measurement Data; (b) MatchOpen Board Model (C_b) to Measurement Data; (c) Schematics Adding Test Board Parameters ($L_b/R_b/C_b$) to Obtain Capacitor Characteristics.

The second step here is to mount the capacitor on the test board to obtain the measured S parameter. The third step is to use L_b, R_b, and C_b as compensation elements in the capacitor data. The compensation values are subtracted from the measured values. The S parameter obtained in Figure 4.12(c) is the capacitor characteristic without the test board characteristics. By converting to a Z parameter, both ESR and ESL are obtained.

See Figure 4.13.

Since ESL is coupled with test board inductance, it may not be perfect. However, when compared with the past studies of electromagneto construction simulation results by FEA, the results of this study gave out to very close ESR values. There could be further discussion, but the data results are a very close approximation.

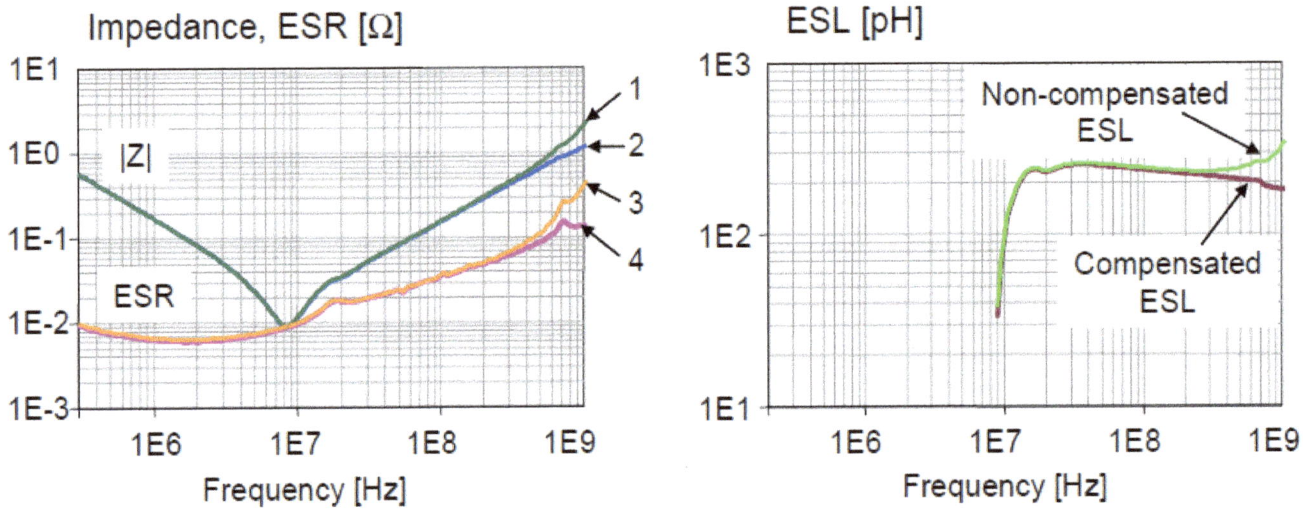

Figure 4.13: Capacitor Characteristics Compensated for Test Board Model. Impedance Magnitude and ESR on the Left (1: Uncompensated |Z|, 2: Compensated |Z|, 3: Uncompensated ESR, 4: Compensated ESR); ESL Correlation on the Right.

Measurement Results

Figure 4.14 and Figure 4.15 show the impedance profiles of the samples prepared by the TDK-controlled ESR method. The test samples are ESR-controlled based upon 0603/1μF and 0805/10μF capacitors. The ESL increased slightly compared with the standard samples. This is caused by the inner electrode series connection. See Table 4.1.

Test Sample	ESR(mohm)	ESL(pH)
STD 0603/X5R/1uF	9.0	258
0603/X5R/1uF /0.2ohm	196	330
0603/X5R/1uF /0.3ohm	336	329
0603/X5R/1uF /0.7ohm	650	358
0603/X5R/1uF /1.2ohm	1230	364
STD 0805/X5R/10uF	4.5	218
0805/X5R/10uF /20mohm	18.0	210
0805/X5R/10uF /35mohm	34.7	214

ESL@300MHZ
ESR@SRF

Table 4.1: ESR and ESL Values.

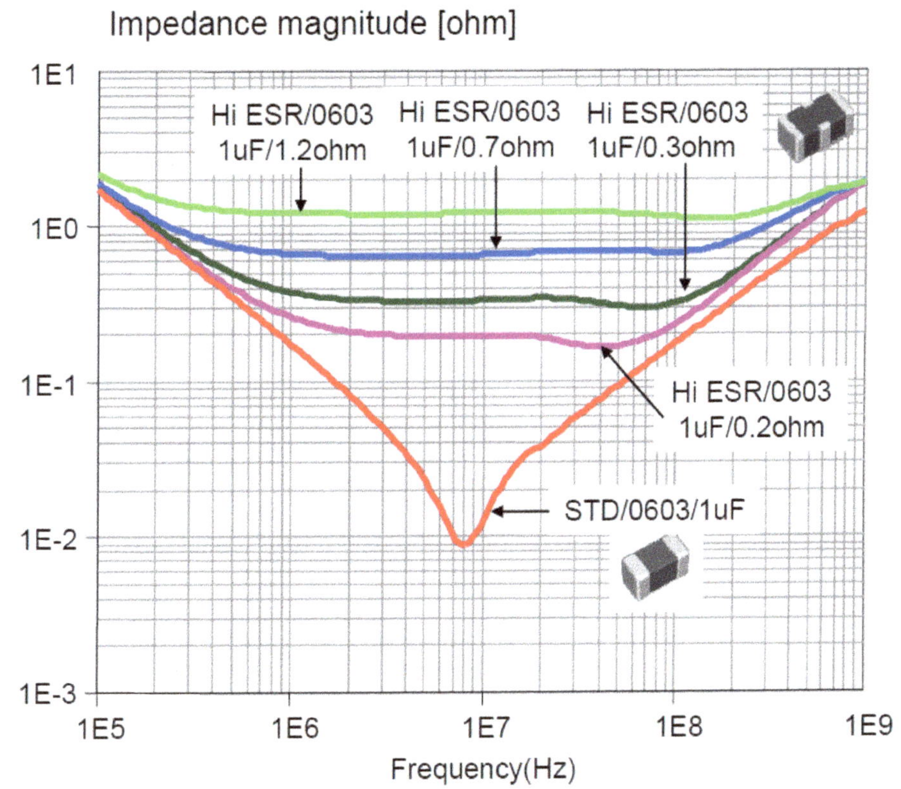

Impedance magnitude [ohm]
1E1
1E0
1E-1
1E-2
1E-3
Hi ESR/0603
1uF/1.2ohm
Hi ESR/0603
1uF/0.7ohm
Hi ESR/0603
1uF/0.3ohm
Hi ESR/0603
1uF/0.2ohm
STD/0603/1uF
1E5
1E6
1E7
1E8
1E9
Frequency(Hz)

Figure 4.14: ESR-Controlled MLCCs versus STD MLCCs in 0603/X5R/1μF Size.

Impedance [ohm]
1E1
1E0
1E-1
1E-2
1E-3
High ESR /0805
10uF/35mOhm
High ESR/0805
10uF/20mOhm
STD/0805/10uF
1E5
1E6
1E7
1E8
1E9
3E9
Frequency (Hz)

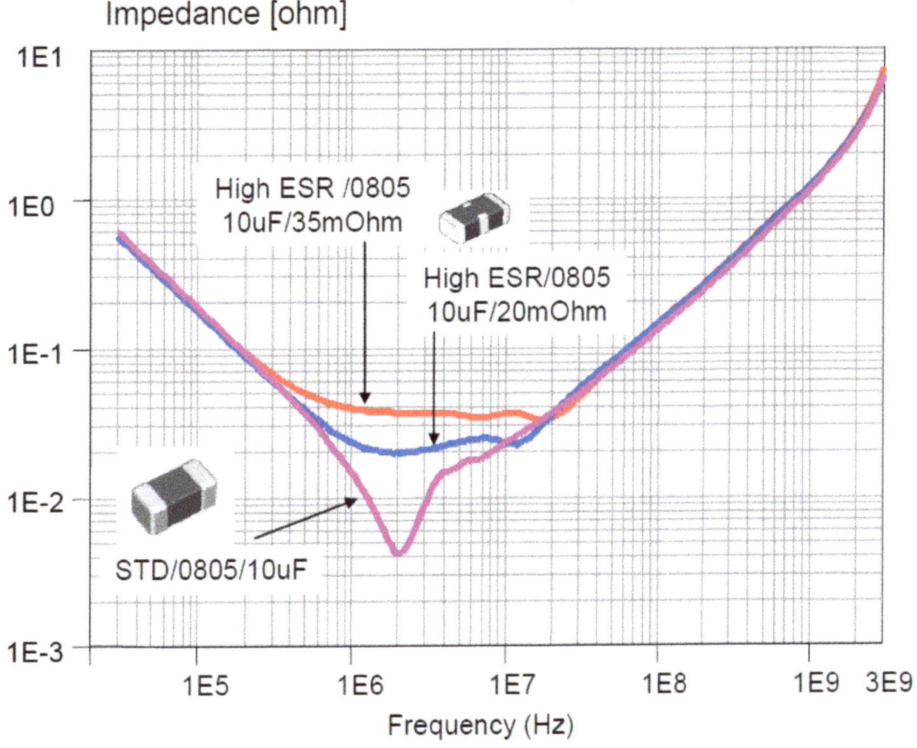

Figure 4.15: ESR-Controlled MLCCs versus STD MLCCs in 0805/X5R/10μF

Circuit Analysis using SPICE Simulation

By simple VR and DCN modeling, the PDN was formed and simulated using the ESR-controlled MLCC. The DCN model was analyzed in the frequency domain and shown in capacitor model Case 1 and Case 2. See Table 4.2. Case 1 was constructed using a standard MLCC, while Case 2 was constructed with an ESR–optimized MLCC using the equivalent circuit model.

Frequency domain analysis results of the PDN model constructed with Case 1 and Case 2 are shown in Figure 4.16. Case 1 illustrates impedance peaks at 300 kHz and at 1.5 MHz due to parallel resonances, exhibiting 3.2 mOhm at 300 kHz and 3.1 mOhm at 1.5 MHz.

On the other hand, Case 2 does not show the impedance peak. In fact, the impedance profile is below 2.5 mOhm up to 100 MHz. Next, the load current response was analyzed in the time domain in the PDN. The load condition was set at 30A to 90A at a 300 kHz switching frequency.

The analysis results of the load current responses are shown in Figure 4.17.

Case 1 shows larger voltage variation than Case 2 due to the transient current from the parallel resonance. Such transient noise is one cause of MPU instability when operated at low voltage.

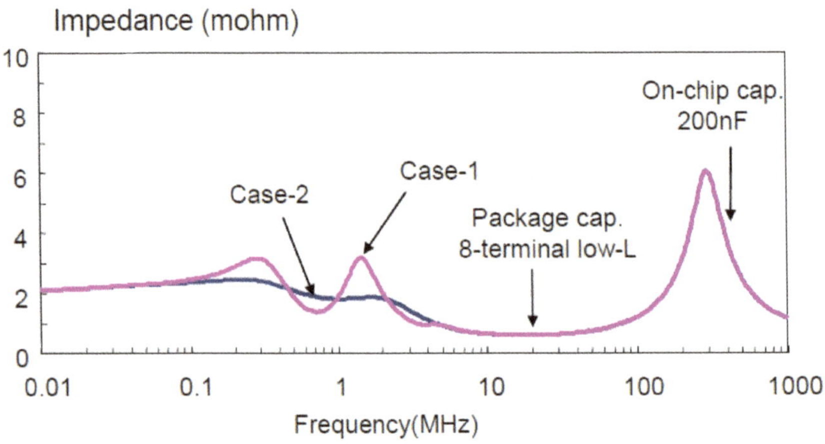

Figure 4.16: Frequency Domain Analysis of Load Current Response Using SPICE Simulation

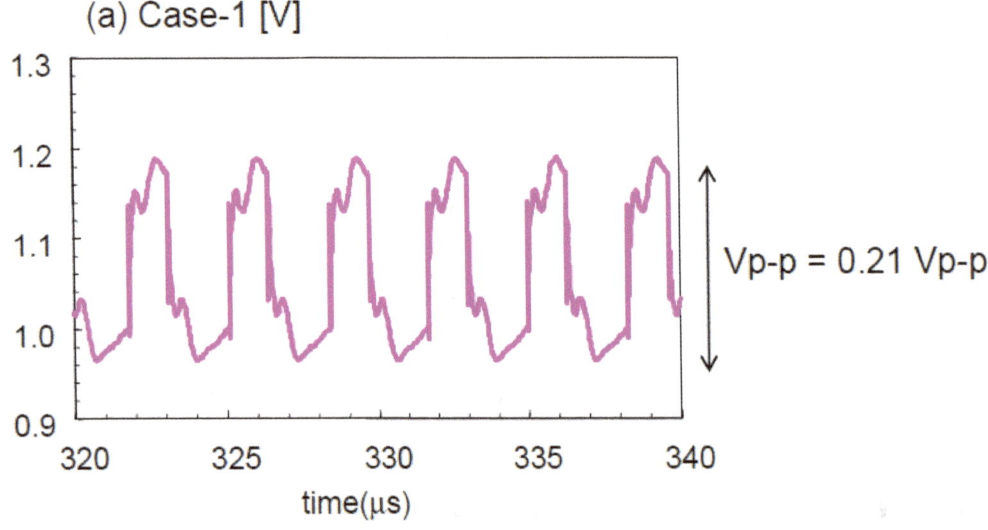

Figure 4.17a: Time Domain Analysis of Load Current Response Using SPICE Simulation

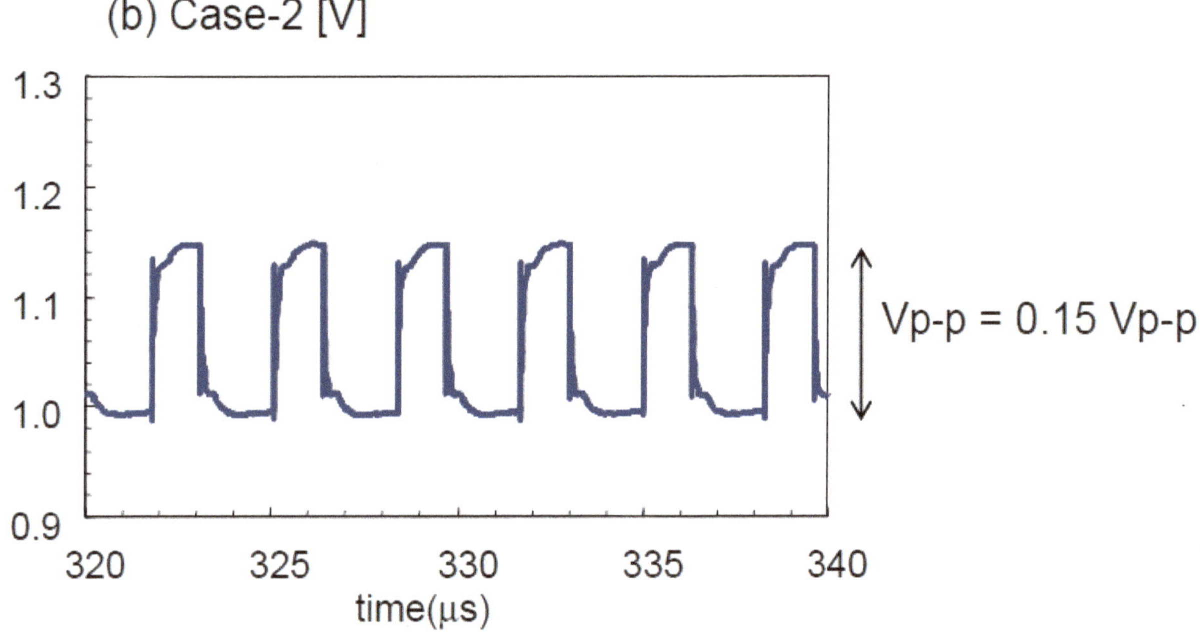

Figure 4.17b: Time Domain Analysis of Load Current Response Using SPICE Simulation

Case-1

	Cap/pcs(uF)	ESR/pcs(mohm)	ESL/pcs(pH)	Quantity(pcs)
Polymer AL Cap	820	7	4000	8
0805 / X5R / 10uF	8	4.5	218	30
0603 / X5R / 1uF	0.8	9	258	30

Case-2

	Cap/pcs(uF)	ESR/pcs(mohm)	ESL/pcs(pH)	Quantity(pcs)
Polymer AL Cap	820	7	4000	8
ESR controlled /0805 /X5R /10uF	8	35	210	30

Table 4.2: Capacitor Model Using SPICE Simulation

Lower ESL Development

Under the same concept, the development of a lower-ESL version is under way. This device has an eight terminal construction, with the middle four terminals having no external electrical contact.

See Figure 4.18.

The optimized ESR and ESL will provide for an improved impedance profile.

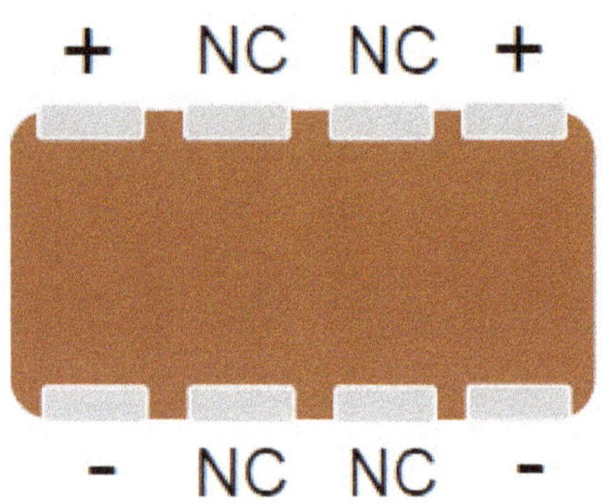

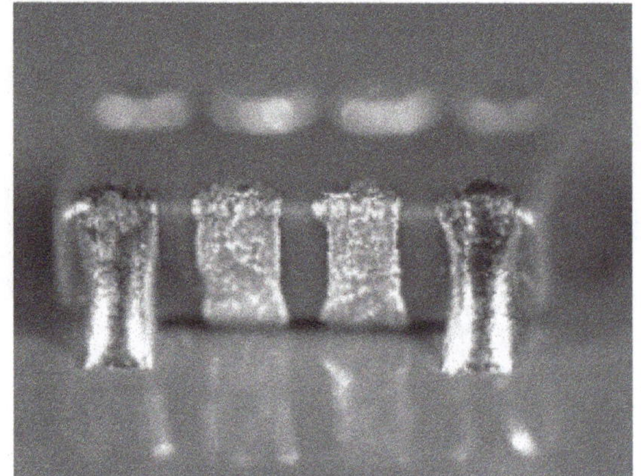

	ESR(mohm)	ESL(pH)
0603 /X5R/1uF	150	166
	600	169

Figure 4.18: Low ESL-Type ESR-Controlled MLCCs.

Conclusion

The multifunctional use of microprocessors, coupled with the increased processing speed, will require the impedance to be further reduced, and the PDN design will continue to face more difficulty.

Since the PDN is required to form small and flat impedances under specified frequency ranges, the ESR-controlled MLCC will play a very important role in PDN designs.

References

[1] Novák, Noujeim, St. Cyr, Biunno, Patel, Korony, Ritter, *Distributed Matched Bypassing for Power Distribution Network*, Manuscript for IEEE CPMT, August 2002.

[2] István Novák, *Frequency-Domain Power-Domain Power Distribution Measurements*, Manuscript for DesignCon East 2003, June 2003, Boston, Massachusetts.

[3] Larry D. Smith, *Frequency Domain Target Impedance Method for Bypass Capacitor Selection for Power Distribution Systems*, TecForum TF-MP3 Comparison of Power Distribution Network Design Methods: DesignCon 2006.

[4] Joseph M. Hock, Andrew P. Ritter, *A Measurement Technique for High Frequency Low Inductance Decoupling Capacitors*, TecForum TF-MP2 *Inductance of Bypass Capacitors, How to Define, How to Measure, How to Simulate*, DesignCon East 2005.

[5] EIA Standard: Electronic Components, Assemblies, Equipment & Supplies Association, *Test Procedure for High–Frequency Characterization of Low Inductance Multilayer Ceramic Chip Capacitors*, PN-4563.

Chapter 5—A Power Distribution System

K. Barry A. Williams, Principal Engineer, Hewlett-Packard

This chapter is an edited and revised version of Chapter 6 of K. Barry A. Williams, *Designing Power Distribution Systems for Electronic Circuits*, Aikman Engineering, 2005.

THE GOAL OF a power distribution system is to have an impedance that is flat over frequency. This goal may or may not be achievable depending on the system requirements. In a case where flatness cannot be achieved, the design will exhibit some pseudo-parallel resonance. The magnitude of the parallel impedance will have to be determined to make sure that the voltage drop, as seen by the load, does not exceed expectations.

An Example of a Power Distribution Design System

The input parameters for the example are shown in Table 5.1.

Description	Min	Nom	Max
Power Supply Voltage	1.400	1.500	1.600
Power Supply Transient Response		10.6 μs	
Load Current Static		10 Amps	
Load Current Dynamic		10 Amps	
Load Switching Frequency		533 MHz	
Pk – Pk Ripple Voltage on Chip		100 mV	
On Die Capacitance (parasitic est.)		40 nF	

Table 5.1: Input Parameters for Design Example

The Problem Schematic

The schematic in Figure 5.1 covers a five-stage system with three stages on the board low-inductance capacitor arrays (LICAs) that may be placed within the package and additional on-die capacitance that may be required. The schematic also indicates several series R and L components placed in series with each capacitor bank. The value of each component cannot be determined until the spacing between the components can be determined. The package inductance and resistance must be determined first because their values significantly affect the design of the power distribution system.

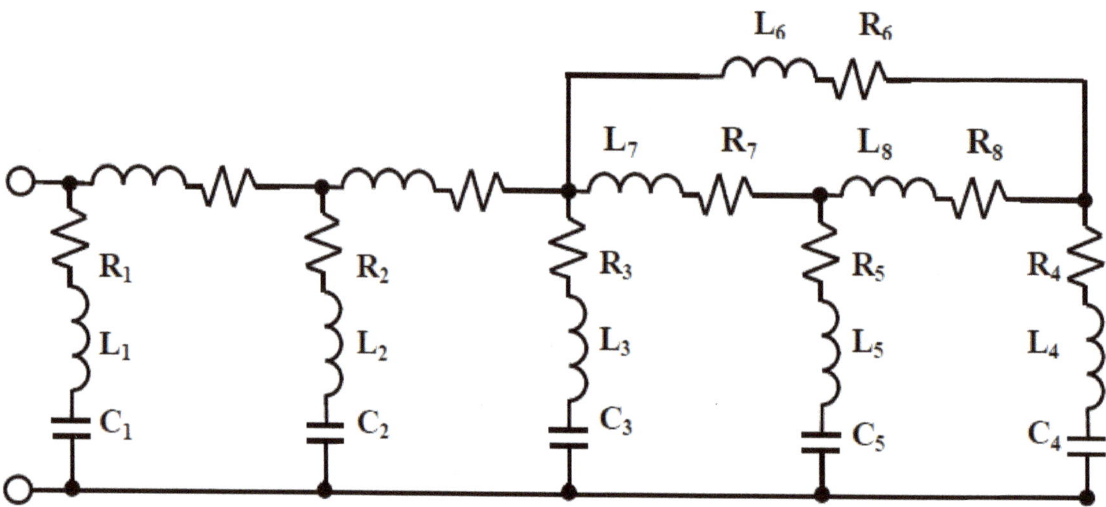

Figure 5.1: Schematic of a Power Distribution System

In Figure 5.1, R4 represents the equivalent load within the chip. C4 represents the amount of on-die capacitance. L7 and R7 represent the chip package inductance and resistance to any LICAs. L8 and R8 represent the inductance and resistance from the LICAs to the die itself. L6 and R6 represent the inductance and resistance of the board under the chip area.

R4 has two values—the first based on the static current and the second based on the sum of the dynamic current and the static current. Using the nominal value of supply voltage, the two values for R4 are 0.150 Ohm, and 0.075 Ohm.

The combined impedance of the board section under the chip and the package represents a series impedance to the load. The combination of the equivalent inductance and resistance creates a significant voltage drop at the load, R4. If the maximum peak-to-peak voltage drop is 100 mV, then the voltage drop due to the series resistance is the static factor. Whereas the voltage drop due to the series inductances is the dynamic factor.

If the chip is a ball grid array or land grid array, the current in the leg represented by L6 should be equal to the current in the leg represented by L7 and L8.

This is of particular importance if there is a LICA capacitor represented by C5. The proper sharing of current in each leg is assured if the current per pin of the chip is equal to the total peak current divided by the number of pins. The board inductance is then determined by the number of pins in each row of the ball grid array (BGA) or land grid array (LGA). Let us subtract the number of power pins in a row from the total current times the current per pin. The difference in current then becomes the total current seen in the next row. Let us repeat the process until the center of the chip has been reached. Moving from row to row represents a strip inductor with current flowing through it. A voltage drop is then created across the strip. Summing the individual voltage drops represents the total drop across the board. This sum when multiplied by time gives volt-seconds. Dividing the volt-seconds by the dynamic current gives the value of inductance L6 and should also be equal to the sum of L7 and L8.

The package inductance is estimated to be 12 pH. The board inductance is estimated to be 8.32 pH. The combined inductance is 4.91 pH.

The package resistance is estimated at 320 µOhm, and the board resistance is estimated at 84 µOhm. The combined resistance value then becomes 67 µOhm.

The Characterization of the Load

Figure 5.2 shows the waveforms. The AC current has a 10 amp peak and a 5 amp average, making the total average of 15 amps. The difference between peak current and average current is 5 amps. The frequency is 533 MHz. Assuming a 50 percent duty cycle, the on-time is 938 ps, Ton. The on-time determines the necessary on-die capacitance as well as the maximum series inductance from the supply voltage.

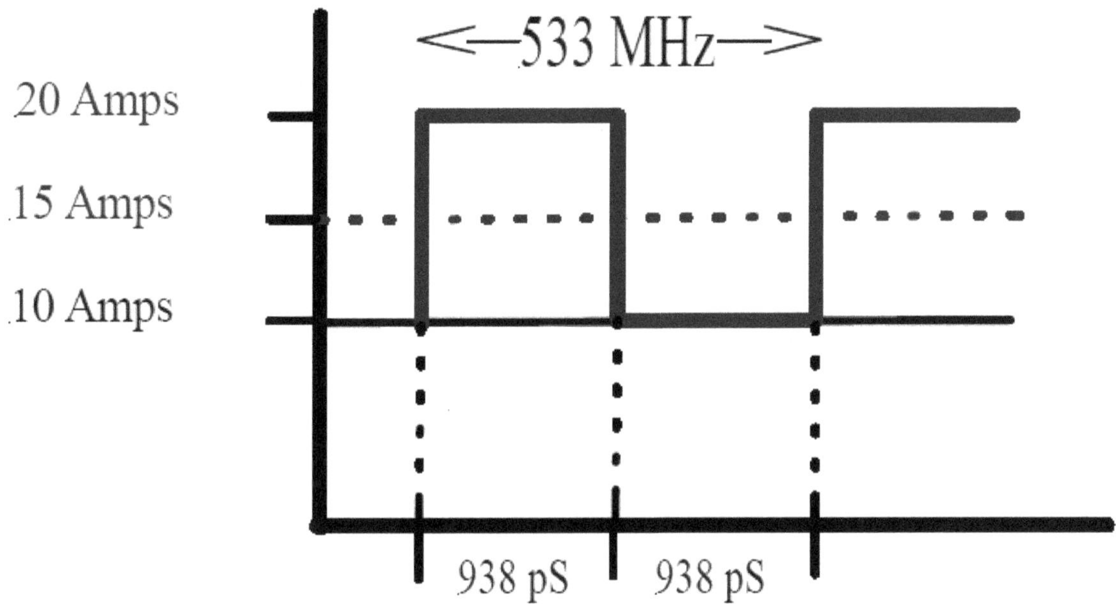

Figure 5.2: Load Current Waveforms

System Bandwidth

The bandwidth and the number of stages of the system should be determined first, because this will give a preview to the number of stages necessary to complete the goal. Table 5.1 gives the loop response of the power supply as 10.6 µs, and this becomes the low-frequency end of the bandwidth. The on-time as given by the duty cycle of the waveform at the load provides the upper-frequency range of the bandwidth. The ratio of these two frequencies gives the bandwidth ratio of the system:

$$b_{RS} = \frac{f_{HS}}{f_{LS}} = \frac{10.6\mu s}{938ps} = 11{,}300$$

The number of stages is then estimated from n ~ Log(bRS) by taking the log of bRS.

From inspection, it is somewhat larger than four. However useful this may seem, it does not give any information about the Q of the system. Further, it may be more useful to have three stages or even five stages rather than four.

$$Q_S = \sqrt[N]{\frac{f_R}{BW}}$$

Q_S can be determined from the two values of time currently known. The first part of the process is to determine the value of the center resonance frequency of the system as indicated by the abbreviation f_R. The value of Q_S can then be determined. If the inversion of the value of time is used to determine the value of frequency, then the series resonant frequency, f_R, becomes 10.03 MHz.

The second part of the process is to determine the bandwidth (BW) and the difference between these two frequencies, or 1,066 MHz. The Q of the system is the ratio between the center frequency and the bandwidth, resulting in a series Q of 0.0094. This value is far lower than what may be expected. The value of Q as calculated is the Q for the system. The value of Q for each stage is found by taking the nth root of the system Q. The initial estimate of four stages results in a value of Q for each stage of 0.311. This is very close to the ideal situation. Figure 5.3 shows four stages and a system Q of 0.0094, which are very close to the third line from top, which represents a stage Q of 0.300.

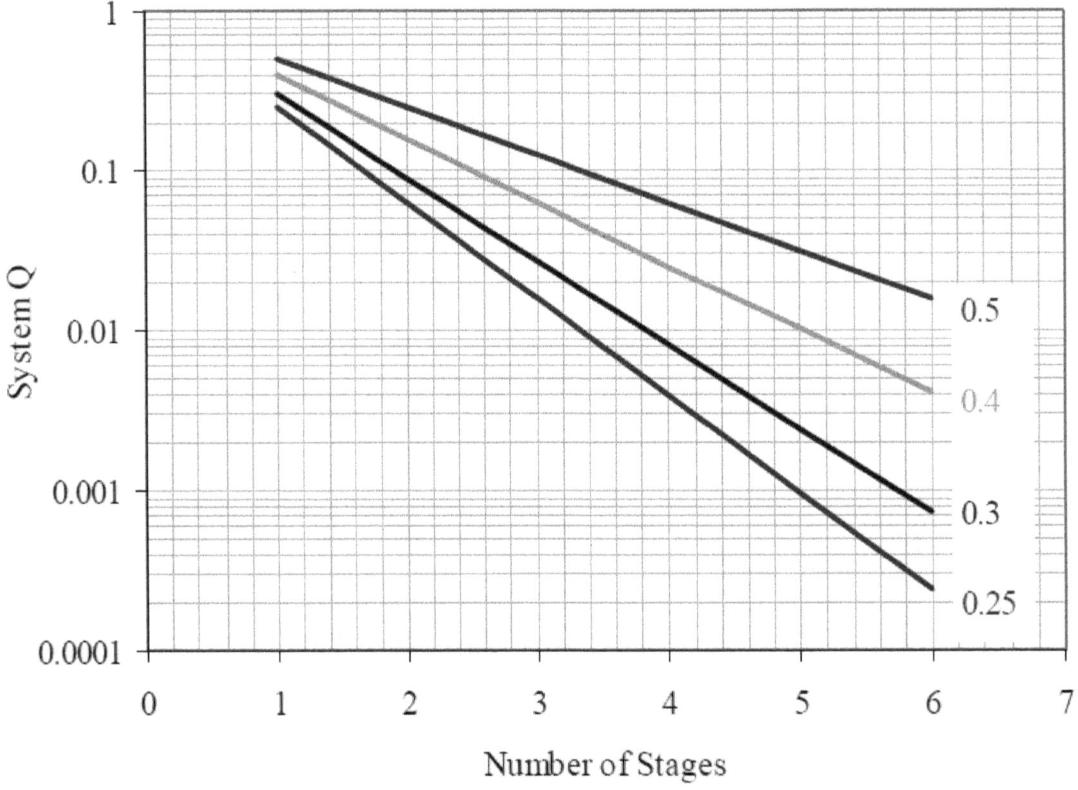

Figure 5.3: Number of Stages Based on System Q

Figure 5.3 also indicates that the same system Q could be accomplished by using five stages. Each stage would then have a value of Q equaling 0.400, as represented by the second line from top.

There are two solutions: the first with a lower value of Q; a system having four stages and a lower parallel resonance impedance. With only four stages, then three would be placed on the printed wire board and the fourth would be the added on-die capacitor.

The second solution requires the addition of package capacitance as represented by C5 in Figure 5.1. The type of capacitor used for this purpose is the LICA. This capacitor is an expensive solution and is usually needed when the package inductance is relatively high.

Determination of the Maximum Impedance

Table 5.1 shows a maximum noise voltage of 100 mV. The maximum impedance can actually be determined from the value of this noise voltage value and the dynamic current. The dynamic current is 10 amps and thus the impedance is 0.010 ohm. The value of impedance is the value at the pseudo-parallel resonant frequencies. The limit line for all frequencies should be less than 0.010 ohm. For parallel resonance, the impedance level should not exceed the limit line. When impedance is plotted against frequency, all values of impedance should stay below the limit line.

Q of the System

For n stages, the series Q_S for each stage is approximated by:

$$Q_S = \sqrt[N]{\frac{f_R}{BW}}$$

To keep the system stable, the value of Q_S for the system is 0.311. The bandwidth ratio has been determined as 11,300. The value of f_{R4} is the switch frequency at the load and the value of f_{R1} is the loop response of the power supply. These values are 533 MHz and 47 kHz, respectively.

Given the approximate value of Q_S, 0.311 for each stage, the bandwidth for the fourth stage is based on the resonant frequency 533 MHz. For this stage, the bandwidth is 1,714 MHz. The value of f_{H4} needs to be determined and is found by dividing the resonant frequency by the series Q.

Since:

$$f_H = \frac{f_R}{Q_S} \text{ and } f_l = f_R\, \mathrm{Q_s} \text{ and bandwidth is defined as: } B_W = f_H - f_L$$

Therefore, by substituting for f_L and f_H, bandwidth is:

$$B_W = \left(\frac{1 - Q_S^2}{Q_S}\right)$$

With Q_S equal to 0.311, the value of f_{H4} is 1,714, thereby making f_{L4} equal to 1,714 minus the bandwidth or 165 MHz. However, f_{L4} is the same point as f_{H3}, making this a parallel resonant point. In addition, at f_{H4}, the value of the parallel impedance and resistance is at the maximum.

Since the maximum impedance is 10 mohm and assuming that this impedance is resistive, then the parallel resistance is the maximum value of 10 mohm. Dividing this value by *(1 + Qs2)* will give the series resistance value. The calculated value of Rs then is 9.11 mOhm. The value of $X_L - X_C$, the difference in impedance at this center frequency, is equal to $Q_S x R_S$ or 2.83 mOhm.

The value of impedance Z then is equal to:

$$Z = \sqrt{R_S^2 + (Q_S R_S)^2}$$

In our example, using Q_S equal to 0.311, the value of R_S is 9.11 mOhm and the value of Z is 9.54 mOhm.

Capacitance and Inductance Determination

The value of f_{R4} is 533 MHz, and the value of f_{H4} is 1,714 MHz. The bandwidth for this stage is found by subtracting f_{R4} from f_{H4}, 1714 MHz-165 MHz = 1549 MHz. Having the bandwidth, the series-resonant frequency and the high-frequency point, f_{H4}, all the other points for the entire system can now be found using a value of Q_S for each stage of 0.311.

Important points to remember are the following:

- The maximum resistance is the maximum parallel resistance, R_P.
- Divide R_P by *(1 + Q2),* where Q is the series Q. This is the equivalent series resistance.
- Multiply the series resistance by the series Q; this is the resistance at the series-resonant frequency.

Given the resonant frequency and impedance, the values of inductance L and capacitance C can now be found.

For the third stage, the value of f_{L4} is the value of f_{H4} minus the bandwidth or 165 MHz. The value of f_{L4} is also the value for f_{H3}. This duality represents the first center frequency f_{C3}, or a parallel resonance point, whose value is now 165 MHz. Having the center frequency, the value of capacitance for stage 4 and the inductance of stage 3 can now be determined. The value of RS has already been determined and the value of capacitance required within the die can now be estimated.

Allocating the die capacitance assumes that there is a certain amount of inductance associated with it and that the series resistance is the value of RS. The risk posed with this assumption is that the actual value of capacitance, inductance, and resistance of the on-die capacitance has been communicated from the original manufacturer. Instead of these required communications, alternate means to estimate these values can be carried out, but at the peril of the user. The package designers all too often do not communicate with those who design the actual die, and then they fail once again by not properly communicating with the user of the device. These parametric elements should be part of the specifications of the device, thereby eliminating unsuccessful prototype builds due to power quality problems as well as signal integrity problems.

First, let us try to estimate the die capacitance that already exists within the logic of the complementary metal oxide semiconductor (CMOS) driver or a similar transistor. This can be accomplished by using the dynamic current and supply voltage to estimate power, where power is:

$$P_C = 0.5CV^2f$$

...where P_C is the power consumption of the on-die capacitance, V is the supply rail voltage, and f is the load-switching frequency. For the example being contemplated, V is 1.5 volts and f is 533 MHz.

The value of P_C is determined knowing that the dynamic current is 10 amperes and the supply voltage is 1.5 volts, giving a value of P_C of 15 watts. From this knowledge, C is determined to be 25.0 nF.

Although conservative, the basic assumption is that all the bits will switch state.

Those that are in a "1" state switch to a "0" state, and those that are in a "0" state switch to a "1" state. Realistically, 25 to 50 percent of the bits will change state, which indicates that the amount of capacitance due to the non-switched bits will increase by a factor of two to four times. The estimated capacitance then could be as much as 100 nF. This value of capacitance can be more than the required need of on-die capacitance. It also could be still far less than the required need of capacitance at the die.

To determine the required value of die capacitance, the use of the characteristic impedance can be determined based on the value of the required series Q.

$$Q_S R_S = \sqrt{\frac{L}{C}}$$

The value of impedance calculated is the impedance of both L and C at resonance. For R_S equal to 9.11 mOhm and Q_S equal to 0.311, the characteristic impedance becomes 2.83 mOhm. At 533 MHz, C is 106 nF and L becomes 0.845 pH. However, the value of impedance, X_C, at f_{L4} is 9.11 mOhm as required at the first parallel resonant frequency point, 165 MHz. The required on-die capacitance from a conservative point of view is far from the required value.

Additional capacitance is required.

The minimum additional capacitance needed is approximately 80 nF, thus bringing the total to 105 nF. The additional capacitance should be increased by the tolerance of the device. If the capacitance increases to 100 nF, the added capacitance of 80 nF would bring the total to 180 nF. The system series Q needs to be maintained, therefore the quotient of L and C must remain constant; if the total capacitance is increased from 106 nF to 180 nF, then the inductance must also increase from 0.845 pH to 1.43 pH.

A particular problem presents itself at this point. The series resonant frequency will now vary based on the actual value of total capacitance on the die itself and the inductance associated with it.

There are two extremes. The first is when the capacitance is a maximum value, 180 nF, and maximum inductance, 1.43 pH. This situation has a self-resonant frequency of 314 MHz. The second situation is when the total on-die capacitance is a minimum and the inductance is a minimum value. The result of this combination is 533 MHz for a self-resonant frequency. The design effort should be based on the minimum valuation of the on-die capacitance and its minimum inductance.

Important points to remember are the following:

- For design purposes, use the minimum total on-die capacitance with a minimum tolerance.
- The inductance should be determined by the system series Q and the switching frequency of the load.
- When it is possible to obtain information about the on-die capacitance, use those parasitic values of L and C.

Stage 3 and Stage 4 Resonances

It has been shown that in a parallel resonance circuit, the resistance factor affects the actual resonant frequency. Therefore, a modification to the presumed resonant frequency must be done to be incorporated to obtain the correct value of on-die capacitance.

As shown in Figure 5.4, the impedance in both branches must be equal for resonance to occur. The conversion is shown in Figure 5.5.

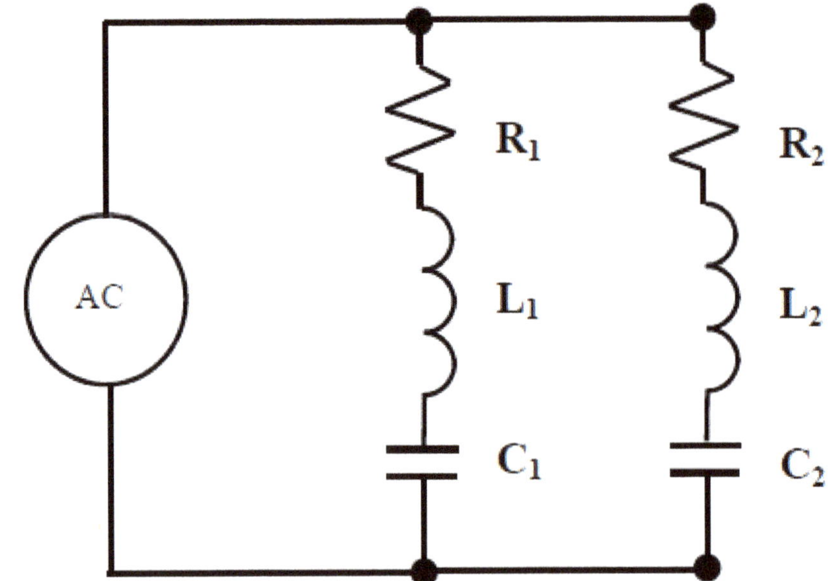

Figure 5.4: The Two-Capacitor Circuit

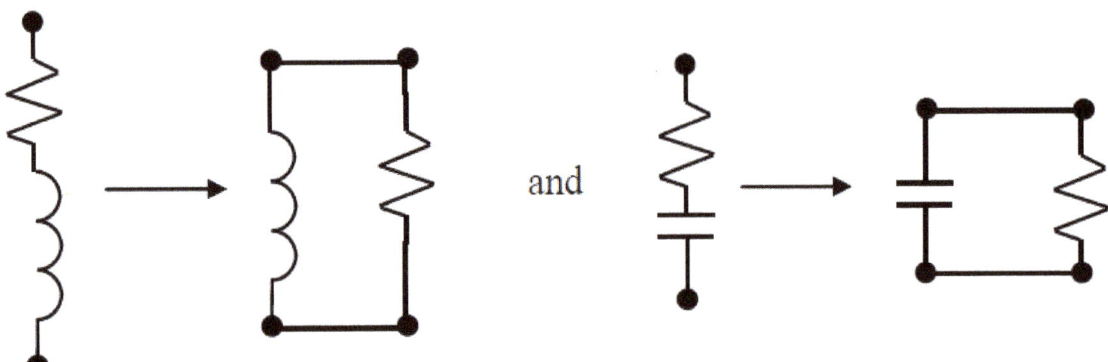

Figure 5.5: Converting Series Components to Parallel Components

Parallel resonance is comprised of the inductance of stage 3 together with its resistance, and the capacitance and its resistance of stage 4. Since each in its own way is a series element, each then should be converted to its parallel equivalency.

Beginning with the series inductance and resistance of stage 3 elements, the impedance is:

$$Z_1 = \sqrt{(R_{S3})^2 + (X_{L3})^2}$$

...and for stage 4, using the series resistance and capacitance, the impedance is:

$$Z_2 = \sqrt{(R_{S4})^2 + (X_{C4})^2}$$

For the parallel equivalency, the conductance is the reciprocal of the resistance, and admittance is the reciprocal of the impedance.

$$R_{S1} = Z_1 Cos\,\theta_1 \text{ and } X_L = Z_1\, Sin\,\theta_1$$

Therefore:

$$G_1 = \frac{Z_1}{Cos\theta_1} \; and \; Y_1 = \frac{Z_1}{Sin\theta_1}$$

In a similar fashion, for stage 4 the conductance and admittance is:

$$G_2 = \frac{Z_2}{Cos\theta_2} \; and \; Y_2 = \frac{Z_2}{Sin\theta_2}$$

Having all the conductance and admittance terms, the elements are in parallel. The circuit now appears in the Figure 5.6.

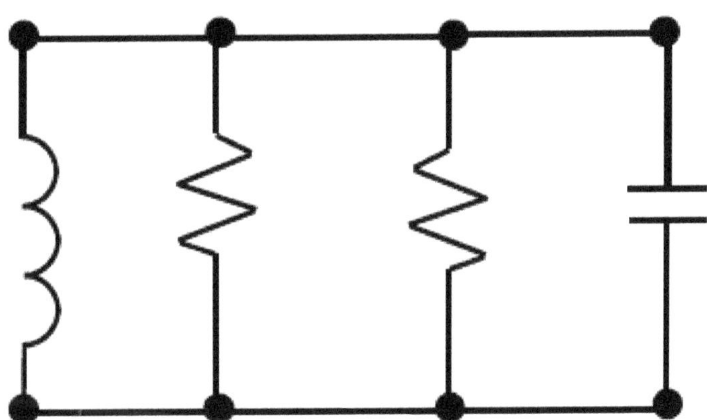

Figure 5.6: Parallel Equivalent Circuit of the Series Elements

Now, since:

$R_{S1} = Z_1 Cos\,\theta_1$, then...

$$Cos\theta_1 = \frac{R_{S1}}{Z_1} \; and = G_1 = R_{P1} = \frac{Z_1}{Cos\theta_1} \text{ then } Cos\theta_1 = \frac{Z_1}{R_{P1}}$$

Therefore:

$$\frac{R_{S1}}{Z_1} = \frac{Z_1}{R_{P1}} \text{ and } R_{P1} = \frac{Z_1^2}{R_{S1}} = \frac{R_{S1}^2 + X_L^2}{R_{S1}}$$

In similar form, then,

$$R_{P2}=\frac{Z_2^2}{R_{S2}}=\frac{R_{S2}^2+X_C^2}{R_{S2}}$$

However, since parallel resistances are in parallel with each other, the equivalent resistance is the parallel equivalent of each R_P. With the two parallel resistors reduced to one resistor, the parallel value of L and C form the classical resonant circuit shown in Figure 5.7. The value of R_P cannot exceed the value of 10 mOhm in the discussed example.

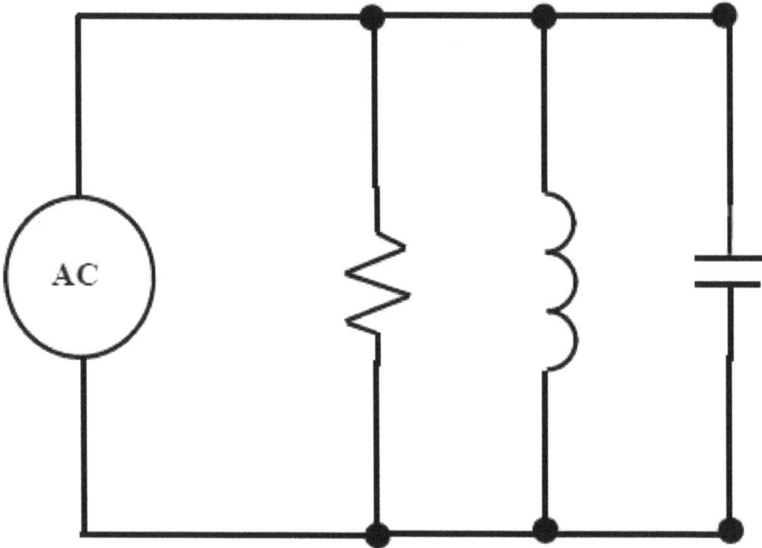

Figure 5.7: Parallel Resonant Circuit

The values of equivalent parallel impedances are found through the same method used to find the equivalent parallel resistance.

Therefore,

$$X_{CP}=\frac{R_{S2}^2+X_{CS}^2}{X_{CS}} \text{ and } X_{LP}=\frac{R_{S1}^2+X_{LS}^2}{X_{LS}}$$

At the point of resonance, X_{LP} is equal to X_{CP}, therefore:

$$X_{CP}=\frac{R_{S1}^2+X_{LS}^2}{X_{LS}}$$

...and determines the value of parallel capacitance at resonance. However, to find the resonant frequency, the above equation needs to be reduced to:

$$\frac{\omega L_S}{\omega C_P}=R_{S1}^2+\omega^2 L_S^2 \; and \; 1=\frac{R_{S1}^2 C_P}{L_S}+\omega^2 L_S C_P$$

or

$$f_R = \frac{\sqrt{\dfrac{1 - R_{S1}^2 C_P}{L_S}}}{2\pi\sqrt{L_S C_P}}$$

This equation has a special situation because if the radical of the numerator is equal to 0, then there is no resonance.

This equation implies that if the value of Q_S is greater than 10, the actual value of the resonant frequency is found in the traditional way—by taking the square root of L_S and C_P and multiplying this result by 2π to get the radial time. The inverse of time is the radial frequency.

In designing a PDN, the value of Q_S is always less than 1 and greater than approximately 0.2.

In the problem being studied, the value of the series Q is 0.311, and the parallel frequency of resonance is determined by L_S and C_P, which is now modified by Equation 1. The modification to the resonant frequency is by the terms that involve Q_S, the other terms define the resonant frequency itself. The series Q terms are equal to 0.297.

Resonant Frequency Points

Figure 5.8 shows the relative series resonant points on the bottom and the parallel resonant frequencies on top. The declining slopes are the capacitive effect with its series resistance and the rising slopes represent the inductive portion and its series resistance. The upper horizontal line represents the maximum value of impedance and the lower horizontal line represents the minimum series value.

The value of frequency is given in MHz. When the series Q is 0.311, then if the switching frequency of the load is 533 MHz; this represents a series resonant frequency. The first parallel resonant frequency that should occur is located at a frequency of $Q_S f_R$. This product evaluates to 165.76 MHz. The next lower-series resonant frequency will occur at a frequency that is the series Q multiplied by the previous frequency, (0.311)(165.76 MHz) or 51.55 MHz.

The process continues until all four stages are defined.

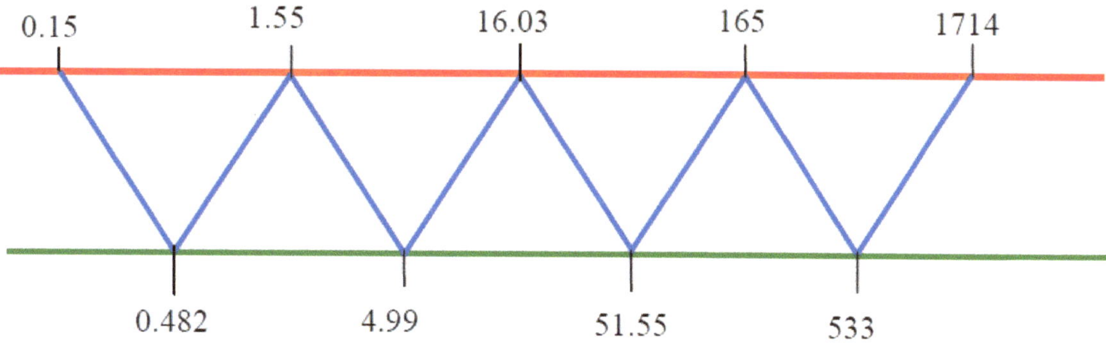

Figure 5.8: Point-to-Point Resonant Frequencies

Keeping with the desire to have a constant value of Q suggests that the value of series impedance should be 2.83 mOhm, which is R_S times Q_S. The value of impedance of 533 MHz is the same value at 51.55 MHz,

4.99 MHz, and 0.482 MHz.

The basic premise in determining the permitted values of inductance as well as the amount of capacitance needed, is based on the fact that the series Q for each stage is identical, thus the series resistance is also equal. Further, at the point of parallel resonance, the impedance of the capacitor and of the inductor is equal and opposite in sign, which when added results in 0. Figure 5.9 shows the basic premise.

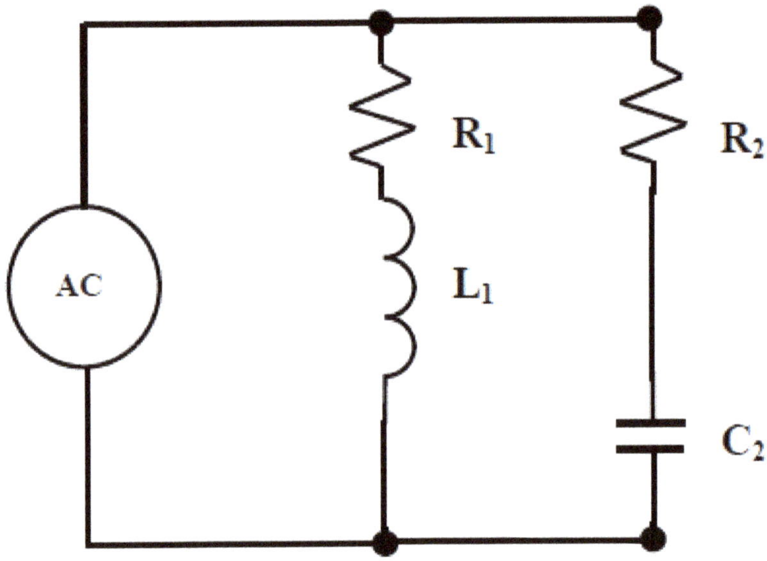

Figure 5.9: An Equivalent Parallel Circuit

If:

$$Z_1 = \sqrt{R_{S1}^2 + X_{LS}^2} \; and \; if \; Z_2 = \sqrt{R_{S2}^2 + X_{CS}^2}$$

...and if $Z_1 = Z_2$ and $R_{S1} = R_{S2}$ and $X_{LS} = X_{CS}$, then it can be stated that with these equalities and with Z_1 and R_{S1} known, the value of X_{LS} and X_{CS} can be found.

Thus,

$$X_{CS}^2 = Z_1^2 - R_{S1}^2$$

Using the present problem, the value of X_{CS} or X_{LS} is 2.83 mOhm at each of their respective series-resonant frequencies. The actual value of capacitance was determined using the fourth resonant frequency of 533 MHz, thus the value of C is 106 nF. This basically answers the question of how much on-die capacitance is required. If a conservative approach is used, then with 25 nF already present, 81 nF will need to be added on the die. If only 30 percent of the bits are switched, then the total capacitance seen will be 187 nF. The value of the series inductance L_S is found at the third resonant frequency of 51.55 MHz and has a value of 8.74 pH.

Using this same method for the other series resonant points yields values of capacitance of 1.1, 11.33, and 117 μF. The inductance for each series point is found in the same way, that is, the impedance is 2.83 mOhm at the points of series resonance. Therefore, these inductances are 8.74, 90.4, 934, and 9,659 pH. These values represent not only the equivalent series inductance of each capacitor stage, but also any board inductances such as vias and planes, interposers, package inductance, and other inductances that may be encountered along the way to the load.

Number of Capacitors

The next task is to select the package style for each stage. Starting with stage 3, the value of capacitance is 1.1 μF and the value of inductance is 8.74 pH. To obtain the value of inductance, the choice is between 0612 capacitors and 0306 capacitors. Both of these are sideways capacitors and have inductance values of 550 pH or 320 pH. To obtain the required inductance for the 0612- capacitor package, 63 capacitors will be needed; for the 0306- capacitor package, 37 capacitors will be needed. The value of capacitance is 0.018 μF for the 0612-capacitor package and 0.033 μF for the 0306 capacitor package. The choice of package type can be easily determined from this data because there are no 0612 capacitors that are that low in value, and therefore the 0306 device of 0.033 μF will be used.

Similarly for stage 2, the value of capacitance is 11.33 μF and the value of inductance is 90 pH. There are two choices for package types—1210 capacitors, which have an inductance value of 1,000 pH, and 0612 capacitors, which have an inductance value of 550 pH. The first package type requires 11 capacitors with a value of 1.0 μF. The second choice is the 0612, which will require six capacitors with a value of 2.2 μF. The 1210 package will be chosen since it is within the range of acceptable capacitance.

Stage 1 requires a capacitance value of 117 μF and an inductance value of 9,659 pH. The power supply requirements are such that the value of capacitance is considerably higher than 117 μF, and as a result, the inductance requirement will be easily met.

Summary of Compiled Results

This completes the analysis of the determination of the die capacitance, the series inductance of each stage, the capacitance of each stage and the frequencies of interest for each stage. Table 5.2 illustrates the values of f_L, f_H and f_R for each stage.

	F_L	f_H	f_R
Stage 4	166	1714	533.0
Stage 3	16	166	51.55
Stage 2	1.55	16	4.98
Stage 1	0.15	1.55	0.482

Table 5.2: Frequency Ranges in MHz for Each Stage of the Design Example

Table 5.3 shows the capacitance needed and the permissible inductance for each stage.

Stage	Capacitance μF	Inductance pH	$\sqrt{\frac{L}{C}}$
Stage 4	0.106	0.84	2.82 mΩ
Stage 3	1.1	8.7	2.81 mΩ
Stage 2	11.33	90	2.82 mΩ
Stage 1	117	934	2.83 mΩ

Table 5.3: Inductance and Capacitance for Each Stage

Table 5.4 summarizes the estimated values of capacitance and their parasitics, quantities, and total equivalent component values.

Stage	*C* µF	*L* pH	*R* mΩ	No. of Devices	Total *C* µF	Total *L* pH	Total *R* mΩ
Stage 4	0.106	0.84	9.11	1	0.106	0.84	9.11
Stage 3	0.033	320	167	37	1.22	8.67	4.51
Stage 2	1.0	1000	16.0	11	11	91	1.45
Stage 1	1000	2450	35	14	14000	175	2.5

Table 5.4: Capacitance and Inductance for the Design Problem

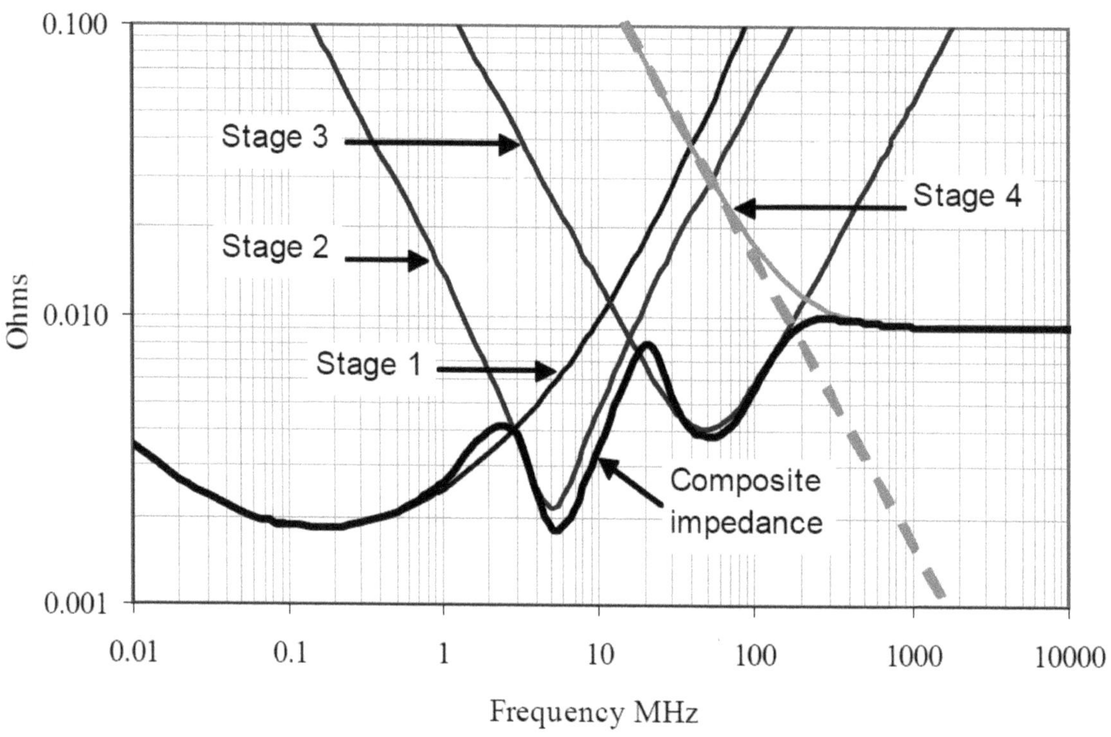

Figure 5.10: Impedance of Design Example

The result of the selection of components and their equivalent parasitics, as well as the overall effective impedance, is shown in Figure 5.10. Although the intended goal is to have a relatively flat impedance across the bandwidth, the actual lack of flatness is due to available component types.

Summary of the Capacitor Selections and Graphical Results

Figure 5.10 shows the results of the design. The leftmost curve represents stage 1, the second from left is stage 2, the third is stage 3, and the rightmost is stage 4. The overall impedance of these four stages is represented by the heavy line. The system will see the impedance of the heavy line because it is the composite of all four stages.

One can make several observations from Figure 5.10. The bandwidth is wider than calculated, which is a result of using a considerably larger number of stage 1 capacitors to meet the power supply requirements. The estimated value of stage 1 capacitance is 117 μF. This is not suitable to maintain the estimated transient response from the power supply. A much larger value is needed to keep the transient response at a reasonable level.

The impedance at each parallel resonant frequency has met the goal of 10 mOhm or less. The parallel resonant frequency of 1.55 MHz and 16 MHz has been achieved. The impedances at the parallel resonant frequencies have met all other calculated expectations.

The resultant composite line at each of the parallel resonant frequency points has a value of impedance less than or equal to that which is represented by the intersection of any two capacitors. The lines of stage 1 and stage 2 intersect at 3 MHz, which is a point of equal impedance. The same situation applies to the intersection point of the stage 2 and stage 3. The intersection of the third and fourth lines is higher in the impedance value than that represented by the composite line in this same area. This indicates that the value of Q_S, although not necessarily the perfect value, is adequate. The stage 1 line also intersects with the stage 3 and stage 4 lines, thus causing the resultant composite line to shift. All of these must be accounted for, although these are secondary affects.

The actual parallel resonance between stage 3 and stage 4 is as predicted. The approximate value of 170 MHz, as shown by the dashed line that crosses the stage 3 line, represents the inductance of stage 3. The dashed line represents the on-die capacitance.

The goal for a reasonably flat impedance across the entire bandwidth has not necessarily been met. The series resistance of stage 3 is considerably less than the goal, as is stage 2 and stage 1. The goal of having a Q_S of less than 0.50 has been met; the expectation of no parallel resonances has not been met. This is due to the selected capacitor for stage 1. The value and number of packages for stage 1 has been increased to meet the requirements of the power supply. The value of capacitance selected for stage 4 is a minimum value and, as can be seen from Figure 5.10, just meets the goal of 10 mOhm.

The overall performance of the impedance is quite acceptable for the entire range of frequencies. Each stage will need additional resistance to flatten the impedance profile in order to see the effects of having the correct value of series resistance. Added resistance is determined by the value of the series Q required for each stage to provide a zero resonant frequency. An interesting fact is that stage 1 needs additional inductance to bring its impedance to 0.009 Ohm. In fact, stage 2 and stage 3 need some small amount of resistance to bring the overall impedance to the desired level.

Graphical Results with Added Resistance

Figure 5.11 shows that with added resistance, the resultant impedance curve improves considerably, almost reaching an ideal. The goal of 10 mOhm maximum for each stage has been met and the overall bandwidth, more than estimated. Figure 5.11 also shows the parallel resonant points, the impedance intersections of each stage and the resultant impedance as represented by the black line.

The summary line shows very little deviation with added resistance. In addition, Figure 5.11 shows the impedance needed for each stage, particularly the amount of the added series resistance needed to obtain a flat impedance.

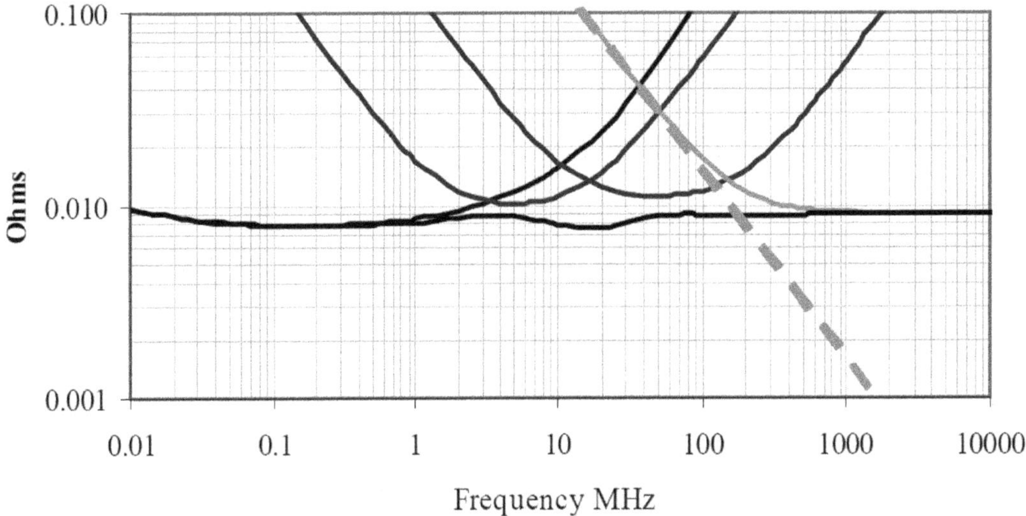

Figure 5.11: Impedance with Added Resistance

The actual ESR value of the capacitor is much lower than required. With a lower ESR value, the temperature rise in each capacitor will be lower, increasing its useful life. Since the power dissipated in each capacitor is related to I^2R, if the resistance is increased by a factor of 5, then the power dissipated within the capacitor is increased by the same factor. The dynamic current is 10 amperes for the system and there are 37 capacitors in stage 3. If the capacitors share the current equally, then each capacitor supports 270 mA. With the ESR of each capacitor being 167 mOhm, the amount of power being dissipated is 12 mW for each capacitor.

The method of adding resistance or impedance to each stage is accomplished by adding resistance to each capacitor for stage 2 and stage 3, and some inductance to stage 1. The effect as shown in Figure 5.11 is the reduction of the overall variation in the series impedance between the power supply and the load or chip. The series impedance in Figure 5.11 varies from about 8.0 to 9.47 mOhm. At specific frequencies, where the impedance is the lowest, there is the lowest series voltage drop to the chip or load. This method does have the problem, however, of finding resistance values low enough, which result in more components, more opportunities for failures, and an increase in both cost and cost of ownership.

Sensitivity—Added Resistance

The next obvious question is: to what degree is the overall impedance degraded with changes in some of the other parasitics? The changes in resistance to stage 2 and stage 3 are shown in Figure 5.11. The value of resistance is based on the number of capacitors in each stage. The value of resistance is divided by the number of capacitors used. Each of the added resistors has some inductance, and this inductance has to be added to the total inductance of each stage. As a result, some additional capacitors may have to be added in order to maintain the proper inductance value.

Sensitivity—Added Inductance

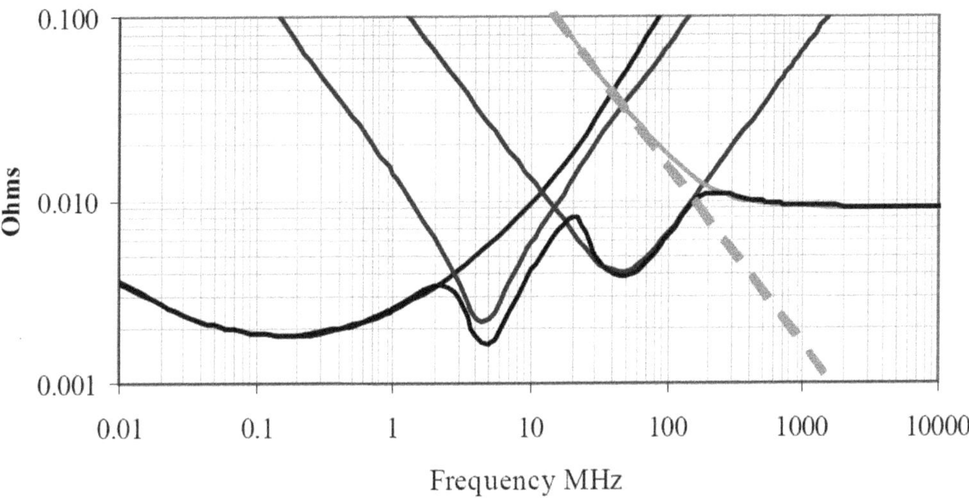

Figure 5.12: Impedance with Added Inductance

The effect of adding more inductance is not necessarily intuitive. Figure 5.12 shows the effects with just 20 percent more inductance in stage 2 and stage 3. The result of the increase in inductance is a shift in the resonant frequencies to the left and an increase in the values of impedance. The circuit now fails to be acceptable in the range of 150 MHz to 350 MHz. The value of impedance in this range is above the 10 mOhm goal. The value of inductance estimated is the maximum value and a lower value should be used, requiring the addition of a few capacitors.

Figure 5.13 shows the results with 20 percent less inductance. The resonant frequencies have shifted to the right just a little and also provide slightly lower impedance. Again, the system is still quite well behaved. The system Q is improved and can be readily seen in the 100 MHz to 1,000 MHz region. This region, as represented by the line of composite impedance, is considerably smoother than in Figure 5.10 or Figure 5.12.

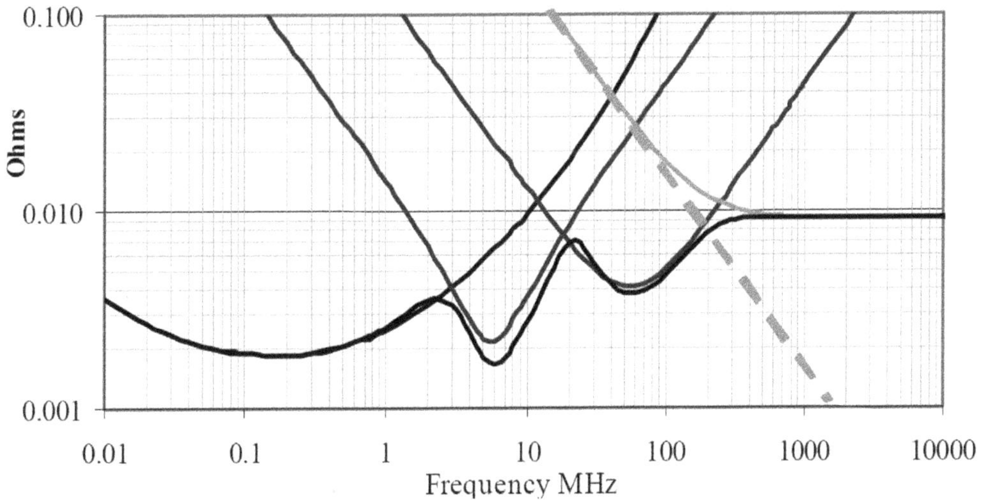

Figure 5.13: Impedance with Less Inductance

Sensitivity—Added Capacitance

The advantage of adding more capacitance to stage 2 is extremely low because it would interact with the inductance of stage 1, which has an impedance that is already lower than required. Adding capacitance to this stage would shift the intersection of the bulk capacitor line to a lower frequency. However, adding more capacitance to stage 3 and stage 4 would produce a lower peak impedance of the parallel resonance between stage 2 and stage 3, and it would shift the resonance to the left.

Adding more on-die capacitance to stage 4 would lower the impedance and shift the parallel resonant frequency to the left. Adding capacitance to just stage 3 alone has some interesting effects. Figure 5.13 shows that with the current capacitance, the impedance relative to stage 1 and stage 2 is much higher. If the number of stage 3 capacitors were to be doubled, then the resonant frequency would remain the same and the value of the total ESR would be halved. The doubling effect would provide an ESR value that is close to the values of stage 1 and stage 2. The doubling effect also produces twice as much capacitance for this stage and half the inductance. The new parallel resonant frequency would be 70 percent of the previous frequency. Looking at Figure 5.13, at the point of about 23 MHz, the new frequency is about 16 MHz and the peak impedance would be about 6 mOhm. Doubling the number of capacitors would also reduce the inductance. The inductive portion that would intersect with the on-die capacitance has a new parallel resonant frequency that is 30 percent higher in value and whose impedance would be reduced at the new resonant frequency.

A reduction in capacitance could have detrimental effects and would significantly increase the impedances of stage 2, stage 3, and stage 4 at their resonant frequencies. The effect of reducing the on-die capacitance will cause the impedance to increase to the point of failure in meeting the required goals. This also points to the fact that the value of capacitance estimated is a minimum value, and therefore the tolerance of these devices should be accounted for.

Typically, a 20 percent tolerance for X5R or X7R capacitors should be used to keep costs to a minimum. The estimated value of capacitance should be increased by the tolerance level and will lower the inductance by the same amount. From a design point of view, a check should be made with minimum capacitance and maximum inductance.

Summary of the Design

Given the design factors that described the input parameters (e.g., the loop response of the power supply; static and dynamic currents of the load; load voltage and its switching characteristics), a certain number of stages and a selection of capacitors were determined. The value of noise voltage was provided and was used to determine the maximum value of series resistance that will satisfy the needs of the design. The value of capacitance and inductance for each stage has been determined. The inductance, resistance, and capacitance requirement of each capacitor within each bank has been determined by the quantity of each capacitor type used.

The design was plotted to see the curves for each capacitor bank and to see if the design met the goals.

It was found, as expected, that the actual impedance was lower than required in some areas. The results were due to the lower series resistance of the capacitors themselves. The design was found to be acceptable and met all the requirement goals except for flatness of the impedance curve. Resistors were added to each stage to bring the series resistance up to 9 mOhm.

The result was plotted, and it was found that the impedance curve was considerably flatter. It was also found that the advantage of adding resistors to each capacitor was that a smaller variation of the impedance resulted over the entire frequency range of the distribution network. In addition, one big disadvantage of adding resistors to each capacitor was that it doubled the number of components, which increased opportunities for failure in the manufacturing process, increased probability of failure over time, and increased cost of ownership and the additional board space to accommodate the resistors with real estate costs that are already considerable.

The design methodology has been proven both theoretically and by simulation results, representing two parts of the Iron Triangle. A practical implementation of the design is achieved by building prototype boards to prove the results obtained by simulation. An aspect of the design that was not incorporated into this simulated system is the inductance and resistance of the planes between the different capacitor banks. The position of each of these capacitor banks to each other will add inductance and both alternating current (AC) and direct current (DC) resistances. In the case of the stage 3 capacitor bank, the distance relative to the package of the chip is fixed.

The components will be placed next to or, as manufacturing will permit, as close as possible to the package.

Stage 1 capacitors are typically located as close as possible to the point of entry to the board of the power supply.

Finally, board inductance, due to the manufacturing processes, will vary by approximately 10 percent. In addition, the inductance of the capacitors and their series resistances is a strong function of the variation in the length, width, and thickness in the capacitor package. Typically, these dimensional variations are between five and 10 percent, depending on the original equipment manufacturer (OEM) supplier. These dimensional variations can be as much as 10 percent at the extreme and should be accounted for. The series resistance of the added on-die capacitor has a tolerance value and the maximum value of resistance should be used. In addition to the tolerance of the series resistor, the operating temperature range effects of the resistance of the on-die capacitor should be factored into the overall tolerance study.

Table 5.1 shows the requirements for this simulated design and has no information about the package inductance or DC resistance of the load. No attempt has been made to estimate its real values.

Although the load may be a memory chip, a central processing unit (CPU), or some other device, the inductance and resistance from its pin(s) to the actual die in the center of the package is very important. Some of these devices have gull wing–type pins around the periphery of the device or short stubs all along its underside. It is to have equal current sharing between the board and the pins or pads of the application-specific integrated circuit (ASIC), CPU, or any other integrated circuit.

The goal is to provide a voltage to the center of the die that has a very small variation due to the switching action of the load.

The value of inductance and resistance of the package has a strong influence on the voltage that the die will actually see. It has not been the practice of OEM suppliers to provide data about the parasitic elements. If the package inductance, resistance, and capacitance were known, then the work of the signal integrity engineer would be much easier and there would be no guessing as to the value of these elements. This information would further the quest to minimize simulation failures, reduce cost, and bring the product to the marketplace sooner.

The value of the signals—that is, the quality of the edges and amplitudes of the signal—is dependent on the quality of power delivered to the load. This is the responsibility of the power integrity engineer, and without power integrity, there is no signal integrity. The purpose of a power distribution system should be quite clear, as its design has been demonstrated to work in the simulator.

However, the evidence clearly shows that there is an urgent need for power integrity which another compelling reason for a good power distribution system.

Important Points to Remember

- With on-die resistance, the value estimated is a maximum value and should be reduced by its tolerance value.
- The inductance estimated for each capacitor stage is a maximum value and should be reduced by adding more capacitors.
- The value of on-die capacitance is a minimum value and should be increased by its tolerance level.
- The value of capacitance for each stage is a minimum value and should be increased by 20 percent.
- Each stage is represented by a different type of capacitor. Each type of capacitor has its own unique characteristics. This essentially prevents the designer from meeting the perfect goal of a maximally flat impedance line.

Chapter 6—Designing Minimum-Cost VRM 8.2/8.3 Compliant Converters

Richard Redl, President, ELFI, S.A.
Brian Erisman, Project Engineer, Analog Devices, Inc.

PROVIDING THE SPECIFIED core voltage to a Pentium®II or other processor throughout all of its dynamic loading conditions is a challenge for which there are many advocates of solutions, but for which few have provided the analysis. Fast voltage-mode architectures have the appeal to the central processing unit (CPU) power converter designer of not requiring a current-sensing resistor and holding the promise of the fastest possible recovery from a transient. However, a worst-case analysis of permissible load transients, operating set-point voltage tolerance, and transient response performance tell another story. The cost-, size-, and performance-optimized solution requires current sensing and does not require a fast transient recovery, yet it effectively provides an instantaneous response. A careful design analysis is required and is provided in this paper. The solution, which is as simple as any other in regards to component count, is realized with the ADP 3152/3 buck converter controller integrated circuit (IC) from analog devices. The cost and size reduction objective is realized in terms of the number of output capacitors required to keep the core voltage within a specified window. This can easily mean cutting that number of capacitors in half.

Introduction

Direct current to direct current (DC–DC) converters to power high-performance microprocessors (e.g., Intel's Pentium® II) must have small transient deviation to a step change in the load current. The step load current change is the result of active power management. The rate of change of load current is on the order of 30 A/µs. Specifications for required performance are provided in a combination of text and table form. The following analysis is focused on the 2.8 V Pentium II processor, but it would apply to other processors with specifications of similar magnitude, including the 2.0 V and below subsequent-generation processors from Intel. Compliance with the specifications should be achieved without unnecessarily increasing the converter cost. One of the biggest cost items in the converter is the output capacitor array. By following the design procedure discussed in this paper, the volume and cost of that capacitor array can be kept at its theoretical minimum.

Objective Specifications

Symbol	Parameter	CPU core freq. (MHz)	Min.	Typ.	Max.	Unit
Vcc_{CORE}	Vcc for proc. core			2.8		V
	Vcc_{CORE} static tol.		–0.060		0.100	V
	Vcc_{CORE} transient tolerance	233 266 300	–0.140 –0.140 –0.130		0.140 0.140 0.130	V V V
Icc_{CORE}	Current for Vcc_{CORE}	233 266 300		6.9 7.8 8.7	11.8 12.7 14.2	A A A
$Icc_{SGNTCORE}$	Icc for Stop-Grant Vcc_{CORE}	233 266 300		0.8 0.9 TBD	1.1 1.2 TBD	A A A
$Icc_{SLPCORE}$	Icc for Sleep Vcc_{CORE}			0.070	0.080	A
$Icc_{DSLPCORE}$	Icc for Deep Sleep Vcc_{CORE}				0.020	A
$dIcc_{CORE}/dt$	Icc slew rate				30	A/µs

Table 6.1: Pentium^{R} II Processor Voltage and Current Specifications

Table 6.1 presents the voltage and current specifications for the Pentium II processor. The output voltage measured at the converter output pins on the system board must never deviate outside the transient limits shown in Table 6.1, including transitions between $Icc_{SGNTCORE}$ and $Icc_{CORE(MAX)}$ at slew rates up to the specified maximum (e.g., 30 A/µsec). Furthermore, even during such a transition, the output voltage may not deviate outside the static voltage limits for longer than 2 µsec. The output voltage ripple must be contained within the static limits. The toggle rate for the output load transition may range from 100 Hz to 100 kHz. Under these conditions, and for all toggle rates, the output voltage must be measured at a 20 MHz bandwidth, and at ambient temperatures between 25°C and 50°C.

Load Transient Performance Limits of the Buck Converter

Figure 6.1 shows the typical output voltage transient (top) and the load current that generates it (bottom).

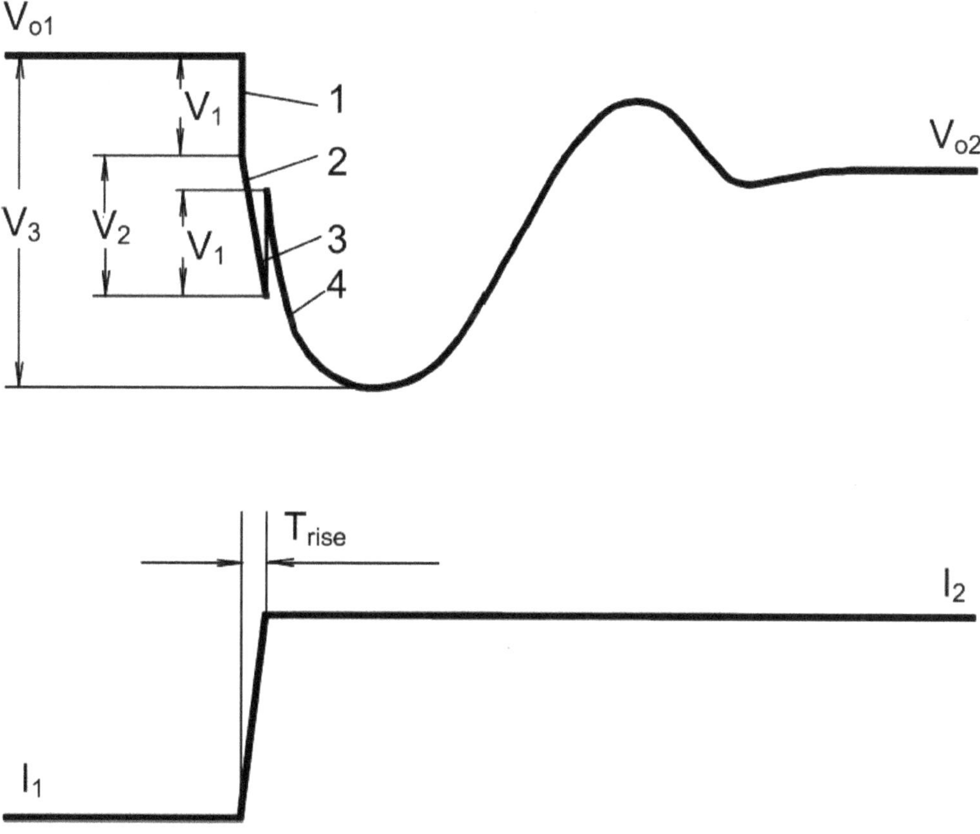

Figure 6.1: Output Voltage (Top) and Load Current (Bottom)

The transient comprises four distinct sections. In section 1, the output voltage steps downward from its steady state value V_{o1} by a voltage V_1. This step is caused by the equivalent series inductance (ESL), or Le, of the output capacitor of the switching regulator. The magnitude V_1 of the step is equal to the product of Le and the rate of change, or di/dt, of the load current. The di/dt is equal to $(I_2 - I_1)$/Trise.

In section 2, the output voltage continues to move downward, but as a ramp function rather than a step function. The magnitude V_2 of the ramp voltage is equal to the product of the load current and the equivalent series resistance, equivalent series resistance (ESR), or Re, of the output capacitor. In section 3, the output voltage steps in the opposite direction (upward). Again, this step is caused by the combination of the change in the di/dt of the load current and the ESL of the output capacitor. The magnitude of the step is V_1. In section 4, the output voltage continues to move but at a much slower rate than in the first three sections. The peak deviation of the output voltage from the initial value is V_3; eventually the output voltage settles to a new value V_{o2}. In section 4, the time function of the output voltage is determined by the dynamic behavior of the converter and the applied control method.

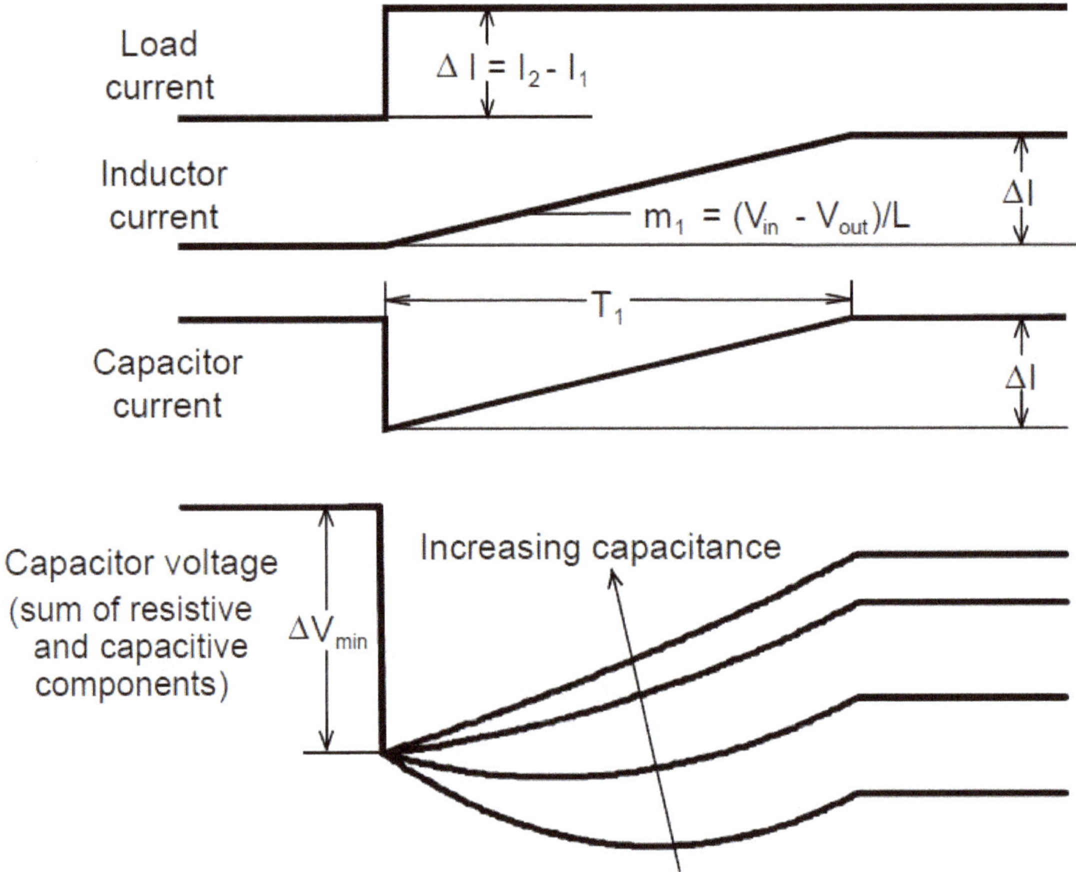

Figure 6.2: Load Transient Waveforms

Figure 6.2 shows the load transient waveforms (neglecting for the moment the ESL of the capacitor, the rise time of the load current, and the effect of the ripple component in the inductor current). It is assumed that the control keeps the power switch on until the inductor current ramps up to the value required by the change in the load current. It is clear from Figure 6.2 that even with an optimal controller, it would not be possible to reduce the peak deviation of the output voltage below the value:

$$\Delta V_{MIN} = \Delta I \ x \ R_e$$

If the capacitance is less than a certain minimum, the capacitor voltage will go below the value determined by the resistive voltage drop. The minimum capacitance required to avoid the dip produces an initial dv/dt, which is equal (but opposite in sign) to the dv/dt generated by the slope of the inductor current and the ESR.

$$C_{MIN} = \frac{\Delta I}{R_e \frac{V_{in} - V_{out}}{L}}$$

If the capacitance is less than C_{min}, the peak deviation can be calculated from the output voltage versus time function:

$$\Delta v_0(t) = t^2 \frac{m}{2C} + t\left(mR_e - \frac{\Delta I}{C}\right) - R_e \Delta I$$

...where $m = (V_{in} - V_{out})/L$ (or $m = -V_{out}/L$, for the case of a downward load current step) is the slope of the inductor current. The result is:

$$\Delta V = -\frac{\Delta I^2}{2mC} - \frac{mCR_e^2}{2}$$

Optimal Load Transient Response

By neglecting the effect of the ESL, a step change of ΔI in the load current causes an initial change in the output voltage of a DC-DC converter that is equal to the product of the ESR and ΔI.

It is not possible to reduce the transient deviation below that value. For a given capacitor technology, the cost of the capacitors tends to be inversely proportional to the ESR—the smaller the ESR is, the more expensive the capacitor will be. Therefore, it makes economical sense to select a capacitor that has a worst-case (i.e., maximum) ESR that is just below the limit determined by the tolerance specifications such as the ones shown in Table 6.1.

Considering the ±130 mV transient tolerance of the 300 MHz Pentium II processor, and assuming infinite dc loop gain, constant-off- time control, ±1 percent accuracy of the reference voltage, and 2 percent peak-to-peak ripple voltage, the maximum allowed resistive deviation is 130m − 0.02×2.8 = 74 mV, both for a step load increase and a step load decrease. With a 14.2 - 1 = 13.2 A step change in the load current, the allowed maximum ESR is 74m/13.2 = 5.6 mOhm.

Figure 6.3 shows the achievable waveform in worst case, i.e., when the reference voltage is 1 percent below the nominal value, and the step load increase happens at the beginning of the off time. In the waveform, the capacitive voltage drop was neglected, which is reasonable if electrolytic capacitors are used. The voltage is assumed to return to the steady-state value within the shortest time that is physically possible.

The Intel specification requires that the output voltage return to within the static limits in less than 2 μs. This sets a lower limit on the switching frequency. For this case, a straightforward calculation yields a minimum switching frequency of 570 kHz and a peak-to-peak inductor ripple current of 10 A.

Unfortunately, the high frequency significantly increases the switching losses both in the control IC and in the power transistor. Reducing the peak-to-peak ripple voltage will not help, since it will lead to even higher switching frequency.

Increasing the ripple voltage, on the other hand, will lead to increased ripple current and more RMS losses, without allowing the frequency to drop below about 500 kHz.

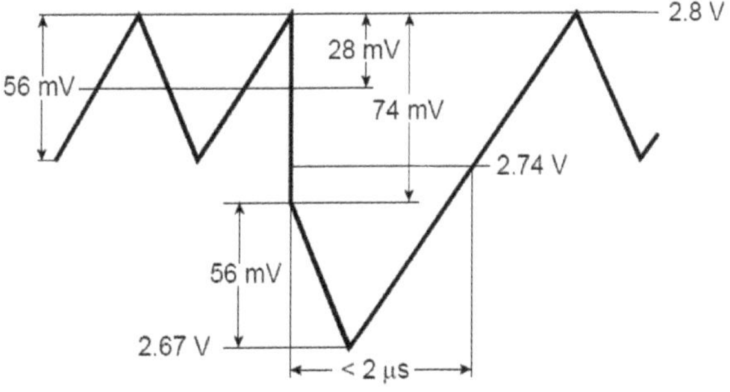

Figure 6.3: Transient Waveform with Infinite DC Gain and Quick Recovery to the Steady-State Value

In order to be able to operate the converter at a reasonably low frequency, i.e., 100 to 200 kHz, a different control strategy should be used. Instead of returning the output voltage to the nominal value as quickly as possible, the control should position the light-load output voltage close to the upper static limit and the full-load output voltage close to the lower static limit. The transition should be step-like. Such a response allows the use of capacitors with increased ESR and, consequently, reduced cost.

Figure 6.4 shows the optimal transient waveform, assuming constant off-time control, ±1 percent accuracy of the reference voltage, 0.5 percent peak-to-peak ripple voltage, and negligible capacitive voltage drop. As Figure 6.4 illustrates, with this control strategy, the maximum allowed resistive deviation is 90 mV, which corresponds to a maximum allowed ESR of 90m/13.2 = 6.8 mOhm.

It is clear, both from this discussion and from Figure 6.4, that for an optimal load transient response, the converter's output impedance must be resistive and equal to the maximum allowed ESR of the output capacitor. If the output resistance is higher than the maximum ESR, the converter will fail to meet the static tolerance specifications. An output resistance that is smaller than the maximum allowed ESR would be acceptable, but it actually costs more than necessary. Frequency-dependent output impedances unavoidably increase the peak-to-peak value of the load transient response, so they also lead to a higher cost.

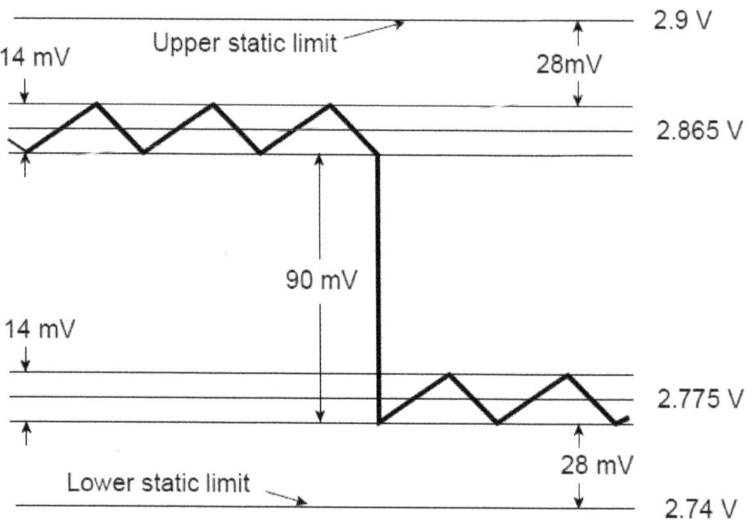

Figure 6.4: Optimal Transient Waveform

One should note that, with the proposed optimal control strategy, the converter can operate at any switching frequency as long as the capacitance of the output capacitor is above the minimum value defined by the C_{MIN} equation.

Commonly Used Control Techniques—Voltage Mode Control

One of the most common control techniques is constant-frequency pulse-width modulation (PWM) voltage-mode control (Figure 6.5). With a properly designed wideband compensation and having a large capacitance at the output, the load transient response of the regulator is a step of Re times ΔI with an approximately linear return to the steady-state value. The slope of the output voltage during the return phase is m times Re, where m is the slope of the inductor current, as previously defined.

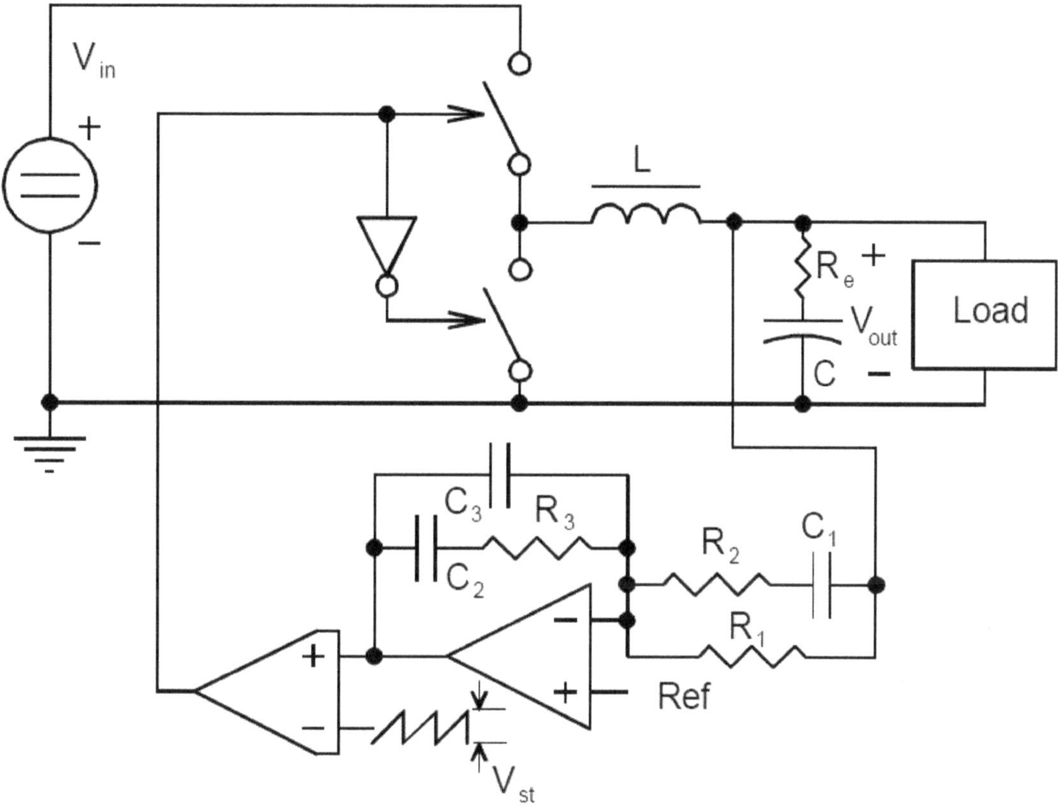

Figure 6.5: Voltage-Mode Controlled Buck Converter

Voltage-mode control cannot produce the optimal output impedance ($Z_{out(opt)} = R_{e(max)}$) due to the inherently small DC output impedance of the buck converter and the lack of direct control of the inductor current.

V^2 Architecture

The recently introduced V^2 architecture shown in Figure 6.6 combines the AC–coupled variation of constant-off-time current-mode control and the feed-forward of the current in the output capacitor for a fast load transient response. The concept of feed-forward of the capacitor current to improve the load transient response was originally presented in [3] and analyzed in detail in [4].

While the output voltage of the V^2–controlled converter rapidly returns to the static value after a load transient, the V^2 technique is not suitable for realizing the optimal output impedance. The basic reason is the same as with the voltage-mode control, the inherently low DC output impedance of the converter and the lack of direct control of the inductor current.

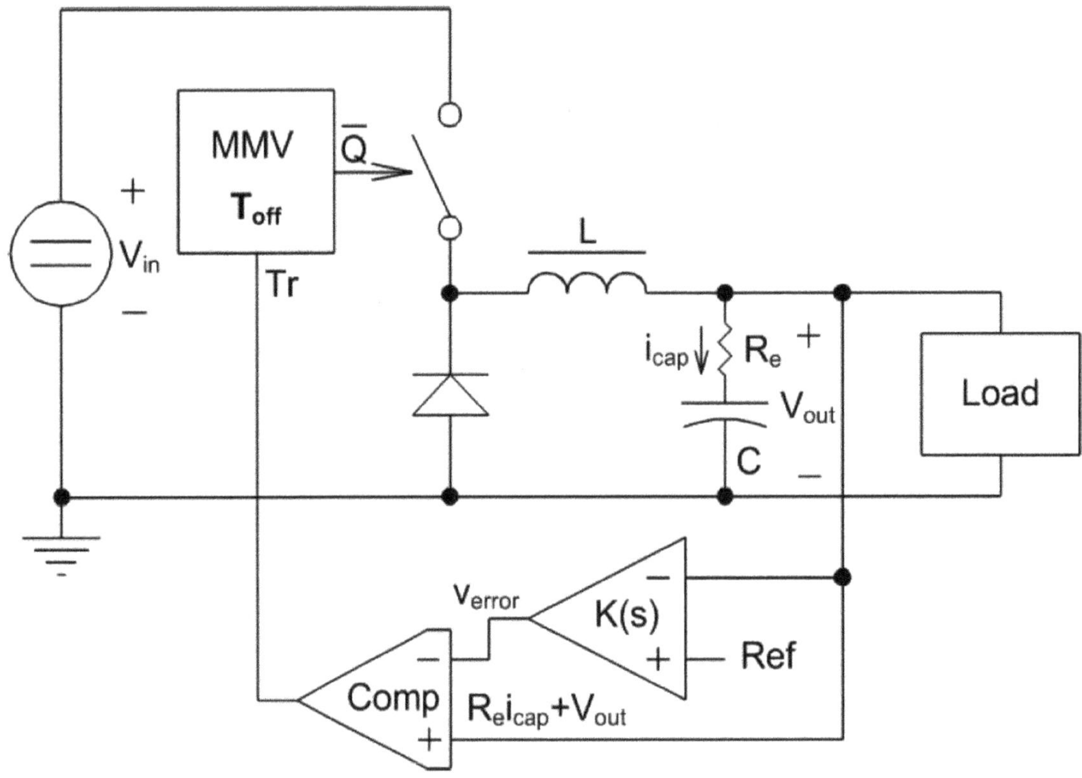

Figure 6.6: Buck Converter with V^2 Control

Droop Resistor

It has been known for some time that the smaller-than-optimal DC output impedance produced by voltage mode control or V^2 technique is undesirable. The proposed solution to increase the output impedance was to add a "droop" resistor between the voltage-feedback node and the junction of the output capacitor and the load.

While the inclusion of the droop resistor certainly helps, the sum of the output converter impedance and the droop resistor resistance only crudely approximates the optimal output impedance. In addition, neither voltage-mode control, nor V^2 control, provides inherent overload protection despite the presence of the droop resistor. The requirement of a reasonably precise droop resistor value that is tuned to produce the optimal voltage droop (which is a function of the converter specification, which may vary) is also impractical.

Design for Optimal Output Impedance

The optimal output impedance can be achieved with current-mode control and suitable compensation of the voltage-error amplifier.

Required Voltage-Error Amplifier Transfer Function

The first step in the design is to determine the output impedance of the feedback-regulated converter. This can be achieved, for example, by using the method of injected/absorbed currents discussed in [5].

Figure 6.7 shows the equivalent circuit of the system.

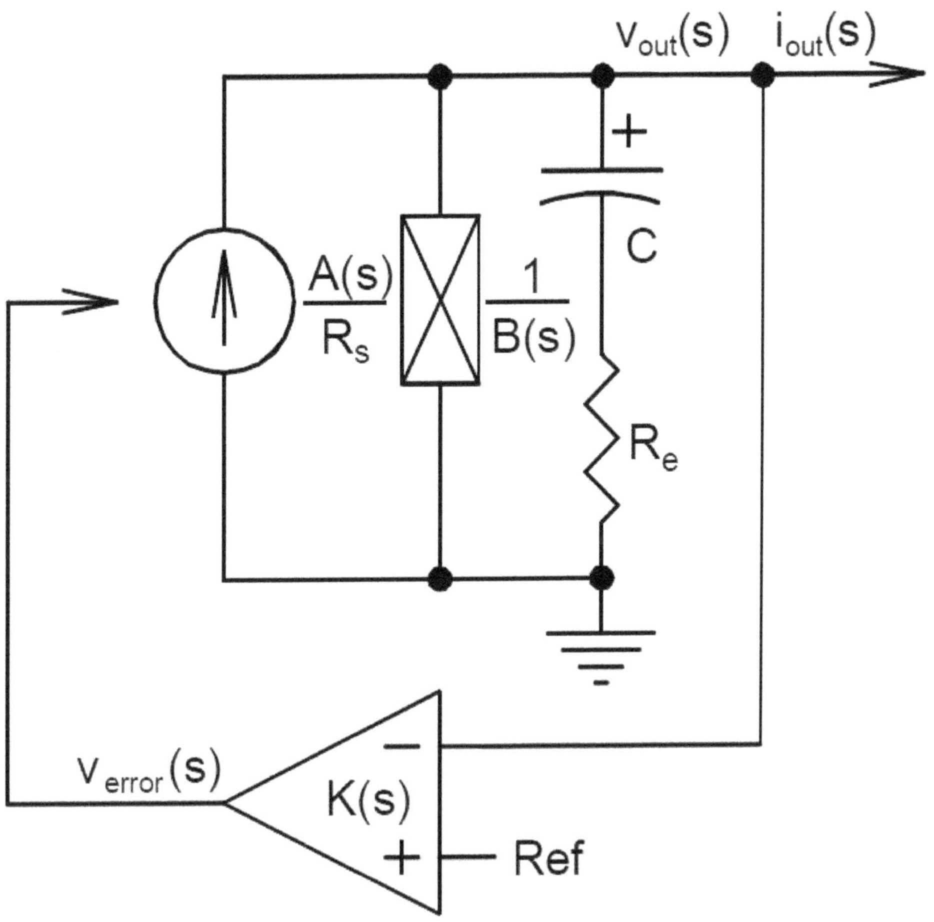

Figure 6.7: Equivalent Circuit of the Feedback-Regulated Converter for Determining the Optimal Output Impedance

In Figure 6.7, A(s) and B(s) illustrate the characteristic coefficients that describe the dynamic behavior of the converter. A(s) is the relationship between the controlled quantity, which is the inductor current for current-mode controlled converter, and the current injected toward the load. B(s) is the measure of the dependence of the injected current from the output voltage. In the equivalent circuit, the effect of the input voltage variations is neglected, which is a reasonable assumption for current-mode controlled buck converters. Rs is the value of the current-sense resistor.

The following equation can be written for the system in Figure 6.7:

$$\left[v_{out}(s)\ x\ K(s)\frac{A(s)}{R_S} - i_{out}(s)\right]\left[\frac{1}{B(s)}\right] || \left(R_e + \frac{1}{sC}\right) = v_{out}(s)$$

From this equation, the output impedance is:

$$Z_{out}(s) = -\frac{v_{out}(s)}{i_{out}(s)} = \frac{1 + sR_eC}{B(s) + sC[1 + B(S)R_e] - K(s)A(s)R_S^{-1}(1 + sR_eC)}$$

Substituting R_e for the output impedance and solving for K(s) yields the following transfer function for the voltage-error amplifier:

$$K(s) = \frac{B(s) + sCB(s)R_e - R_e^{-1}}{A(s)R_S^{-1}(1 + sCR_e)}$$

The best practical choice for controlling the converter is constant-off- time current-mode control (Figure 6.8). This control technique features high rejection ratio for input voltage variations and frequency-independent response from the control input to the inductor current; it also allows the use of only the switch peak current information.

The characteristic coefficients of the constant-off-time current-mode-controlled buck converter are as follows:

$$A(s) = 1$$

$$B(s) = -\frac{T_{off}}{2L}$$

Substituting the characteristic coefficients in *K(s)* yields:

$$K(s) = -\frac{R_s\left(\frac{1}{R_e} + \frac{T_{off}}{2L} + sCR_e\frac{T_{off}}{2L}\right)}{1 + sCR}$$

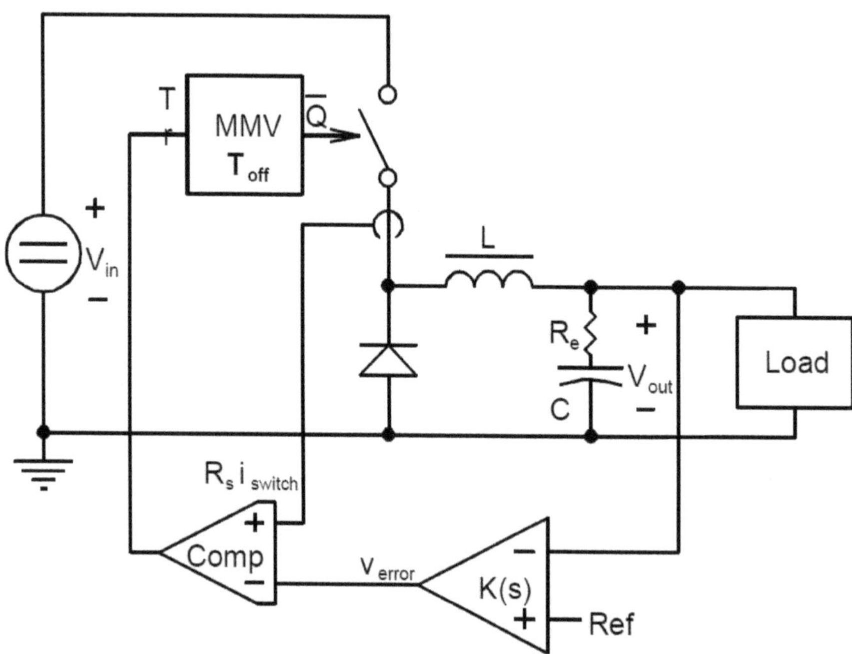

Figure 6.8: Buck Converter with Constant Off-Time Current-Mode Control

One should recognize that the zero in the K(s) equation is typically well above the switching frequency, so the last term in the numerator can safely be neglected.

Also, in most practical applications, R_e is usually much smaller than $2L/T_{off}$, which means that the required transfer function can be written as:

$$K(s) = -\frac{R_S}{R_e}\frac{1}{1 + sCR_e}$$

This transfer function can be easily realized with an operational amplifier as shown in Figure 6.9a or a transconductance amplifier as shown in Figure 6.9b. The ratio R_2/R_1 in Figure 6.9a must be equal to R_s/R_e, and the product R_2C_1 must be equal to the time constant R_eC of the output capacitor. In the gm-amplifier realization of Figure 6.9b, the product gmR_3 must be equal to R_s/R_e, and the product R_3C_2 must be equal to R_eC.

Let us note that if the R_eC time constant is commensurate with, or smaller than, the switching period, the pole of the transfer function can be neglected, and C_1 (in Figure 6.9a) or C_2 (in Figure 6.9b) can be omitted from the error amplifier. Note also that the dc shift of the output voltage is proportional to the value of the sense resistor. A large tolerance in the sense resistance (e.g., using the RDS(ON) of the MOSFET, the winding resistance of the inductor, or the resistance of a copper trace on the board) would greatly reduce the available dc shift, which would increase the minimum dc gain requirement and would require an output capacitance with substantially lower ESR.

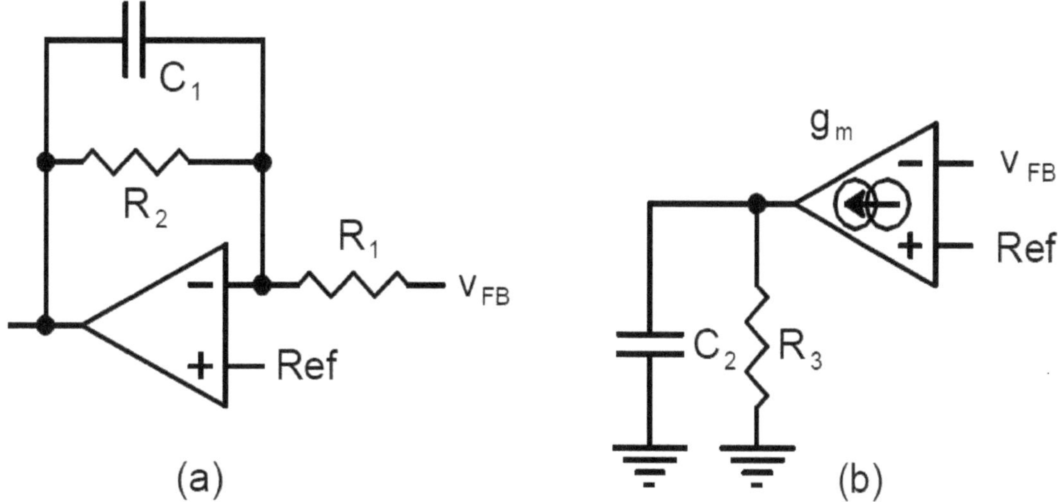

Figure 6.9: Realizing the Transfer Function of (10) with Operational Amplifier (a) or with gm Amplifier (b)

Design Procedure

The steps of a typical design for meeting the load transient requirements of the Pentium II application are as follows:

1. Determine the maximum acceptable ESR of the output capacitor from the allowable dynamic deviation and the magnitude of the load current step.

$$R_{e(max)} = \frac{\Delta V_{out}}{\Delta I} = \frac{90m}{13.2} = 6.8m\Omega$$

2. Calculate the minimum inductance of the energy-storage inductor from the ESR, off time, DC output voltage, and peak-to-peak output ripple voltage. Assuming an off time of 2.4 μs, and a peak-to-peak ripple of 14 mV, the minimum inductance is:

$$L_{min} = \frac{V_{out}T_{off}R_e}{V_{ripple,pp}} = \frac{2.8\ x\ 2.4\mu\ x\ 6.8m}{14m} = 3.26\mu H$$

3. Determine the minimum capacitance of the output capacitor from the requirement that the output be held up while the inductor current ramps up (or down) to the new value. The minimum capacitance would produce an initial dv/dt, which is equal (but opposite in sign) to the dv/dt generated by the di/dt in the inductor and the ESR of the capacitor.

$$C_{min} = \frac{\Delta I}{R_e\left(\frac{di}{dt}\right)} = \frac{13.2}{6.8m\ x\ \left(\frac{2.2}{3.26\mu}\right)} = 2.88mF$$

4. Here di/dt is the rate of rise or fall of the inductor current, whichever is smaller. We assume that the minimum voltage across the inductor is 2.2 V.

5. Select a capacitor that has more capacitance and less ESR than the values calculated from the C_{min} and $R_{e(max)}$ equations.

6. Select the input and feedback resistors for the voltage-error amplifier (or the terminating resistor in the case when the error amplifier is of the trans-conductance type) such that the DC gain is equal to the ratio of the current sense resistor and the ESR of the output capacitor.

7. Select a compensating capacitor such that it provides a pole frequency in the transfer function of the error amplifier that is equal to $1/(2\pi R_e C)$.

Computer Simulations and Experimental Results

Computer simulations were conducted with a complete switched model using the demo version of PowerSIM (PSIM), a circuit-analysis program developed for power electronics applications. Figure 6.10 shows the output voltage and the inductor current for the power component parameters L = 3 μH, Re = 7 mOhm, C = 3 mF, and with the compensation that provides the optimal output impedance.

The off time was 2.4 μs.

We also took experimental data on a test circuit controlled by a new constant-off-time current-mode controller IC, the ADP3152 from analog devices. This IC features the VID output voltage programming and status monitor signal required in the Pentium II guidelines of [1]. Another version of the IC (ADP3153) includes an additional controller for a low-dropout (LDO) linear regulator.

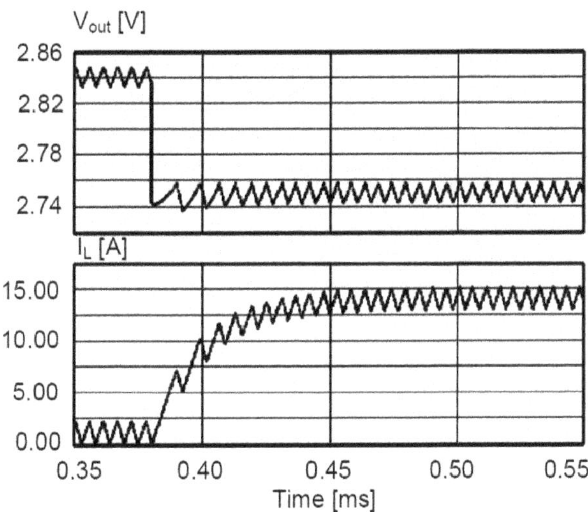

Figure 6.10: Simulated Load-Transient Response with Optimal Compensation

Figure 6.11 shows the schematic of the test circuit. In order to demonstrate the optimal response theory, we chose the compensation not based on the worst case parameters, but on the measured parameters. We used six pieces of 560 μF, 25 V, FA series Panasonic aluminum electrolytic capacitors in parallel. The ESR of the six parallel capacitors was 6.15 mOhm. Thus the 13.2 A load step produced an approximately 81 mV step in the output voltage. The total capacitance of 3.36 mF and ESR of 6.15 mOhm limited the inductance to less than 3.44 μH [calculated from (13)]. We selected 3.3 μH.

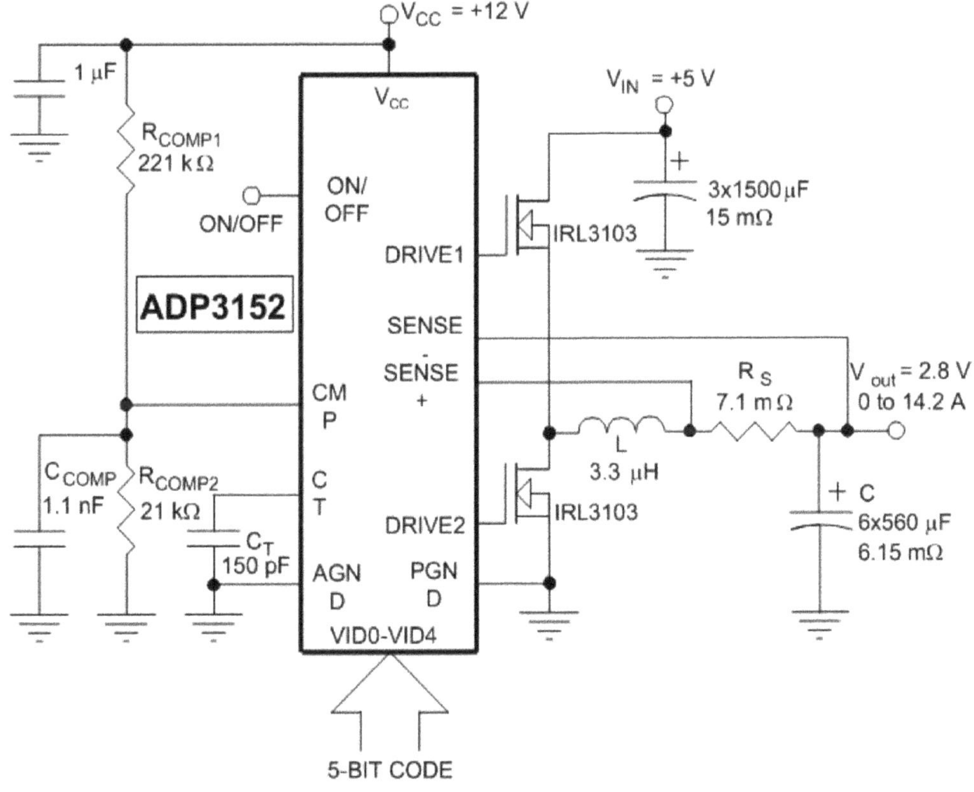

Figure 6.11: Test Circuit Schematic

To determine the compensation and to create an intentional output voltage offset, it is necessary to understand the components of the feedback network. The sensed output voltage is divided by three, is measured against a reference voltage by a transconductance amplifier with a transconductance of 2.2 mS, develops a voltage via the compensation termination, and is divided by 12, and it is with that voltage the peak current is programmed. Those are the gain factors. The offset is determined by noting that the output of the transconductance amplifier commands zero current when it is at approximately 0.7 V, and that voltage also corresponds to a balanced input. Thus, when the output voltage equals three times the reference voltage, the transconductance amplifier will have a net output of zero current.

The compensation components for the ADP3152 were chosen as follows. The equivalent current sensing resistor was chosen to be 7.1 mOhm. This produces a nominal 100 mV drop at 14 A load—allowing sufficient headroom for ripple current below the peak current limiting threshold of 130 mV. According to the text, we should set the DC gain (from output voltage to current programming voltage) equal to the ratio of the current sense resistor Rs and the ESR, R_e. The loop transconductance Gm, from output voltage to inductor current, will be the transconductance amplifier gain, 2.2 mS, divided by both 3 at the front end and 12 at the tail end, for a net transconductance of 61 μS. Therefore, the terminating resistance should be set equal to Rs/(Re Gm), or 7.1m/(6.15m x 61μ) = 19 kOhm. Since the IC is powered by a regulated 12 V supply, we can calculate two resistors to create a divider from 12 V, which yields approximately 1 V, and which has an impedance at the divider tap of 19 kOhm. That yields the values that we used in the test circuit: 221 kOhm (top) and 21.0 kOhm (bottom).

Since the amplifier termination time constant should equal that of the output capacitance and ESR, the amplifier termination capacitance should be 1.1 nF. That value was also used in the circuit.

Figure 6.12 shows the measured transient response, with two different time scales.

We were obviously able to achieve the desired and predicted response. The response of the output current is balanced nearly perfectly against the output capacitor impedance to prevent both the overshoot, which would widen the dynamic regulation tolerance, and the undershoot, which if the load step were to revert at the point where the undershoot is highest, would also widen it.

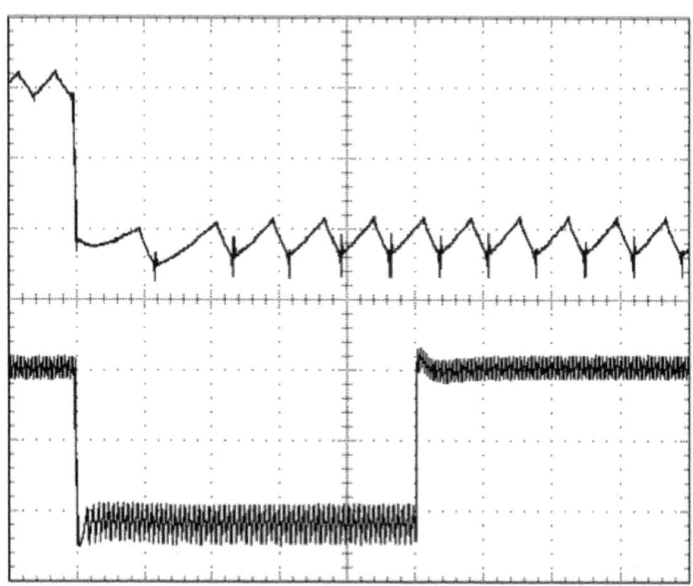

Figure 6.12: Measured Load Transient Response Both Traces Show the Output Voltage Response to a 13-A Change in the Load Current, but with Two Different Time Scales, 10 μs/div. (Top Trace) and 100 μs/div. (Bottom Trace). Vertical Scale: 40 mV/div.

Summary

Switched mode power converters are limited by current technology to switch at speeds slower than the 2 μs that is allowed for dynamic regulation of a Pentium II processor. Thus, the output capacitor impedance sets the limit of achievable response capability for a power converter. By first choosing that capacitor according to the minimum impedance required by the load specification and then tailoring the compensation to match that impedance according to the guidelines set forth in this paper, the converter can have an instantaneous response (180 degrees of phase margin) to a load step change. The regulation performance is limited only by the actual output capacitors.

References

[1] Intel document, *VRM 8.1 DC-DC Converter Design Guidelines*, Order number: 243408-001, May 1997.
[2] D. Goder and W.R. Pelletier, *V^2 Architecture Provides Ultra-Fast Transient Response in Switch Mode Power Supplies*, HFPC Power Conversion, September 1996 Proceedings, pp. 19-23.
[3] R. Redl and N. O. Sokal, *Near-Optimum Dynamic Regulation of DC-DC Converters using Feed-Forward of Output Current and Input Voltage with Current-Mode Control*, IEEE Transactions on Power Electronics, vol. PE-1, no. 3, July 1986, pp. 181-192.
[4] G. K. Schoneman and D. M. Mitchell, *Output Impedance Considerations for Switching Regulators with Current-Injected Control*, PESC '87 Record, pp. 324-335.
[5] A. S. Kislovski, R. Redl, and N. O. Sokal, *Dynamic Analysis of Switching-Mode DC-DC Converters*, Design Automation, Inc., Lexington, Massachusetts, 1996.

Appendix—Maximum ESR, Maximum Inductance, Minimum Switching Frequency, and Ripple Current of Converters with Zero DC Regulation

The maximum ESR of the output capacitor is:

$$R_e = \frac{V_1 - V_{err} - \frac{V_{ripple,pp}}{2}}{\Delta I}$$

...where V1 is the maximum transient deviation from the nominal voltage (130 mV), V_{err} is the maximum setpoint tolerance in one direction (typically 28 mV, corresponding to 1 percent), $V_{ripple,p\text{-}p}$ is the peak-to-peak output ripple (e.g., 56 mV, corresponding to 2 percent), and ΔI is the step load current change (13.2 A).

The following equation can be written based on the waveform in Figure 6.3:

$$T_1 = \frac{V_2}{\frac{V_i - V_O}{L} R_e} + \frac{V_{ripple,pp}}{\frac{V_O}{L} R_e}$$

...where T_1 is the maximum duration for which the output voltage can stay outside the static tolerance limit (2 μs), V_2 is the difference between the transient and static tolerance limits (70 mV), V_i is the input voltage (5V), V_o is the output voltage (2.8V), and L is the maximum inductance.

From the T_1 equation, the maximum inductance is calculated as:

$$L = \frac{T_1 R_e}{\frac{V_1}{V_i - V_O} + \frac{V_{ripple,pp}}{V_O}}$$

The minimum required switching frequency is:

$$f = \frac{R_e V_O \left(1 - \frac{V_O}{V_i}\right)}{V_{ripple,pp} L}$$

The peak-to-peak ripple current can be calculated as:

$$I_{ripple,pp} = \frac{V_{ripple,pp}}{R_e}$$

Substituting the numbers given in parentheses, the results are:

R_e = 5.6 mΩ, L = 216 nH, f = 570 kHz, $I_{ripple,pp}$ = 10A

Chapter 7—Frequency Domain Target Impedance Method for Bypass Capacitor Selection for Power Distribution Systems

Larry D. Smith, Principal Signal Integrity Engineer, Altera Corp.

This chapter is from a revised paper originally presented in TF-MP3, *Comparison of Power Distribution Network Methods: Bypass Capacitor Selection Based on Time Domain and Frequency Domain Performances*, DesignCon 2006, Santa Clara, California, February 6-9, 2006.

THERE HAS BEEN much discussion in the industry on how to choose bypass capacitors for power distribution systems (PDSs). This paper defines and discusses the frequency domain target impedance method (FDTIM). The power system should meet target impedance across a broad frequency range, from direct current (DC) up to the highest frequency of interest. Simulation is used to select capacitor values and quantities from a menu. The number of capacitors required to meet the target impedance up to a corner frequency is estimated by a simple formula. High-impedance resonances are avoided by providing sufficient system damping. The relationship of printed circuit board (PCB) capacitors to the voltage regulator module (VRM) at low frequency and the mounted die at high frequency is discussed. Capacitor equivalent series resistance (ESR) is an important consideration. The FDTIM is compared to the Big "V" and decade methods for choosing bypass capacitors, and it is found to have superior performance and cost characteristics.

Introduction

Bypass capacitors, otherwise known as decoupling capacitors, are an important part of the PDS. There has been much controversy over how to choose the optimum set of capacitors to accomplish a reliable PDS. The primary figures of merit for a PDS include the noise performance, cost, reliability, and surface area consumed by the capacitors.

The primary components of the PDS include the VRM, bulk capacitors, high-frequency ceramic capacitors, PCB power planes, capacitor mounting pads and vias, mount for the electronic package (load, power consumer), package capacitors, package power planes, die mount and internal die capacitance. Many of these components are shown in Figure 7.1, roughly in the frequency range where they are effective.

This paper concentrates on the FDTIM to select bypass capacitors that will work with the rest of the system to provide clean power to the load, usually one or more silicon die. The method has been previously discussed in [1]. A key concept is the determination of a target impedance. By meeting the target impedance from DC up to the highest frequency of interest at each level of assembly, the optimum compromise between performance, cost, reliability, and board area is reached.

Target Impedance

As part of the die design process, engineers assume a nominal power supply voltage with a tolerance, often plus or minus 5 percent. Circuits are simulated to guarantee timing and performance specifications assuming that the power supply at the circuit terminals meets a well-defined specification. This simulation usually involves SPICE analysis with ideal voltage sources at three "corners" to provide power at nominal and nominal ±5 percent voltage conditions.

In the real world, there are no ideal SPICE power supplies because real voltage sources have series source impedance. This series impedance gives rise to load regulation as the magnitude of the current drawn by the load changes during normal operation. The cost of a power supply is inversely proportional to the series impedance, with an ideal zero impedance power supply being infinitely expensive. The optimum power supply delivers the specified power requirements at the minimum cost. By knowing the nominal power supply voltage, maximum and minimum load currents, and allowable voltage tolerance, the target impedance is easily calculated from Ohm's law.

$$T_{target} = \frac{Vdd\ x\ tolerance}{I_{max} - I_{min}} = \frac{1.0V\ x\ 0.05}{100A - 50A} = 1mOhm$$

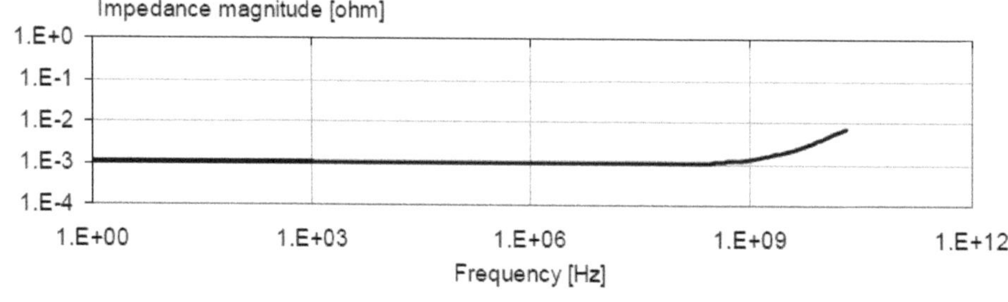

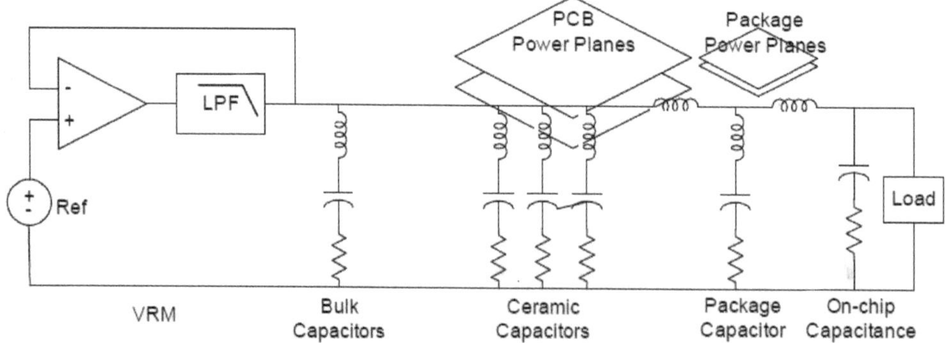

Figure 7.1: Components for PDS Design The Impedance of the PDS Should Be Flat across a Broad Frequency Range.

In this example, the nominal power supply is 1.0V; the tolerance is 5 percent; and the maximum and minimum load currents are 100A and 50A respectively. A 1V power supply with 1 mOhm of source impedance will vary by 0.05V when the maximum and minimum currents are drawn by the load. The difference between maximum and minimum load current is often called the transient current. Both the magnitude and rise (fall) time of the transient is important. The magnitude determines the target impedance; the rise time determines the upper-corner frequency where the target impedance should be met.

It may appear that this calculation is off by a factor of two because plus and minus 5 percent voltage tolerance is allowed. However, the regulation loop of a power supply VRM will bring a 5 percent low voltage, sensed at the load and back to nominal voltage after the VRM time constant has past. When the load goes away, the voltage at the load will bounce to plus 5 percent until another VRM time constant has past, thus using up the entire ±5 percent tolerance.

The target impedance calculation applies at all levels of assembly including a single circuit, block of circuits on a die, full die, package, power plane on a PCB that supplies several loads, the DC source for VRMs (often 12V or 48VDC), and even at the AC system supply level.

The target impedance is useful to size the amount of capacitance necessary at each level of assembly to store sufficient charge and energy for the load. When a fast load draws a power transient, charge is stored in small capacitors near the load supply and near the current for a short period of time, but they are soon depleted.

Current must then come from slower time constant capacitors at the next level of assembly. Typical PDS charge storage areas include on-chip capacitors, on-package capacitors, PCB ceramic and bulk capacitors, and capacitors associated with the VRM input and output. Each level of assembly provides transient current long enough for the time constant of the next level to expire so that energy can begin to come (or stop coming) from the next stage.

A PDS that meets target impedance across the entire frequency span has sufficient stored charge to supply clean power at all frequencies. On the other hand, a power system that is higher than the target impedance in any frequency band will usually fail to meet the specified voltage tolerance under some load condition. If the measured PDS impedance is substantially below target impedance in any frequency range, the cost of the PDS can usually be reduced while still meeting the power specifications. It is referred to as a "target" because it gives the best cost and performance solution for the PDS.

Impedance in the Frequency Domain

The target impedance should be met not only at DC but at all frequencies up to some corner frequency, fc. There are mechanisms at the die and system level that can cause power supply transients in all frequency bands, not just the clock frequency. The circuits on the die have rise times on the order of 10 ps. Using the common formula that relates frequency content to rise time $fc = \frac{0.35}{t_{rise}}$, the largest frequency of interest for an individual circuit is 35 GHz.

A die operating with a clock frequency of 1 GHz can suddenly begin to draw (or cease drawing) current from the power supply in about 1 nSec, giving a corner frequency of about 350 MHz. Integrated circuits can suddenly draw transient current after a key stroke or cease to draw current when operations stall while waiting for data from memory. It may take 40 ns to retrieve data from DRAM and a die could repetitively draw pulses of power current in 100 ns cycles putting a 10 MHz load on the PDS. Or, the central processing unit (CPU) could be waiting for data from a hard drive with a 1 ms time constant leading to kHz loading on the PDS. It is possible for customer code to cause repeating current transients at virtually any frequency from DC to 35 GHz, therefore, the PDS should meet target impedance in all frequency bands in the respective levels of assembly. The circuits on the die should look out and see an impedance that is close to target impedance throughout the entire frequency range.

Simulation is used to select the best mix of capacitors. The impedance can be measured by using a 1 amp current source in the die position of the circuit. The voltage across the current source is numerically the same as the self-impedance of the PDS measured from the die position. Voltages at other positions of the circuit are numerically the same as the trans-impedance, which is the voltage across any two nodes of the circuit divided by the current forced at some other position of the circuit. By selecting optimum values and quantities of bypass capacitors to be placed at various locations of the circuit, the PDS can be designed to have a relatively flat impedance profile across a broad frequency range. Figure 7.1 shows a 1 mOhm target impedance up to the corner frequency associated with the rise time of a silicon gate along with the components that are effective in each frequency range.

Cap Value	Size	Diel.	Measured Value	Units	ESR (m Ω)	L intern (nH)	L mount (nH)	SRF (MHz)	Q	ESR div. by 1 m Ω	Meet 1 m Ω Ztarget
100uF	1812	X5R	80.3	µF	1.8	2.112	0.600	0.341	0.7	2	2
47uF	1210	X5R	42.1	µF	1.9	1.487	0.600	0.537	1.1	2	3
22uF	1210	X5R	17.7	µF	2.5	1.300	0.600	0.867	1.3	3	7
10uF	0805	X5R	7.26	µF	3.6	0.773	0.600	1.60	1.6	4	9
4.7uF	0805	X5R	4.12	µF	4.2	0.544	0.600	2.32	2.1	4	5
2.2uF	0805	X5R	1.98	µF	6.1	0.413	0.600	3.55	2.2	6	8
1.0uF	0603	X5R	0.79	µF	9.1	0.391	0.600	5.69	2.3	9	12
470nF	0603	X5R	404	nF	13	0.419	0.600	7.85	2.3	13	16
220nF	0603	X7R	172	nF	19	0.438	0.600	11.9	2.3	19	28
100nF	0603	X7R	75	nF	29	0.443	0.600	18.0	2.3	29	30
47nF	0603	X7R	39	nF	38	0.451	0.600	24.7	2.4	38	40
22nF	0603	X7R	17	nF	64	0.492	0.600	36.6	2.1	64	53
10nF	0603	X7R	8.9	nF	80	0.518	0.600	50.4	2.4	80	60
								Totals		273	273

Table 7.1: Discrete MLC Capacitors That Are Useful for PCB Bypass. Measured Values for Capacitance and Inductance Are Shown Together with the Series Resonant Frequency and Q for 600 pH Mounting Inductance.

The major difficulty with PDS design is that the system is built from a network of inductances and capacitances that attempt to mimic the flat impedance of a 1 mOhm resistance. A conductor always has parasitic inductance that has a plus 20 dB per decade slope on a log-log impedance plot. Intentional capacitance from discrete capacitors and parasitic capacitance from power planes has a slope of minus 20 dB per decade.

Resonance occurs where the two slopes cross. Parallel combinations of capacitance and inductance form impedance peaks, while series combinations form impedance dips. For power loss reasons, resistance is often designed out of power systems and can lead to high Q resonances if the crossings are not well controlled. A major objective in choosing decoupling capacitors is to insure that resonant peaks that exceed the target impedance are avoided.

This is achieved through the use of accurate frequency domain models for each of the PDS components.

PCB Bypass Capacitor Selection

Much board space in modern computer systems is dedicated to the power distribution components, mostly bypass capacitors. There is cost associated with the components, assembly process, vias, and the PCB area occupied by the capacitors. It is important to get the most benefit out of each mounted capacitor in order to minimize cost. Important considerations include capacitance value, ESR, internal inductance, and mounting inductance.

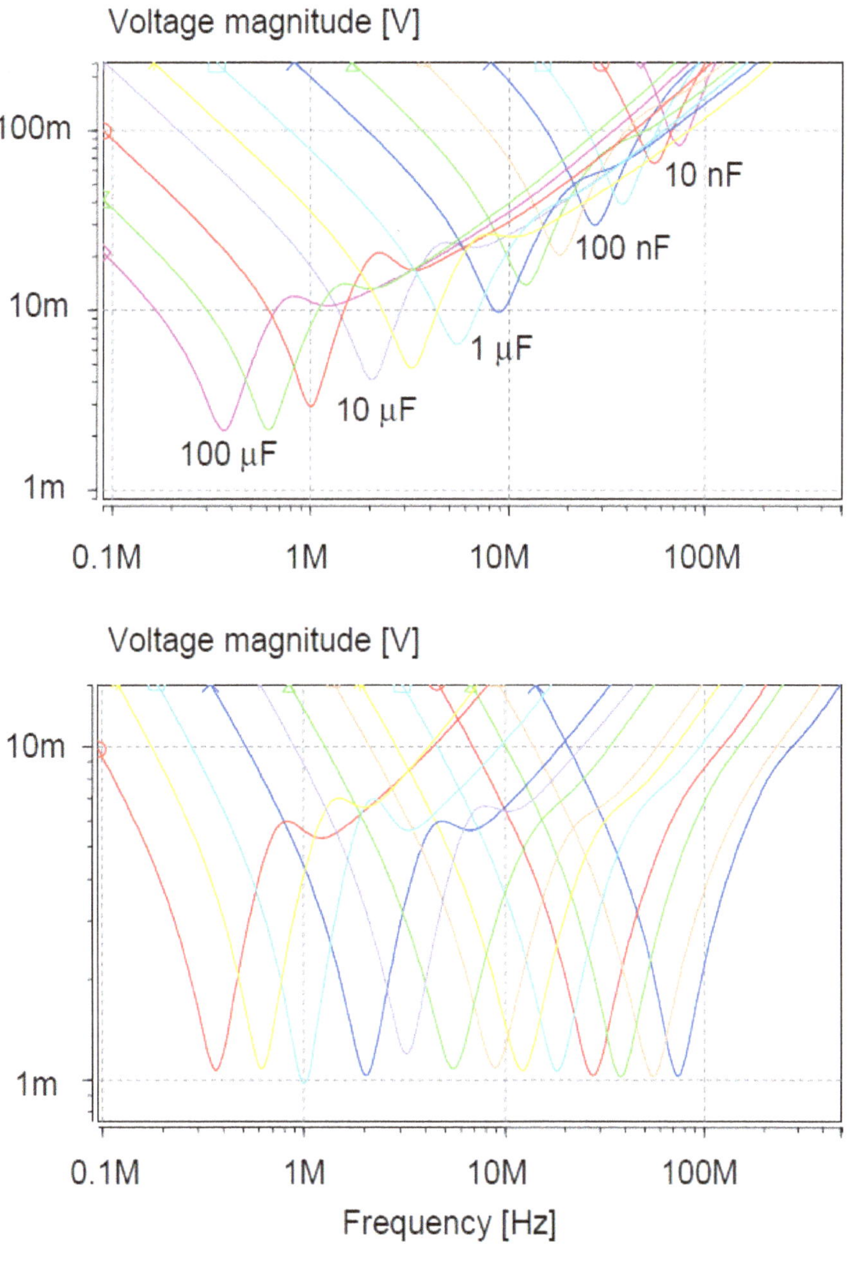

Figure 7.2: Simulated Impedance for Several Bypass Capacitor Values. Top: Single Capacitors; Bottom: Quantity to Reach 1 mOhm. Capacitors Are Selected From This Menu to Build a Power Distribution System. The Vertical Scale of Volts Should Be Interpreted as Ohms Because of the 1 Amp Current Source Used in Simulation.

Discrete PCB capacitors are used to cover the frequency range where the VRM leaves off up to where the inductance of the mounted electronic package dominates. This is typically the 10 kHz to 100 MHz band, about four decades. The upper part of the frequency range is covered by multilayer ceramic (MLC) capacitors and the lower part is covered by aluminum electrolytic, tantalum oxide, or tantalum polymer bulk capacitors.

Advancements in VRM technology have increased the bandwidth and decreased response times to the point where the bulk capacitors may not be necessary, but this usually involves a more expensive VRM. Table 7.1 gives the size, dielectric type, and ESR associated with several ceramic capacitor types that are useful for bypassing. The FDTIM for PDS design involves choosing the number of capacitors of each value such that the parallel combination of capacitors approximates a flat line at the target impedance. The approximate number of capacitors for each value can be calculated by dividing the target impedance into the ESR of that capacitor.

The top graph of Figure 7.2 shows the simulated impedance of the capacitors of Table 7.1 using the transmission line model topology for the capacitors discussed in [2]. The bottom graph of Figure 7.2 shows simulation results for the number of capacitors in parallel needed to reach 1 mOhm as calculated in Table 7.1. Three capacitor values are used per decade: 10 μF, 22 μF, 47 μF, etc. Each capacitor curve has three distinct portions: capacitance at -20 dB/decade, inductance at +20 dB/decade, and the bottom of the curve at the ESR value for the capacitor. This simulation includes only the internal inductance for capacitors (does not include mounting inductance). The family of curves shows that decreasing valued capacitors have a series resonance at increasing frequency.

The ESR of smaller valued capacitors is higher than that of lower valued capacitors. This is the menu of capacitor components that may be used in a PDS to meet a target impedance within this frequency range.

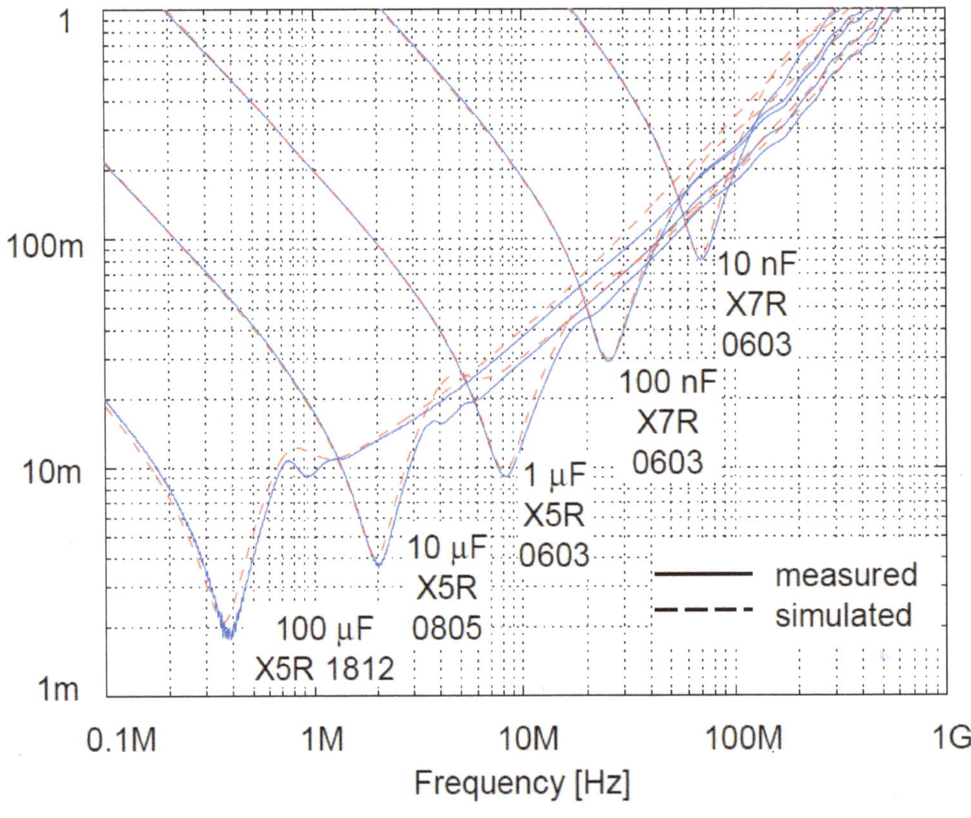

Figure 7.3: Model to Hardware Correlation for Several Ceramic Capacitors

Simulation parameters for capacitor models are chosen for best fit with capacitors measured using two port techniques with a VNA [3, 4]. Figure 7.3 shows model-to-hardware correlation.

Figure 7.4 shows simulation results for 273 total capacitors in parallel compared to a 1 mOhm target impedance. The specific quantities of capacitors are given in the column of Table 7.1 named "quantity for 1 mOhm." The simulated impedance drops below 1 mOhm because of the influence of capacitors at nearby resonant frequencies. The bottom graph of Figure 7.4 shows the simulation results for the same number of capacitors, but value quantities are chosen to meet the 1 mOhm target impedance in the lower frequency range and are given in the final column of Table 7.1.

Capacitor Sizing from Target Impedance and Corner Frequency

50 MHz was chosen as the corner frequency, above which the target impedance will not be met at the PCB level of assembly. It is actually chosen from knowledge of the mounted inductance of the load. There is little point in pushing the low impedance on the PCB much more than 2x the frequency supported by the package mounting inductance. The corner frequency is the frequency where the target impedance equals the parallel combination of all capacitors that have become inductive.

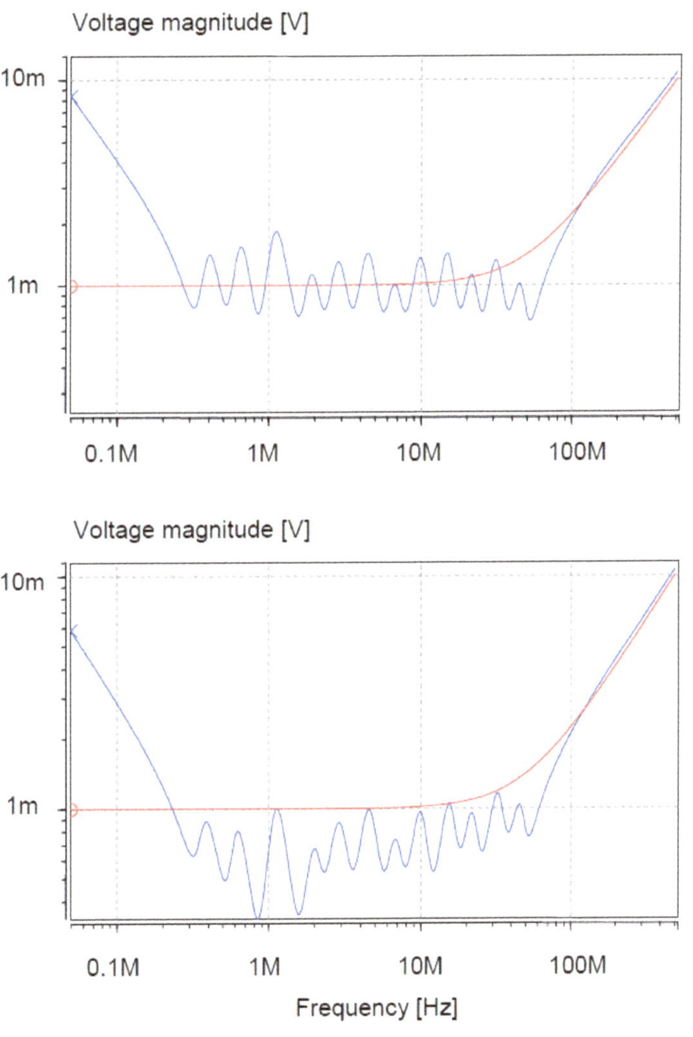

Figure 7.4: Top Graph: Simulation Results for Parallel Capacitors Calculated to Reach 1 mOhm. Bottom Graph: Capacitors Chosen to Stay below 1 mOhm Target Impedance.

$$Z_{target} = \left| j\omega_c \frac{L_{cap} + L_{mount}}{n} \right| = 2\pi f_c \frac{L_{cap} + L_{mount}}{n}$$

Where:

$$f_c = \frac{n \; x \; Z_{target}}{2\pi(L_{cap} + L_{mount})}$$

n is the number of capacitors in parallel
L_{cap} is the internal capacitor inductance
L_{mount} is the mounting inductance

After all the capacitors have become inductive, the location of the rising slope is simply the impedance of all the inductances in parallel.

The top graph of Figure 7.5 shows curves with several values of mounting inductance added in series with each capacitor: 200, 400, 800, and 1,600 pH. This is the expected range depending on whether great or little priority is given to achieving low mounting inductance in the PCB geometries.

As shown in Figure 7.5, the resonant minimum of each individual capacitor moves up and to the left with increasing mounting inductance. The bottoms of the individual capacitor curves are initially rounded, then become sharper as the mounting inductance is increased. As expected, the corner frequency of the entire set of capacitors in parallel moves to the left with increased mounting inductance.

The bottom graph of Figure 7.5 shows simulation of a reduced number of capacitors by eliminating the highest-frequency capacitors first. The same target impedance is met but only up to a reduced corner frequency.

From the earlier examples and figures, it is estimated that the number of ceramic capacitors—with ESRs that are commonly available today—required to meet a target impedance in the mOhm range with a corner frequency in the 50 MHz range is:

$$n = k \frac{f_c \; x \; (L_{cap} + L_{mount})}{Z_{target}}$$

k is about 6
average L_{cap} is 0.4 nH
L_{mount} is the mounting inductance in nH
Z_{target} is the target impedance in mOhms
f_c is the corner frequency in MHz.

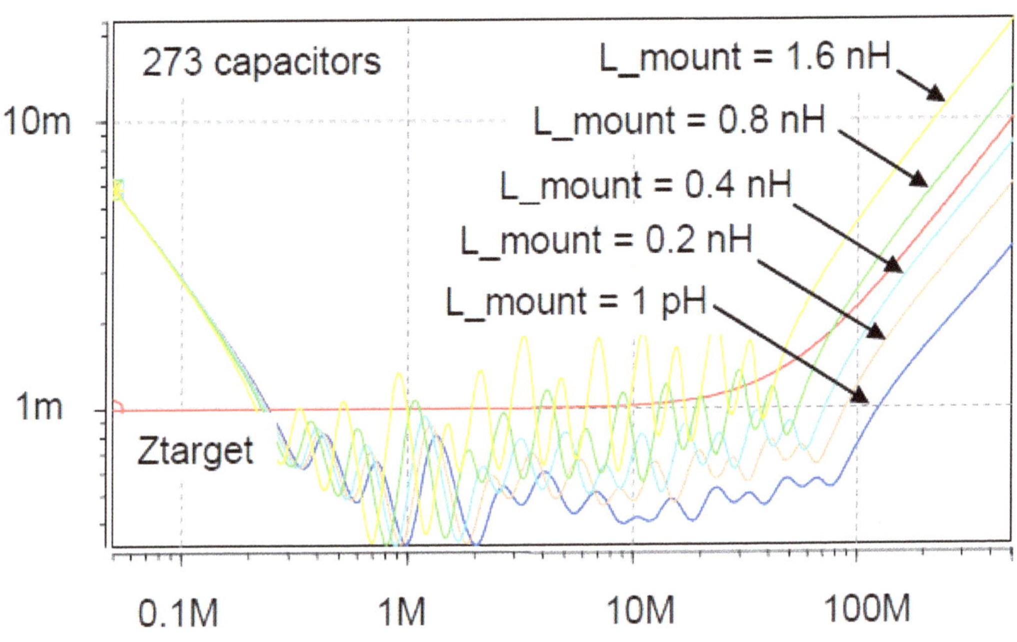

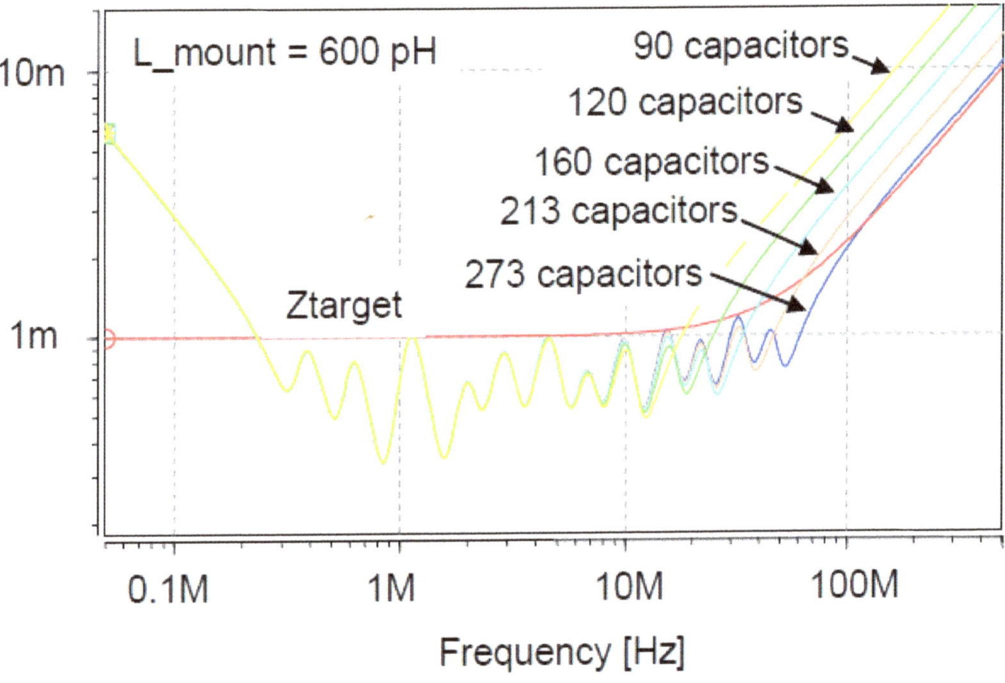

Figure 7.5: Top Graph: Simulation Results for Capacitors with Several Values of Mounting Inductance. Bottom Graph: A Reduced Number of Capacitors Is Able to Meet Target Impedance, but only up to a Lower Corner Frequency.

ESR Considerations

There is apparently a relationship between capacitor ESR and the number of capacitors required to meet a target impedance. The initial estimate for an individual capacitor quantity was ESR/Ztarget, and it would appear that a lower ESR would result in fewer capacitors required. However, the low ESR that creates deep resonant dips may also create high resonant peaks unless the Q (mounting inductance) is well controlled.

The ESR in commonly available X7R and X5R ceramic capacitors combined with reasonable (600 pH) mounting inductance gives a $Q = \omega L/R$ of about 2 as shown in Table 7.1. Three capacitor values per decade, as earlier simulated, appears to be optimal. If ESRs were lower, more capacitor values per decade would be required to keep the resonant peaks from getting out of hand. If ESRs were higher, fewer capacitor values per decade would be required, but a greater quantity of each value would be needed in parallel to reach the target impedance. The FDTIM will work in any case.

There have been industry proposals for a controlled, higher ESR for ceramic capacitors [5]. This has merit since it would reduce the number of different part number capacitors on a PCB and reduce manufacturing complexity. But what would the optimum value of ESR be? It would most likely turn out that the optimum ESR for several different products would be several different ESRs for the same capacitance value (e.g., the optimum ESR for a 1 µF capacitor would be 10 mOhm, 20 mOhm, and 40 mOhm for three different products), leading to multiple part numbers for each capacitance value. This would be difficult to manage from a component and inventory standpoint.

Several years ago, the metallurgy for MLC capacitors was silver-palladium. ESR values varied by a factor of four from supplier to supplier and sometimes within a single lot from a single supplier, therefore making it very difficult to simulate with SPICE models and obtain consistent measured results from product boards. The industry has moved to nickel base metal, resulting in very consistent ESRs from supplier to supplier for each capacitor value and somewhat lower ESR than the previous metallurgy.

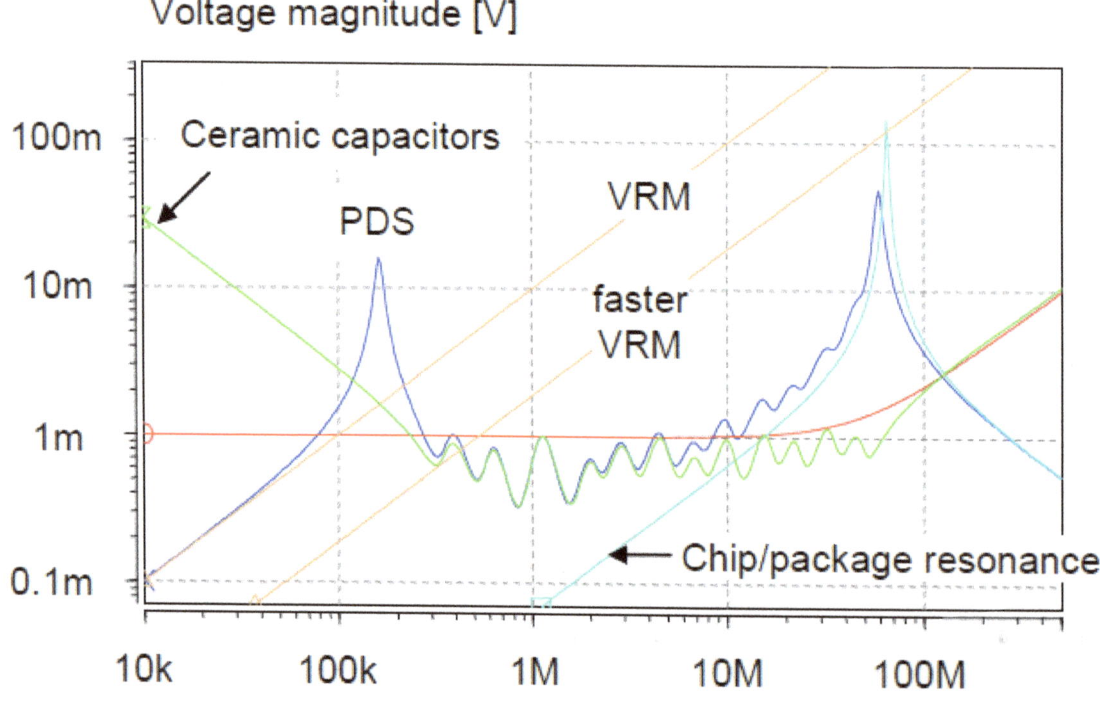

Figure 7.6a: PDS Impedance Peaks Due to VRM at Low Frequency and Chip/Package Resonance at High Frequency.

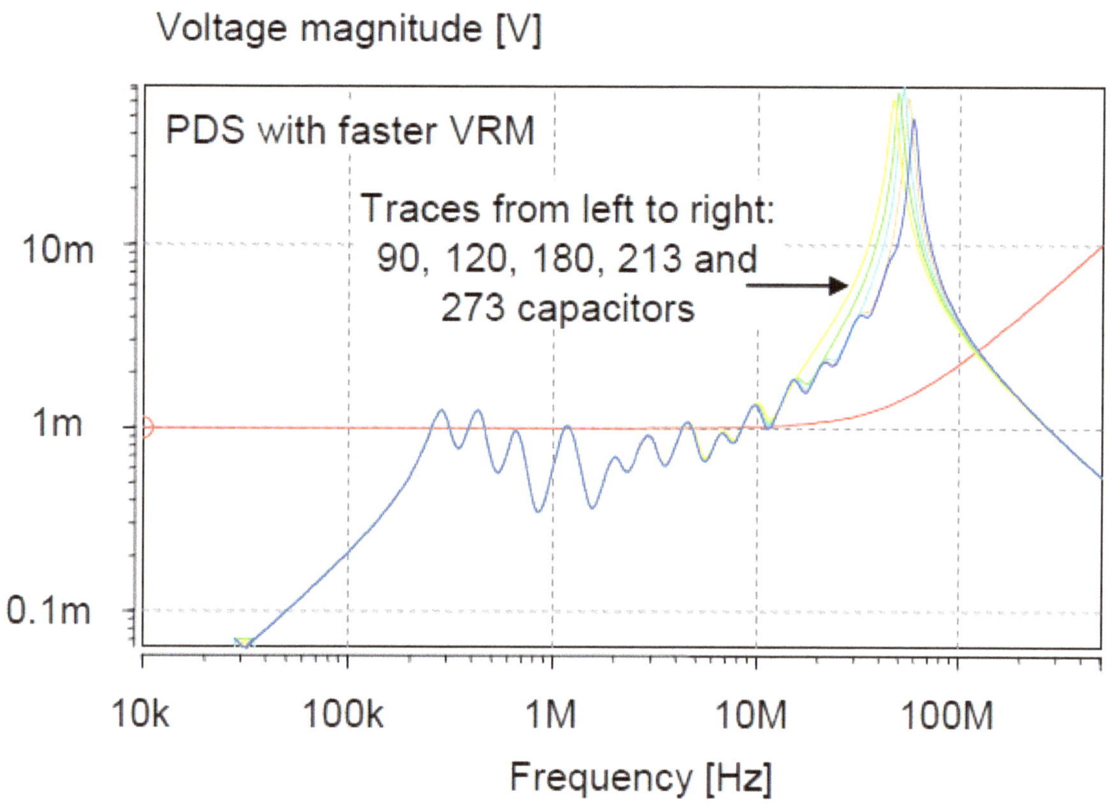

Figure 7.6b: Faster VRM Improves Low-Frequency Performance. Chip/Package Resonance Is Not Improved by Any Combination of PCB Capacitors.

The capacitor suppliers should, in fact, develop the most consistent and cost-effective process possible for manufacturing ceramic capacitors. Today's X7R and X5R capacitors work well. Good PCB design techniques enable a mounting inductance that gives a Q below 3, which is manageable from a parallel resonance standpoint with three capacitor values per decade. Any attempt to increase ESR to a specified value may result in undesirable cost, complexity, size, or inductance for the capacitor and complexity in component selection and inventory.

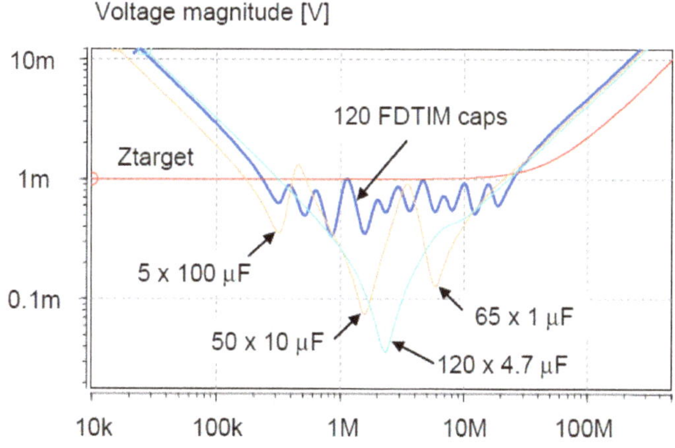

Figure 7.7a: 120 Capacitors Selected by Deep V and Decade Method Compared to FDTIM. PCB Capacitors by Themselves.

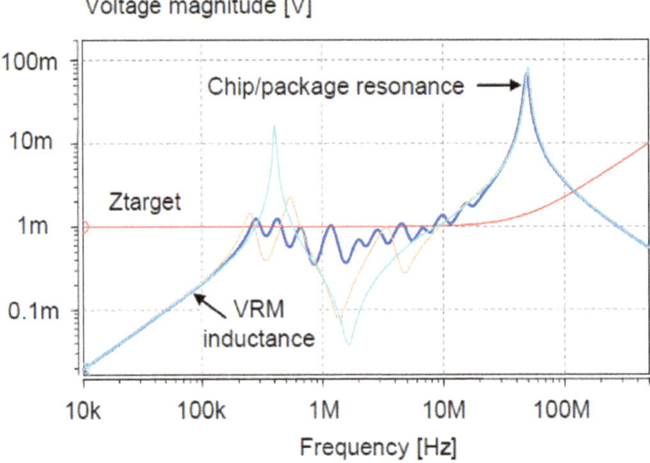

Figure 7.7b: 120 Capacitors Selected by Deep V and Decade Method Compared to FDTIM. Same Capacitors Together with VRM and Chip.

Problems at High and Low Frequency

FDTIM results in an optimal set of ceramic capacitors that meet a target impedance across a broad frequency range. The low end of this frequency range must transition to the VRM output impedance, which is usually inductive. The chip/package resonance impedance is near the target impedance from about 300 kHz to 30 MHz.

The bottom graph of Figure 7.7 shows system simulation along with the VRM and chip/package resonance.

The "big V" method is particularly vulnerable at the VRM crossing. There is very little system resistance in providing damping and, as a result, a large peak develops. The decade method does better because the 100 μF capacitors are nicely matched with the VRM output inductance, but there are still some peaks above the target impedance.

The component cost of large valued and sized capacitors is greater than for the smaller values. The three methods are nearly equivalent in the chip/package resonance band of frequencies.

The FDTIM capacitors are smaller on the average and use a higher percentage of low cost capacitors.

Conclusion

The FDTIM for selecting bypass capacitors, which has been discussed previously, uses quantities of several capacitor values to develop a flat impedance versus frequency profile. A major advantage is the damping provided for resonant peaks that may develop as capacitive or inductive components are added to the system.

This method usually produces the lowest cost set of capacitors that will meet power quality specifications. There is, however, some additional complexity in manufacturing with several part number capacitors being assembled onto the PCB. The number of capacitors required to meet the target impedance up to a corner frequency is proportional to the corner frequency and inversely proportional to the target impedance.

A formula has been given to estimate the number of PCB capacitors required.

References

[1] L. D. Smith, R. E. Anderson, D. W. Forehand, T. J. Pelc, T. Roy, *Power Distribution System Design Methodology and Capacitor Selection for Modern CMOS Technology*, IEEE Transactions on Advanced Packaging, Vol. 22, No. 3, August 1999, p. 284.

[2] L. D. Smith, D. Hockanson, K. Kothari, *A Transmission Line Model for Ceramic Capacitors for CAD Tools based on Measured Parameters*, Proc. 52nd Electronic Components & Technology Conference, San Diego, California, May 2002, pp. 331-336.

[3] L. D. Smith, *MLC Capacitor Parameters for Accurate Simulation Model*, Design Con 2004.

[4] I. Novak, *Measuring Milliohms and PicoHenrys in Power Distribution Networks*, Design Con 2000.

[5] I. Novak, S. Pannala, J. R. Miller, *Overview of Some Options to Create Low-Q Controlled-ESR Bypass Capacitors*, EPEP2004, October 25-27, 2004, Portland, OR.

[6] L. D. Smith, R. Anderson, T. Roy, *Chip-Package Resonance in Core Power Supply Structures for a High Power Microprocessor*, ASME Proceedings of Interpack'01, July 2001.

[7] S. Weir, *Bypass Filter Design Considerations for Modern Digital Systems, A Comparative Evaluations of the Big 'V,' Multi-pole, and Many-pole Bypass Strategies*, Design Con East 2005.

Chapter 8—Resonant-Free Power Network Design Using Extended Adaptive Voltage Positioning Methodology

Alex Waizman, Principal Engineer, Intel Corp.
Chee-Yee Chung, Principal Engineer, Intel Corp.

EXTENDED ADAPTIVE VOLTAGE positioning (EAVP) is a new robust methodology for the design and analysis of a low impedance resonant-free power delivery network, which utilizes and extends on the AVP theory that is often used in voltage regulator module (VRM) design and operation. Through EAVP, uncertainties in design guard-band noise budget can be removed, which result in significant performance bin-split improvement and cost reduction. Design optimization of decoupling capacitors with EAVP will be illustrated by using both time and frequency domain analysis.

Introduction

It is a well-known fact that power supply noise induced by the switching core logic of microprocessors can limit their performance [1, 2]. The distribution of power within the system must be done in a manner that will provide a low impedance power and ground connection for current path. However, there is no documented methodology in place to designing a low impedance resonant free power delivery network and the ways to achieve the goal are not well understood.

For illustration purposes, we have used an example of a central processing unit (CPU) core power delivery design. However, the methodology is generic to any power network design: e.g., input/output (I/O) power delivery design, which deals with issues of simultaneous switching output (SSO) noise of output buffers.

The paper will first address the motivation for doing a lumped power delivery model and will provide an explanation of the various elements in the circuit. Through the lumped power delivery model, an explanation of the AVP theory that is used widely in VRM application is provided.

Next, we will discuss the theory and application of EAVP. Using the theory of EAVP and the analysis technique presented, analysis on the impact of imperfect power delivery network on CPU performance caused by different decoupling capacitors on package and mother board (MB) or incorrect design in interconnect planes are discussed.

Lumped Power Delivery Model

For the benefit of this paper, we use a simple one-dimensional lumped model. This model is fairly easy to construct, and it requires minimal computational time. This model's limitation is due to its inability to model any non-uniform physical behavior, either hotspot current consumption on die or non-uniform interconnects layout. Only a distributed power delivery model analysis could accurately predict all the artifacts described above on nonuniformity.

However, for the purpose of explaining EAVP theory, we will use the lumped model to illustrate the interactions between the different parasitic components in the power delivery system.

During the actual layout implementation, we also attempt to maintain symmetry and uniformity of components placement such that the approximation of simple one-dimensional model can be considered fairly accurate. For brief illustration, the block diagram of lumped power delivery model is shown in Figure 8.1.

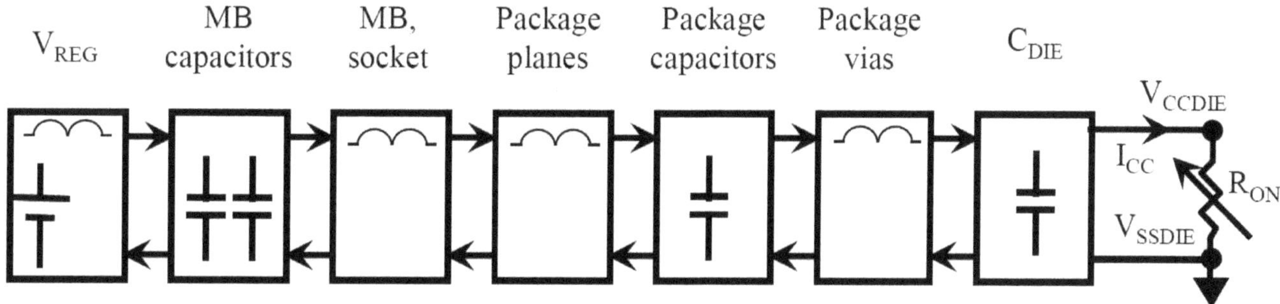

Figure 8.1: Power Delivery Block Diagram

The power delivery model includes serial elements labeled with the inductor symbol that, in reality, also have some amount of series resistance while the parallel elements are shown as capacitors. These capacitors, in fact, have some amount of series parasitic inductance and resistance.

The power delivery starts with a VREG voltage source, which consists of mainly a series inductor representing the inability of the VREG to respond to quick changes of current consumption.

Next, MB capacitors comprised of parallel combination of bulk (BLK) electrolytic and mid-frequency (MF) ceramic capacitors are followed. The MB and pin-grid array socket pins (SKT) mostly add series inductance and resistance.

Power feeds into the PKG CAP high frequency (HF) capacitors located on the pin side of flip chip pin grid array (FCPGA) package through the package planes (PKG PLN) [4, 5].

Finally, PKG vias connect the power delivery network to the silicon, where some amount of C_{DIE} decoupling capacitance is integrated. The R_{ON} variable resistance represents the time-dependent current consumption of the CPU. The value of the resistor R_{ON} is simply R_{ON} = supply voltage (VCC)/supply current (ICC).

A different number of switch resistors in parallel can be used to model the behavior of different linear current waveform (e.g., di/dt). In this model, on-die inductance effect is neglected.

Adaptive Voltage Positioning—Time Domain

AVP [3] theory was originally developed by VRM vendors to minimize the cost of bulk decoupling capacitors that is required for the module to function within the output voltage specifications. The old method of VRM design is shown in Figure 8.2.

For the sake of discussion, the inductance due to the bulk capacitors is ignored for the moment.

Waveforms in Figure 8.3 illustrate that the voltage source is set to be at the VCCNOM (nominal value) and bulk capacitors are added such that the noise on VCC due to ICC current consumption steps is within VCC ± TOL, where TOL is the tolerance specified by the VRM to function.

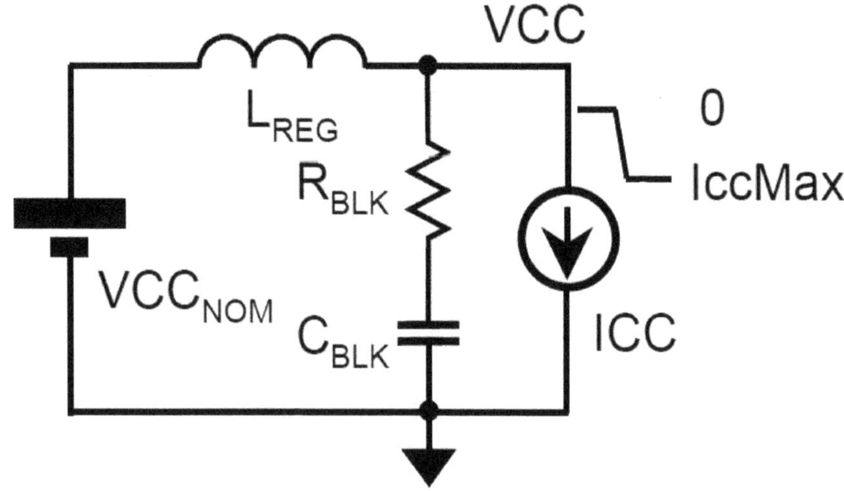

Figure 8.2: Conceptual VREG Circuit

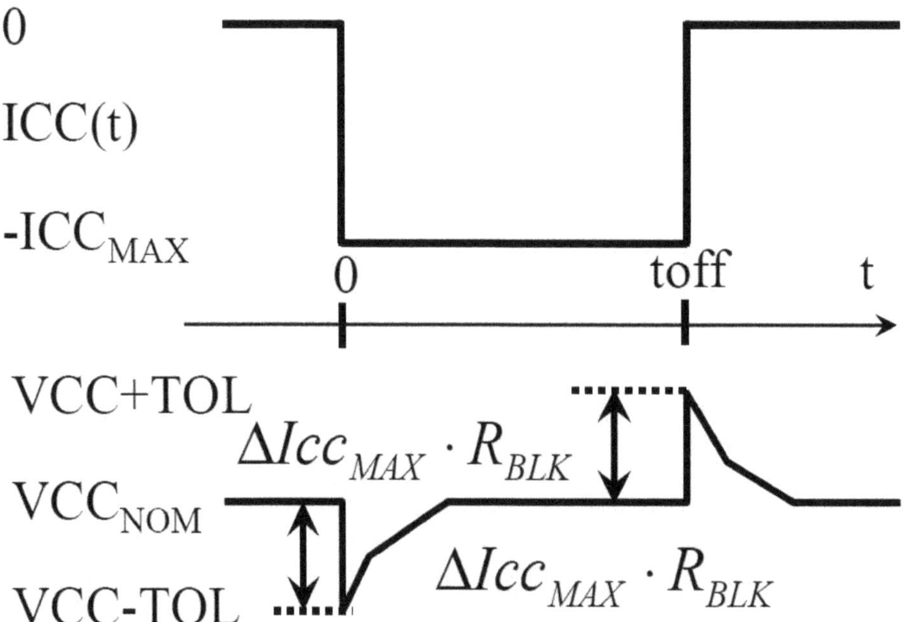

Figure 8.3: Conceptual VREG Waveforms

AVP, on the other hand, utilizes the circuit shown in Figure 8.4. Waveforms in Figure 8.5 show that AVP sets the voltage to VCC+TOL when no current is consumed. When all current is consumed, the voltage is then simply VCC-TOL. By following certain design criteria of the parasitic parameters, to be discussed next, the AVP circuit was able to completely remove the overshoot and undershoot in the power supply voltage when current switches ON and OFF.

In both Figure 8.4 and Figure 8.5, we observe that ICC transient current at t = 0 is provided purely by the bulk capacitance initially thereby setting the initial voltage drop to ICCMAX*RBLK. On the other hand, the steady state ICC current is solely provided by the VREG thereby the steady state voltage drop is ICCMAX*RREG. In order to have identical voltage droops at t = 0 and in the steady state the condition of Equation 1 has to be met:

$$R_{BLK} = R_{REG}$$

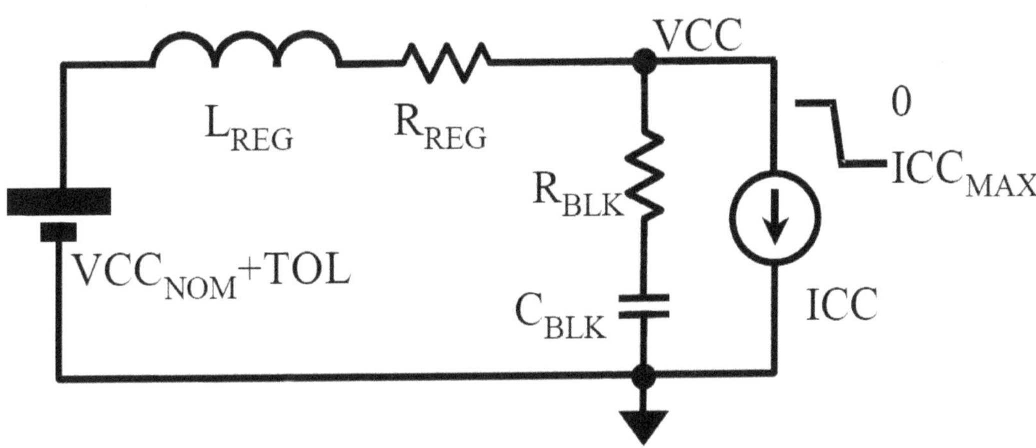

Figure 8.4: AVP Conceptual Circuit

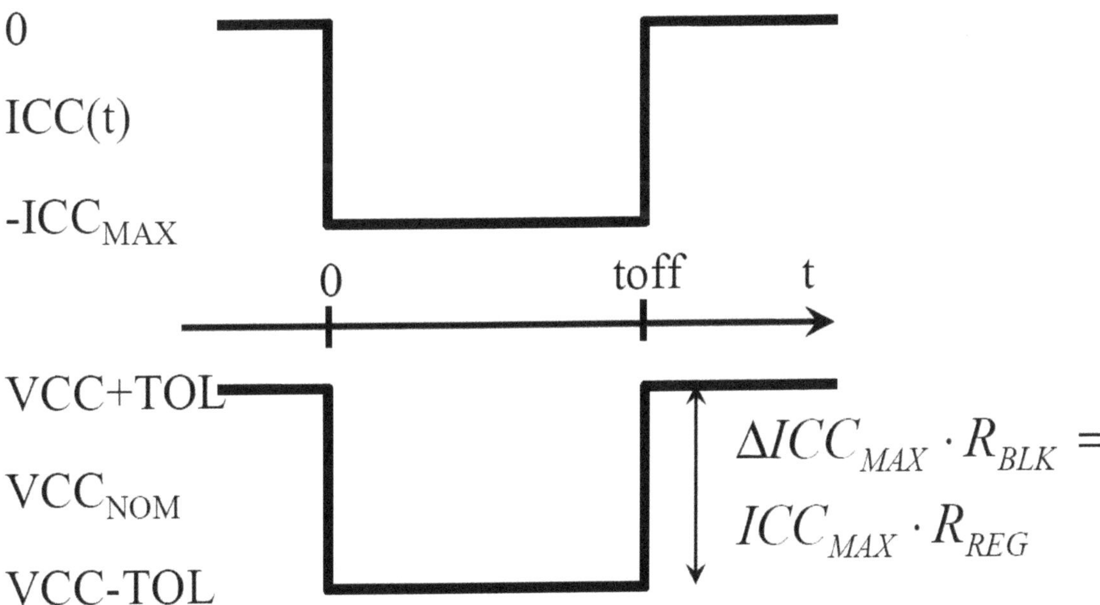

Figure 8.5: Adaptive Voltage Positioning Waveforms

To achieve zero overshoot and undershoot response, another condition of equal inductive and capacitive time constants has to be met as shown by the $R_{BLK}C_{BLK}$ equation.

Some intuition of the validity of this equation will be shown through subsequent sections.

$$R_{BLK}C_{BLK} = \frac{L_{REG}}{R_{REG}}$$

Notice that in order to achieve the TOL specifications for a regular VREG design, R_{BLK} had to meet the constraint of the R_{BLK} equation.

$$R_{BLK} = \frac{V_{CC}TOL}{I_{CCMAX}}$$

However, for the AVP design method to meet the same tolerance window, R_{BLK} requirements are relieved by a factor of two—as expressed by the revised R_{BLK} equation:

$$R_{BLK} = 2\frac{V_{CC}TOL}{I_{CCMAX}}$$

The revised R_{BLK} equation demonstrates the cost-saving benefits of AVP. Specifically, comparing AVP to regular VREG design, only half the total bulk capacitors are needed in AVP, since R_{BLK} target is two times bigger to meet the same $\pm$TOL requirements.

Assuming TOL = $\pm$ 40 mV ($\pm$ 2.5 percent of 1.6V VCC) and CPU ICC = 16.3A, this will result in R_{BLK} = 4.89 mOhm.

Using six Nichicon bulk capacitors (ESL = 9 nH, ESR = 29.3 mOhm and C = 2.2 mF), we were able to achieve this requirement. The value of C_{BLK} (13.2 mF) will be set from the original R_{BLK} equation to the value given by:

$$L_{REG} = 2200\mu F \; x \; 6 \; x \; 4.89m\Omega = 315nH$$

Appendix A describes how a VRM error amplifier can be represented as an inductor, and how L_{REG} value is determined by AVP.

Simulated waveforms of this design example are shown in Figure 8.6, where a clean transient is observed in VCC supply even when ICC changes from 0 to 16.3A in 25 ps. Figure 8.6 shows the current provided by C_{BLK} and VRM going through a smooth transition of supplying the current to the core switching.

As expected, C_{BLK} supply the initial current demand and then, gradually, the VRM takes over.

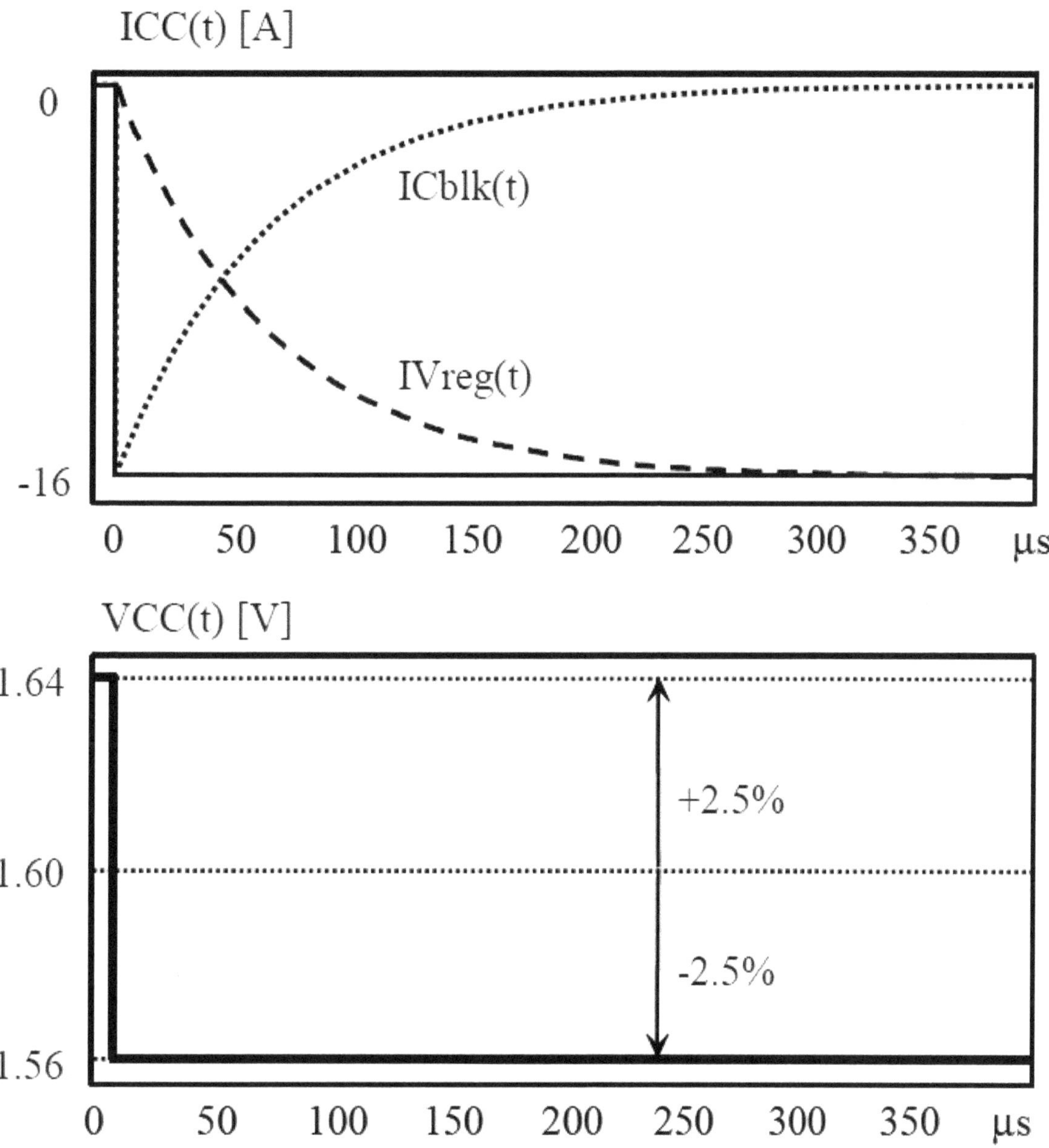

Figure 8.6: AVP Simulated Waveforms

Figure 8.7 illustrates the simulated waves for the same ICC(t) as before but for the various cases where the constraints of Equation 1 and Equation 2 are not met.

Some of them show clear violation of design targets for VCC where they exceed the ± TOL window of 2.5 percent.

Others seem initially to be harmless, e.g., the case of C_{BLK} *2.

However, a more careful observation of this case, as shown in Figure 8.8 manifest the issue when ICC(t) current turns OFF at about t = 90 μs. Then, the voltage VCC(t) does exceed the specified range of ± 2.5 percent.

Adaptive Voltage Positioning—Frequency Domain

Replacing ICC(t) current source of Figure 8.4 with a sine wave 1A current source with variable frequency, offers analysis in frequency domain, as shown in Figure 8.9 and Figure 8.10. The Z(f)-profile behavior of bulk capacitors, VRM and both in parallel can be clearly observed in Figure 8.10. The intuition behind Equation 1 and Equation 2 now becomes obvious. Equation 1 guarantees that the high-frequency impedance determined by R_{BLK} is the same as the low impedance determined by R_{REG}. Equation 2 takes care of matching the transition frequency between the two networks, resulting in flat, purely resistive frequency response with impedance of Z = 4.89 mOhm. This is frequently referred in the literature as the perfect pole-zero cancellation in the Z(f) impedance profile.

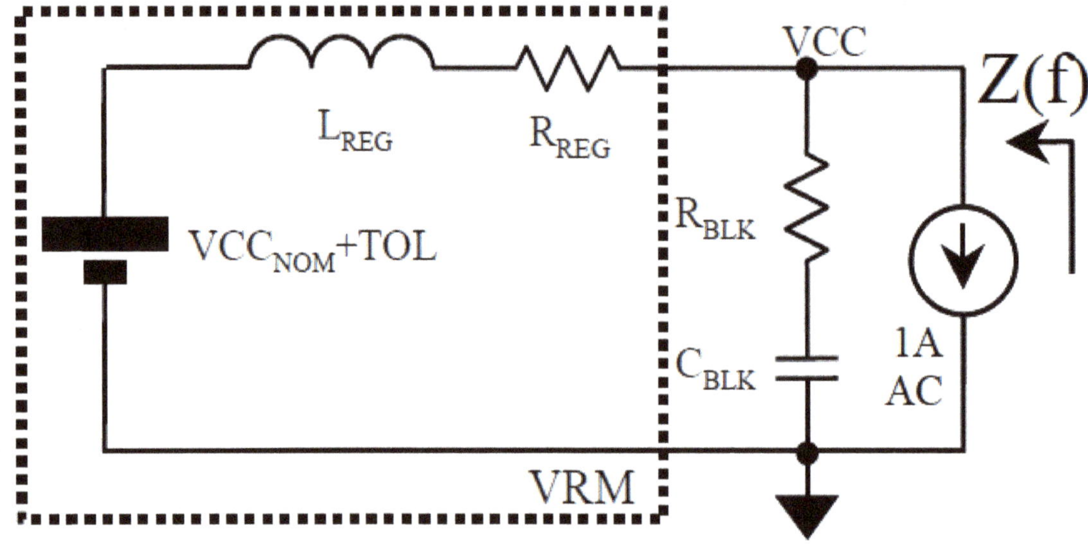

Figure 8.9: AVP Z(f) Simulation Circuit

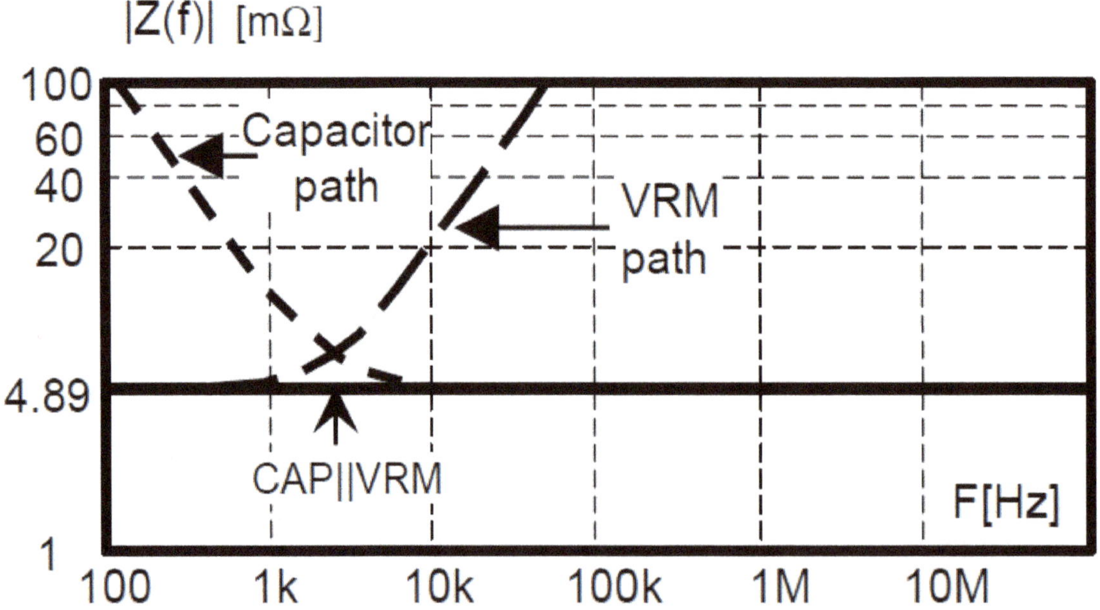

Figure 8.10: AVP Z(f) Impedance Profile (L_{BLK} = 0)

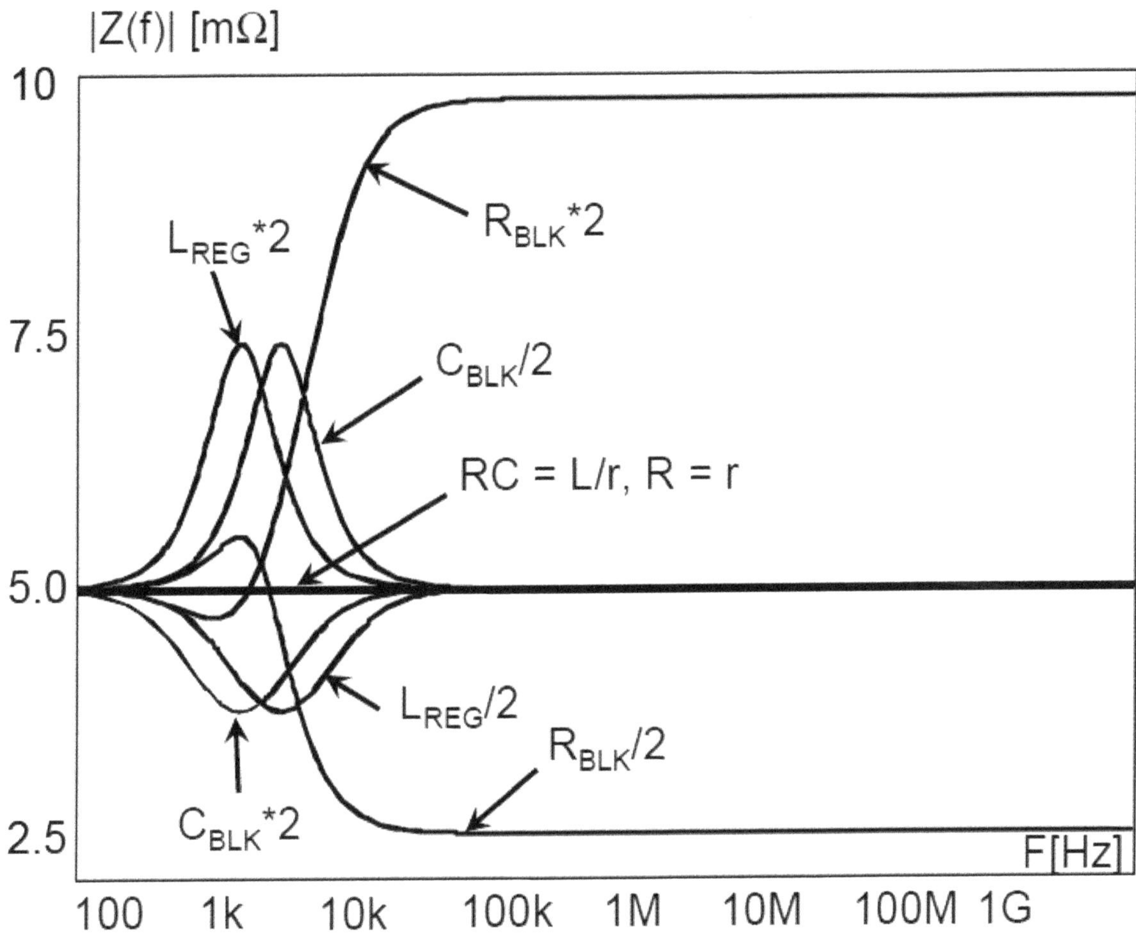

Figure 8.11: Z(f) for Non-Optimized AVP (L_{BLK}=0)

Figure 8.11 illustrates the impact in Z-profile when the constraints of Equation 1 and Equation 2 are not met. The important conclusion is that for AVP non-optimized cases, when ICC(t) has a repetitive ON/OFF current consumption, power supply noise can grow significantly.

For example, for the case when L_{REG} *2 and if ICC(t) would toggle ON/OFF at about 1 kHz repetition rate, we observed that the power supply network will exhibit resonating behavior with the peak noise increasing more than 1.5x compared to the optimized case that maintains the constraints of the $R_{BLK}C_{BLK}$ and R_{BLK} equations.

Effect of Non-Ideal Bulk Capacitors on AVP

Unfortunately, the effectiveness of bulk capacitors at higher frequencies is limited by their parasitic inductance (L_{BLK}). The modified simulation circuit is shown in Figure 8.12 with the addition of L_{BLK}, while Z(f) impedance profile for this case is shown in Figure 8.13.

The presence of L_{BLK} causes the impedance to start increasing again beyond 1.4 MHz.

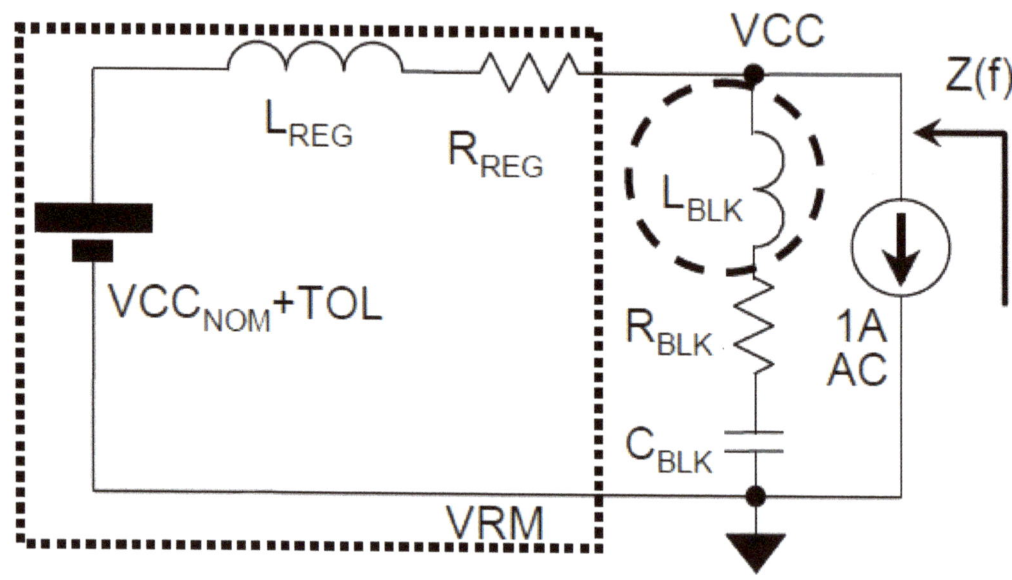

Figure 8.12: AVP Circuit with L_{BLK}

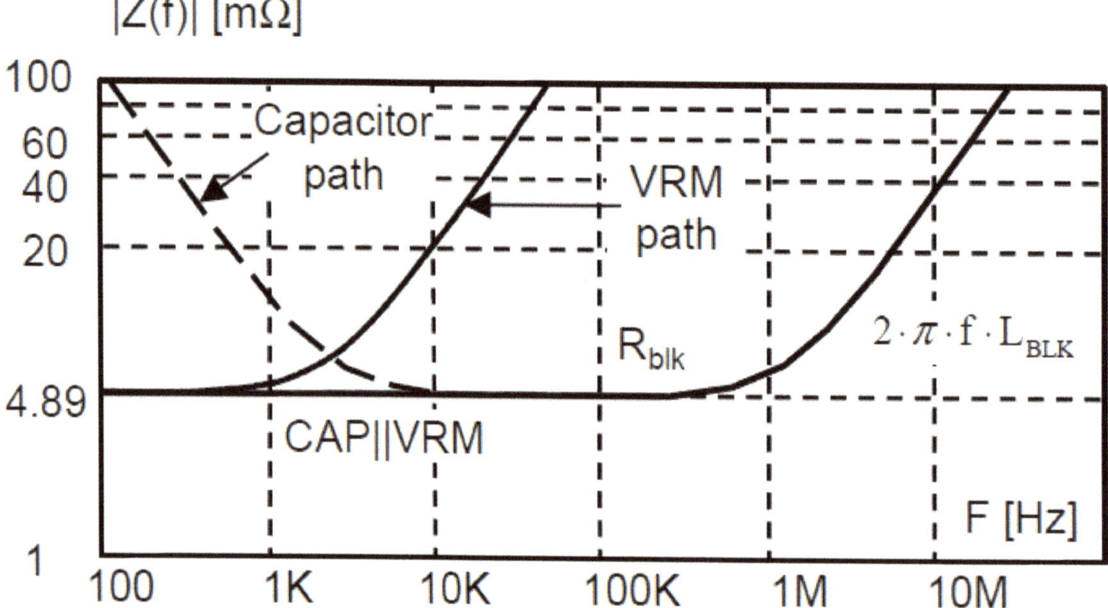

Figure 8.13: AVP Z(f) Impedance Profile with L_{BLK}

The common method of dealing with this impedance increase is with addition of MF decoupling capacitors on the motherboard next to microprocessor socket.

In the following sections on the EAVP method, we will discuss the optimal method of selecting the rest of the decoupling capacitors and interconnects toward the switching source.

EAVP—Principle

Rule 1: Select the ESR of decoupling stage N to be equal to the effective series R of decoupling stage N-1
Rule 2: Select RC time constant of decoupling stage N to match L/R time constant of stage N-1

—or—

Select the Z(f) 3-dB point of decoupling stage N to the Z(f) +3-dB point of stage N-1

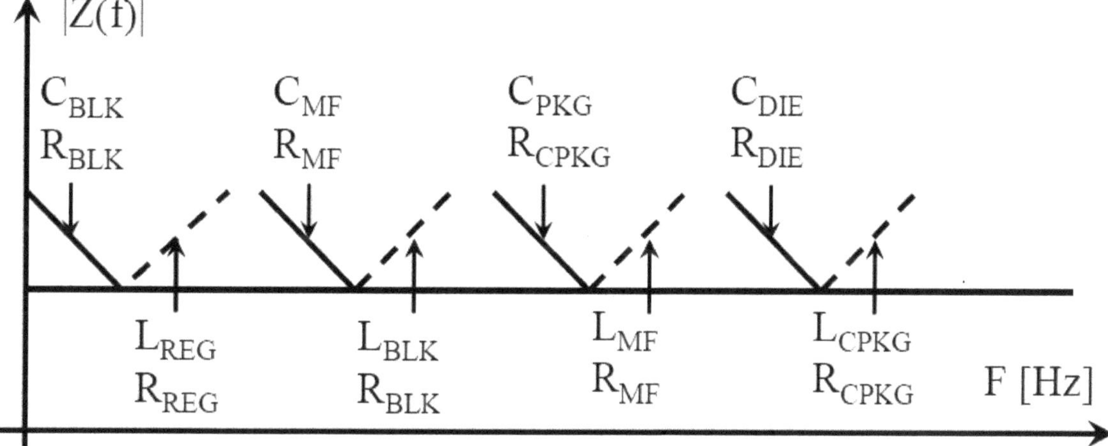

Figure 8.14: EAVP Conceptual Principle

EAVP basically extends the method that was used for optimized AVP design of bulk capacitors to the successive design of MF, package, and C_{DIE} capacitor parameters. The conceptual EAVP principle is as illustrated in Figure 8.14.

The conceptual principle turns out to be somewhat inaccurate when the different stages of decoupling are too close together in frequency and requires some second order effect correction. However, the basis of EAVP will provide the adjustment of second order effect.

EAVP Design of MF Capacitors

For the specific layout in the MB in this case, the MF ceramic decoupling capacitors are placed in parallel with the C_{BLK} capacitors. This is illustrated in the simulation schematics of Figure 8.15. The resulting Z(f) profile is shown in Figure 8.16. EAVP design of MF capacitors is as follows: First, we select according to Rule 1:

$$R_{BLK} = R_{MF} = 4.89m\Omega$$

For Rule 2, a graphical method is used to select the value of C_{MF} capacitors.

First, we find the +3 dB frequency [the frequency where the impedance is increased by $\sqrt{2}$ where the impedance of the circuit without the MF capacitors is:

$$Z(fcr) = 4.89\ x\ \sqrt{2} = 6.913m\Omega$$

The +3 dB frequency turns out to be 526 kHz. Then, we compute for C_{MF}.

$$C_{MF} = \frac{1}{2\ x\ 4.89m\ x\ 526K} = 61.91\mu F$$

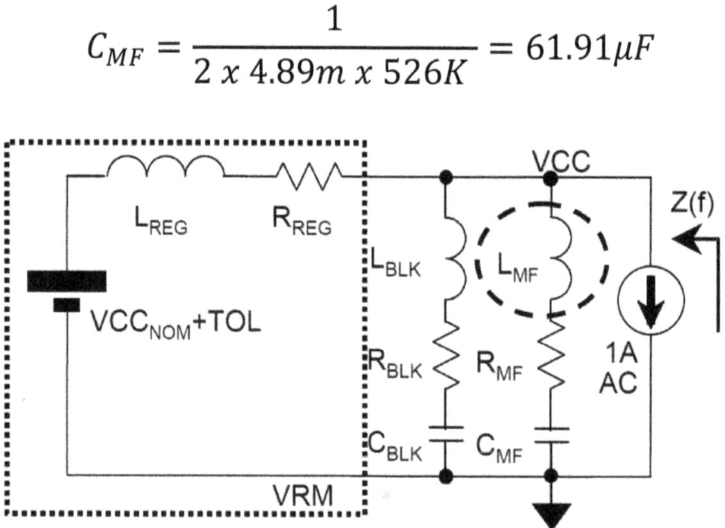

Figure 8.15: MF Capacitor Optimization

Using nine multi-layered ceramic chip (MLCC) MF capacitors in 1206 form factor in parallel yields a convenient motherboard layout and a value of 6.90 µF with ESR of 44.0 mOhm per capacitor. Figure 8.16 illustrates that, for these calculated values, fairly flat Z(f) frequency response results.

However, when we account for the 562 pH as well (extracted from 3D electromagnetic modeling) loop inductance of 1206 capacitors, we observed some frequency response flatness degradation. The degradation of flatness results from the interaction between the poles and zeros of the Z(f) frequency response.

If L_{MF} were much smaller, the flatness degradation would have been very small and unnoticeable. An optimization routine script was written, where several values of C_{MF}, R_{MF} yielded the optimized response. The final results of nine MF capacitors with 6.816 µF and ESR of 47.0 mOhm were not significantly different from the originally calculated values of 6.90 µF and ESR of 44.0 mOhm, as shown in Figure 8.16.

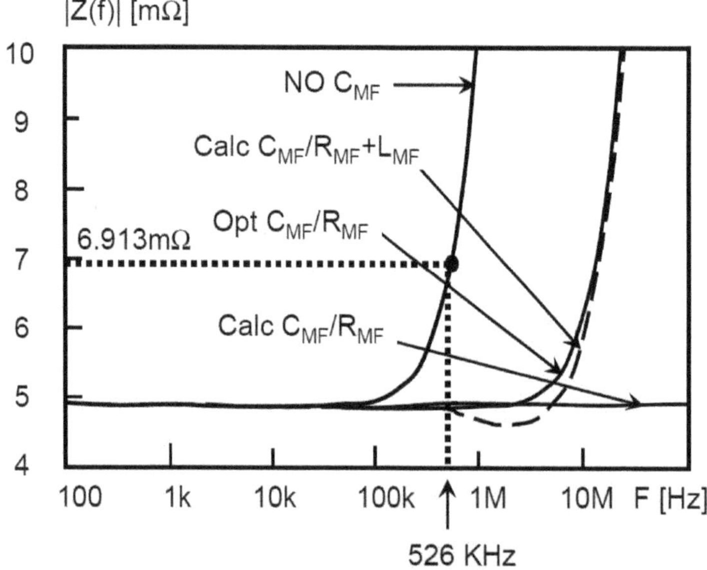

Figure 8.16: MF Optimization Z(f) Profile

Connection of MF and BLK Capacitors to Socket Pins

Figure 8.17 illustrates the cross-section of the motherboard layout that shows the parallel placement of the bulk and MF capacitors and the way they feed the socket pins through the 4.5 mil separated VSS island on the top layer of the motherboard and VCC internal plane.

This optimized connection results in a very low loop inductance.

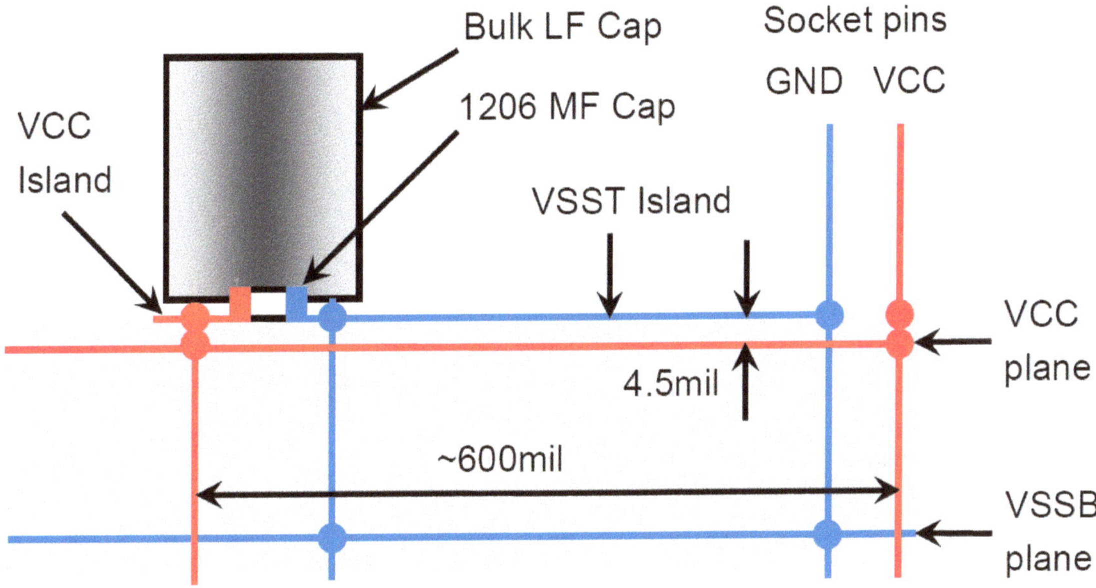

Figure 8.17: MF and Bulk Caps Parallel Connection

RLC parasitic matrices are extracted using an electromagnetic field solver at 1.68 MHz (critical frequency for this part of the circuit), and the results summarized are in Table 8.1.

	L[nH]			R[mΩ]			C[pF]		
	VCC	VSSB	VSST	VCC	VSSB	VSST	VCC	VSSB	VSST
VCC	0.233	0.179	0.224	0.095	0.027	0.030	643.8	-54.6	-589.3
VSSB	0.179	0.234	0.174	0.027	0.103	0.021	-57.2	70.0	-12.8
VSST	0.224	0.174	0.232	0.030	0.021	0.070	-586.7	-15.4	602.1

Table 8.1: RLC Parameters of MB Connection between MB Capacitors and CPU Socket

Package Capacitors Optimization

A simulation circuit for package capacitor optimization using EAVP is shown in Figure 8.18.

In addition to the previous VRM, BLK, and MF capacitors, we now append the MB connection, socket inductance and resistance parameters, and package capacitors (CCPKG) and their resistive and inductive parasitics, RCPKG and LCPKG. The FCPGA package design also has a 2" x 2" sandwiched power/ground planes (30 μm separation between the metal planes with relative dielectric constant of 3.4) that connects socket pins to the CCPKG capacitors.

These planes are represented as almost ideal CPKGPLN capacitors with capacitance of about 4 nF and resistance of 1 mOhm.

Rule 1 is applied for the selection of RC_{PKG} but with the additional series resistance added by socket pins and MB planes that connect the MF and BLK capacitors to CC_{PKG}. This results in RC_{PKG} = 5.747 mOhm.

Next, we employ Rule 2 to design the value of CC_{PKG} to get the Z(f) profile in Figure 8.19. The +3dB impedance and frequency are 8.127 mOhm and 5.67 MHz, respectively, which in return yielded a calculated CC_{PKG} = 4.88 μF. For the specific layout we have, using six 0612 MLCC package capacitors, the effective loop inductance is 72 pH per capacitor [4], and 12 pH for the parallel combination. However, even this small inductance results in some flatness degradation as shown in Figure 8.19.

Again, optimization routine script results in optimized values of six 0612 capacitors, where each capacitor is 0.77 μF with ESR of 39.1 mOhm.

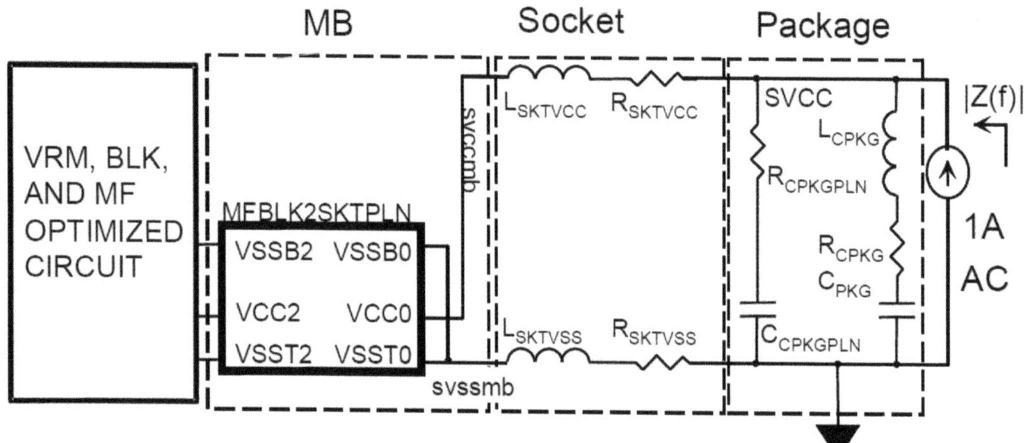

Figure 8.18: Package Capacitors Optimization Circuit

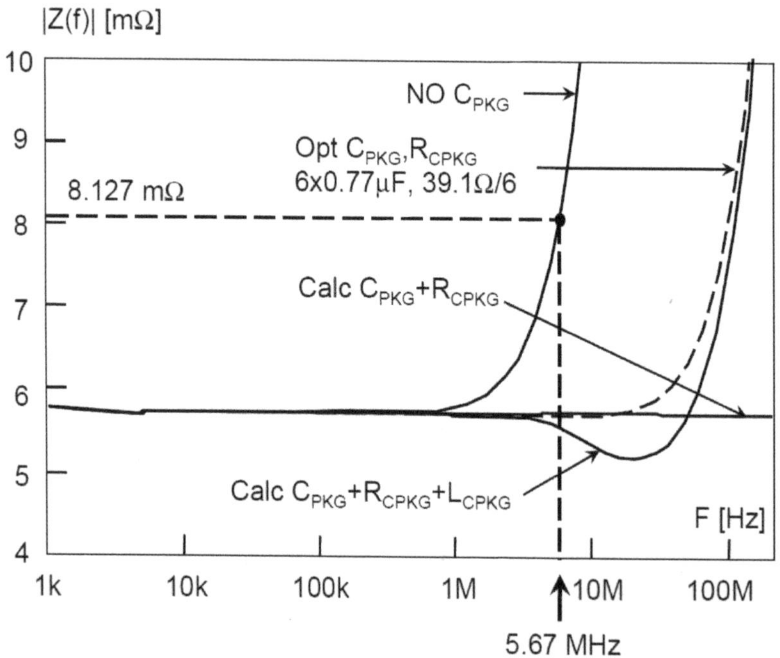

Figure 8.19: Package Capacitors Optimization Z(f)

C_{DIE} Optimization

Last, we will use the EAVP method to optimize the value of C_{DIE} and R_{DIE} with the simulation circuit of Figure 8.20.

We append to the previously optimized circuit, the C_{DIE} and R_{DIE} that are connected to the rest of the circuit with R_{PKG} VCC/VSS and L_{PKG} VCC/VSS, which reflect the inductance and resistance of package vias that connect the pin side package capacitors to the die. It is also assumed that C_{DIE} has no noticeable inductance effect in this case, which is true if the capacitors on the die are laid out in an effective manner. Simulation results are shown in Figure 8.21.

According to Rule 1, R_{DIE} is selected to be 6.43 mOhm, while C_{DIE} is graphically found by Rule 2 from Figure 8.21 (+3 dB point of 9.093 mOhm and 33.21 MHz) to be 745 nF.

This completes the EAVP design of this power network.

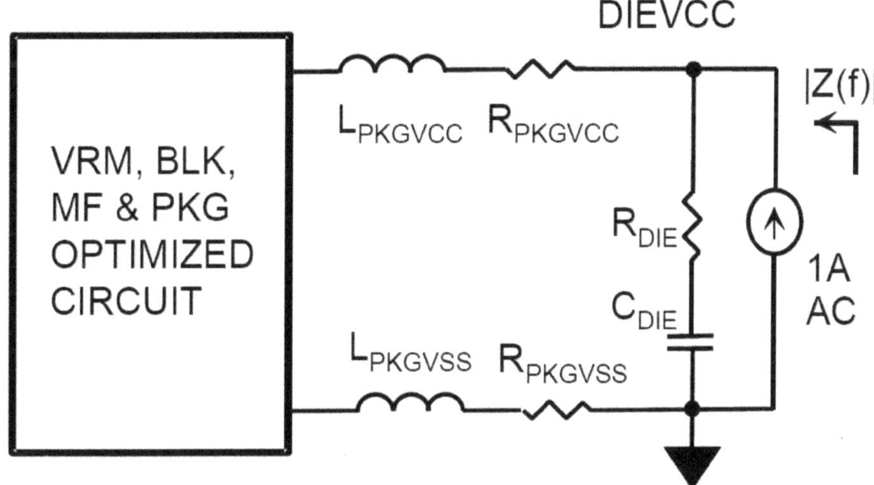

Figure 8.20: C_{DIE} Optimization Circuit

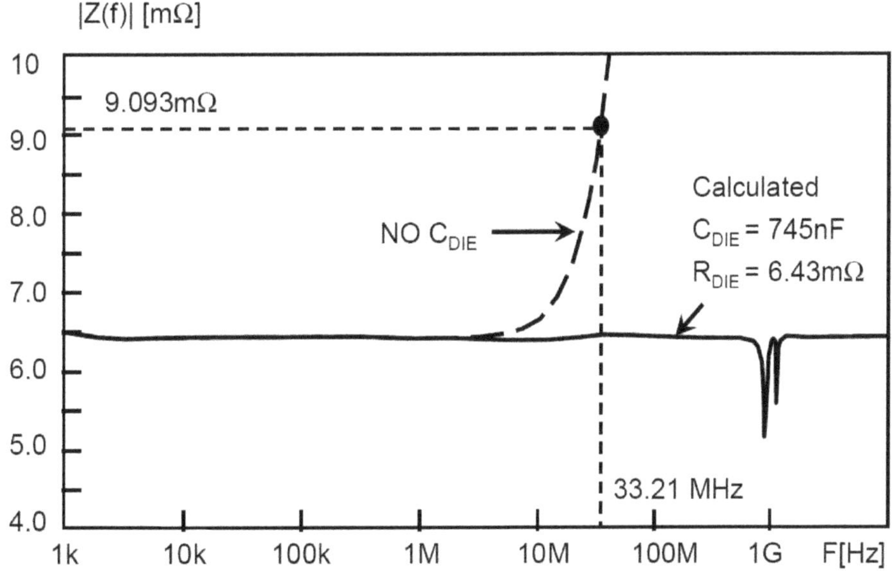

Figure 8.21: C_{DIE} Optimization Z(f) Profile

Frequency Domain Current Profile

Figure 8.22 illustrates the frequency domain current profile of each decoupling capacitors and the VRM.

In Figure 8.22, we observe that VRM is effective in providing current only up to 2.4 kHz (-3 dB point), which is in a good agreement with the 140 kHz typical switching frequency of the switching regulator. The low switching is also vital for the simple Laplace-transform analysis performed that allows the simple representation of VREG operation with L_{REG} & R_{REG}.

Bulk capacitors' effectiveness gradually reduces starting at about 0.5 MHz, while the peak current for MF capacitors is at 1.68 MHz. This is why the extraction of RLC parasitics for the MB connection between the capacitors and the socket in the previous section was done at 1.68 MHz.

Peak current of the package capacitors is provided at 13.75 MHz and above ~33 MHz, most of the current is provided by C_{DIE} with the exception of about 900 MHz series resonance of package planes that provide the current in this narrow band.

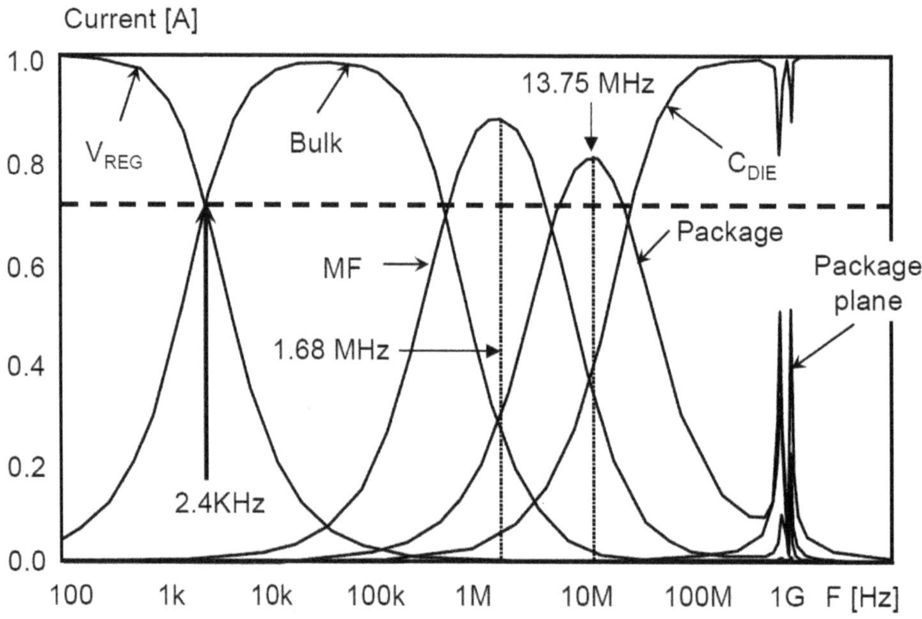

Figure 8.22: Frequency Domain of Current Profile of Each of Capacitors

Figure 8.22 also illustrates the fact that current spectral content gradually shifts from one capacitor to another, each one in its frequency band, without any shared resonating peaking.

This efficiency is the result of EAVP optimized design.

Time Domain Results

Next, the EAVP-optimized design is stimulated at the die by a 0-12.4A transient within 25 ps. This is shown at different time scales on Figure 8.23, Figure 8.24, and Figure 8.25.

A perfectly flat voltage step response, monitored at the die nodes, shows that it is well within the target-specified window of 80 mV.

Issues with Optimized EAVP Values

Up to this point, the assumption was that any value resulting from EAVP analysis is readily achievable by commercially available components. Table 8.2 summarizes the critical calculated and optimized parameters for the decoupling capacitors.

Cap Type	Form Factc	N	Ind	Calc C	Calc R	Opt C	Opt R	Actual C	Actual R
			nH/Cap	uF	mΩ	uF	mΩ	uF	mΩ
Bulk	Electr	6	9	2200	29.33	2200	29.33	**2200**	**29.33**
Mid Freq.	1206 Cer	9	0.562	6.879	44	6.816	47	6.8	3
PKG.	0612 Cer	6	72	0.813	34.5	0.7714	39.1	0.68	8
Cdie	Gate/Diff		0	0.745	6.43	0.745	6.43	**0.16**	**0.625**

Table 8.2: Summary of Critical Optimized Values

In reality, ceramic capacitor vendors are not aware of the need to have controlled ESR capacitors.

On the contrary, the common belief until now was that lower ESR is always better. For example, for this design, we need MF capacitor with ESR of 47 mOhm, while the common component's ESR is about 15 times lower.

A similar situation exists with package capacitors where the standard components' ESR is about five times lower than those needed by an EAVP-optimized design. The authors have contacted several capacitor vendors for specialized capacitors with integrated and specified ESR. At the time of this paper, we have received samples of such capacitors from one of the vendors and are evaluating their performance.

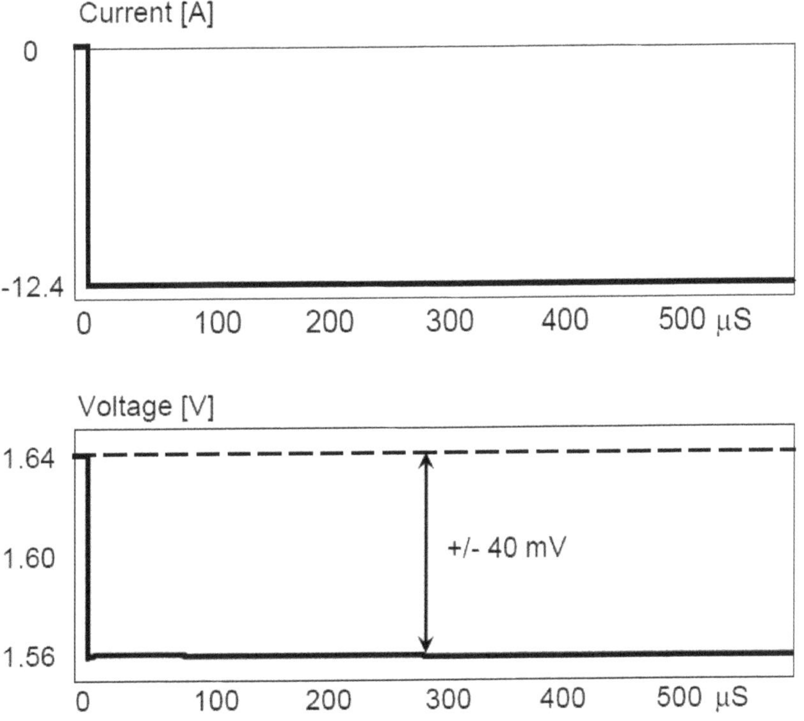

Figure 8.23: Optimized Time Domain Results

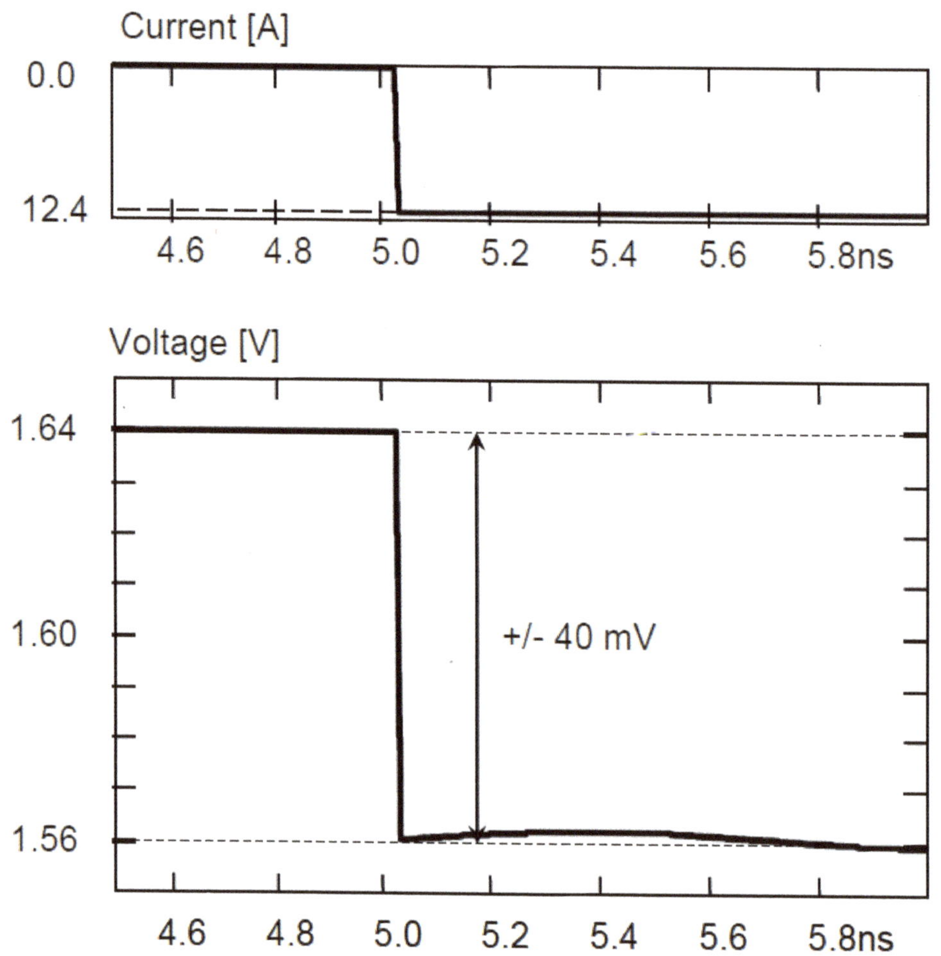

Figure 8.24: Optimized EAVP Time Domain Response Zoom in View

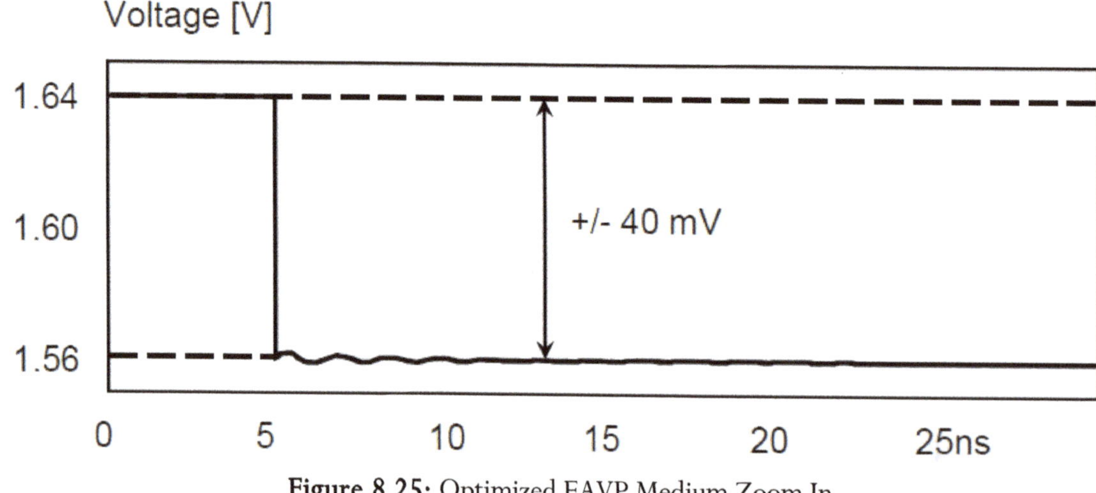

Figure 8.25: Optimized EAVP Medium Zoom In

Similarly, EAVP requires C_{DIE} of 745 nF with R_{DIE} of 6.43 mOhm, but for this product, they are 160 nF and 0.625 mOhm, significantly lower than the optimized values.

At this point, we do not know how to bring the values closer to the target values without causing silicon area growth of more than 1.5x to accommodate the needed C_{DIE}. We also need further exploration of the effect of higher R_{DIE} on switching of on-die bus drivers that might demand quite high current in sub 100 ps windows. A recent paper by Sun Microsystems[6], however, suggests an area-efficient implementation of active circuit-decoupling capacitors that significantly improve area utilization efficiency for on-die decoupling capacitors.

Resonant Power Network Effect Due to Realistic C_{DIE}/R_{DIE}

Figure 8.26 illustrates the resonating behavior of Z(f) profile for the realistic case of C_{DIE}.

Power network impedance increases up to about 5x at the resonance frequency of 72 MHz. A 72 MHz repetitive current with transients from $I_{CCMAX}/2$ to I_{CCMAX} is simulated and shown in Figure 8.27. The EAVP–optimized design results in VCC that does not go below 1.56V, which is the lower 2.5 percent limit.

However, the case where C_{DIE} = 160 nF and R_{DIE} of 0.625 mOhm results in about 200 mV p-p ringing that significantly exceeds the tolerance window defined by the initial design. Probably not available today, the concern is the possibility that one may write an application or a virus program that will stimulate a repetitive wave as shown by Figure 8.27, thereby creating the significant power supply ringing.

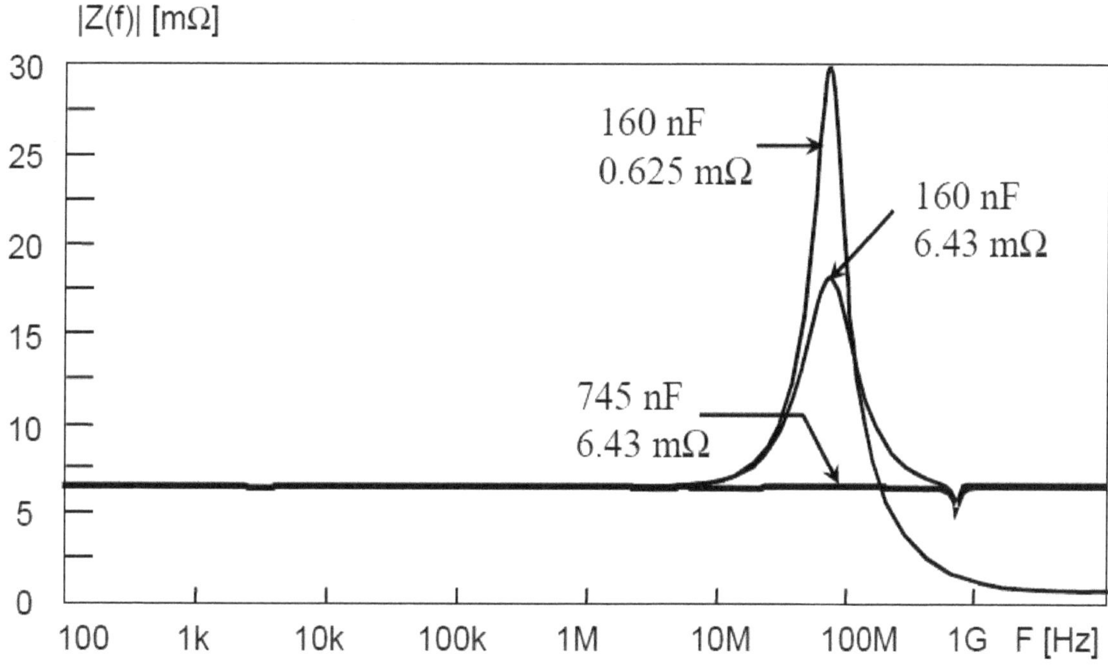

Figure 8.26: C_{DIE} Effect in Frequency Domain

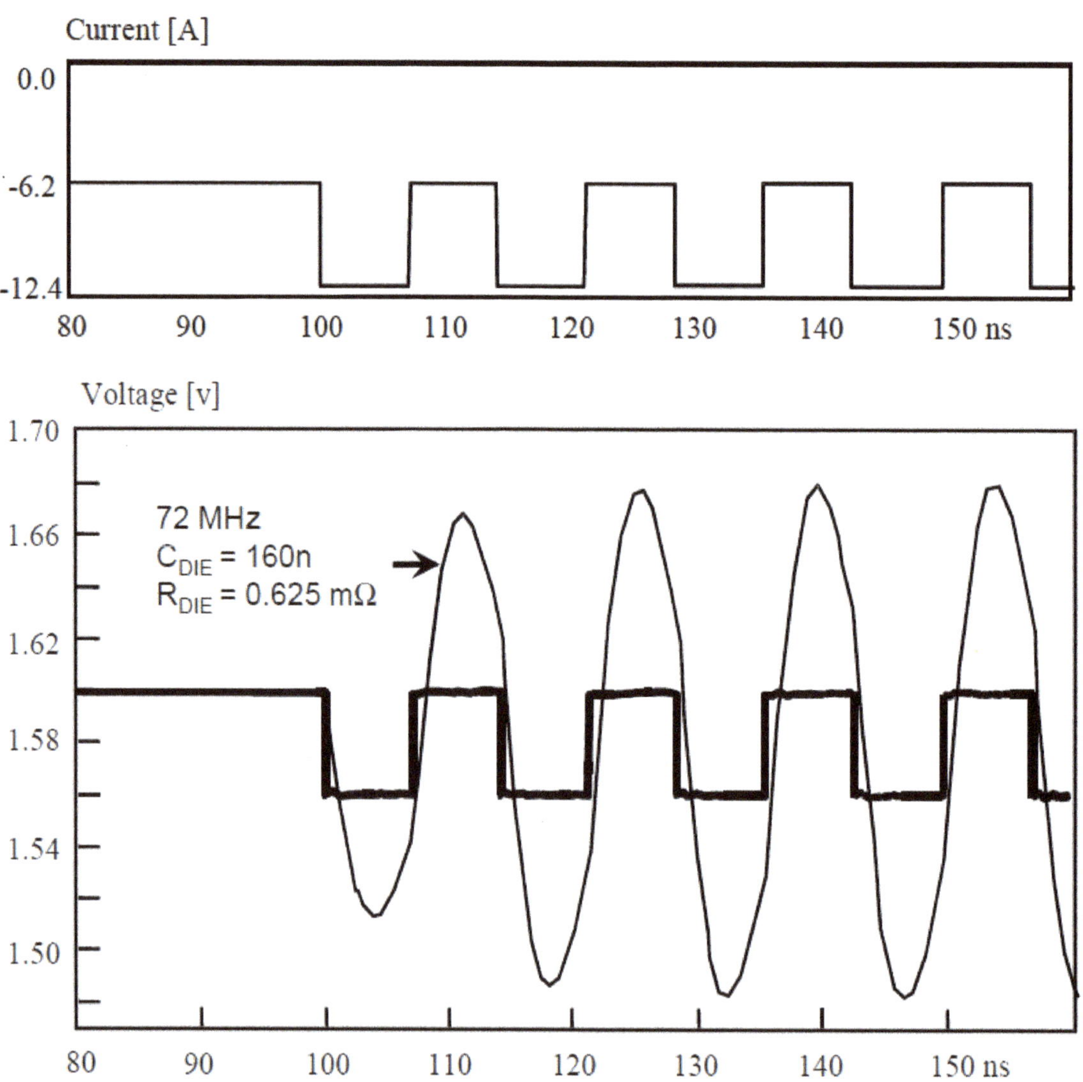

Figure 8.27: Application of Repetitive Current Pattern to Non-Optimized Power Delivery

For input/output (I/O) operations, the I/O power network resonance could be fatal when the I/O output buffers repetitive ON/OFF switching frequency coincides with the resonant frequency of the power network.

Stimulation of repetitive ON/OFF stimulus of I/O operation is usually more likely to happen.

Future Work

As part of next steps, the authors plan to complete the validation of EAVP with the higher and controlled ESR capacitors on this design platform.

Several improvements in EAVP methodology are required in respect to simulation capability development: modeling of frequency dependent resistance, and performing Monte Carlo analysis to understand the statistical risk of manufacturing variation.

Summary

A new approach of performing power delivery design and analysis is presented. Using the EAVP method, it was demonstrated how to design for a power delivery Z-profile that is constant with regards to frequency. EAVP also shows how an optimized design is obtained. Analysis method using time-domain simulations as well as frequency domain simulation are presented to illustrate the robustness of this design methodology. An application design is used for illustration and verification results clearly show the correctness and importance of this design and analysis approach.

Acknowledgments

The authors would like to thank their managers for the support, and the following individuals from Intel for their technical help during the development of this theory: YL Li, I. Bary, D. F. Figueroa, D. Ayer, T. Arabi, A. Birr, N. Yosef, M. Stapleton, and G. Hackendorn.

References

[1] R. Downing, P. Gebler, and G. Katopis, *Decoupling Capacitor Effects on Switching Noise,* IEEE Trans. CPMT vol. CHMT-16, no. 5, pp. 484-489, August 1993.

[2] K. Lee and A. Barber, *Modeling and Analysis of Multichip Module Power Supply Planes*, IEEE Trans. CPMT, Part B, vol. 18, no. 4, pp. 628-639, November 1995.

[3] R. Redl, B. P. Erisman, and Z. Zansky, *Optimizing the Load Transient Response of the Buck Converter*, APEC '98. Conference Proceedings 1998., Thirteenth Annual, Volume: 1, 1998 pp. 170-176 vol. 1.

[4] Y. L. Li, T. G. Yew, C. Y. Chung, and D. F. Figueroa, *Design and Performance Evaluation of Microprocessor Packaging Capacitors using Integrated Capacitor-via-Plane Model*, IEEE Trans. Advanced Packag., vol. 23, No. 3, pp. 361-367, Aug 2000.

[5] A. Birr and C. Y. Chung, *Electrical Characterization of Flip Chip Pin Grid Array (FCPGA) Package Technology*, Intel Assembly and Test Technology Journal, Chandler, AZ, vol. 2, pp. 111-116, 1999.

[6] M. A. R. Salem and A. Taylor, *An On-Chip Voltage Regulator using Switched Decoupling Capacitors*, ISSCC, Feb 9, 2000, San Francisco California.

Appendix

Figure 8.28 illustrates conceptual schematics of the VREG circuit comprising an error amplifier with output impedance of R_O and open loop gain given by:

$$A_o = \frac{1}{1 + st_o}$$

$$s = j2\pi f$$

Usually the DC gain A_O of the error amplifier and can be made very large (105-106). The open loop gain pole $1/\tau_o$ is controlled by the CC compensation capacitor. V_{BG} is an accurate voltage reference source with a nominal voltage of 1.225v applied to the positive input of the error amplifier.

Negative feedback comprising the R_1 and R_2 resistors is fed back from the output voltage node V_O to the negative input of the error amplifier. The ratio of R_1 and R_2 resistors, as shown by Equation 10, controls the steady state output voltage of the V_{REG} circuit. Thus, setting the voltage to V_{CCNOM} + TOL when using AVP can be easily achieved by changing the R_2/R_1 resistors ratio.

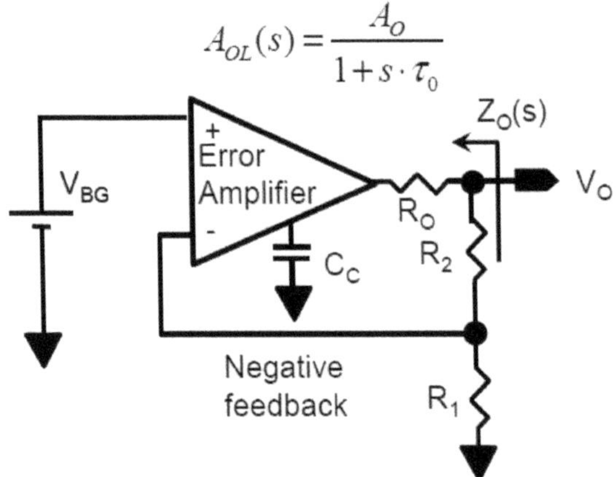

Figure 8.28: VREG Error Amplifier

$$\boldsymbol{V_O = V_{BG}\left(1 + \frac{R_2}{R_1}\right)}$$

Due to the negative feedback, output impedance Z(s) of the VREG circuit is given by:

$$Z_O(s) = \frac{R_O}{1 + \beta A_{OL}(s)}$$

Substituting the DC gain (A_o) equation into the output impedance $Z_O(s)$ equation and rearranging and multiplying the nominator and denominator by $(1 + s\tau_O)$ results in:

$$Z_O(s)=\frac{R_O}{1+\beta\frac{A_O}{1+s\tau_O}}=\frac{R_O}{1+\beta A_O}\frac{1+s\tau_O}{1+s\frac{\tau_O}{1+\beta A_O}}\approx\frac{R_O\tau_O}{1+\beta A_O}s$$

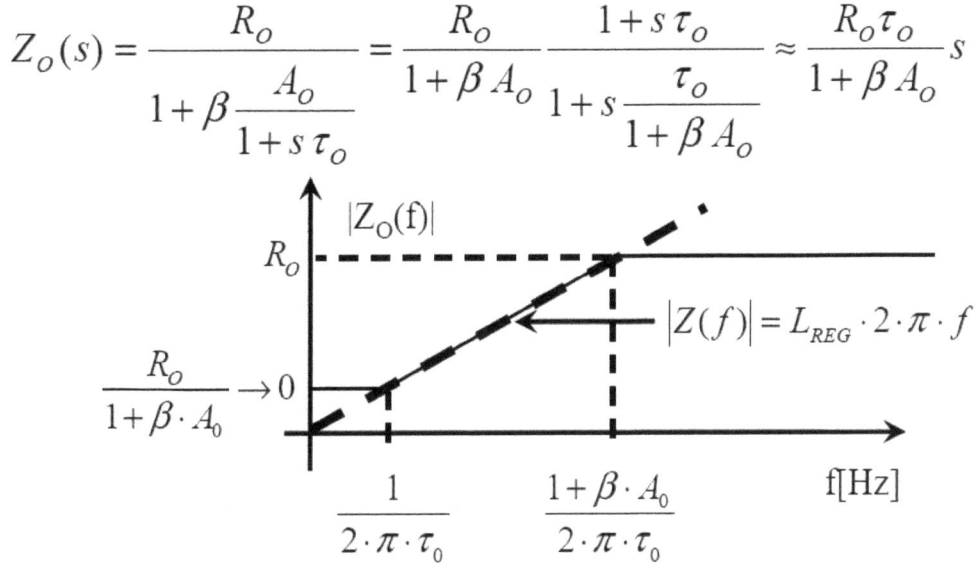

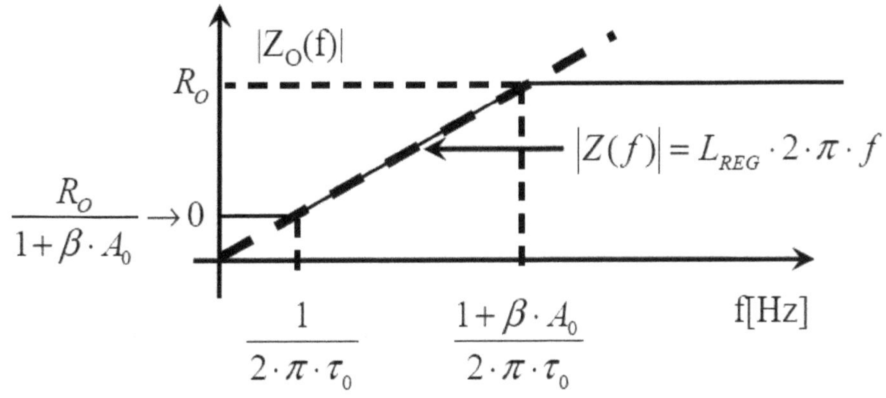

Figure 8.29: Graphical Illustration of the $\boldsymbol{Z_O(s)}$ equation $Z_O(f)$ Profile of VREG Circuit

The reconfigured output impedance equation ($Z_O(s)$) is illustrated in Figure 8.29.

At low frequency where $s\rightarrow 0$, the impedance is $R_O/(1+\beta A_0)$ and is very small since A_0 is very large.

At high frequency, ($s >> (1+\beta A_0))/(2\pi\tau_0)$), the impedance value becomes R_O, which is very high (several Ohms) in comparison to the impedance of typical power delivery system (several milliohms).

In the mid-frequency range the impedance linearly increases with frequency given by:

$$Z_O(s)\approx\frac{R_O\tau_O}{1+\beta A_O}s=2\pi f L_{REG}$$

Thus for the frequency range of interest the VREG circuit can be approximated by a simple inductor L_{REG} with the value of L_{REG} given by:

$$L_{REG}=\frac{R_O\tau_O}{1+\beta A_O}$$

By changing the value of the CC compensation capacitor of Figure 8.28, we can tweak the value of the time constant τ_0 in the L_{REG} equation.

This allows for the easy control of L_{REG} in order to perform the optimization of the AVP design.

Chapter 9—Distributed Matched Bypassing for Board-Level Power Distribution Networks

István Novák, Senior Staff Engineer, Sun Microsystems
Leesa Noujeim, Staff Engineer, Sun Microsystems
Valérie St. Cyr, Supply Base Development Manager, Sun Microsystems
Nick Biunno, Principal Engineer, Sanmina-SCI
Atul Patel, Process Engineer, Sanmina-SCI
George Korony, Senior Member of Technical Staff, AVX Corp.
Andy Ritter, Senior Member of Technical Staff, AVX Corp.

This chapter is a revised and shortened version of the paper that appeared with the same title in the IEEE Transactions on Advanced Packaging, Vol. 25, No. 2, May 2002, pp. 230-243.

POWER DISTRIBUTION NETWORKS (PDNs) need to provide impedance response with specified shape and value over a wide-frequency band. Bypass capacitors with different values, along with capacitors and planes, may create resonance peaks unless the capacitor parameters are selected properly. Distributed matched bypassing (DMB) is suggested to create a smooth impedance profile. DMB requires components with Q<<1, which in turn requires user-defined equivalent series resistance (ESR). Different options are shown to set (increase) the ESR of bypass capacitors. We introduce bypass quality factor (BQF) and bypass resistor (BR).

Introduction

There has been considerable interest in recent years to improve the PDN of high-end computer and networking equipment. At the module level, on printed circuit boards (PCBs), full-area conductive layers over regular thickness or thin dielectric laminates provide low impedance for high-frequency decoupling [1] and a low-inductance conduit among PDN components. This is complemented by at times several thousand capacitors for mid and lower-frequency bypassing and decoupling [2, 3]. Conductive plane pairs in PCBs may exhibit multiple resonances [4, 5, 6], which can be suppressed by proper damping of the structure [7, 8].

Bypass capacitors with different values connected to the conductive planes also may exhibit resonances either between different capacitor banks [9] or between capacitors and planes. The application and benefits of higher-ESR bypass capacitors are mentioned in several papers (e.g., [8, 9, 11-15]).

One universal approach to reduce the resonance peaks is to minimize the inductance connecting the parts. With discrete surface-mount capacitors, the loop inductance is several hundred pH, and usually the dimensions of PCB and capacitor do not allow us to lower it below 100 pH, which is still too high a value in some applications to suppress resonance peaks. The ESR of bypass capacitors could also be selected to provide a flat impedance response, however, the ESR parameter for today's capacitors is not user-definable.

ESR of tantalum and electrolytic capacitors is usually in the Ohm range, whereas ESR of multi-layer ceramic capacitors is usually in the milliohm range. Tantalum and electrolytic capacitors are usually considered as low-frequency bulk capacitors, and as such their construction and geometry are not optimized for low-inductance connection to the PCB.

This paper introduces the BQF parameter of bypass capacitors, followed by the description of the board-level DMB design methodology for PDNs, with possible ways to set (increase) the ESR of bypass capacitors.

Distributed Matched Bypassing of Power Distribution Networks

The cumulative impedance of all board-level bypass capacitors should be a basin-shape impedance profile. For a lumped equivalent model, as shown in Figure 9.1, the cumulative capacitance and inductance values are C_{tot} and L_{tot}. At low frequencies it is complemented by the (inductive) impedance of the power source, usually a voltage regulator module (VRM). At high frequencies, it is met by the capacitance of power planes or by the package/die capacitance (C_p). The goal is to create a cost-effective design that meets the system's impedance requirement.

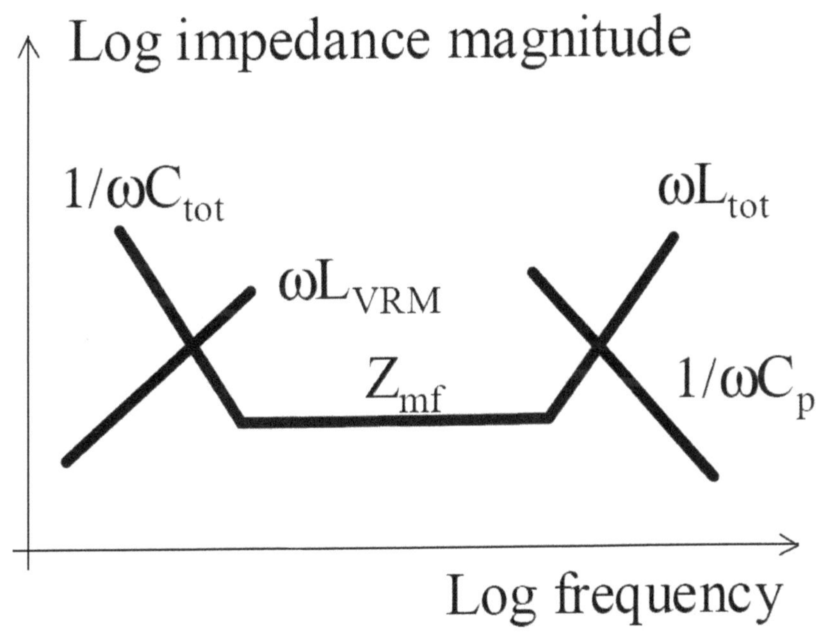

Figure 9.1: Typical Impedance Curve of a Board-Level Power Distribution Network. The Cumulative ESR Values of Parts in One or More Bypass Capacitor Banks Create the Flat Bottom with Value of Z_{mf}.

For PDNs, the requirement is to guarantee that the peak-to-peak transient noise stays within specified limits for all possible combinations of noise-source activities. With some exceptions, this is usually translated to a wideband, resistive impedance requirement of PDN. Whenever the flat, resistive impedance profile cannot be maintained, or rather due to other constraints, it is not the optimum solution; the requirement toward the PDN's impedance profile can be expressed as having either overshoot-free transient response or overshoot-free impedance-magnitude response.

The board-level bypass capacitors are typically surface-mount ceramic parts for capacitance values at or below 10 μF, and tantalum, electrolytic, or organic through-hole or surface-mount parts for large capacitance values. In its simple form, the equivalent circuit of a bypass capacitor is a series capacitor-resistor-inductor (C-R-L) network.

The Bypass Quality Factor of Capacitors

Let us assume lumped models for multiple bypass capacitors, and frequency-independent equivalent-circuit elements for each capacitor. If we start with one single capacitor value (either a single capacitor piece or multiple capacitors with the same value, what will be referred to as a capacitor bank) with C-R-L values for the nominal capacitance, ESR and ESL parameters, respectively, we get the impedance profiles shown in Figure 9.2. The top graph of Figure 9.2 shows the impedance of a capacitor with C=1 μF, L=1 nH, and two different values of R (R = 0.1 and R = 0.01 Ohm).

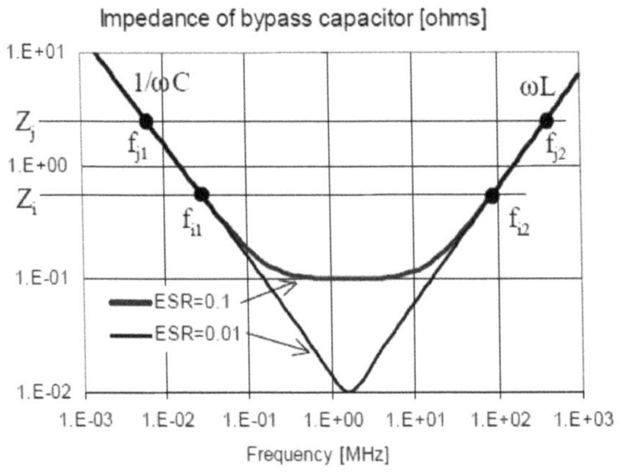

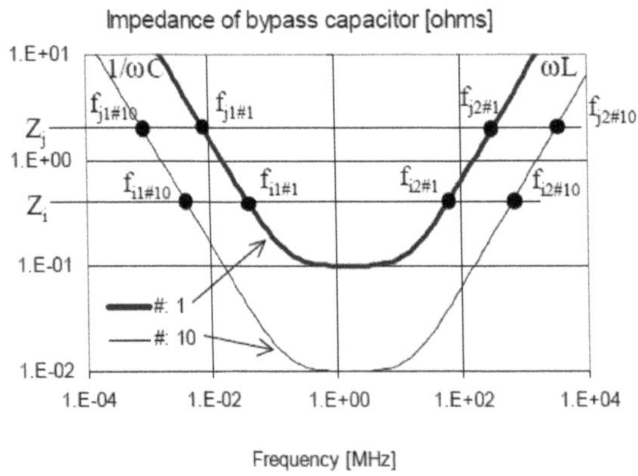

Figure 9.2: Illustration of Bypass Quality Factor for a Capacitor Bank with C = 1 μF, L = 1 nH. The Top Chart Shows the Effect of ESR; the Bottom Chart Illustrates the Effect of Number of Capacitors in the Bank. Let Us Note That as Long as $Q < 1$ and $Z > ESR$, $BQF = f_{i2}/f_{i1}$ is the same for All Z and ESR Values. One Should Also Note the Expanded Frequency Scale on the Bottom Chart.

The series resonance frequency is independent of R and is:

$$f_0 = \frac{1}{2\pi\sqrt{LC}} = 1.59Mhz$$

The quality factors (Q) of the capacitor for the two different R values are:

$$Q = \frac{2\pi f_0 L}{R} = \frac{\sqrt{\frac{L}{C}}}{R} = \begin{cases} 1 \text{ for } R = 0.01 \text{ ohms} \\ 0.1 \text{ for } R = 0.1 \text{ ohms} \end{cases}$$

If we need to achieve an impedance profile in a given frequency range at or below a specified value of Z, as long as Z > R, there are two frequencies (f_1 and f_2) where the impedance curve of the capacitor intercepts Z: f_1 is where the C-R-L impedance is still capacitive, and f_2 where the impedance of capacitor is already inductive. For Q<1, these frequencies can be approximated by:

$$f_1 = \frac{1}{2\pi CZ}, \qquad f_2 = \frac{Z}{2\pi L}$$

In traditional radio-frequency applications, the above definition of Q is a good indication of the losses in the capacitor. In power distribution applications, however, a bypass capacitor is more effective if it covers a bigger f_2/f_1 ratio of frequencies, within which frequency range its impedance is below a required value of Z. We call the f_2/f_1 ratio the bypass quality factor (BQF). For Q < 1, as illustrated in Figure 9.2, BQF depends only on the ratio of capacitance and inductance of the individual capacitors, and can be also expressed as the inverse square of the product of Q and R.

For a bank of N (N ≥ 1) identical capacitors, it is:

$$BQF = \frac{\frac{f_2}{f_1}}{(ZN)^2} = \frac{C}{L} = (QR)^{-2}$$

The top graph of Figure 9.2 illustrates the concept of BQF for one piece of capacitor with C = 1 μF, L = 1 nH, and with two ESR and two Z values.

For these parameters:

$$BQF = \frac{\frac{f_{j2}}{f_{j1}}}{Z_j^2} = \frac{\frac{f_{i2}}{fi1}}{Z_i^2} = \frac{C}{L} = 10^4$$

The bottom graph of Figure 9.2 illustrates the concept of BQF for two capacitor banks, both having capacitors with C = 1 μF, L = 1 nH, one bank having one piece of capacitor (N = 1), the other bank having 10 pieces of the same capacitor (N = 10).

$$BQF = \frac{\frac{f_{j2\#1}}{f_{j1\#1}}}{(1Z_j)^2} = \frac{\frac{f_{j2\#10}}{f_{j1\#01}}}{(10Z_j)^2} = \frac{\frac{f_{i2\#1}}{f_{i1\#1}}}{(1Z_i)^2} = \frac{\frac{f_{i2\#10}}{f_{i1\#01}}}{(10Z_i)^2} = \frac{C}{L} = 10^4$$

Let us note that the above definition of BQF yields a parameter, which is independent of Z, N, and R (ESR). The BQF parameter in bypass applications is in line with the design methodologies where only the highest capacitance value of any given case style is used, thus maximizing BQF of the parts.

To create the specified PDN response, the DMB uses concepts similar to those described in the adaptive voltage positioning [10], extended adaptive voltage positioning [11], and dissipative edge termination [12] concepts. Recently bypass capacitors with increased ESR values have also been proposed to suppress capacitor-capacitor and capacitor-board resonances ([13–15]).

It should be noted that while the capacitance is a sole attribute of the capacitor piece, the inductance presented by the capacitor to the PDN depends not only on the shape and size of the capacitor body (which sets its partial self-inductance), but also on its connection to the PDN. It has also been shown ([16, 17]) that the inductance and resistance may be a noticeable function of frequency.

Adaptive Voltage Positioning

The adaptive voltage positioning is a low-frequency concept, referring to the proper selection of output resistance of VRMs. It suggests that with a given mid-frequency impedance requirement (Z_{mf}), the optimum (lowest) value of bulk capacitance for a given peak-to-peak transient noise can be achieved with an output resistance of VRM matching the mid-frequency impedance value:

$$R_{out\_VRM} = Z_{mf}$$

This is contrary to the practice when a high direct current (DC) gain in the control loop sets the output resistance of VRMs to very low values.

Extended Adaptive Voltage Positioning

The extended adaptive voltage positioning suggests that not only the VRM–to-bulk capacitor interface, but also the bulk capacitor-to-bypass capacitor interface should follow the above procedure by setting the ESR of the lower- and higher-frequency capacitor banks the same and to match the square root of inductance-to-capacitance ratio of the adjacent capacitor banks.

Dissipative Edge Termination

The dissipative edge termination provides resistive termination to the power-distribution planes, by matching the real part of termination elements to the characteristic impedance of planes.

The DMB combines and extends the above concepts, as described below. The DMB methodology provides matching between and among the impedances of the low-frequency VRM, mid-frequency bypass capacitors, and high-frequency PCB and package planes and/or silicon elements, and thus creates a controlled impedance response over the entire frequency range of interest. The methodology gives simple rules and conditions for the parameters of PDN components, both for the same required impedance value throughout the entire frequency range and for situations when the required impedance varies with frequency.

The Unit Cell of DMB

The DMB concept uses elements with Q<1 (Q<<1 is preferred). The Q<1 condition creates a shallow flat bottom on the impedance curve of each capacitor bank. As a result, capacitor banks that are adjacent on the frequency axes can be described by a lower frequency R-L and a higher-frequency R-C element in parallel.

This R-L-R-C network approximation is valid only in the vicinity of the crossover frequency of the adjacent component banks, as this first-order analysis neglects the interaction of the capacitance of the lower-frequency bank and the inductance of the higher-frequency bank. The equivalent circuit and frequency response of the unit cell is shown in Figure 9.3.

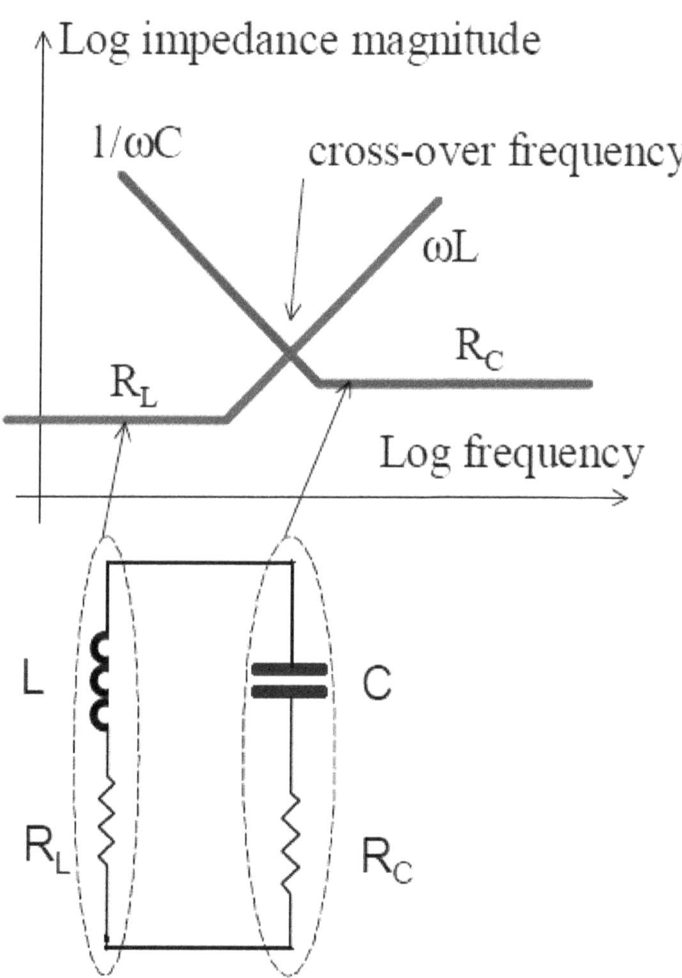

Figure 9.3: Equivalent Circuit and Bode Diagram of Unit Cell of Distributed Matched Bypassing. The Equivalent Circuit Is Valid Near the Cross-Over Frequency, Where the Lower- and Higher-Frequency Capacitor Banks Are Represented by an RL-L and RC-C Circuit, Respectively.

As long as the serial losses interconnecting the DMB elements are not significant, the PDN can be represented by a number of such unit cells lumped in parallel, each one covering different portions of the required frequency range, while still not having interaction among the elements other than within each unit cell. This means that each bank in the PDN is represented by two adjacent unit cells: one represents the lower-frequency R-C signature, whereas the other represents the higher-frequency R-L signature of the same bank of elements.

The frequency-dependent impedance of the unit cell (Z_{DMB_u}) is a second-order complex function:

$$Z_{DMB\_u} = \frac{\left(R_C + \frac{1}{j\omega C}\right)(R_L + j\omega L)}{\left(R_C + \frac{1}{j\omega C}\right) + (R_L + j\omega L)}$$

With s = jω, and rearranging, this becomes:

$$Z_{DMB\_u} = R_C + \frac{s(L - CR_C^2) + (R_L - R_C)}{s^2 LC + sC(R_C + R_L) + 1}$$

The characteristic equation in standard form is:

$$s^2 + 2\zeta\omega_n s + \omega_n^2;\ \omega_n = \frac{1}{\sqrt{LC}};\ \zeta = \frac{(R_C + R_L)}{2}\sqrt{\frac{C}{L}}$$

When $R_L = R_C = R$, the impedance expression can be simplified:

$$Z_{DMB\_u} = R + \frac{s(L - CR^2)}{s^2 LC + s2CR + 1}$$

If, moreover, we set:

$$R_O = \sqrt{\frac{L}{C}} = R$$

...the unit-cell impedance reduces to a frequency-independent resistive impedance of R:

$$Z_{DMB\_u} = R = R_O$$

When capacitor banks with different capacitance values are connected together along the frequency axis, keeping this condition yields the optimum solution, as the resistive impedance provides overshoot-free transient response, and no peaking in the frequency response. This is the default solution for DMB: the adjacent DMB unit cells both have the same resistive flat impedance response.

Under some circumstances, the entirely flat impedance response is either not desirable or not feasible to maintain. In the general case:

$$R_C \neq R_L$$

...and the second-order impedance function yields a complex impedance. The impedance has to transition from R_L at low frequencies (or DC in this simplified DMB unit-cell model) to the R_C value at higher frequencies (or infinite frequency in this simplified DMB unit-cell model). Of the various possibilities, three special conditions will be considered here:

- Case 1: overshoot-free transient response
- Case 2: overshoot-free impedance-magnitude response
- Case 3: impedance-magnitude response with a fixed amount of small overshoot (peak)

Case 1

Overshoot-free transient response for a step-like current excitation is achieved when the characteristic equation has at least critical damping and yields two real roots. This means that the ζ damping factor must be equal to or bigger then one. At the boundary, the damping factor should be unity, which sets the following condition among the DMB unit-cell parameters:

$$\zeta^2 = 1; (R_C + R_L) = 2R_O = 2\sqrt{\frac{L}{C}}, \frac{(R_C + R_L)^2}{4} = \frac{L}{C}$$

Case 2

Sometimes it is useful to set the requirement for the impedance magnitude as not having a peak, or:

$$|Z_{DM\_u}| = \max(R_L, R_C)$$

This means that for a sinusoidal current excitation, the response magnitude will stay within the bounds of the responses at DC and infinite frequency. In order to achieve overshoot-free impedance response, the following condition should hold among the DMB unit-cell parameters:

$$R_L R_C = R_O^2 = \frac{L}{C}$$

Case 3

The condition for case 2 becomes very restrictive and pessimistic for:

$$max\left\{\frac{R_L}{R_C};\frac{R_C}{R_L}\right\} \gg 1$$

If we allow for a small, approximately 1 percent overshoot in the impedance-magnitude response, the conditions become more realistic and favorable.

Numerically evaluating the DMB unit-cell impedance, the condition for the unit-cell parameters becomes:

$$(b_2r^2 + b_1r + b_0)R_{LC}^2 = \frac{L}{C}$$

Where:

$$R_{LC} = max(R_L, R_C)$$

$$r = \frac{R_L}{R_C}\ if\ R_L < R_C and\ r = \frac{R_C}{R_L} if R_C < R_L$$

$$b_1 = 0.4831$$

$$b_2 = -0.0139$$

A few typical applications of these cases and conditions are listed and described in the following:

1. **VRM–to-Bulk-Capacitor Interface**

We assume that the active control loop of the VRM and the inductance connecting the VRM to the bypass capacitors have a simple one-pole R-L equivalent network representation, with a DC output resistance of $R_L = R_{VRM}$ and inductance of $L = L_{VRM}$.

We also assume that the ESR of bulk capacitor(s) meets the mid-frequency impedance requirement:

$$R_C = ESR_{BULK} = Z_{MID}$$

The DC output resistance of the VRM and the required mid-frequency impedance of the PDN are determined by the system partitioning and are not necessarily the same.

For instance, if the VRM supplies several independent boards, its output resistance should be lower than the mid-frequency impedance required by each of the boards.

Depending on how many loads the VRM has to service and what value of series DC loss in the PDN has to be assumed, the DC output resistance of the VRM is equal to or lower than the mid-frequency impedance requirement:

$$R_{VRM} \leqq Z_{mf}$$
$$R_L = rRC = rZ_{mf}$$

From these conditions, we can solve either for C or for L. It is usual that we estimate the affordable inductance (L) of the VRM output and its connections, and solve for the required total low-frequency (bulk) capacitance. The required interrelation among the four DMB unit-cell parameters:

$$Case\ 1: \frac{L}{CR_C^2} = \frac{(1+r)^2}{4}, C = \frac{L}{Z_{mf}^2}\frac{4}{(1+r)^2}$$

$$Case\ 2: \frac{L}{CR_C^2} = r, C = \frac{L}{Z_{mf}^2}\frac{1}{r}$$

$$Case\ 3: \frac{L}{CR_C^2} = b_2r^2 + b_1r + b_0, C = \frac{L}{Z_{mf}^2}\frac{1}{b_2r^2 + b_1r + b_0}$$

For all of the above cases, the lowest value of bulk capacitance is required if r = 1. For r < 1 values, the normalized capacitance requirement is shown in Figure 9.4.

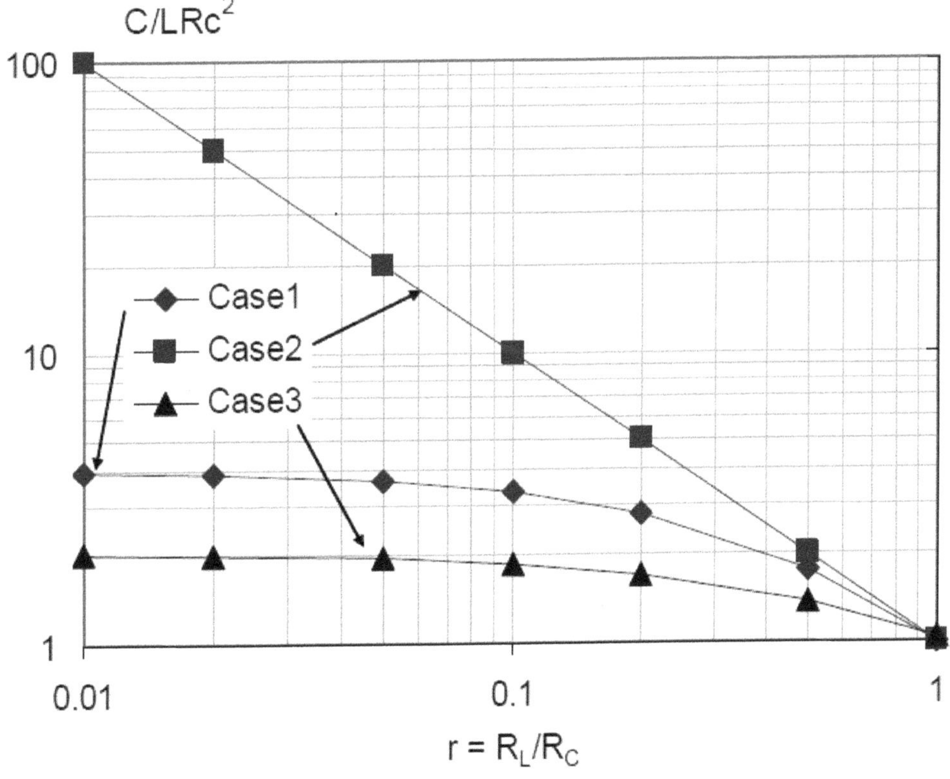

Figure 9.4: Normalized Bulk-Capacitance Requirement versus $r = R_L/R_C$

One should note that as r decreases, the capacitance value necessary to maintain the specified condition monotonically increases; the highest value is required for overshoot-free frequency response (case 2), and the lowest value is needed if we allow for a small amount of peaking (case 3).

2. Bulk-to-Mid Frequency Capacitor Interface

In case of capacitor-capacitor interfaces, selecting the resistance values in adjacent DMB unit cells to be equal will guarantee a smooth continuation of impedance from one DMB cell to the next and creates a flat and resistive impedance profile. As shown in Figure 9.4, for $R_C = R_L$, the optimum solution is defined for all three cases by $R_L R_C = L/C = R_{o2}$. If, for any reason, R_L does not equal R_C, we get either the situation of VRM–to-bulk-capacitance described earlier, or the situation of mid-frequency-to-package/die situation described in the following.

3. Mid-Frequency-to-Plane Interface

If packages and active electronics are assumed to have no noticeable influence on the PDN impedance, the mid-frequency bypass capacitor banks interface with the power-distribution planes. A pair of power/ground planes with dimensions of x*y and separation of h, with ε_r relative dielectric constant of the insulating laminate, can be chosen to substitute for all of the power plane pairs the stack-up may have on the particular supply rail. The approximate characteristic impedance of the equivalent plane pair then becomes:

$$Z_p = \frac{266(\frac{h}{x})}{\sqrt{\varepsilon_r}\left(1+\frac{y}{x}\right)} = \frac{532\,h}{\sqrt{\varepsilon_r}\,P}$$

...where P is the perimeter of the rectangular plane shape, P = 2*(x+y). A bedspring equivalent circuit is used to simulate the impedance profile of the planes with their assumed bypass capacitors [25]. Figure 9.5 shows the simulated self-impedance profiles at the center of a pair of 25.4 x 25.4 cm (10"x10") plane pair with 50 μm (2-mil) separation, and N pieces of DMB components are placed uniformly around the edge of the planes, each having an ESR of R_L and ESL of L.

In the following page, the top chart shows the impedance profile for the case when the R_L/N cumulative ESR of the mid-frequency DMB cells matches the Z_p plane impedance. The parameter is the L/N cumulative inductance of all of the DMB parts. For a total inductance of 10 pH, the impedance profile is smooth; for 100 pH and 1 nH inductance values, there is an increasing peaking in the impedance profile due to the resonance peak of the static plane capacitance and DMB inductance. The bottom chart shows the same scenario, but the ESR of parts is 10 times lower. One should note that even with the lowest inductance there is a noticeable peaking in the impedance profile. Detailed simulations show that sufficiently smooth impedance profile can be achieved for a 10" x 10" size under the following conditions:

$$\frac{R_L}{N} = Z_p, \frac{L}{N} = \frac{\mu_0 h}{5}$$

One should note that these matching conditions do not assume thin power and ground laminates, and they are valid for a wide range of laminate thicknesses. There are situations when additional damping is necessary: on planes with large open areas, planes with few silicon chips attached (e.g., double data random access memory [DDRAM] termination planes), or when socket/package inductances isolate the silicon from the planes at the resonance frequencies.

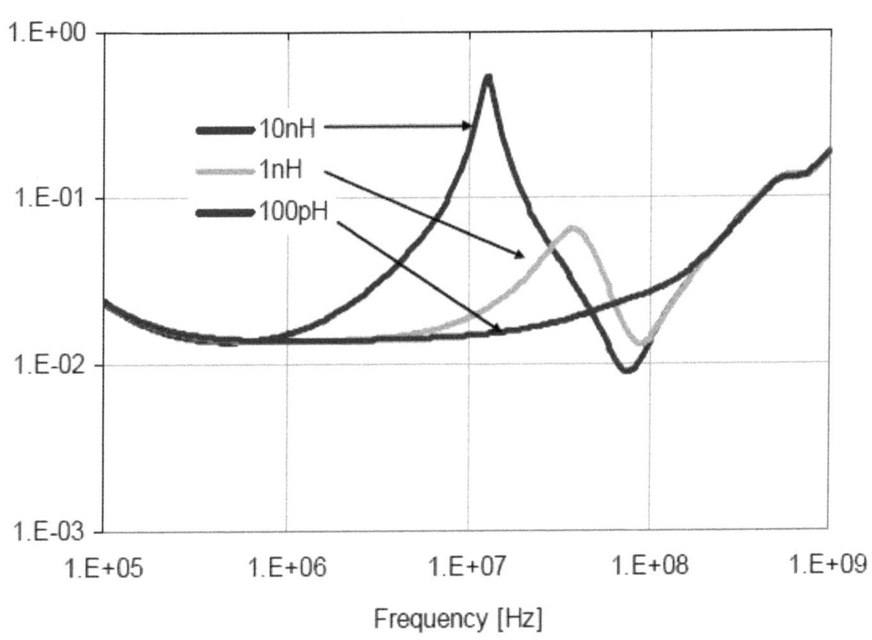

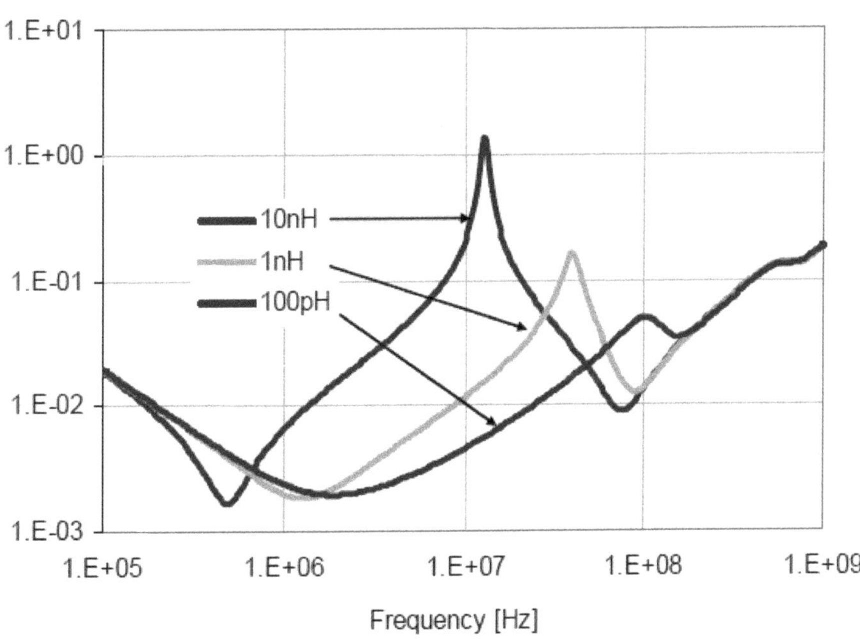

Figure 9.5: Self-Impedance Magnitude at the Center of One Pair of Power/Ground Planes with h = 2 Mil Separation, $\varepsilon_r = 4$, with DMB Elements along the Board Periphery. Top: $R_L/N = Z_p$, Bottom: $R_L/N = 0.1Zp$. Parameter; L/N Inductance of DMB Elements

4. Mid-Frequency-to-Package/Die Interface

In PDNs, by feeding a large integrated circuit, the connection may be point-to-point, where the path goes through a package, with optional package capacitors, and ends on the silicon. The low equivalent resistance of silicon usually creates a package resonance [19]. The DMB unit cell on the package and silicon boundary has the mid-frequency target impedance (Z_{mf}) and package inductance (L_{pkg}) in the R-L leg, and the die capacitance (C_{die}) and die equivalent resistance (R_{die}) in the R-C leg.

In this situation, usually $R_C < R_L$, $r = R_C/R_L$, and the task may be to find the maximum permissible value of L (L_{pkg}). Expressing L in cases 1 through 3 yields:

$$Case\ 1: \frac{L}{CR_C^2} = \frac{(1+r)^2}{4}, L = \frac{C}{Z_{mf}^2}\frac{(1+r)^2}{4}$$

$$Case\ 2: \frac{L}{CR_C^2} = r, L = \frac{C}{Z_{mf}^2}r$$

$$Case\ 3: \frac{L}{CR_C^2} = b_2r^2 + b_1r + b_0, L = \frac{C}{Z_{mf}^2}b_2r^2 + b_1r + b_0$$

For all of these cases, the highest value of inductance is allowed if $r = 1$. For $r < 1$ values, the normalized inductance requirement is shown in Figure 9.6. Let us note that similar to the VRM–to-bulk capacitance interface, as r decreases, the inductance value necessary to maintain the specified condition monotonically decreases; the lowest value is required for overshoot-free frequency response (case 2), and the highest value is allowed if we accept a small amount of peaking (case 3).

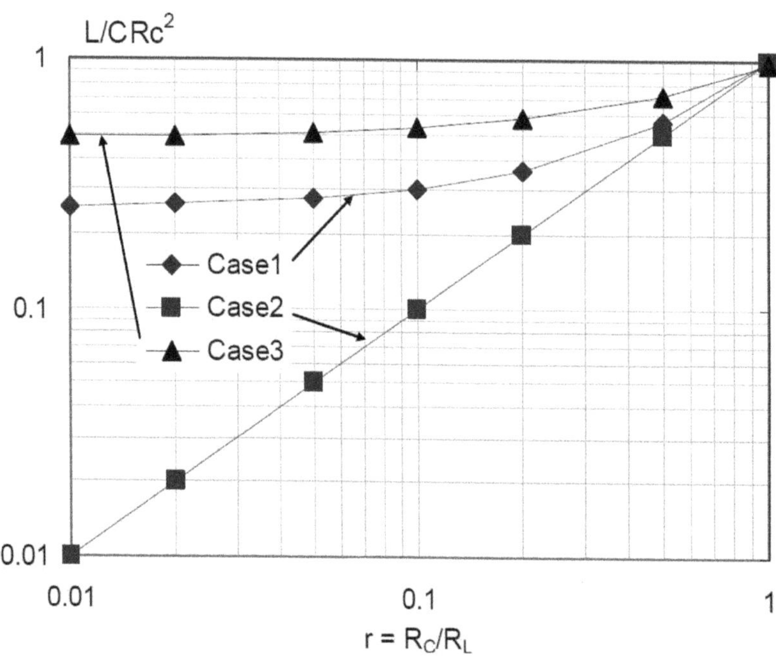

Figure 9.6: Normalized Inductance Requirement versus $r = R_C/R_L$

As a summary, the DMB methodology follows a few simple steps that involve:

- Determining the number of high-frequency capacitors from the required total inductance.
- Selecting the highest available capacitance in the given size.
- Calculating the required ESR of each capacitor. If the total capacitance of high-frequency bypass capacitors and the achievable connecting inductance of VRM would still create a resonance peak, additional (lower-frequency) capacitor banks are selected, similar to the process described in [14].
- Optionally, if suppression of plane resonances is also required, the inherent plane dimensions should be selected to match the mid-frequency impedance requirement.

By following the above procedure, the possibility of inter-capacitor and plane-to-capacitor resonances can be minimized. One should note that requesting the highest available capacitance in a given package style also conveniently minimizes the inductance of the capacitor body. As shown in Figure 9.7, higher capacitance in the same package size often comes with a thinner cover layer, thus lowering the inductance.

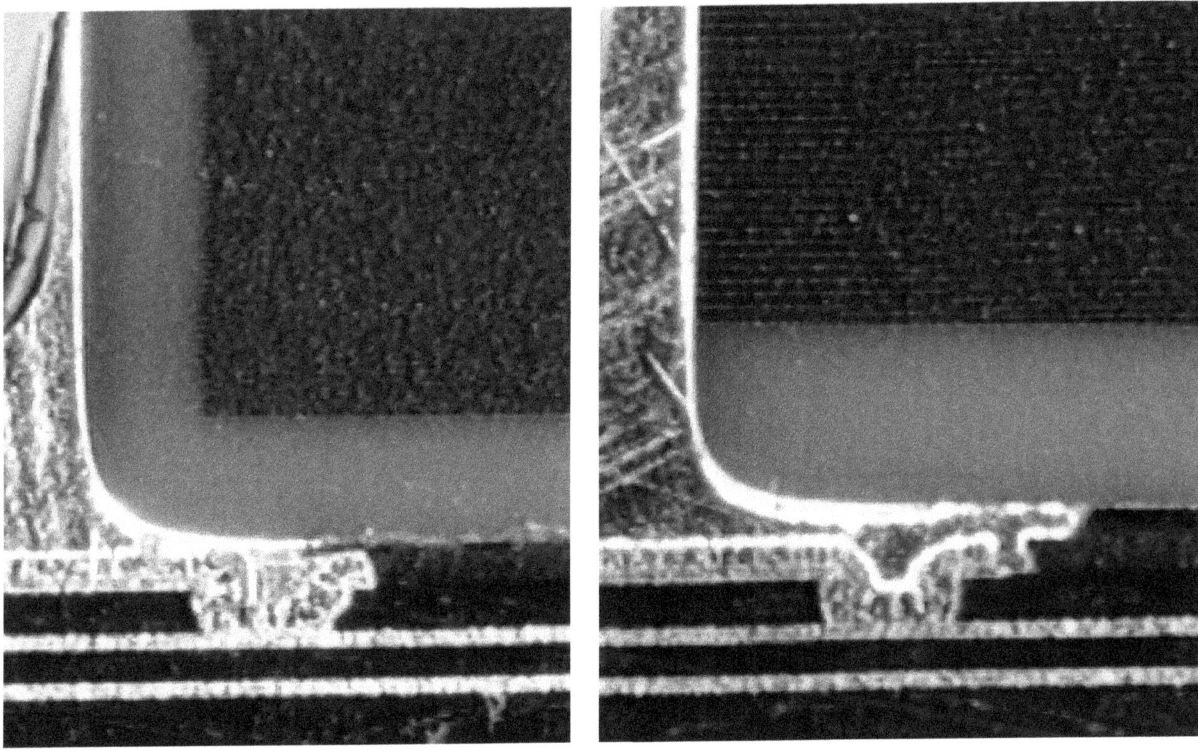

Figure 9.7: Cross-Section of Mounted Capacitors Showing the Bottom Cover Thickness in 0612-Size IDC Package: 2.2 μF Part on the Left (7.2 Mils Bottom Cover) and 1 μF Part on the Right (10 Mils Bottom Cover). Layer 2-Layer 3 Laminate is 2-mil Dielectric with One-Ounce Copper on Either Side

5. Sensitivity to Component Tolerances

The Q<<1 condition of the DMB components yields a lower sensitivity to component tolerances.

Contrary to the case when the bypass components have Q >= 1, when the impedance magnitude of peaks and valleys at each frequency transition between adjacent components depend on at least three parameters (C, L, and R), the Q << 1 condition decouples the L and C values within each unit cell, thus leaving effectively only the tolerance of R itself.

Implementation of Distributed Matched Bypassing—ARIES Implementation

The annual resistive interstitial element screened-in (ARIES) solution is based on the annual buried resistor (ABR)™ (Figure 9.8) process [20, 21], where the ESR of a ceramic capacitor is increased by adding a series resistor element, created in an annular void between a conductive pad and its surrounding anti-pad.

In bypassing applications, one terminal of the resistor should be on one of the power rails, usually located on large metal areas or full planes, conveniently eliminating the need for an anti-pad ring connection, thus maximizing the available density.

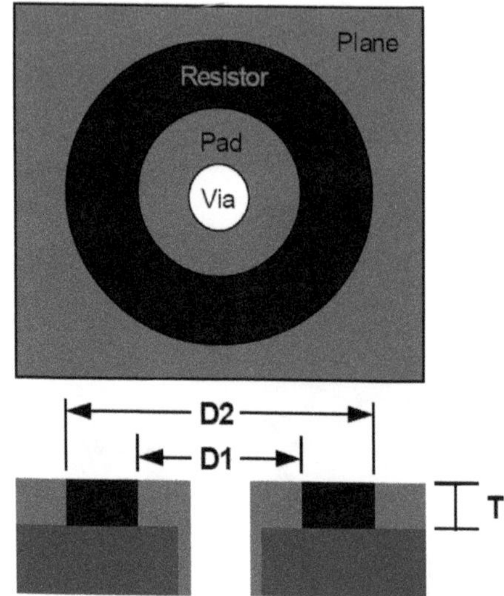

Figure 9.8: Construction of Annular Buried Resistor (ABR™)

In a series R-C connection, assuming linear components, the sequence of the two parts does not affect the resulting impedance. The external resistance can be either on the ground side, on the power side, or split in any ratio between the ground side and power side. To reduce the number of required components in the ARIES solution, resistors are inserted only between the capacitor terminals and the power plane.

To minimize the required footprint and the loop inductance, as shown in Figure 9.9, the ARIES solution uses a multi-terminal capacitor (an eight-terminal capacitor is shown in the figure) with blind vias connecting to the PCB planes. The capacitor sits on the outermost (L1) metal layer, and four of the capacitor's terminals are connected to the second (L2) ground plane. Four other blind vias connect the remaining terminals of the capacitor to the third (power) layer (L3). Having the ground layer outside with direct connection to the capacitor has several advantages:

- The ground layer provides electromagnetic interference (EMI) shielding.
- The capacitor body is tied to ground; it does not "float" on the noise voltage across the series resistance.
- Larger-diameter ABR™ components can be used, which can extend underneath the ground blind vias.

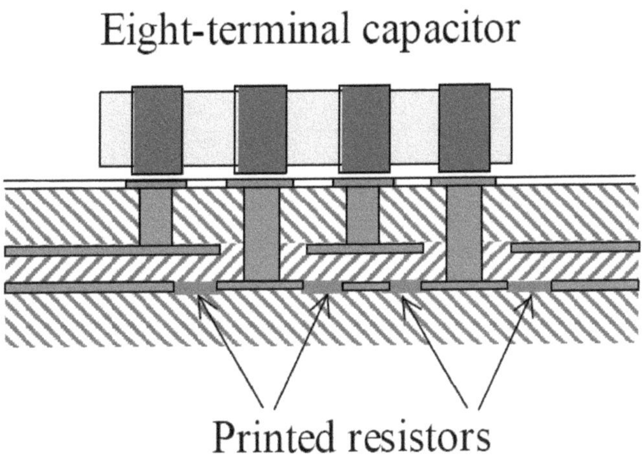

Eight-terminal capacitor

Printed resistors

Figure 9.9: ARIES Construction, Shown in a Cross-Sectional Side View. The Multi-Terminal Capacitor Is Connected with Blind Vias to the Power and Ground Planes Below. Four ABR™ Embedded Resistors on the Lower Plane Set the ESR to the Required Value

The direct physical connection of the embedded resistors to one of the planes also increases their power rating.

In one ARIES element, there is either one eight-terminal capacitor (inter-digitated capacitor [IDC]) or one array of four capacitors with eight independent terminals (integrated passive component [IPC]), plus four embedded resistors. The proximity of the embedded resistors in one block provides good current sharing among the four legs, thus reducing the current difference due to resistance tolerances.

One should be aware that the same geometrical concept can be extended to capacitor packages with more terminals [22], or fewer terminals, including the regular two-terminal straight or reversed geometry capacitors. The tolerance of embedded printed resistors depends on the resistive ink composition, printing consistency, and long-term behavior of the part. Without trimming, around 25 percent of the tolerances can be achieved, which can be further improved, if necessary, by trimming processes. In bypassing applications, though, a tolerance tighter than around 20 percent is rarely required.

DMB Implementation with Controlled ESR Capacitors

Designs of high-CV, low-inductance capacitors typically include a large number of internal layers, multiple parallel external termination contacts, and minimized inactive margins, all of which combine to also reduce the ESR, relative to standard multilayer chip capacitors (MLCCs).

This trend of lowering ESR is contrary to the system need of controlled, increased ESR. Thus, in leading edge decoupling applications, capacitor equivalent circuit parameters must be fully specified and controlled, including series capacitance, inductance, and now, resistance. The challenge for capacitor designers is to develop ways to control and increase ESR without compromising hard-won gains in lower ESL and higher CV ([22-24]). AVX Corporation is actively developing capacitors with controlled, selectable ESR while preserving the low-inductance structure of the capacitor device. Efforts to date yield capacitors with deliberate ESR variations over three orders of magnitude in 0306, 0508, and 0612 LICC and 0508 and 0612 IDC (Figure 9.10).

Figure 9.11 shows the impedance profiles of a conventional (low-ESR) and a controlled-ESR 0612 1μF ceramic capacitor. The parts were soldered on two pads, connecting with one pair of blind vias from the pads to a plane pair one and two layers below.

Capacitor type	max. cap.	inductance	BQF	ESR min.	ESR max.
	[μF]	[pH]	[xE+03]	mOhm	mOhm
0612 LICC	10.00	325	31	6	1000
0508 LICC	6.80	250	27	5	1000
0306 LICC	2.20	200	11	4	1000
0612 IPC	4.00	150	27	6	N/A
0612 IDC 8 terminals	10.00	110	91	5	500
0508 IDC 8 terminals	6.80	90	76	4	500
0612 IDC 10 terminals	10.00	75	133	4	500
0508 IDC 10 terminals	6.80	65	105	3	500
1818 IDC 32 terminals	68.00	15	4533	1	TBD
LICA	0.13	25	5	15	N/A
HiFLI ™ 8x8	0.22	10	22	4	200
Ta capacitor	1000.00	3600	278	25	N/A
Best cap	4.00E+05	1400	285714	50	N/A

Figure 9.10: Maximum Capacitance, Typical Attached Inductance, BQF and ESR Range of Some AVX Capacitor Families

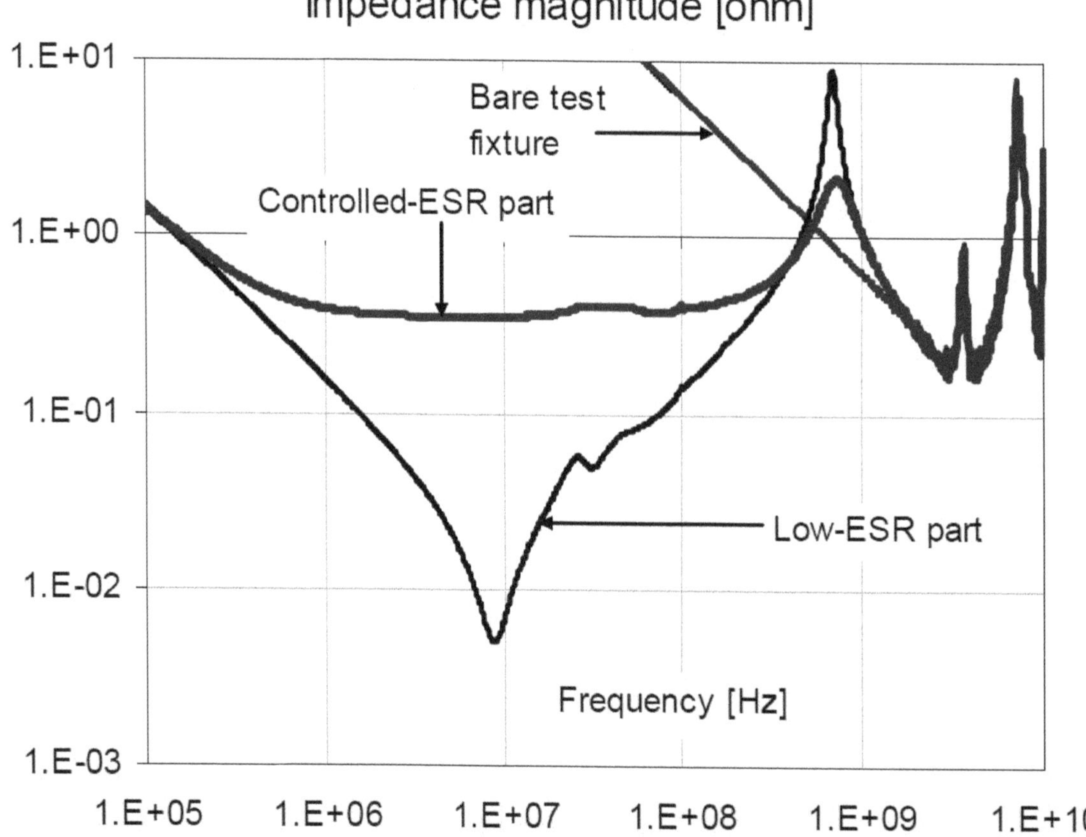

Figure 9.11: Measured Impedance Profile of AVX 1μF 0612-Size Low-ESR and Controlled–ESR LICC Capacitors

The figure also shows the impedance of the small bare fixture board. Let us note that:

- The impedance profile of controlled-ESR part is flat over more than two decades of frequencies.
- The first anti-resonance peak is at about 750 MHz for both parts, indicating that ESL is not increased by controlling ESR.

Test Board Implementation

Twenty-layer 25.4x12.7 cm (10" x 5") test boards with 50 μm (2 mil) FR4 dielectric layers between power and ground planes were designed and built.

The stackup is shown in Figure 9.12.

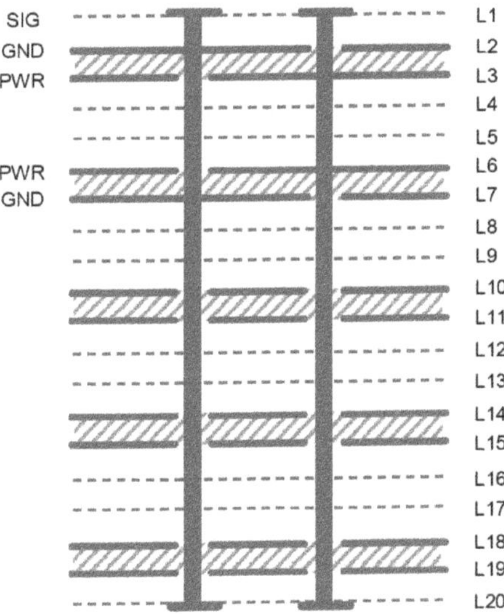

Figure 9.12: Stackup of Test Boards. In the Measured Networks, Only the Upper Two Plane Pairs (L2-L3 and L6-L7) Are Connected to Test Vias. The Other Six Planes and Eight Signal Layers Were Left Floating

The boards had two 2.54 cm (one-inch) grids with 1.27 cm (0.5") offset with respect to each other. On one of the grids, nominally starting at the lower left corner of the board, test through-holes were placed to allow the measurement of the impedance profile.

Because the length and width of the board are integer multiples of an inch, there would be 30 test points exactly falling on the periphery of the board. These 30 test points were pulled back 0.63 cm (0.25") from the board edge.

The grid points of the second square grid start at a 1.27 cm x 1.27 cm (0.5" x 0.5") offset from the lower left corner, and they accommodate surface pads for 1206 eight-terminal capacitors with the appropriate blind vias and embedded resistors (for the locations of test vias and capacitor pads, see Figures 9.13 and 9.14).

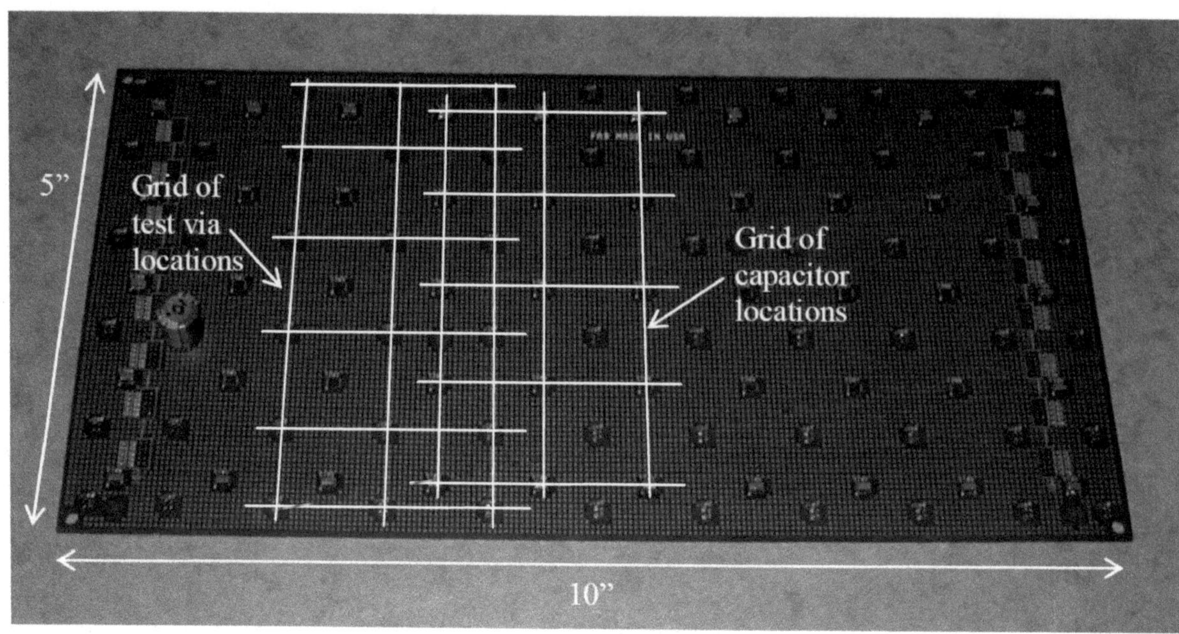

Figure 9.13: Photo of 10" x 5" Test Board, with the Test Via and Identified Capacitor Grids. Besides the Single Bulk Capacitor, There Are 26 Pieces of 2.2 μF IDC Parts on the Outer Ring of ARIES Positions

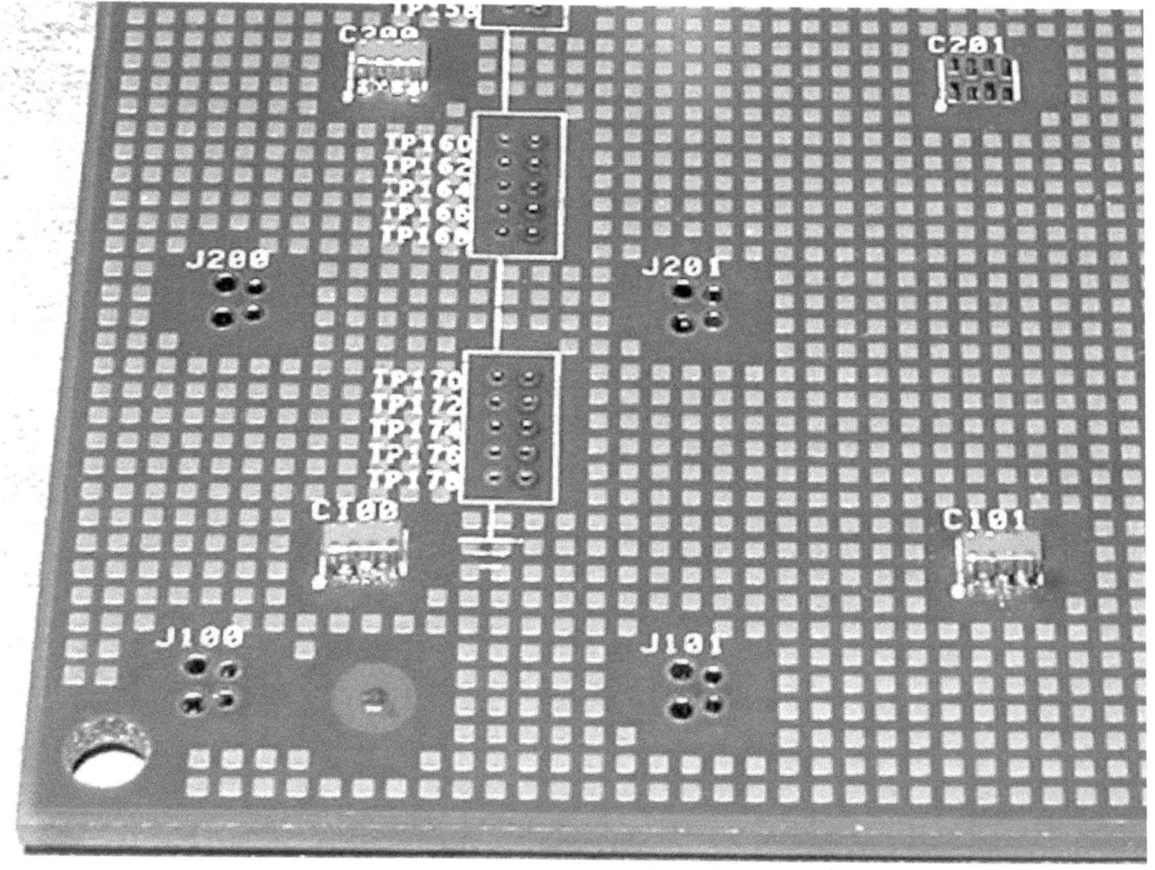

Figure 9.14: Close-up of Test Points and Capacitor Elements on the Test Board

A second set of boards with the same stackup and dimensions, except with solid copper connection in place of the embedded resistors, serve the purpose of reference structure to measure and simulate the PCB itself. The same reference boards with no embedded resistors were also used to measure controlled-ESR capacitors.

The test structure was characterized by measurements and simulations in several steps:

1. Bare Board Parameters

The bare board was characterized by detailed SPICE-grid simulations and measurements. Uniform and homogeneous cross section and materials were assumed. Considering the 1.27 mm (50 mils) pullback of planes from the board edge, the plane dimensions were a = 25.15 cm (9.9") and b = 12.44 cm (4.9"). There are two parameters, however, the dielectric constant and the plane separation, which on a finished board cannot be measured directly without destructive probing.

To obtain the dielectric constant and the plane separation, we can use the expressions of static plane capacitance and first modal resonance frequency:

$$C = \varepsilon_0 \varepsilon_r \frac{ab}{s}$$

$$f_{res} = \frac{1}{2a\sqrt{\varepsilon_0 \varepsilon_r \mu_0}}$$

...where C is the static capacitance of the plane pairs, a and b are the length and width of planes, s is the separation of planes, f_{res} is the first modal resonance frequency of the planes, and ε_0, ε_r, μ_0 are the dielectric constant of free space, relative dielectric constant of laminate material between conductive planes, and permittivity of free space, respectively.

By rearranging, we get:

$$s = \varepsilon_0 \varepsilon_r \frac{ab}{C}$$

$$\varepsilon_r = \frac{1}{4a^2 f_{res}^2 \varepsilon_0 \mu_0}$$

From the above expressions, with the measured values of C = 44.5nF, and f_{res} = 291 MHz, the dielectric constant and plane separation were calculated:

s = 52.3 μm (2.06 mils) per pair, and $\Sigma_r = 4.217$

The good correlation of simulated and measured self-impedance of the bare test board, defined in Figures 9.12 through 9.14, at the test point of x = 10.16 cm (4") and y = 7.62c m (3") from the lower left corner, is illustrated in Figure 9.15. The SPICE grid used for the simulation had 0.635 cm (0.25-inch) uniform grid of lossy transmission-line segments, the topology and simulation parameters were as described in [25].

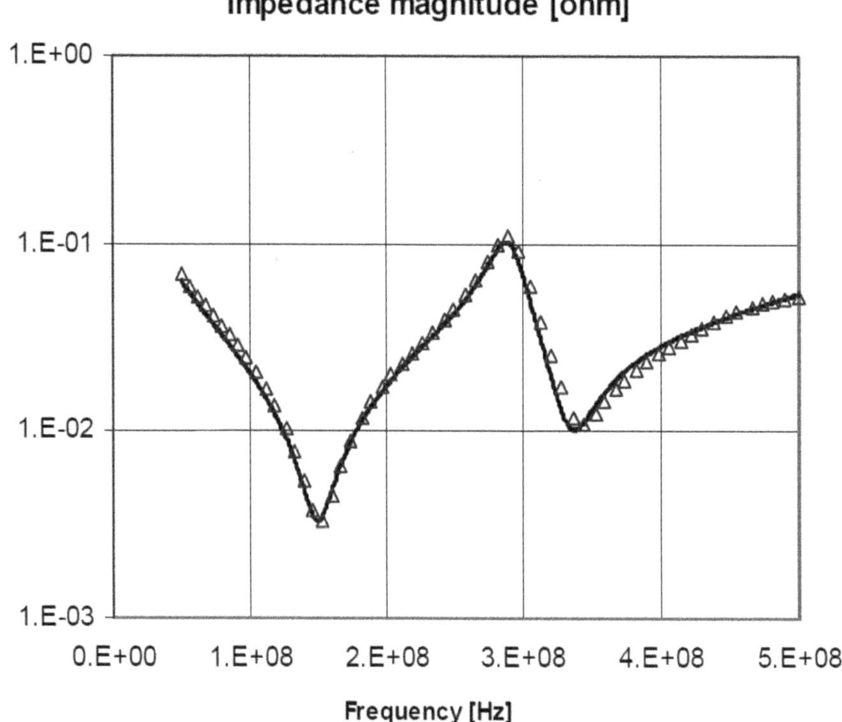

Figure 9.15: Simulated (Solid Line) and Measured (Triangles) Self- Impedance Magnitude of the Bare Test Board, Measured at the Test Point at x = 4", y = 3" from the Lower Left Corner

2. Capacitor Pad and Blind via Connection Parameters

The electrical parameters of the eight-terminal capacitor connection were derived by shorting all eight pads of one capacitor location on the surface with a copper sheet. The self-impedances and transfer impedances at several locations were measured and simulated. Three different capacitor locations were shorted, one at a time. Using the parameters of the bare board, the elements of a series R-L circuit were selected during the simulations to fit the measured impedance profile of the shorted board.

The attached impedance of the via/pad combination was matched with the following elements:

$$L_a = 48.5 \text{ pH}, R_{dc\_a} = 5\text{E-4 ohms}, R_{sk\_a} = 1.6\text{E-7 ohms*sqrt(f)}$$

Let us note that this R-L equivalent circuit represents the pads and vias only, these values are independent from and do not contain the impedance contribution from the planes and capacitors.

The correlation of the simulated and measured self-impedance of the shorted test board is illustrated in Figure 9.16, having a short at the capacitor location x = 11.43 cm (4.5") and y = 6.35 cm (2.5"), the impedance being measured at test point x = 7.62 cm (3") and y = 10.16 cm (4").

The peak at 90 MHz corresponds to the resonance of the static plane capacitance with the shorting inductance.

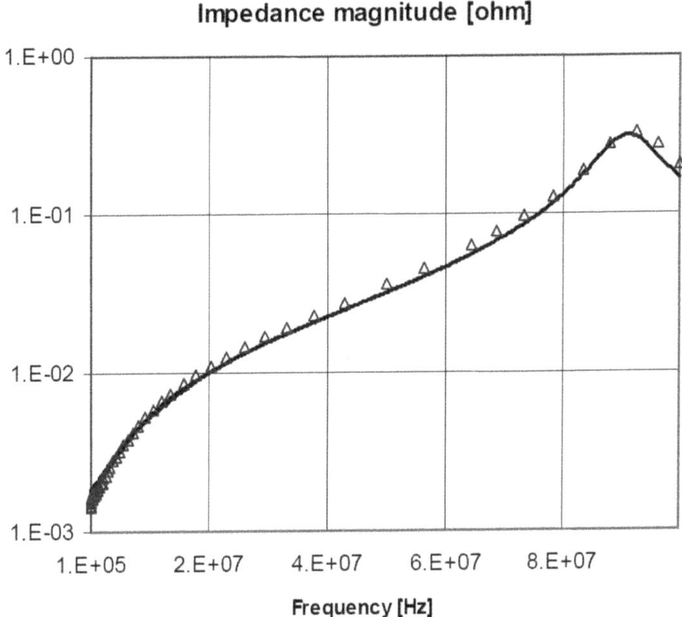

Figure 9.16: Illustration of the Degree of Correlation Obtained with the RL Via-Pad Model. Solid Line: Simulated, Triangles: Measured. The Same Via Model Yields Similar Good Correlation at All of the Tested Plane Locations, for Both Self- and Transfer Impedances.

3. Capacitor Parameters

Similar to the extraction of via-pad parameters, the capacitor parameters were also extracted by soldering one piece of eight-terminal capacitor on the board, and they correlate with the measured self- and transfer impedances on the board. Data in this paper is shown for the AVX 2.2 μF X7R IDC part (W3L16C225MAT). The equivalent circuit is a series C-R-L network, however, as it was shown in [16] and [17], the inductance and resistance associated with the capacitor body are both frequency-dependent.

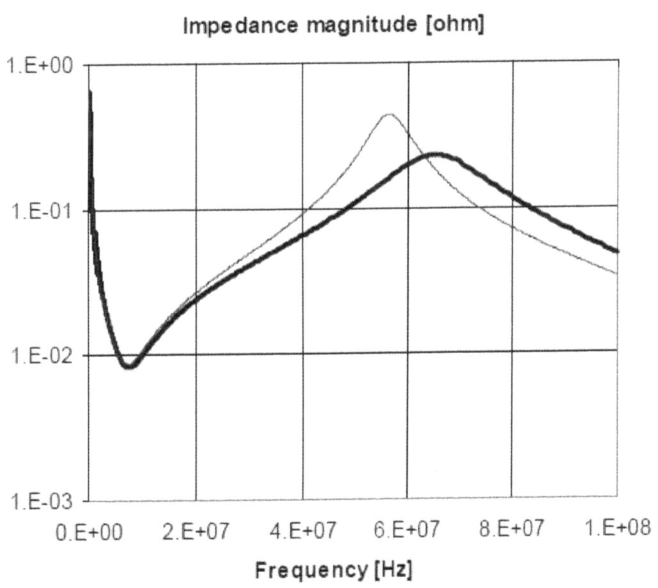

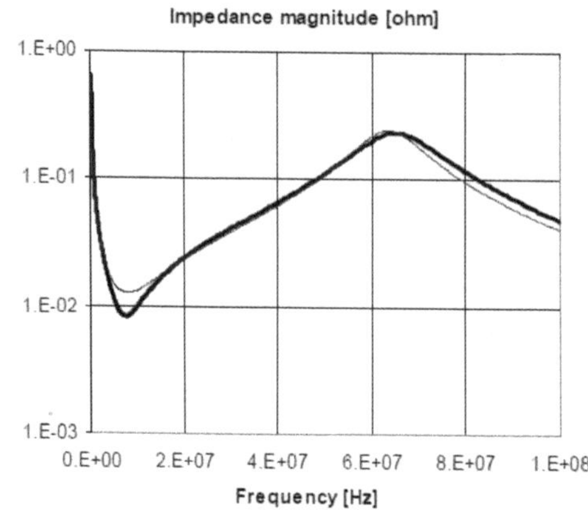

Figure 9.17: Correlation at 8 MHz (Top) and 62 MHz (Bottom) between Measured (Heavy Line) and Simulated (Thin Line) Impedance of Test Board with One Piece of Capacitor Attached

The correlation and curve fitting was performed at two characteristic frequencies, namely, the series and first parallel resonance frequencies. In this particular case, the series resonance was around 8 MHz, the first parallel resonance was at 62 MHz. The extracted R and L parameters at these two frequencies were 6 mOhm and 160 pH at 8 MHz, and 10.5 mOhm and 120 pH at 62 MHz. One should note that as was shown earlier, the via-pad combination at these frequencies represent 1 mOhm and 48.5 pH at 8 MHz, and 1.8 mOhm and 48.5 pH at 62 MHz, therefore the extra resistance and inductance due to the capacitor piece itself is 5 mOhm and 111.5 pH at 8 MHz, and 8.7 mOhm and 71.5 pH at 62 MHz.

The correlation of measured and simulated impedances with the above parameters is illustrated in Figure 9.17.

4. Fully Populated DMB Test Board

The test boards with and without embedded resistors were also measured with a full population of low-ESR capacitors.

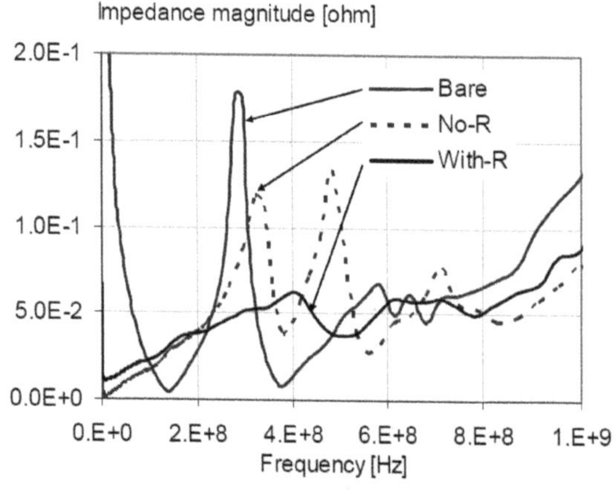

Figure 9.18: Self-Impedance Measured on the 10” x 5” Test Board with 26 Pieces of Low ESR IDC Parts with and without Embedded Resistors. The Three Traces: a) Thin Continuous Line: Impedance of the Bare Board (for Reference), b) Dashed Line: Impedance of the Same Board without Embedded Resistors, and c) Solid Heavy Line: Impedance with Embedded Resistors

Though other allocations are also feasible, the population described in this paper was a full ring of capacitors along the periphery of board, a total of 26 pieces. Figure 9.18 compares the self-impedance magnitude of the populated board with and without embedded resistors and the self-impedance of the bare board.

The three traces: a) thin continuous line: impedance of the bare board (for reference), b) dashed line: impedance of the board populated with low-ESR parts without embedded resistors, and c) solid heavy line: impedance with embedded resistors.

The graph above shows that by adding the embedded resistors, the impedance at 324 MHz is reduced from 0.119 ohms to 0.052 ohms, and at 480 MHz from 0.133 ohms to 0.038 ohms.

5. Module Card Implementation

The left of Figure 9.19 shows a photo of the embedded resistor on an un-laminated inner layer.

The photo on the right shows the cross-section at the printed-resistor element.

Figure 9.20 illustrates the savings in component count and board area on a module card.

Figure 9.21 shows the distribution of measured resistance values on two separate DMB layers on the finished board.

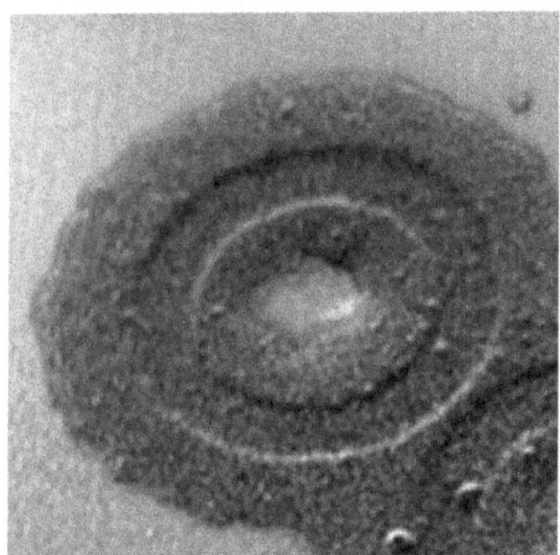

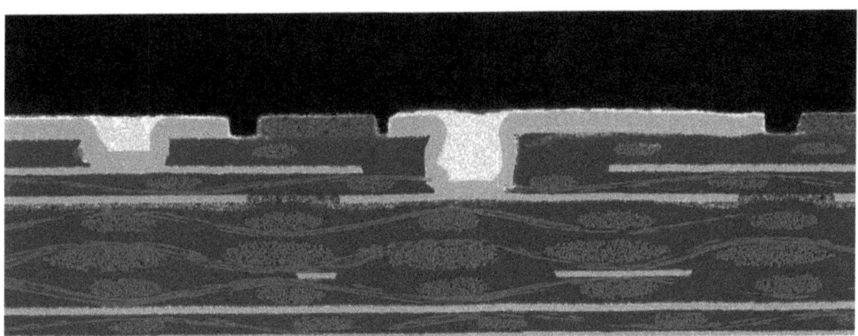

Figure 9.19: Left: ABR™ Embedded Resistor on Inner Layer; Right: Cross-Section at the Printed Resistor of the Finished Module Card

The Concept of the Bypass Resistor

The DMB relies on the controlled ESR of capacitor parts to create a wide-band resistive impedance profile. It also assumes capacitors with Q<<1, so that there is a pronounced flat resistive-like impedance bottom of their impedance curve.

This reduces the interaction of capacitance and inductance tolerances, creating a less sensitive network.

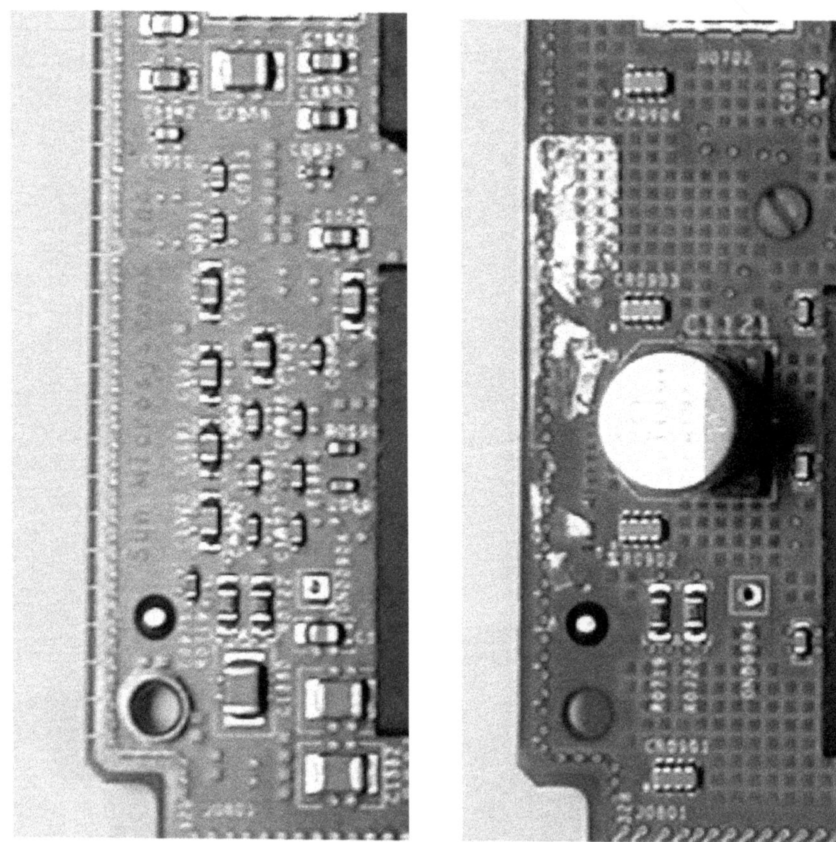

Figure 9.20: Illustration of Savings in Component Count and Board Area on a Module Board. Left: Board Detail of Module with More Than 250 Pieces of Mid-Frequency Bypass Capacitors, and Right: Same Module with 50 Distributed Matched Bypassing Capacitors

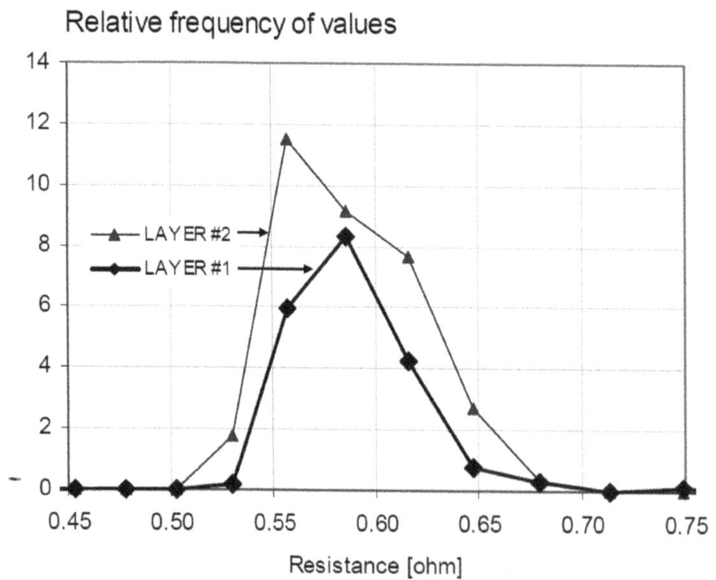

Figure 9.21: Relative Frequency of Embedded Resistance Values Measured on the Two Separate DMB Layers of Board Shown in Figure 9.20

To achieve a design goal with the lowest number of parts, we also need the highest available capacitance in a given package style.

Realizing that for bypass applications the resistance (e.g., ESR) is of primary importance, we can define a new part, called BR. It is a virtual R-C-L component, where the R and C are specified, and L depends both on the geometry of package(s) and the geometry of usage. It can be either one piece: controlled-ESR capacitor; or separate pieces: a low-ESR capacitor in series to resistor(s) forming a low-inductance geometry. The suggested specification items for the bypass resistor:

- Capacitor part: all specification items that are used for bypass capacitors, except the capacitance, which is always the maximum available for the particular case style and material
- Resistor part: all specification items that usually go with a resistor

The available resistance values should match the resistance tolerance and stability. For a 30 percent resistance tolerance, the E3 series is adequate: 1, 2.2, 4.7, 10. For a 20 percent resistance tolerance, the E6 series is suitable: 1, 1.5, 2.2, 3.3, 4.7, 6.8, 10.

Assuming 20 percent tolerance (E6), the 10 mOhm to 10 Ohm range can be covered with just 19 ESR values for each case size. As a comparison, the ±10 percent E12 series covers the four decades of 100 pF to 1μF range with a total of 49 entries.

Conclusion

It has been shown that the adaptive voltage positioning, extended adaptive voltage positioning, and dissipative edge termination concepts can be merged to provide a user-defined optimum PDN impedance profile over a wide band of frequencies. Distributed matched bypassing requires low Q ($Q<1$) bypass capacitor elements. Two solutions and their implementations are described how to set the ESR of bypass capacitors. It is also shown that for DMB applications, the BQF is a useful representation of the expected capacitor performance. The BR concept is introduced. Simulated and measured results on test boards and module cards illustrate the feasibility of DMB.

Acknowledgments

The authors would like to acknowledge the support and contribution of the following people: Paul Baker, Karl Sauter, Merle Tetreault, Michael Freda, Ram Kunda, Marc Foodman, Paul Sorkin, and Sreemala Pannala of SUN Microsystems; George Dudnikov and Greg Schroeder of Sanmina Corporation; and John Galvagni of AVX Corporation.

References

[1] Charbonneau, *An Overview of the NCMS Embedded Capacitance Project*, NCMS Embedded Capacitance Conf., Tempe, Arizona, February 28-29, 2000.

[2] Garben, McAllister, *Novel Methodology for Mid-Frequency Delta-I Noise Analysis of Complex Computer System Boards and Verification by Measurements*, Proc. of EPEP conference, October 23-25, 2000, Scottsdale, Arizona.

[3] L. Smith et al., *Power Distribution System Design Methodology and Capacitor Selection*, IEEE Trans. Adv. Packag. , Vol. 22, No. 3, August 1999, pp. 284-291.

[4] Hubing et al., *Power Bus Decoupling on Multilayer Printed Circuit Boards*, IEEE Trans. Electromagnetic

Compat., Vol. 37, No. 2. May 1995.
[5] Eged, Balogh, *Analytical Calculation of the Impedance of Lossy Power/Ground Planes*, Proceedings of the Instrumentation and Measurement Technology Conference, May 21-23, 2001, Budapest, Hungary.
[6] Carver, Mink, *Microstrip Antenna Technology*, IEEE Transactions on Antennas and Propagation, AP-29, 1981, pp. 2-24.
[7] Morris et al., *AC–Coupled Termination of a Printed Circuit Board Power Plane in its Characteristic Impedance*, U.S. Patent 5,708,400, Jan. 13, 1998.
[8] Zeef, Hubing, *Reducing Power Bus Impedance at Resonance with Lossy Components*, Proceedings of EPEP2001, October 28-31, 2001, Boston, Massachusetts.
[9] Brooks, *ESR and Bypass Capacitor Self Resonant Behavior—How to Select Bypass Caps*, www.ultracad.com.
[10] Redl et al., *Voltage Regulator Compensation Circuit and Method*, U.S. Patent # 6,229,292.
[11] Waizman, Chung, *Extended Adaptive Voltage Positioning (EAVP)*, Proc. of EPEP conference, October 23-25, 2000, Scottsdale, Arizona.
[12] Novák, *Reducing Simultaneous Switching Noise and EMI on Ground/Power Planes by Dissipative Edge Termination*, IEEE Tr. CPMT, 22, No. 3, pp. 274-283, August 1999.
[13] Archambeault, *Power Ground Reference Plane Decoupling Analysis of Design Alternatives Using Measurements and Simulations*, Proceedings of the 2001 IEEE EMC Symposium, August 13-17, 2001, Montreal, Canada.
[14] Waizman, Chung, *Package Capacitors Impact on Microprocessor Maximum Operating Frequency*, Proceedings of the 51st Electronic Components and Technology Conference, May 29-June 1, 2001, Orlando, Florida.
[15] Peterson et al., *Investigation of Power/Ground Plane Resonance Reduction Using Lumped RC Elements*, Proceedings of ECTC2000, May 21-24, 2000, Las Vegas, Nevada.
[16] Y. L. Li, *Distributed Models for Multi-Terminal Capacitors—Using 2D Lossy Transmission-Line Approach*, Proceedings of the 51st Electronic Components and Technology Conference, May 29-June 1, 2001, Orlando, Florida.
[17] Smith, Hockanson, *Distributed SPICE Circuit Model for Ceramic Capacitors*, Proceedings of the 51st Electronic Components and Technology Conference, May 29-June 1, 2001, Orlando, Florida.
[18] Kim, Swaminathan, *Analysis of Multi-Layered Irregular Distribution Planes with Vias Using Transmission Matrix Method*, Proceedings of EPEP2001, October 28-31, 2001, Boston, Massachusetts.
[19] Mandhana, *Design Oriented Analysis of Package Power Distribution System—Considering Target Impedance for High-Performance Microprocessors*, Proceedings of EPEP2001, October 28-31, 2001, Boston, Massachusetts.
[20] *Capacitor Laminate for Use in Capacitive Printed Circuit Boards and Methods of Manufacture*, U.S. Patent 5,079,069.
[21] *Annular Circuit Components Coupled with Printed Circuit Board Through-Hole*, U.S. Patent 5,708,569.
[22] Prymak, *Advanced Decoupling Using Ceramic MLC Capacitors*, AVX Technical Information, www.avxcorp.com.
[23] Korony et al., *Controlling Capacitor Parasitics for High Frequency Decoupling*, Proceedings of IMAPS2001, 2001 October 9-11, 2001, Baltimore, Maryland.
[24] Galvagni et al., *Method of Forming Thin Film Terminations of Low Inductance Ceramic Capacitors*, U.S. Patent 4,842,318, Aug. 29, 1989.
[25] Novák et al., *Lossy Power Distribution Networks with Thin Dielectric Layers and/or Thin Conductive Layers*, IEEE Tr. CPMT, 23, No. 3, pp. 353-360, August 2000.

Chapter10—Comparison of Power Distribution Network Design Methods, an Approach to System-Level Power Distribution Analysis

Dale Becker, Senior Technical Staff Member, IBM Corp.

This chapter is taken from a paper was presented as part of TF-MP3 *Comparison of Power Distribution Network Methods: Bypass Capacitor Selection Based on Time Domain and Frequency Domain Performances* at DesignCon 2006, Santa Clara, California, February 6-9, 2006.

THE HIGH CURRENT drawn by the integrated circuits in modern system designs create a design challenge for the power distribution of these systems.

The high-frequency circuits, high-leakage circuits, and process and functional variability are creating a current density that presents a challenge for the designer to meet voltage regulation specifications. The direct current (DC) and alternating current (AC) voltage variations are a challenge to control by design and a challenge to verify by measurement. I will present one approach to the power distribution system design with today's demands.

Introduction

From the high-end computers to the desktop systems, the need for power and noise control becomes greater as the level of integration and the speed of the operation evolves. In this chapter, I would like to challenge your perspective on how to approach the analysis and design of a power distribution. Where the designer sets its priorities depends on the details of the design, but a good understanding of the design is needed to know where to spend the design and analysis efforts.

In addition, one should note that, traditionally, we split the power distribution from the signal distribution analysis and design. We can no longer afford to do this with the speeds at which our signal interfaces are running.

Complexity of the Power Distribution Network

The complexity of a computing system is quickly increasing. The design and analysis addresses a broader frequency range and the physical design complexity due to greater partitioning of the voltage domains. The increasing level of circuit and function integration with the leakage of the current silicon technology results in an increased DC current demand in a system.

This increased current requires more robust power planes to deliver the current and more sophisticated voltage regulation to manage the voltage variations the circuit terminals will see. Since managing the DC voltage is truly a system-level design, the DC analysis tools are the most advanced of the power distribution tools to incorporate a full system and customized fast solvers of the resistance network. Although the rate of increase of the operation clock frequency is leveling off in microprocessors, it is still increasing and the tools to analyze the high-frequency effects are maturing. The end result is that, with the increased focus on DC analysis and the increasing clock frequency, the system designer has an increasing frequency bandwidth requirement to meet design requirements. To manage the power of the devices, more sophisticated power management schemes are employed, which involve clock gating and sleep modes. The current deltas introduced into the power distribution network (PDN) by these controls ensure that noise is introduced across a full frequency band.

While the clock frequency of the microprocessors may be leveling off, the chip-to-chip interconnect is going through a rapid rate of change. The frequency or bit rate of the signal interconnect is increasing rapidly, and in addition, the number of interconnects per device is increasing and the compliance standards and protocols are changing, requiring the tools and analysis techniques to keep up.

Such quantities as simultaneous switching noise (SSN) and the jitter introduced by the SSN need to be controlled to manage the bit error rate (BER) of the interfaces. This is where the design of the PDN and the signal distribution network meet and no longer can be treated as separate disciplines.

For these reasons, the design and analysis tasks of the power distribution designer become more difficult due to the increasing frequency bandwidth of the switching noise. The physical complexity of the system is also increasing, most strongly because of more voltage domains to optimize the power savings techniques designed into the systems. This physical complexity introduces voltage tolerance requirements between voltage domains and a physical structure that further stresses the modeling capacity of the power distribution tools.

Motivation for Analysis of the Power Distribution

It is important to understand the motivation behind analyzing the response of the PDN. The motivation is to design a power distribution that enables robust and reliable system operation. As difficult as it is to model the physical structure of the power distribution, it is even more difficult to predict the currents in the system that are the source for the switching noise. The amplitudes of changes in current demand from cycle to cycle or across some number of cycles are dependent on the chip architecture performance. The spatial distributions of the current demand on a chip or from chip to chip are also architecture-dependent.

The analysis of the PDN starts by understanding what power distribution disturbances will interrupt the robust operation of the system. For example, one should understand what voltage collapse can be tolerated without causing timing failures of static logic, the amplitude of voltage swings that cause sense amplifiers to not operate as designed or the filtering on the voltage input to a phase-locked loop (PLL) that needs to be done to meet a jitter budget.

A Computer System

To illustrate an approach in designing and analyzing the power distribution, let us consider a high-end server as shown in Figure 10.1. In the middle of the frame are four nodes of 16 processors together with dynamic random access memory (DRAM) and adaptor chips to communicate to the outside world. The four nodes are interconnected by wires in a center board, which also have slots on the back for the voltage regulators providing the logic voltage levels for the nodes. The voltage variation for this system is managed from the voltage regulation plugged into the back of the center board to the chip terminals on the multi-chip modules on the node board in the front.

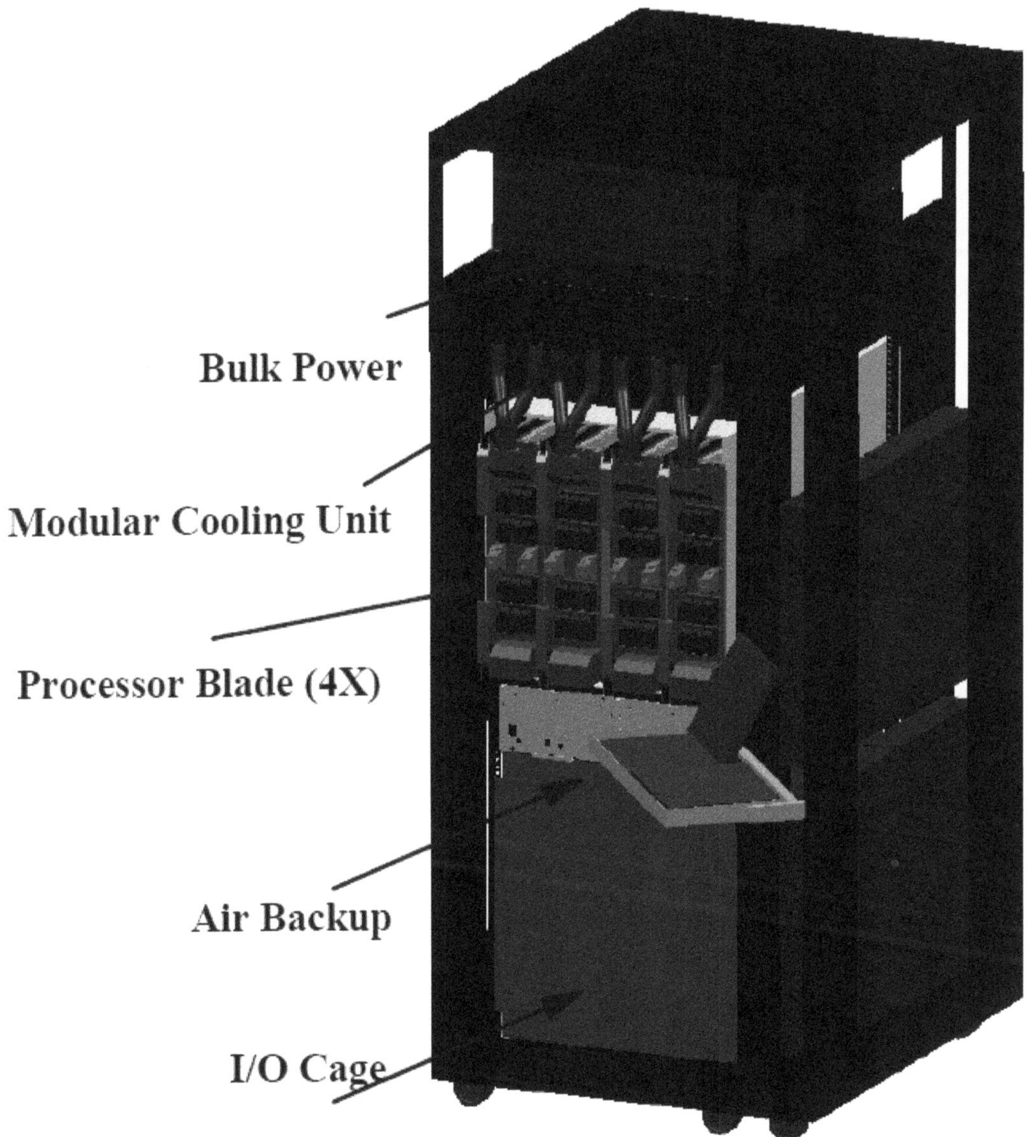

Figure 10.1: An IBM eServer (Source: Winkel, et. al., *First- and Second-Level Packaging of the z990 Processor Cage*, IBM Journal R&D, May/July 2004. www.research.ibm.com/journal/rd/483/winkel.pdf)

Historically, we treated interfaces with static timing and static noise margins. We were able to separate out the timing criteria and noise criteria. Furthermore, we were able to define noise margins based on minimum voltage-level requirements. Over time, this has progressed to the point where we need to consider the noise impact on timing and eye masks, including both timing and noise relationships. The noise margins are not made without knowing the net characteristics.

As an example, let us look at IBM processor board designs over time in Table 10.1. You will see that we have gone from 0.166 Gbps or a 6 ns single data rate cycle time where the transfer happened in the latency of one bit 10 years ago to more than a 10 times reduction in the bit rate today. Then, as the voltages dropped, we needed to contain the switching noise of the chips. Eventually, we discovered that it was not good enough to do a high-level design analysis of the IR drop; we needed to analyze the actual design and use that to determine if we met the voltage tolerance specs. At this time, we are working on incorporating checking techniques that look at the interaction of SSN and the jitter the power distribution disturbances create on the signal interfaces.

Power Distribution Noise Analysis

To illustrate the analysis of the power distribution of such a computer system, let us take a look at one way of partitioning the analysis of the system. By necessity, we break up the analysis into different frequency bands. The DC analysis involves the path from the voltage regulation to the circuit terminals, the largest physical dimensions for any of the analysis.

The mid-frequency noise or noise pulses that are created in the 1-500 MHz range have a board and package model with the decoupling on those package levels.

The high-frequency noise is created by the chip-to-package interaction, requiring a model with a smaller volume representation but much more detailed to reflect the higher frequencies and the local interactions of the circuits.

A fourth set of models is needed to reflect the impact of return path discontinuities on the signal path. These discontinuities used to be the power and ground splits encountered in the routing of the signals. Nowadays, the break in the return path is due to the package or board vias or the component connector pins and the planes where the vias and connectors are terminated. If the via and connector termination do not allow for direct current return paths, this creates a split. Modern commercial tools are able to model these discontinuities and our analysis techniques need to quantify the jitter that can be introduced.

Year	Bit Rate	Evolving Constraints	Evolving Design Verification at tape-out
1996	0.166 Gb/s	Bound Timing Limits AC Noise Budget	Spreadsheet timing SSN determine VDDmin
1998	0.250 Gb/s	ISI In Noise Budget	Multi-pkg level Crosstalk Core Noise included in SSN
2000	0.450 Gb/s	Clock-Data Timing	Noise feedback to timing Spatial Vdd variance spec'ed
2003	0.650 Gb/s	Freq. Dependent Loss	Fast simulation of each Net IR Analysis Required
2005	1.9 Gb/s	Equalization Design	Timing and Noise Combined SSN induced jitter on Gb links

Table 10.1: *A Historical Perspective of Off-Processor Signal Frequency and How the Noise and Timing Margins Were Set* (Source: DesignCon 2004 Panel, Establishing Pass-Fail Criteria for High-Speed Digital Interfaces)

DC Analysis

Figure 10.2 shows part of the processor packaging for a node of our system under consideration. The power is fed from the midplane board to the node board, which holds the multichip module (MCM) and memory.

Once the size of the problem has been determined, the model needs to be created. Figure 10.3 shows the representation of the power distribution gridding needed to get the desired electrical quantities from the model.

One set of the desired electrical quantities is the voltage variation in the system. From this model, the absolute voltage drop from the regulator to the circuit can be determined. The operation of the circuits is more interested in the voltage variation from the regulator set point to the circuit terminals that define the minimum and maximum voltage that the chip designs must tolerate, and the voltage variation between the sending and receiving circuits that impact the noise immunity of chip-to-chip interfaces.

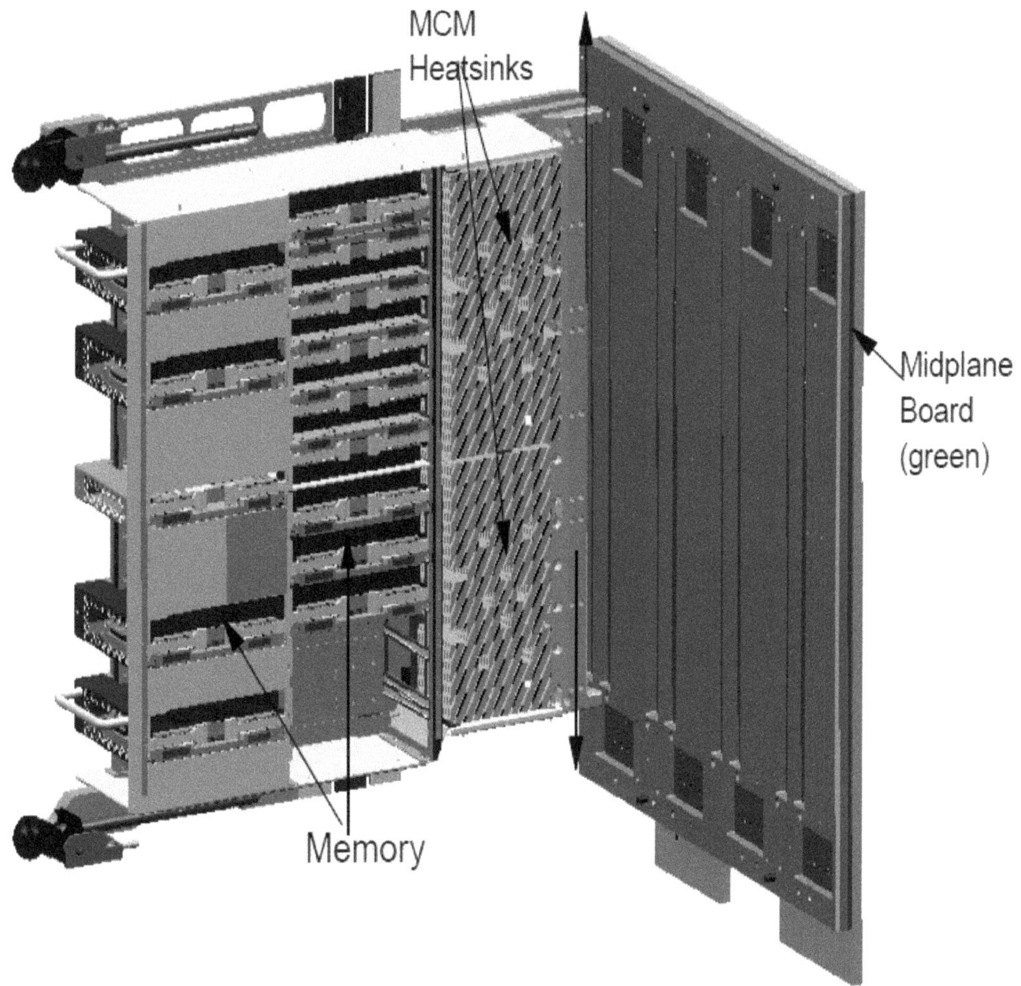

Figure 10.2: *The Midplane Board and Node Board for the DC Power Distribution Mode* (Source: A. Huber et al., *Power Distribution Analysis for IBM eServer System Integration Optimization*, EPEP, October 2004)

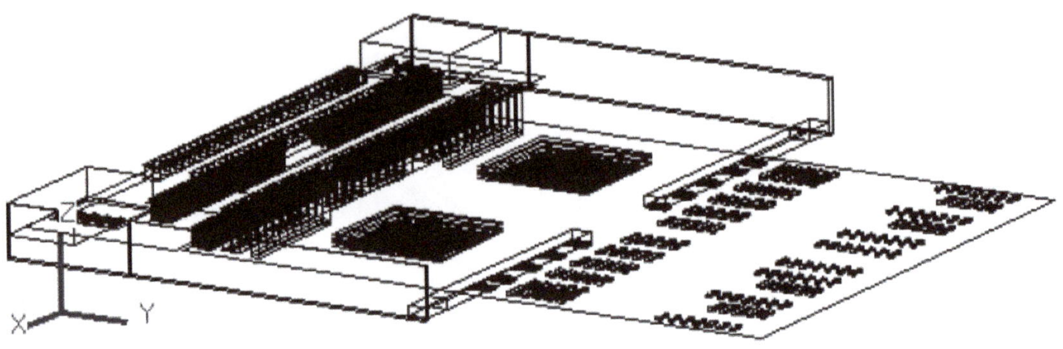

Figure 10.3: The MCM, Node, and Midplane Are Represented in a Model Involving 10,000 Current Sources and 100,000 Elements

The current densities in the system and the local heating and electromigration limits that must be met are actually equally important as the current (amperes) demands of today's systems. The model of Figure 10.3 allows the current plane densities to be modeled and visualized. In addition, the currents through the vias and connectors can be accurately predicted with the large number of elements and current sources that are included in the model.

Figure 10.4 shows a two-dimensional gradient plot of the voltage variation on the processor board and the current density on the board that comes from the DC analysis. Design tradeoffs and decisions can actually be made by focusing on the areas with high voltage drop and high current density.

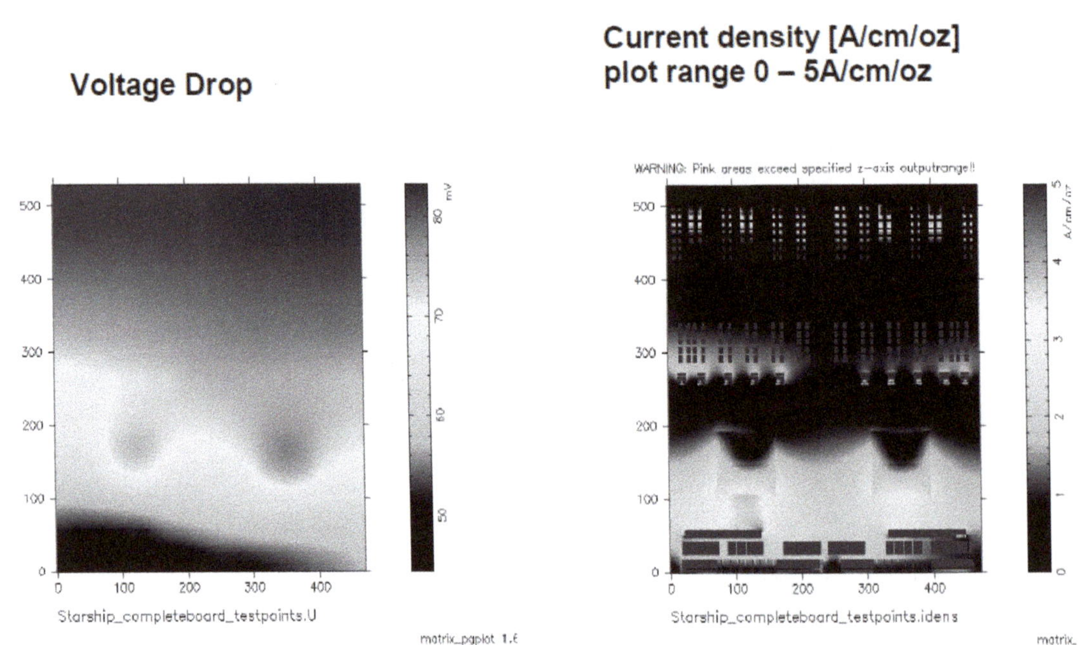

Figure 10.4: The Voltage Gradients (Left) and Current Density Gradient (Right) for One Analysis Involving the Processor Board

Mid-Frequency Analysis

The mid-frequency analysis is directed to understand where the peaks in the power distribution impedance occur and what design and decoupling can be done to manage those peaks. The midfrequency range is roughly from 1 MHz to 500 MHz. The peaks in the frequency impedance response are typically due to loops of the inductance of the package and capacitance of the chip, package, and added decoupling capacitors.

This loop of inductors and capacitors makes a resonant loop with a peak in the impedance at a frequency or frequencies related to the loop parameters. For example, in one of the servers, two primary loops were observed as seen in Figure 10.5. The higher frequency loop is the L1_eff between the on-chip capacitance and the inductance between the on-chip capacitance and the on-module capacitors. The lower frequency loop, L2_eff, is between the on-module capacitors and the board capacitors.

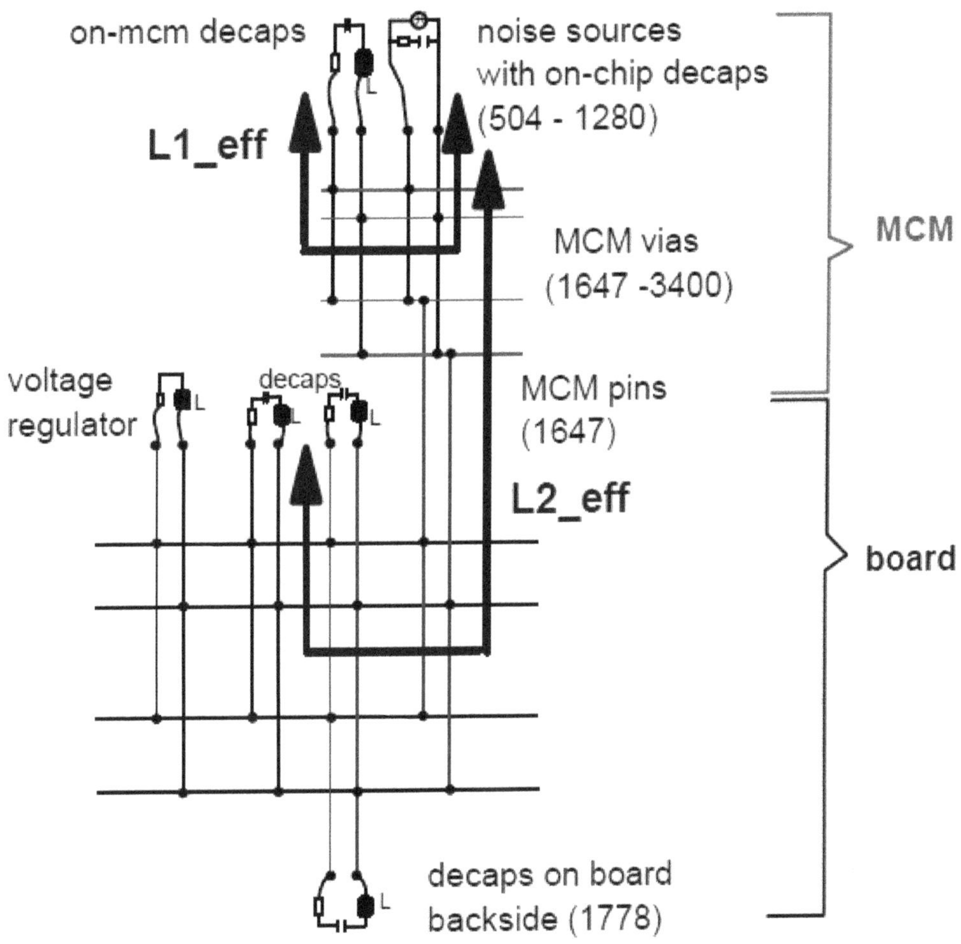

Figure 10.5: The Two Primary Current Loops Creating Peak Impedances in the System (Source: B. Garben et al., "Mid-Frequency Delta-I Noise Analysis of Complex Computer System Boards with Multiprocessor Modules and Verification by Measurements," IEEE Trans. on Adv. Pkg., 294-303, August 2001)

Another set of resonances that needs to be accounted for are the parallel plate ones due to the physical dimensions of the planes.

The resonant modes that are set up depend on both the loading on the plane and the source location.

The tools and techniques to analyze the mid-frequency noise are the most mature of the power distribution modeling tools. The challenge for analysis is that instead of the upper end of the frequency range being approximately a cycle or bit time of the interfaces, it can be 10 bit times or more at current system design points.

High-Frequency Analysis

The high-frequency analysis ideally covers the frequency range from the highest mid-frequency resonance created by the loop between the on-chip capacitors and the off-chip inductance to the frequency content of the edge rates of the current transitions.

Figure 10.6 shows a particular chip in the system where the circuit layout, switching sources, and capacitance distribution are modeled together with the power grid supporting the chip. The multiple power domains together with the multiple clock domains create a complexity that must be included in the analysis. Current analysis tools allow one to model a full chip and visualize the gradient of the voltage variations across the chip as shown in the figure.

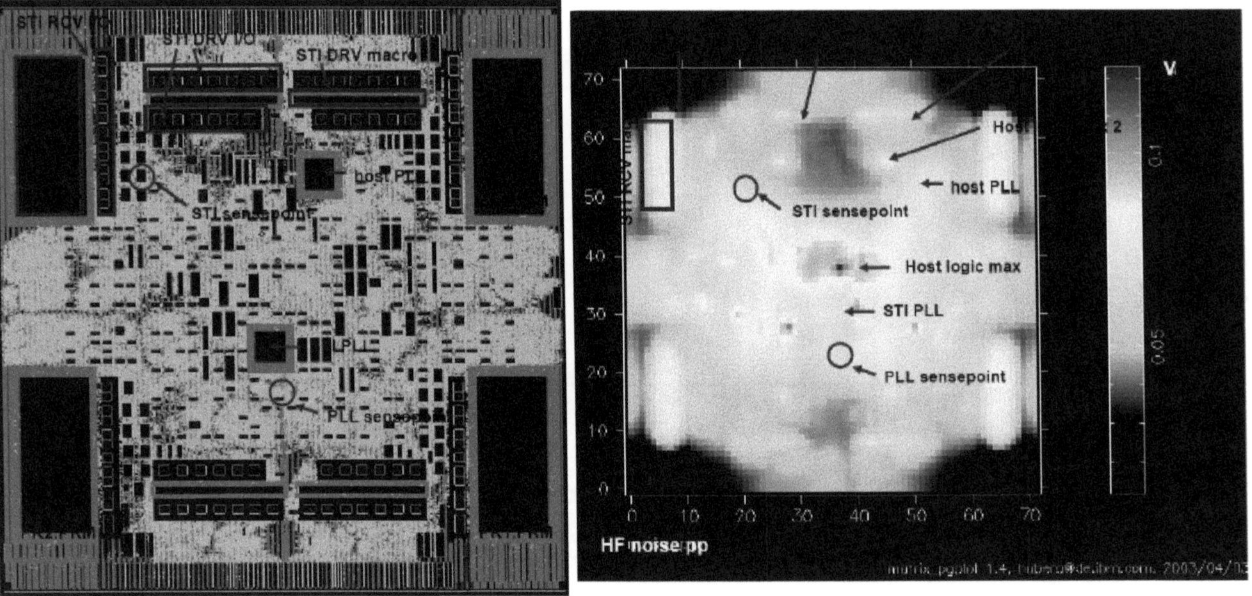

Figure 10.6: The Modeling of On-Chip High Frequency Voltage Variations (Source: A. Huber et al., *Sensitivity Analysis of Generic On-Chip Delta-I Noise Simulation Methodology*, 2004 Workshop on Signal Propagation on Interconnects [SPI 2004])

As higher frequencies are encountered, the high-frequency models will need to include signal interconnect to model the effect of nonideal reference planes on the communication just as we model board-level interconnect nowadays.

This will also lead to the need for better frequency-dependent representation of the resistances, capacitances, and inductances that make up the on-chip power grid.

High-Frequency Signal Return Paths

In the previous section, we discussed signal interaction and power distribution. The signal quality is very dependent on the quality of the return paths for those signals. The impact of power plane splits has been long understood and many papers have been written on modeling the impact of splits.

In today's designs, differential pair signaling is often used, which reduces the impact of discontinuities. In addition, the impact of routing over power plane splits is widely understood by now and usually avoided to preserve the quality of the signal.

In the introduction, the impact of non-ideal termination of vias and connector pins was mentioned. This is illustrated in Figure 10.7 where a via and a connector makes the connection between a trace on card 1 and card 2.

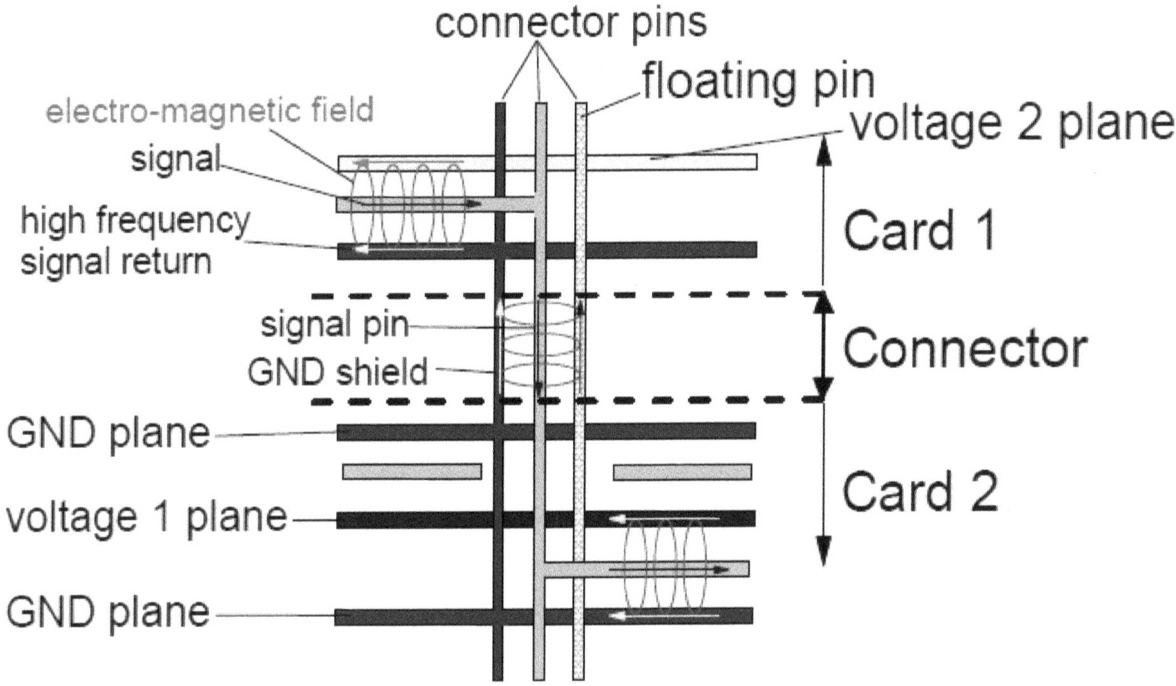

Figure 10.7: An Illustration of a Non-Ideal High-Frequency Return Path on a Connector (Source: Winkel et al., *First- and Second-Level Packaging of the z990 Processor Cage*)

On card 2 in Figure 10.7, the reference planes are of the global information grid (GIG) network defense (GND) potential and connected to a via that also connects to the GND potential in card 1.

However, on card 1, only the second plane denoted voltage 2 is not connected to a via and the return current from the card 1 trace does not have a direct path to follow the signal line and depends on displacement current.

In addition, there is a pin not connected to the reference planes of the trace and for the purposes of the high frequency return path is floating. The impact of this non-ideal path may or may not be tolerable depending on the connection, but the impact needs to be taken into account.

Tools are available to analyze the behavior of these connections.

Summary

We have addressed partitioning the modeling and analysis of a computer system's power distribution network. The DC analysis has become a very critical part of the analysis given the current demand of present integrated circuits. Managing the mid-frequency AC analysis in the MHz frequency range depends on a low-inductance power distribution and a consistent decoupling capacitor placement to manage the resonant loops created by the chips and the PDN.

The high-frequency noise at the chip level is maturing as meeting the on-chip signaling specifications and noise margins becomes increasingly challenging with increasing integration and on-chip interface speeds. The high-frequency return paths are also an important part of power distribution design.

The analysis is done to ensure that the design meets the specifications or objectives to maintain the voltage in an acceptable range that is above a minimum voltage, V_{min}, and below a maximum voltage, V_{max}. The analysis must also ensure that the voltage variations between transmit and receive circuits are maintained at a certain level to achieve the desired BER on the interface.

The jitter quantification introduced on the interface due to power supply variations is also a key part to understand the reliability and robustness of the system variation.

References

[1] Winkel et al., *First- and Second-Level Packaging of the z990 Processor Cage*, IBM Journal R&D, May/July 2004. www.research.ibm.com/journal/rd/483/winkel.pdf.

[2] A. Huber et al., *Power Distribution Analysis for IBM eServer System Integration Optimization*, EPEP, October 2004.

[3] B. Garben et al., *Mid-Frequency Delta-I Noise Analysis of Complex Computer System Boards with Multiprocessor Modules and Verification by Measurements*, IEEE Trans. On Adv. Pkg., 294-303, August 2001.

[4] W. D. Becker et al., *Modeling, Simulation, and Measurement of Mid-Frequency Simultaneous Switching Noise in Computer Systems*, IEEE Trans. Comp., Packaging, and Manuf. Tech. Part B: Advanced Packaging, 157-163, May 1998.

[5] A. Huber et al., *Sensitivity Analysis of Generic On-Chip Delta-I Noise Simulation Methodology*, 2004 SPI.

[6] Weekly et al., *Optimum Design of Power Distribution System via Clock Modulation*, 2003 EPEP.

[7] Budell et al., *Accurate HSPICE Modeling of Arbitrary Package Geometries Using Transmission-Line Equivalent Techniques*, 2003 ECTC.

[8] T. Zhou et al., *On-Chip Circuit Model for Accurate Mid-Frequency Simultaneous Switching Noise Prediction*, 2005 EPEP.

Chapter 11—Bypass Filter Design Considerations for Modern Digital Systems, a Comparative Evaluation of the Big "V," Multipole, and Many Pole Bypass Strategies

Steve Weir, Consultant, Teraspeed Consulting Group

This chapter is a revised version of the paper presented in TF-MP3 *Comparison of Power Distribution Network Methods: Bypass Capacitor Selection Based on Time Domain and Frequency Domain Performances*, DesignCon 2006, Santa Clara, California, February 6-9, 2006.

CERAMIC CAPACITOR BYPASS filter networks may be constructed by any of three well-known methods popularly differentiated by the number of capacitor values used: one, one capacitor per decade, or multiple capacitors per decade. In fact, support of a given method seems to be nearly religious in fervor. We have developed a general model for all three methods that provides insights into the strengths and weaknesses of each method from all: power delivery, signal integrity, electromagnetic compatibility (EMC), and manufacturing perspectives. Our goal is to reduce religious debate to practical choices based on application criteria. Finally, we demonstrate efficient network synthesis for each technique.

What Does the Bypass Network Do?

A bypass filter network applies a shunt across the power rails of sufficiently low impedance to maintain rail voltage in the presence of switching currents. In a modern system, we are concerned with switching currents that range:

1. One-time surge as the power rails initially charge. Large spikes can sustain for a few μs to a millisecond or more as the rails transition through a range that biases complementary metal-oxide semiconductor (CMOS) field effect transistors (FETs) in their linear regions. The overall duration is independent of rise-time, which can be in the low ns.
2. Repetitive surges as power-managed devices enable or disable large functional blocks.
3. Pulsating core currents associated with large state machine or memory block operations.

4. Pulsating input/output (I/O) currents associated with signaling.
5. Where supply-rail (VDD) planes act as signal return image planes, bridges signal switching currents.
6. The ceramic capacitors in an application provide the necessary shunt impedance from a cut-off point where the voltage regulator module impedance rises above the system target, and either:

   1. For systems that do not use bypass capacitors to bridge signal return currents between planes—a point somewhat above the package low-pass filter cut-off of the integrated circuits (ICs)
   2. For systems that do use bypass capacitors to bridge signal return currents between planes—the lesser of:

      - Signal knee frequency
      - Parallel resonant frequency (PRF) where plane cavity net capacitive reactance crosses the discrete capacitor network net inductive reactance

IC package power low-pass cut-off rarely extends above 100 MHz and is fundamentally limited by package size. However, signal currents routinely exceed 1 GHz today and are headed up. It is reasonable to question the prudence of relying on the bypass capacitors and plane cavities to convey image return currents between multiple reference planes.

Despite the low-pass nature of IC packages, massive modern IC currents still translate to very significant currents between package and printed circuit board (PCB) at high frequency. These currents can become real EMC headaches when coupled onto networks. Despite the low-pass nature of IC packages, massive modern IC currents still translate to very significant currents between package and printed circuit board (PCB) at high frequency.

These currents can become real EMC headaches when coupled onto networks and/or structures that resonate or just provide high impedance (read: efficient antenna) near the excitation frequency.

Multilayer Chip Capacitors Bypass Basics

Capacitor impedance magnitude follows a familiar "V" shape. Impedance falls from a theoretical value of infinity at DC to a minimum representative of the equivalent series resistance (ESR) at device-mounted self-resonant frequency where the capacitive reactance and inductive reactance are equal.

At higher frequencies, the inductive reactance dominates.

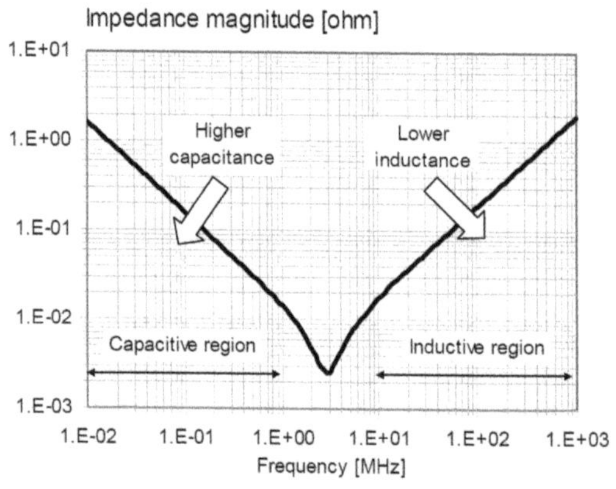

Figure 11.1: Multilayer Chip Capacitors Impedance versus Frequency (Lumped Model)

Multilayer chip capacitors (MLCCs) typically span capacitance ranges of 1000:1 or more in a given case size. Typically available parts in 6.3V ratings:

- 0402 100 pF to 470 nF
- 0603 180 pF to 2.2 μF
- 0805 180 pF to 10 μF

Within any case size and dielectric composition from a given manufacturer, device ESR follows capacitance as:

$$ESR = K1\ x\ C^{K2}$$

...where -0.5 < K2 < -0.3.

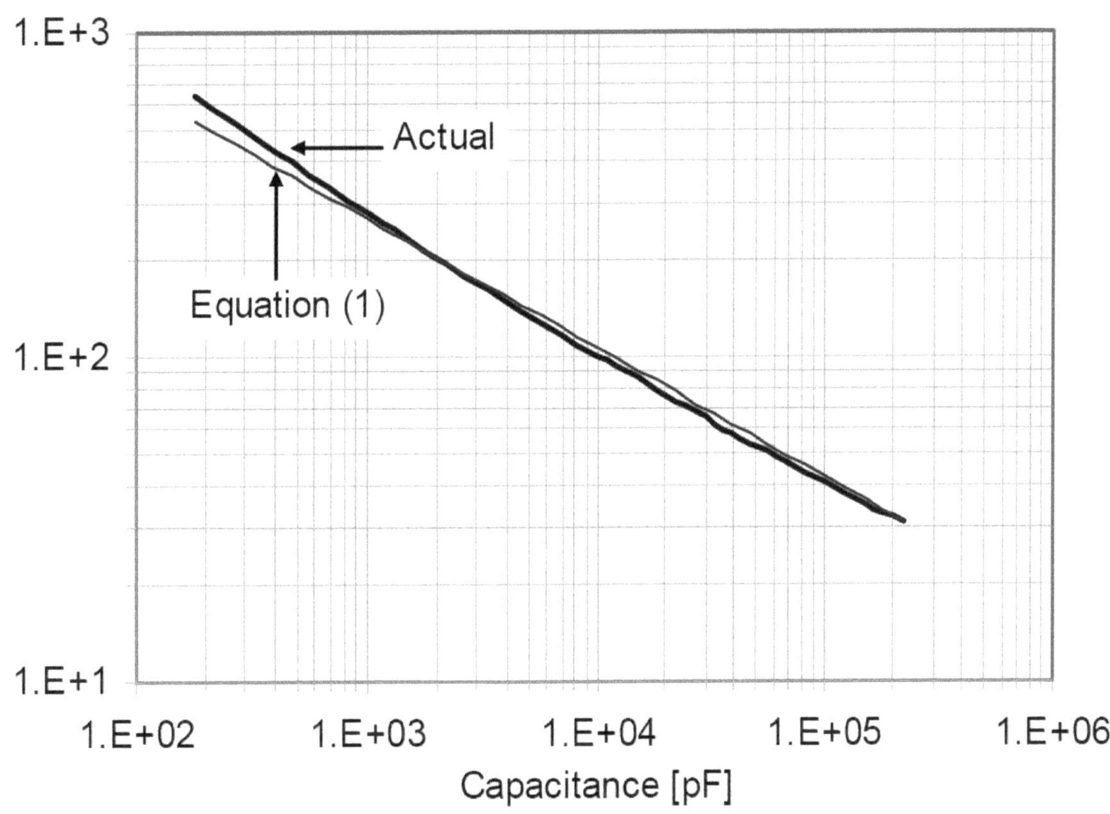

Figure 11.2: ESR versus Capacitance, AVX 0603 X7R Values from SpiCap 3.0™ Shown. SpiCap 3.0 ©2003 AVX Corporation

Parallel Resonance between Capacitor Networks

The phase angle difference between the two networks determines the severity of impedance peaking clearly at the intersection of the two impedance magnitudes networks. This is of concern for MLCC bypass networks in three regions.

- Transition from the voltage regulator module (VRM)/bulk capacitor network to MLCC network
- Transition from the MLCC network to board cavity capacitance
- With multipole MLCC networks: transitions between each capacitor value network

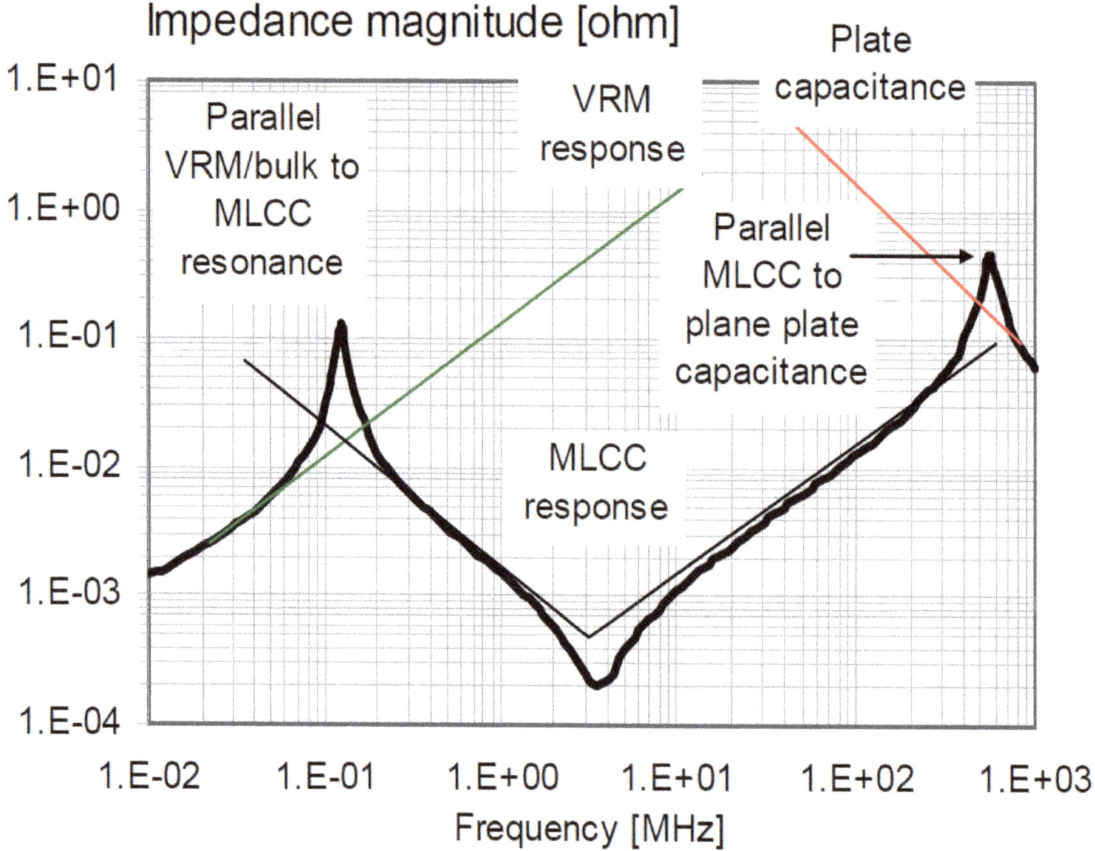

Figure 11.3: Bypass Network Impedance versus Frequency

Due to a series of capacitors with widely spaced pole pairs, the impedance magnitude crossover points occur at near 180 degrees phase difference, resulting in the following relationships:

$$\omega_{PRF} \approx \frac{1}{\sqrt{L_{LF\_NET} C_{HF\_NET}}}$$

$$|Z_{LPRF}| \approx \omega_{PRF} L_{LF\_NET} = \sqrt{\frac{L_{LF\_NET}}{C_{HF\_NET}}} = |Z_{CHAR}|$$

$$Q \approx \frac{|Z_{LPRF}|}{ESR_{HF\_NET} + ESR_{LF\_NET}}$$

$$|Z_{PEAK}| \approx Z_{CHAR}Q \approx \frac{|Z_{CHAR}|^2}{ESR_{HF\_NET} + ESR_{LF\_NET}}$$

If the capacitance of a given network is set as the sum of a capacitor group, i.e., $C_{HF\_NET} = C_{HF\_CAP} * N$ then:

$$ESR_{HF\_NET} \approx \frac{ESR_{HF\_CAP}}{N}$$

$$|Z_{CHAR}| \approx \sqrt{N\frac{L_{LF\_NET}}{C_{HF\_CAP}}}$$

$$Q \approx \frac{\sqrt{N\frac{L_{LF\_NET}}{C_{HF\_CAP}}}}{ESR_{HF\_CAP} + N\,ESR_{LF\_NET}}$$

$$|Z_{PEAK}| \approx |Z_{CHAR}|Q \approx \frac{L_{LF\_NET}}{C_{HF\_CAP}(ESR_{HF\_CAP} + N\,ESR_{LF\_NET})}$$

$$|Z_{PEAK}| \approx \frac{L_{LF\_NET}}{K1C_{HF\_CAP}^{(1+K2)}}$$

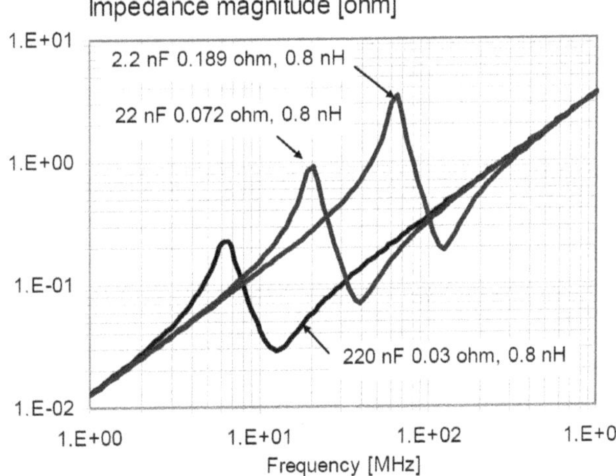

Figure 11.4: Example Parallel Resonance of a Capacitor with a 2 nH Inductor

Since K1 and K2 are constant values that depend on the capacitor manufacturing process, and LLF_NET is set by the lower-frequency network, the |ZPEAK| equation implies that, in order to contain peak impedance, we should seek the largest capacitance value in a given package and geometry combination. Figure 11.4 illustrates this principle as it applies to a single capacitor. Figure 11.5 shows that scaling capacitor quantity provides no improvement to |ZPEAK| when the result fails to improve phase margin.

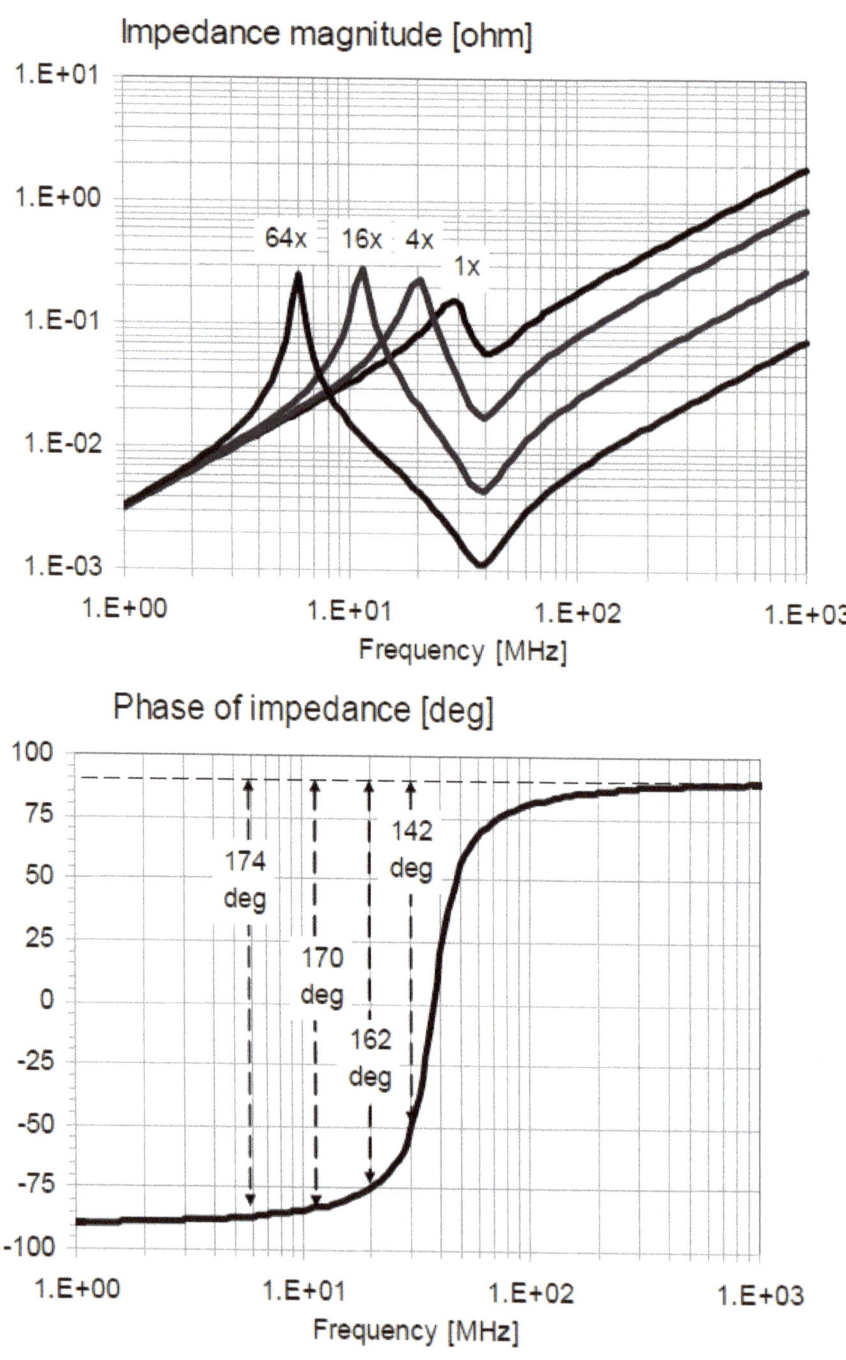

Figure 11.5: Example Peak |Z| versus Capacitor Count, Low Phase Margin. One, Four, 16, and 46 Identical Capacitors of 22 nF 0.072 ohm 0.8 nH Are Connected in Parallel to a 0.5 nH VRM/Bulk Inductance

The |ZPEAK| equation also reinforces the fact that our network impedance is driven by inductance—inductance of all the elements in the network—not just the high-frequency capacitors. At equal cost, capacitors with lower mounted inductance afford benefits across the frequency spectrum.

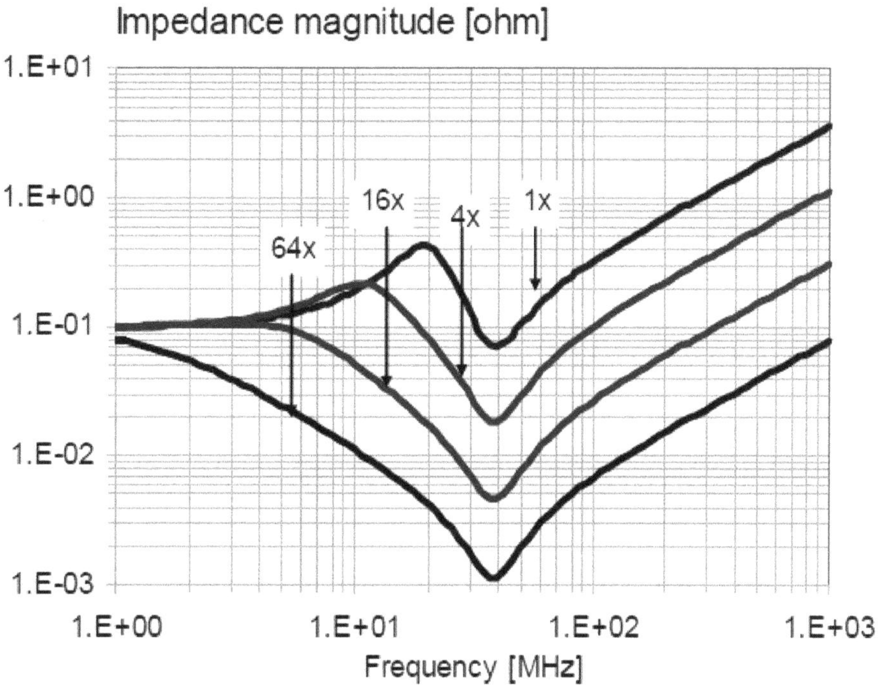

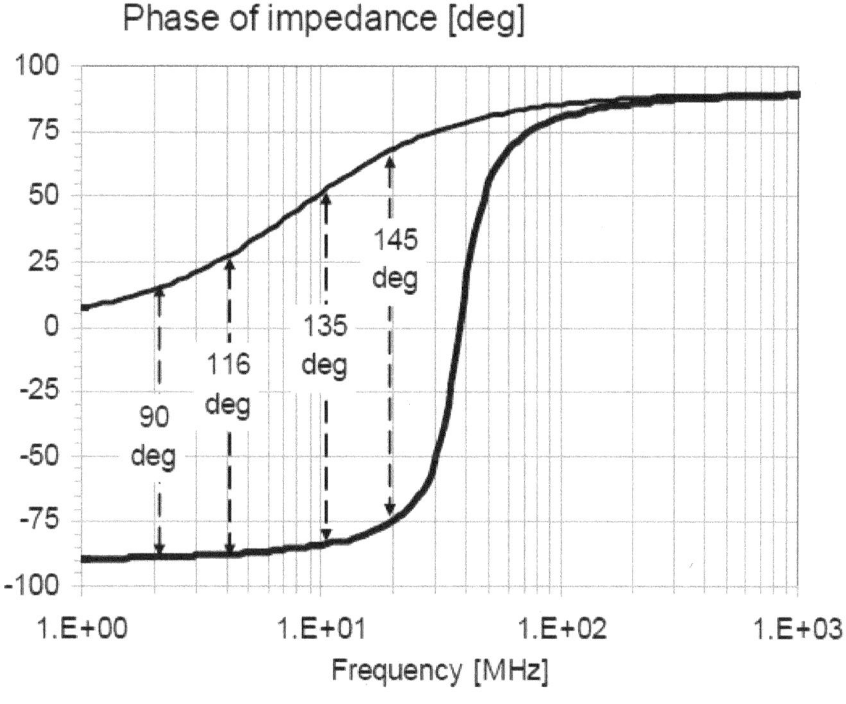

Figure 11.6: Example Peak |Z| with 0.1 ohm 2 nH R-L Impedance versus Capacitor Count of 22 nF 0.072 ohm 800 nH MLCC, High to Low Phase Margin

Figure 11.5 illustrates that when the impedance magnitudes intersect with a phase difference close to 180 degrees, that peak impedance remains essentially constant, while Q increases markedly with increasing capacitor count. Figure 11.6 illustrates that we can bring parallel resonant peaking under control by establishing the magnitude crossings with substantial phase margin versus 180 degrees.

In order to manage peak impedance, we need to provide phase margin at the impedance magnitude intersections. Figure 11.6 illustrates the tremendous difference phase margin makes.

In this example, the 0.1 Ohm resistor limits the phase shift of the L/R network until the 10 MHz region. So long as we cross impedance magnitude with at least 45 degrees phase margin, impedance peaking will be limited to 25 percent or less of |Z| or either network alone at the transition.

Bypass Strategies—Three Methods, Three Faiths?

Today, the following three methods of bypass strategy may be considered popular:

- Finely spaced multipole (SUN Microsystems, Larry Smith et al., flat response)
- Coarsely spaced multipole ("capacitors by the decade")
- Single value ceramic capacitor (big "V")

Multipole Filters, F=K*X$_N$ versus Single Value Big "V"

Beginning in the late 1990s, researchers at SUN Microsystems published a series of papers [1] on a novel method of bypass network design that concatenates many tightly spaced filters. The method emphasizes derivation of high-frequency network performance dictated more by ESR than by ESL.

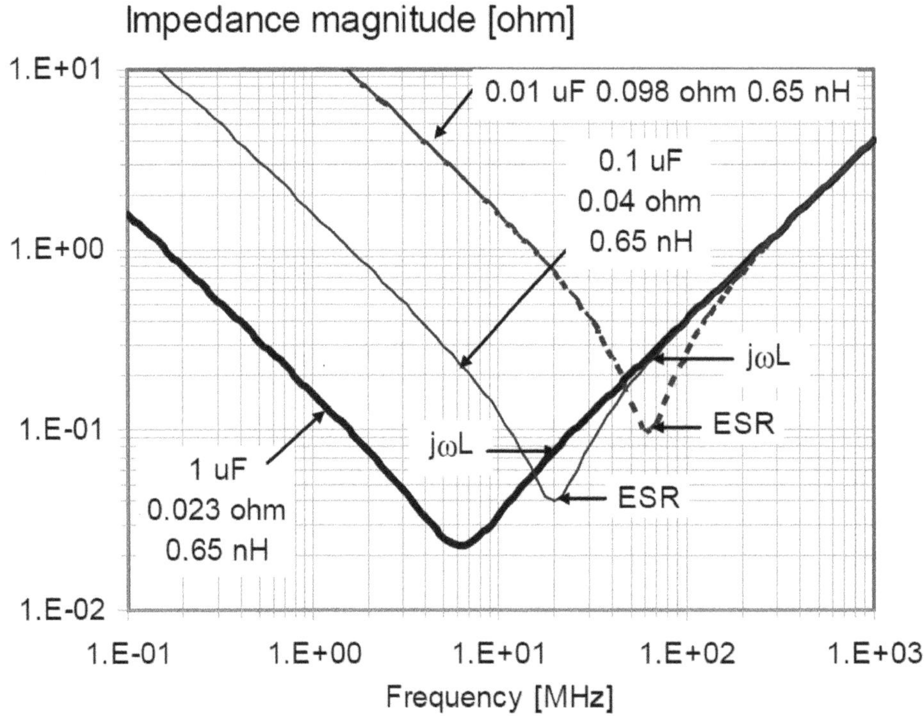

Figure 11.7: ESR Impedance at SRF versus Mounted Inductance |Z|=jwL is Fixed Significantly Beyond Capacitor SRF

As can be seen from Figure 11.7, MLCC capacitors, particularly in smaller capacitance values, exhibit high *Qs*: the ESR at the self-resonant frequency (SRF) for each capacitor is much lower than jωL. Sun researchers reasoned that, given a particular impedance target and bandwidth, one should be able to capitalize on the high Q of these devices and develop networks dominated by the ESR value rather than jωL. The resulting networks should have the following desirable characteristics when compared to other methods:

- Reduced component count
- Higher ESR, which would reduce peaking at parallel resonance with PCB plane cavities
- Faster transient response

The method takes good advantage of low relative phase angles that result with closely spaced SRFs to limit the parallel resonant peaking between networks. If one wished to go to extremes, the entire E12 capacitor series could be used.

However, some balance between benefits of large phase margins and manufacturing complexity is called for. Currently, the method is often advocated with three capacitor values in each decade (e.g., 2.2 nF, 4.7 nF, 10 nF, 22 nF) [6].

In order to maintain a fairly flat impedance floor, the number of capacitors required in each successively smaller capacitance value increases commensurate to ESR. Given a network primarily composed of capacitors all in the same package, and therefore all with the same mounted ESL, the SRF of each network uses the following form:

$$\omega_{SRF(N+1)} = \omega_{SRF(N)}\sqrt{\frac{C_N}{C_{N+1}}}$$

$$X = \sqrt{\frac{C_N}{C_{N+1}}} = 10^{\frac{1}{6}}$$

For example, we have a series of networks each with a self-resonant frequency of $F = K*X^N$.

For the case of three capacitors per decade, X = 1.47. Assuming for a moment that K2 from the $ESR = K1\ x\ C^{K2}$ equation exactly equals -0.5, then we would further have:

$$ESR_{(N+1)} = X\ ESR_{(N)}$$

$$j\omega L_{SRF(N+1)} = j\omega X L_{SRF(N)}$$

...with the immediately observable result that $Q_{(N)}$: $j\omega L_{SRF(N)}/ESR_{(N)}$ is a constant.

The number of capacitors required to hold a constant ESR and a constant |ZCHAR| in each successive value bin then reduces to the following:

$$Count_{(N+1)} = Count_{(N)}\frac{ESR_{(N+1)}}{ESR_N} = Count_{(N)}X$$

Given a capacitor count fixed by frequency, for a given capacitor mounted inductance, case size and chemistry, filter bandwidth simply extends by adding more high-frequency capacitor values in quantities proportionate to the resulting SRFs. The only remaining unknown is how many capacitors to seed into the lowest-frequency bin.

The seed quantity problem on paper is dictated by the extent of loading interaction between the parallel networks. Due to the constant Q, we can show that for K2 = 0.5, a simple constant is sufficient to scale ESR.

This math works independently of the value of X. In the case of the fine-grained networks advocates, X = 1.47, whereas for capacitors on decade values, X = 3.16.

Sample Networks

An illustration of a hypothetical one-dimensional network using fractional capacitor counts, K2 = 0.5, and ESR_{220nF} = 30 mOhms:

Table 11.1 summarizes the network.

Capacitor	ESR [Ω]	Qty.
220 nF	30m	1
100 nF	44.1m	1.47
47 nF	64.8m	2.16
22 nF	95.3m	3.18
10 nF	140m	4.67
4.7 nF	206m	6.86
2.2 nF	300m	10.0
1 nF	441m	14.7
Net	3.75m	44.04

Table 11.1: Example of a Perfect 1.47^N Network

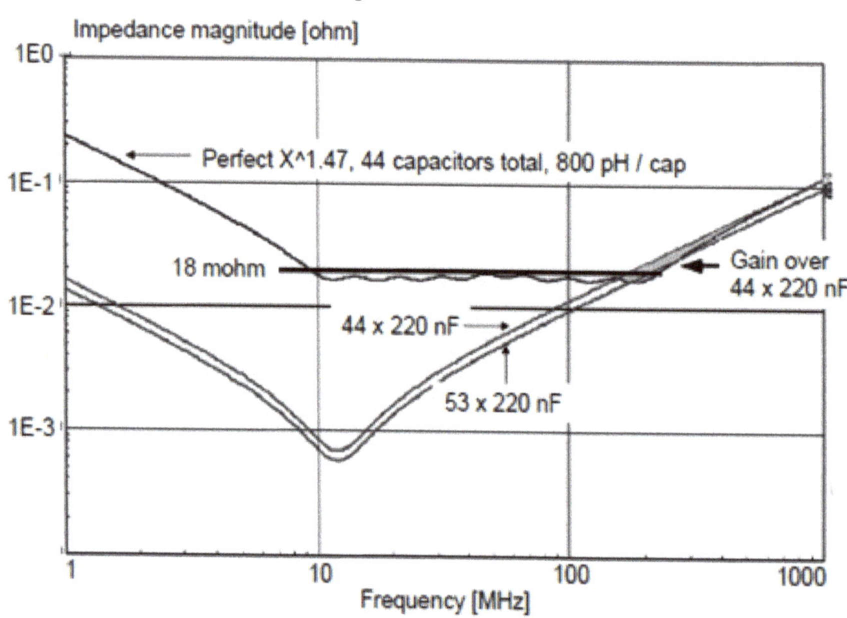

Figure 11.8a: Perfect 1 nF, 2.2 nF, 4.7 nF ... 220 nF Network versus 220 nF big "V"

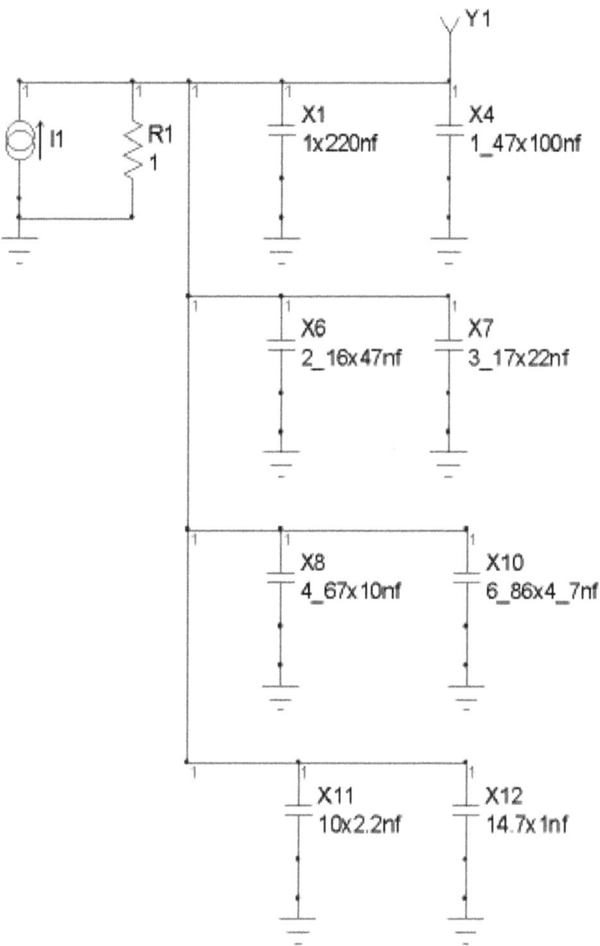

Figure 11.8b: Perfect 1 nF, 2.2 nF, 4.7 nF ... 220 nF Network versus 220 nF big "V"

We see that the impedance floor peaks occur at 18 mOhms—60 percent of the ESR of each parallel capacitor network. The entire parallel ESR is 3.75 mOhms. From 160 MHz to 500 MHz, this network model delivers lower impedance than a network composed only of 44 pieces of 220 nF capacitors.

To reach parity at the higher frequencies, an additional nine capacitors, or approximately 20 percent more, are needed, for a total of 53 using only 220 nF capacitors. In each case, the single-value networks exhibit ESRs only about one-fourth that of complex networks.

Referring to the |ZPEAK| equation:

$$|Z_{PEAK}| \approx Z_{CHAR}Q \approx \frac{|Z_{CHAR}|^2}{ESR_{HF\_NET} + ESR_{LF\_NET}}$$

The higher ESR of this network versus either of the two big "V" networks results in a substantially lower impedance peak at the impedance crossover to the plane cavities for any crossover substantially above the big "V" network SRF. This is perhaps the greatest strength of the fine multipole approach.

This property of higher ESR is potentially of significant importance in systems where the bypass network bridges signal return currents between image planes of both power rails or where coherent signals such as clocks align close to the parallel resonant frequency.

Figure 11.9 illustrates the comparison when plane cavities are included.

But if all of the following three conditions are met:

- IC power low-pass cutoff occurs well before the parallel resonant frequency
- Signals do not excite near the resonant frequency sufficiently to present an EMC issue
- The bypass capacitor network does not comprise a substantial portion of signal return paths

...then under these combined conditions, the higher impedance peaks at the ceramic capacitor to plane cavity transition of big "V" implementations is of no consequence.

Conversely, when either the peak impedance and/or settling time requirements are stringent, the improved ESR of the multipole approach is still woefully inadequate. In these cases other means must be found to either damp the network or push the resonant frequency out of the signal band.

What about capacitors on decade spacing? If we synthesize a network to most closely match the performance of the 1.5N using 3.16N, we get the results shown in Figures 11.9 and 11.10. To match the bandwidth of the one-dimensional multipole filter model, a total of 53 capacitors—the same as for the big "V"—are required.

What we buy for the extra complexity over the big "V," as well as the extra component count over the multipole case, is the best impedance from 110 to 160 MHz, slightly better impedance than the multipole case at the low end, and much better impedance at the parallel resonance with the plane cavity than the big "V" implementations.

However, impedance is markedly higher than the big "V" over a wide range of frequencies, including the low-frequency limit where we need to transition from the VRM/bulk capacitors.

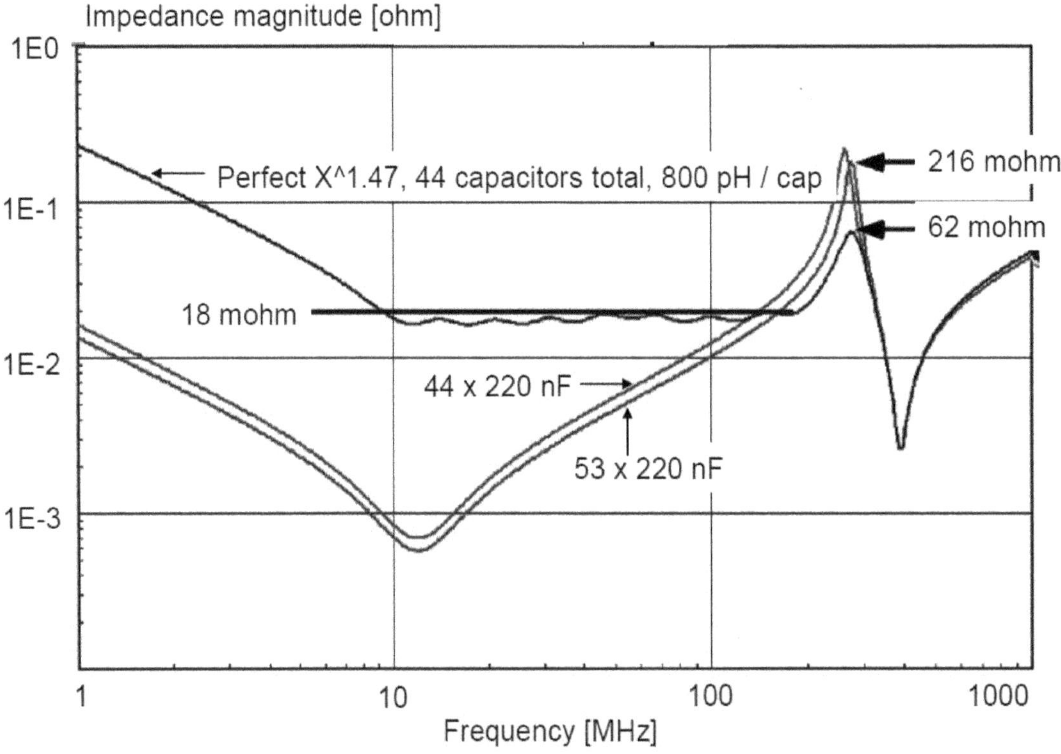

Figure 11.9: Perfect 1.47^N Network versus Big "V," Including 6" x 6" 3-mil Planes

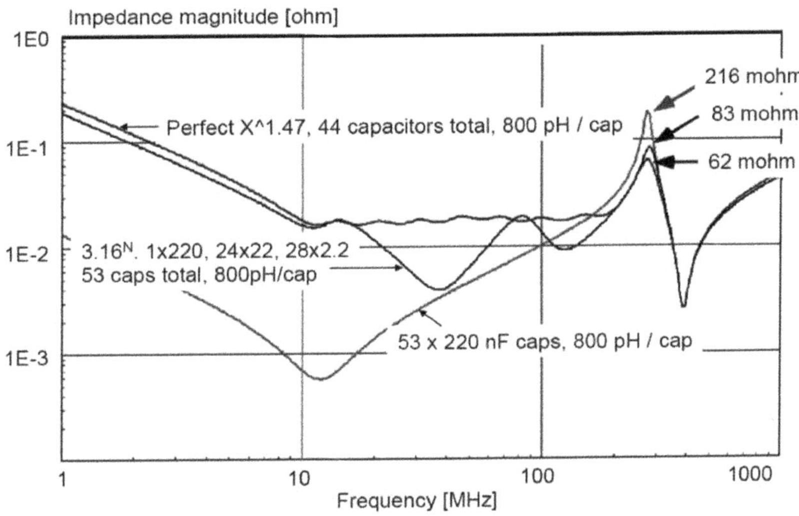

Figure 11.10: Comparison with Decade-Spaced Capacitors, Including 6" x 6" 3-mil Planes

The World in 2D

The foregoing one-dimensional models neglect spatial distribution of the capacitors and plane transmission line effects. Both are very significant and, at high frequencies, they dominate power delivery performance. The component count advantage of fine multipole often fades when we account for spatial distribution of the capacitors and transfer impedance to the ball grid array (BGA) balls. The effect is particularly acute for ICs with center power slugs and/or relatively thick power-plane cavities in the IC package.

Here, we model planes using a 2D bedspring model. A 34 x 34 ball 1 mm pitch BGA attaches at the center of a 4" x 4" 3 mil thick plane represented by the bedspring array. The capacitors are located 0.16" from the outside device balls. We modeled 1.5^N array as hypothetical fractional capacitors evenly distributed around the capacitor periphery.

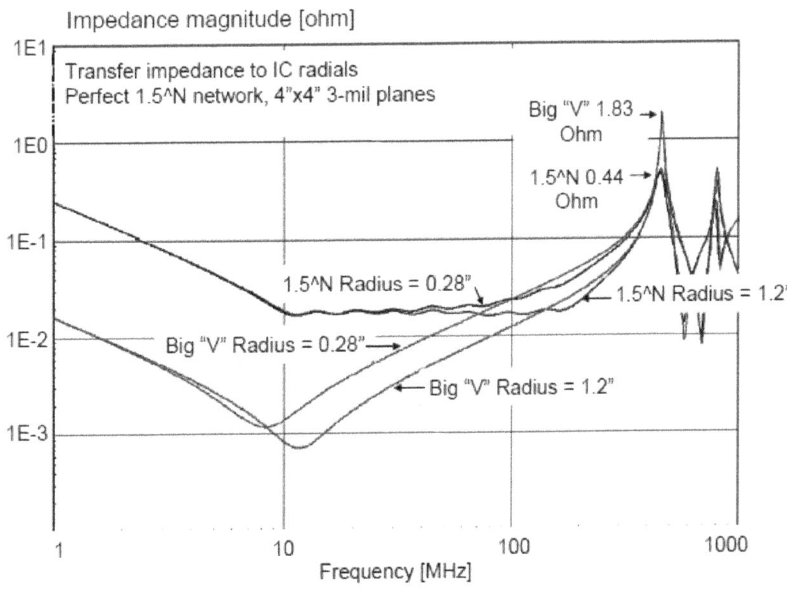

Figure 11.11a: Example 1.5^N versus Big "V" at IC Attachment. Hypothetical Fractional Capacitors Equally Distributed: 44 Capacitors Total versus Big "V" 44 Discrete Capacitors. Logarithmic Vertical Scale.

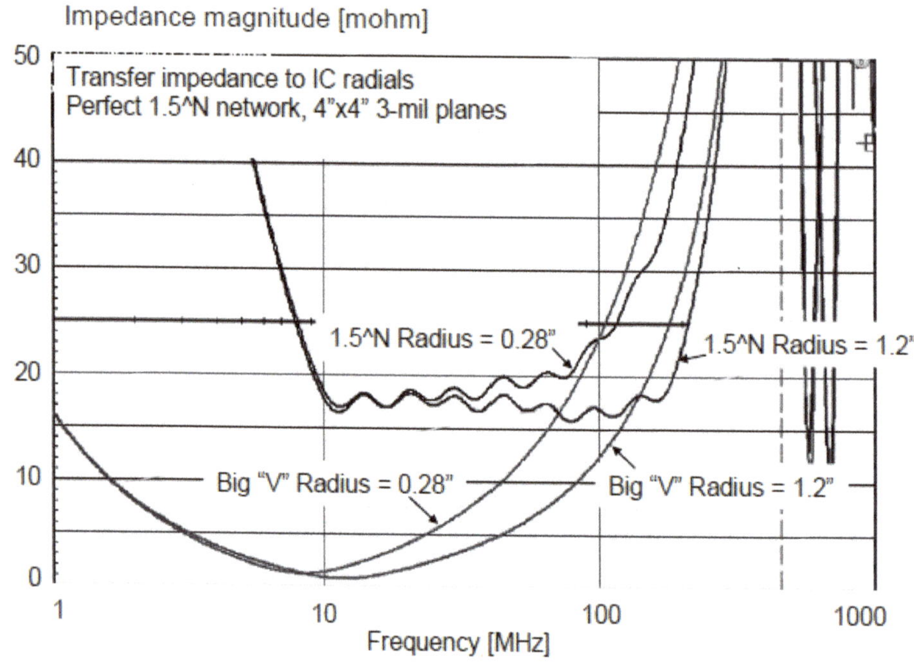

Figure 11.11b: Example 1.5^N versus Big "V" at IC Attachment. Hypothetical Fractional Capacitors Equally Distributed: 44 Capacitors Total versus Big "V" 44 Discrete Capacitors. Linear Vertical Scale

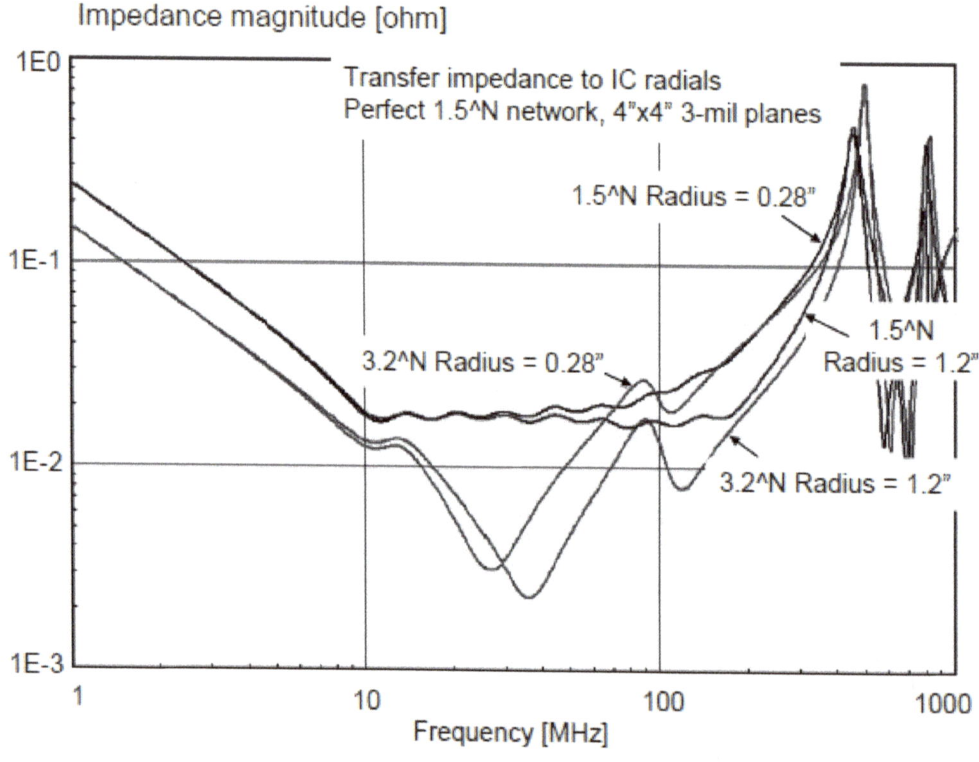

Figure 11.12a: Example 1.5^N versus 3.2^N at IC Attachment. Hypothetical Fractional Capacitors Equally Distributed versus 53-Capacitor 3.2^N Network

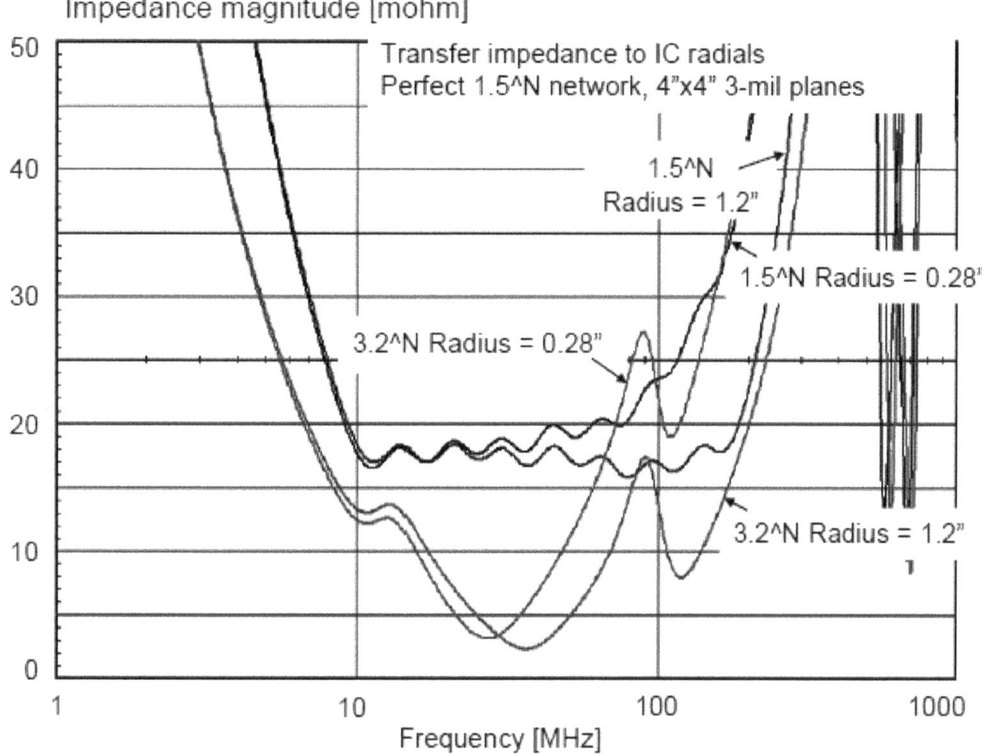

Figure 11.12b: Example 1.5^N versus 3.2^N at IC Attachment. Hypothetical Fractional Capacitors Equally Distributed versus 53-Capacitor 3.2^N Network

The fine multipole approach retains its component efficiency advantage best when used with packages that do not crowd the power connections into a center slug configuration and include very thin planes for internal power distribution to the die.

Common BGA packages with center slug configurations, including the 0.28" radius examples shown in Figure 11.11, diminish the component advantage of fine multipole against big "V" when applied with nominal plane separations. Spreading inductance takes an even bigger toll on X3.2 networks, all but eliminating any advantage over fine multipole, even when using 20 percent more parts.

The 2D models show that fine multipole retains significant advantage in peak impedance, 450 mOhms in this example, at network/plane PRF versus either: big "V," 1.8 Ohms, or X3.2, 800 mOhms.

Transient Responses

Applying a steady stimulus of a 150 MHz trapezoidal waveform with 2 ns Tr/Tf to the networks above, we should expect that the peak amplitudes are similar.

1D simulation bears out that they are as follows.

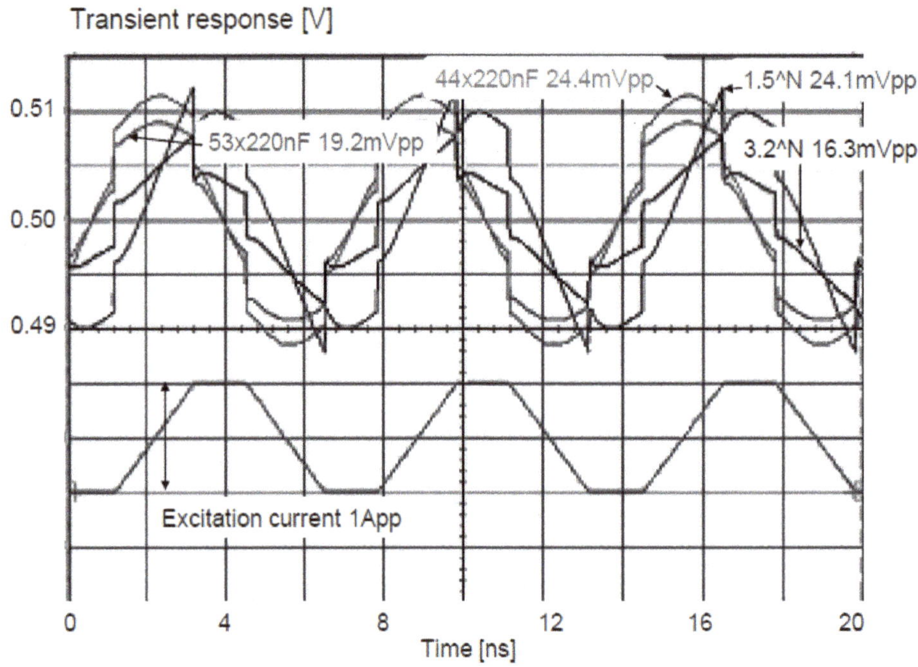

Figure 11.13: 150MHz 2 ns Rise and Fall Time Trapezoid, Steady State Transient Response

Figure 11.13 demonstrates that, for this frequency stimulus, the $X^{3.2}$ network affords the lowest noise levels owing to its lower impedance in this region. The remaining networks remain very close in response as predicted by the impedance magnitude plots.

As a final measure of transient response, we observe the behavior in response to an isolated pulse where Tr and Tf exceed the parallel resonance of the capacitors and the planes, both undamped and damped.

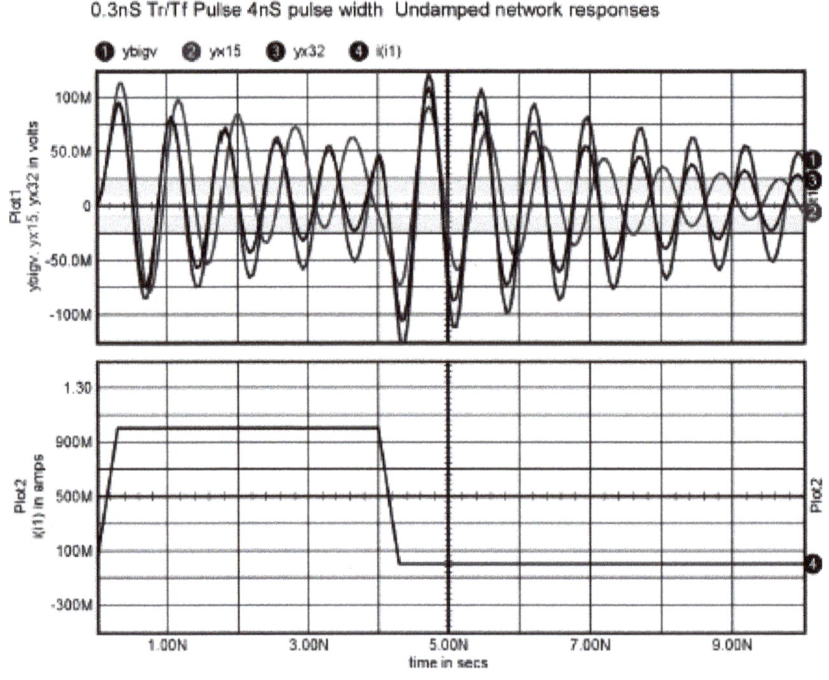

Figure 11.14a: Transient Response, Isolated Pulse, Undamped Network

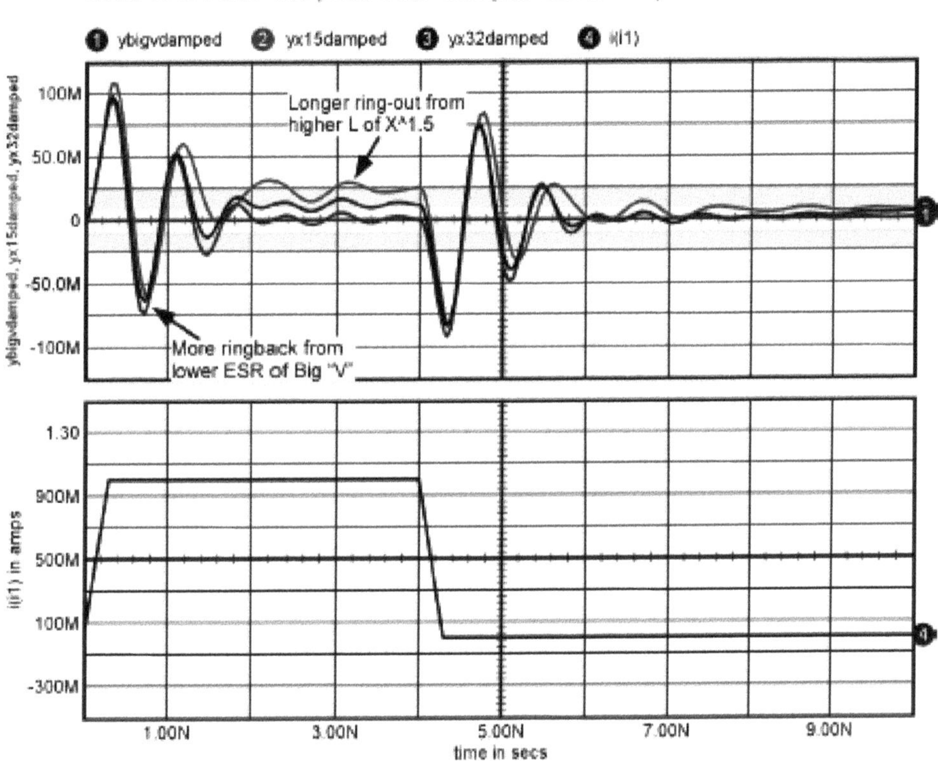

Figure 11.14b: Transient Response, Isolated Pulse, Damped Network

The 1.5^N network exhibits the highest initial amplitude due to the higher inductance from fewer capacitors. However, the big "V" has a longer settling time and higher resonance.

Absent damping, none of the methods as shown are suitable for high data rates. A repetitive signal excitation source such as a "1-0-1-0" DDR2 533 Mbps pattern could excite any of these networks into steady oscillation. The situation is far worse for the big "V" networks than the multipole networks. Clearly, corrective measures to provide damping are required for each of the approaches in fast edge/high signaling rate applications.

The damped results demonstrate a much more suitable picture for I/O supplies. Again, the lower capacitor count of 1.5^N results in a higher initial peak, while the higher ESR results in a higher sustained offset. Whether the offset is an asset or a hindrance depends on whether the next pulse is in the same or opposite direction. The big "V" continues to show slightly higher ringback amplitude due to lower ESR.

However, it has the fastest stable recovery to 25 percent maximum excursion. 3.2^N recovery time is nearly identical.

For I/O rails where power supply bandwidth is wide (e.g., greater than 1GHz), the following is true:

- Peak noise amplitude is determined by the network inductance. The theoretical component gain of the $X^{1.5}$ network does not apply past the reject band floor.
- Undamped, each of the methods rings out for multiple bit periods. Some additional method of damping is required to support high signaling rates.

VRM/Bulk Capacitor Transition

Whereas plane characteristics are fairly easy to define, the VRM/bulk capacitor combination can take on several forms. We concentrate here on the most difficult case, which is a commercial converter using MLCCs as the output capacitors. The resulting device has high Q, which challenges containment of parallel resonance at the transition to the PWB bypass capacitors.

The $|Z_{PEAK}|$ equation dictates that we need substantial phase margin to limit parallel resonant peaking. To achieve this, we need to cross with the converter MLCCs close to their own mounted SRF. This is a parameter that we will most likely have to find by measurement.

Once the parasitics have been determined, we can solve for the left-hand side of our filter response.

Figure 11.15 illustrates an example case where, despite much higher ESRs, the X^N networks peak badly, while the big "V" networks are well behaved. This should make clear that in order to manage peaking, we really are managing the phase margin at the impedance magnitude crossover. Because the X^N networks in this example cross at a much higher frequency where the VRM/bulk phase has reached nearly 90 degrees, the higher ESR of the X^N networks is of little help containing the substantial resonance.

The big "V" works well in this example, because despite being at almost -90 degrees phase the impedance magnitude crosses the VRM/bulk capacitors where they are at a low phase angle.

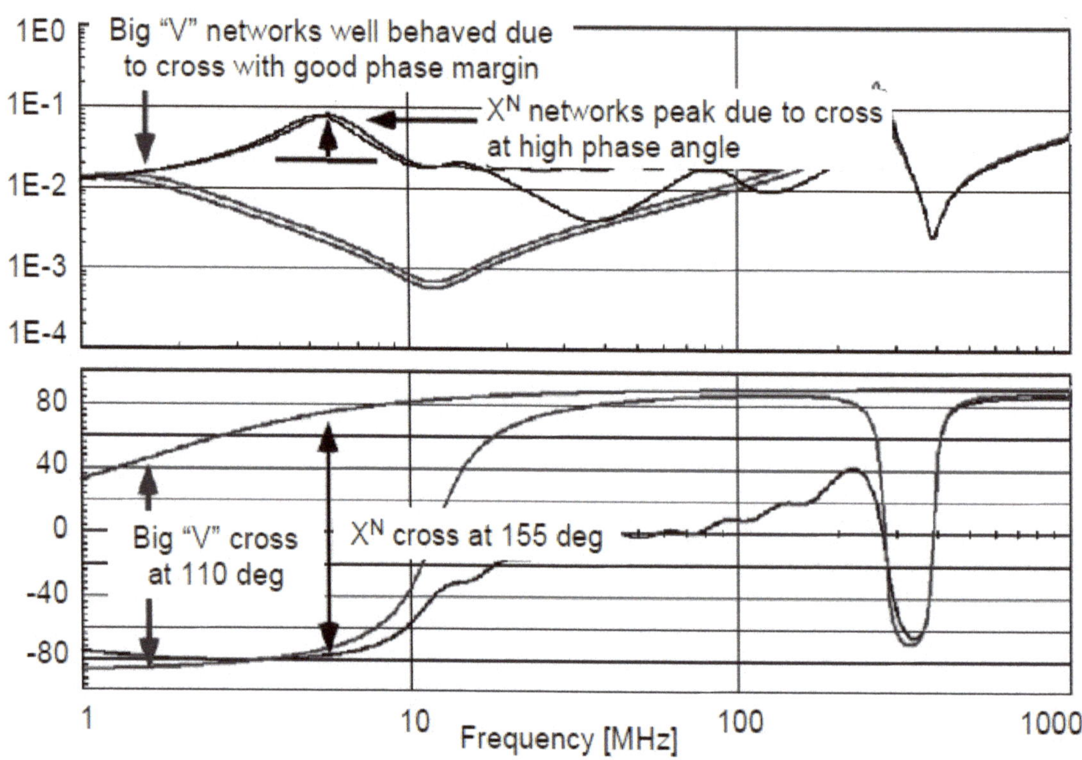

Figure 11.15: Example with VRM Bulk Filter Transition

Given either an X^N or a big "V" network design, and a model for the VRM/bulk network, it is a straightforward matter to check the phase margin and peaking. If sufficient phase margin is not available, then additional, intermediate network(s) will be required to realize an acceptable response.

With the big "V," we have the option of increasing the individual capacitor values until we hit a price point inflection. Currently, this tends to occur at the largest one or two values of capacitance for a given case size, chemistry and voltage rating combination. Since application of the big "V" is built on the premise that parallel resonance with the planes is not a design issue, the sage advice of Dr. Howard Johnson to employ the largest (we suggest the largest value before price inflection) capacitance in a given capacitor case size/chemistry combination is well taken. This approach affords the highest probability that the transition from the VRM/bulk capacitors occurs with sufficient phase margin to meet the impedance profile at little or no cost impact.

For X^N networks, limitations in available capacitor values dictate practical combinations of new filter networks. It is observable that for large-capacitance MLCCs, the component cost is fairly proportionate to capacitance.

Some modest gains in placement and via drilling costs are to be had by using the largest capacitors, e.g., fewest additional poles necessary to satisfy the impedance profile.

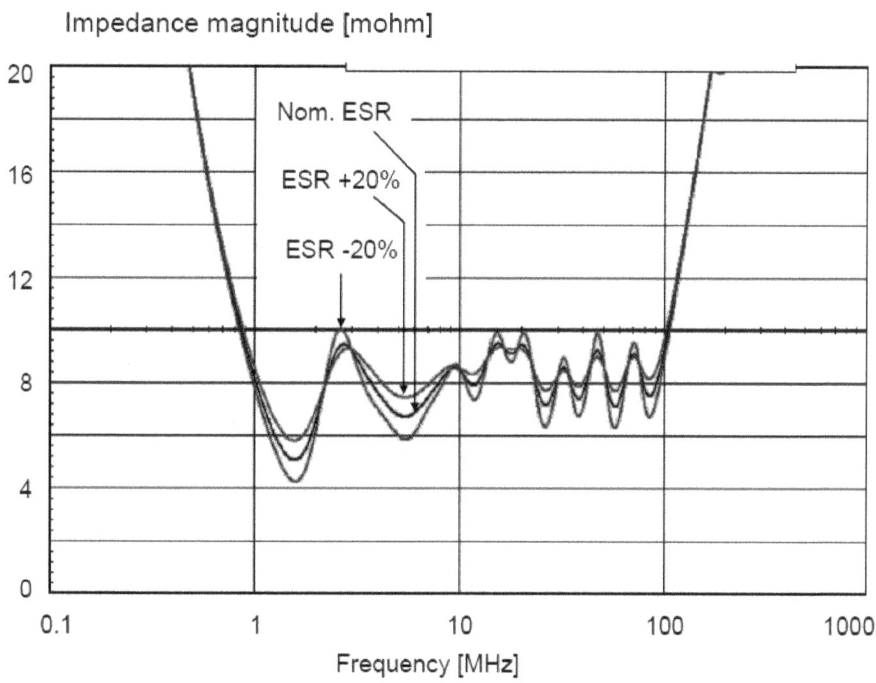

Figure 11.16: Example ESR Sensitivity of 1.5^N Network with Nominal 10 mOhms Target Impedance, 39 Capacitors

X^N networks exhibit increased transfer impedance for lower component ESR values. The sensitivity rises with larger values of X.

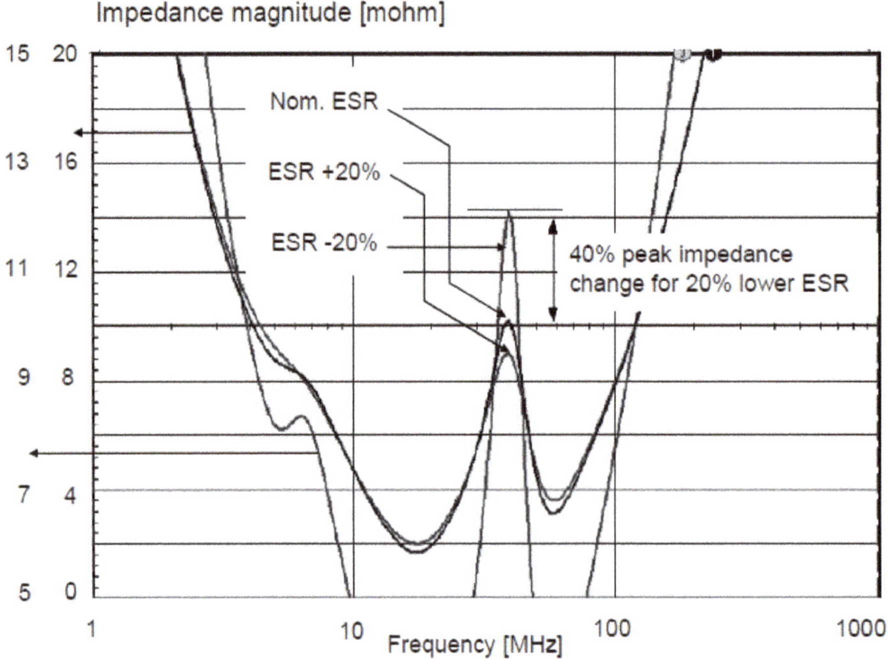

Figure 11.17: Example 3.2^N Network with 53 Capacitors. Note the Different Vertical Scales.

On the other hand, big "V" networks demonstrate little sensitivity to ESR.

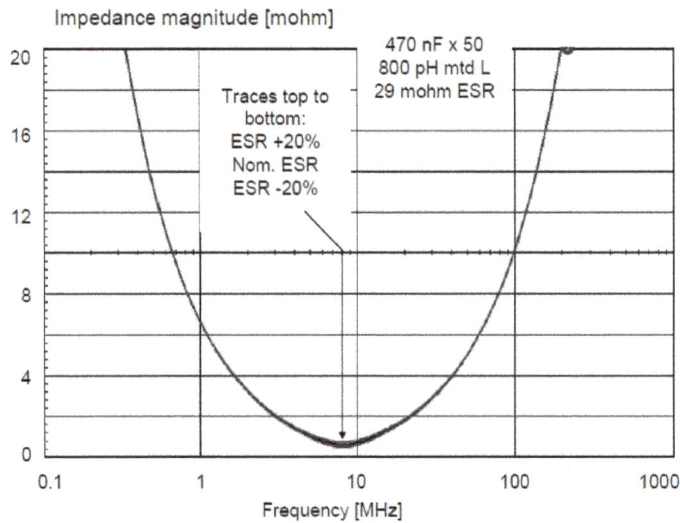

Figure 11.18: *Big "V" ESR Sensitivity*

Mis-Stuffed Component Sensitivity

X^N networks depend upon carefully matched quantities of each capacitor value. It should come as little surprise that minor stuffing errors cause big impedance perturbations.

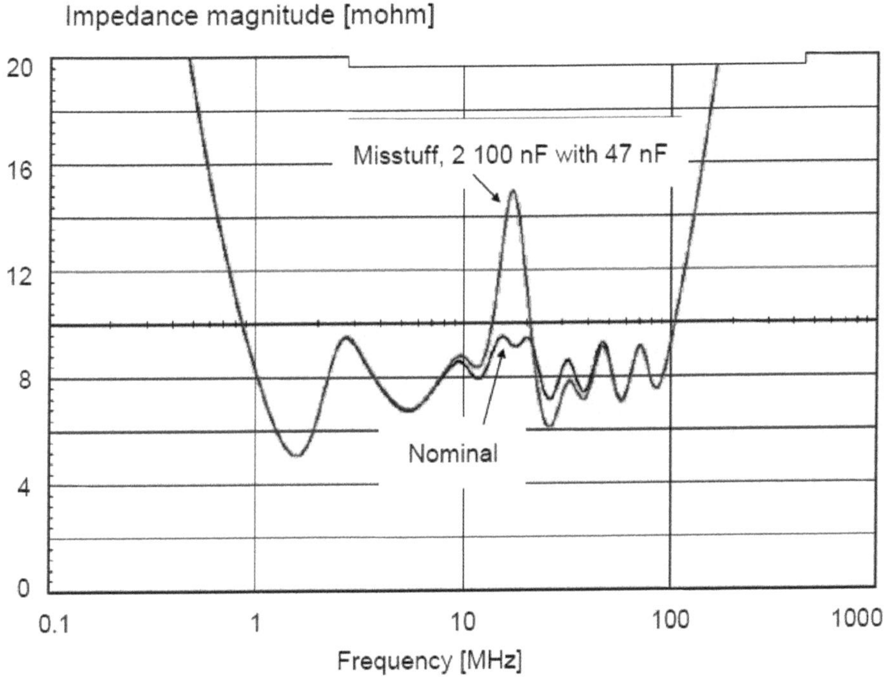

Figure 11.19: 1.5^N with 10 mOhm Target Impedance, 39 Capacitors, 5 Percent Mis-Stuffs

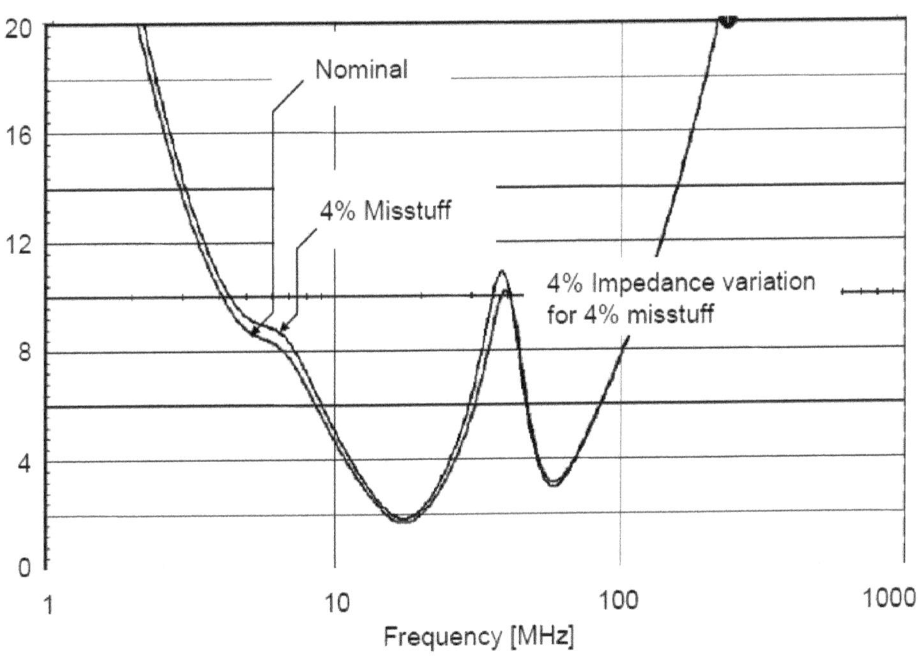

Figure 11.20: 3.2^N 4 Percent Mis-Stuffs

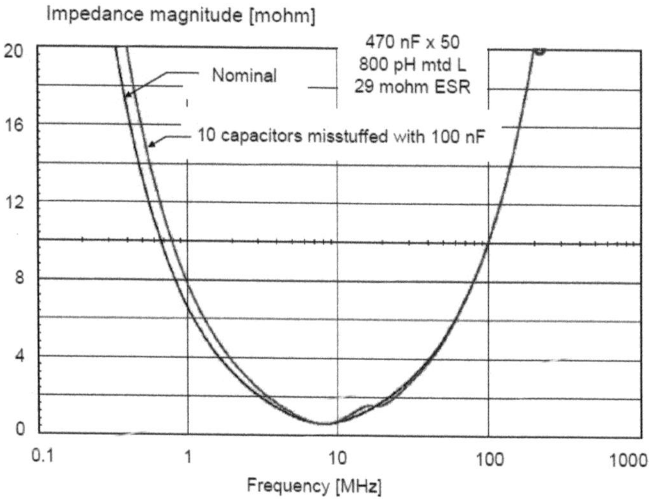

Figure 11.21: Big "V" with 20 Percent Mis-Stuffs

The good news for the multipole cases is that each capacitor has a clear frequency signature. The bad news is that board quality assurance (QA) really needs a frequency sweep. If an error is found, most likely all capacitor locations for the offending value will need to be removed and replaced.

3.2^N networks have much greater redundancy than 1.5^N networks and demonstrate a greater tolerance to mis-stuffed values. Big "V" networks enjoy greater redundancy, and even 20 percent mis-stuffs have little effect on performance.

Open or Missing Component Sensitivity

Sensitivity to open components is worst in the X^N networks at the low-frequency end, where the number of components in a given value is few. Smaller X suffers more than larger X. Big "V" networks show very little sensitivity to open or missing components.

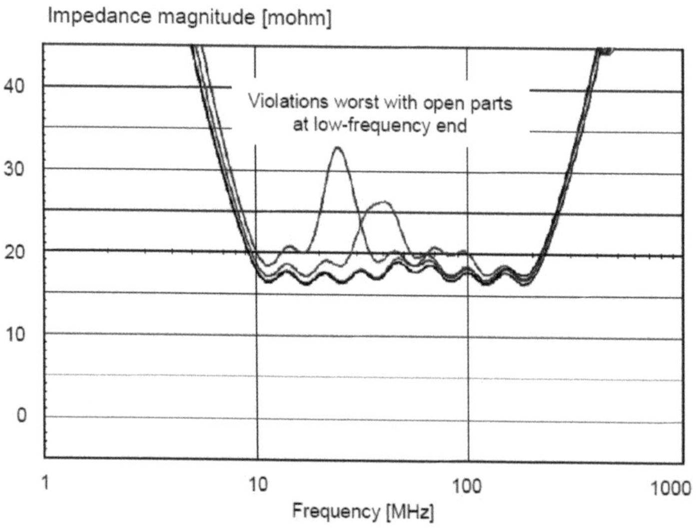

Figure 11.22: 1.5^N Open Component Sensitivity

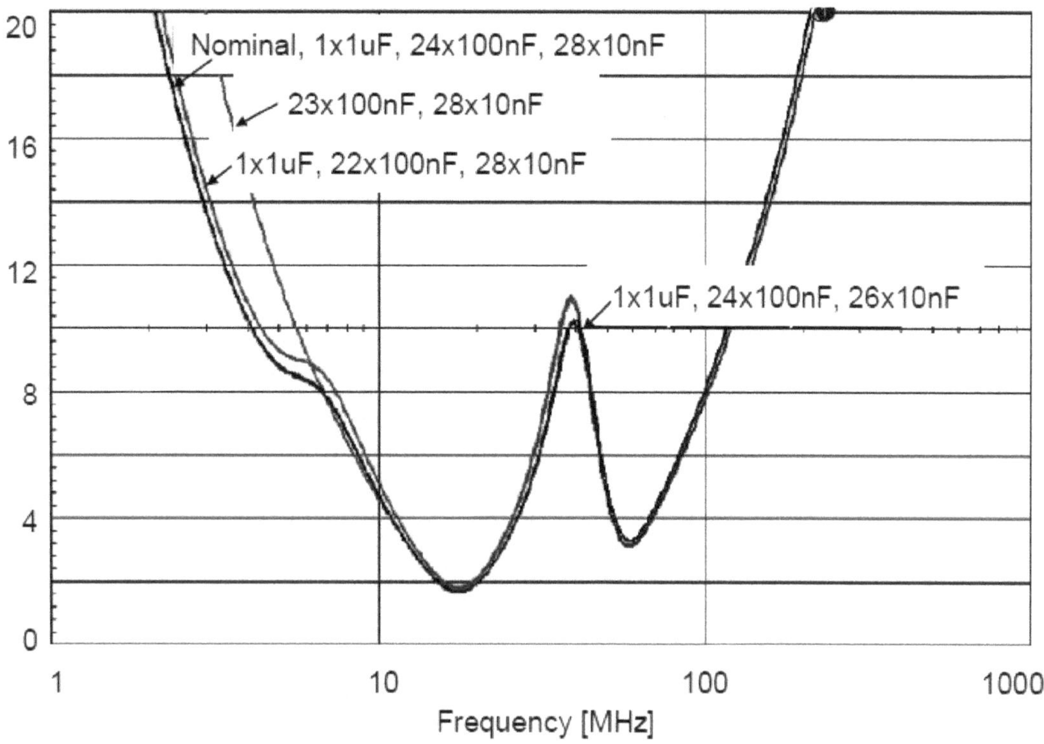

Figure 11.23: 3.2^N Open Component Sensitivity

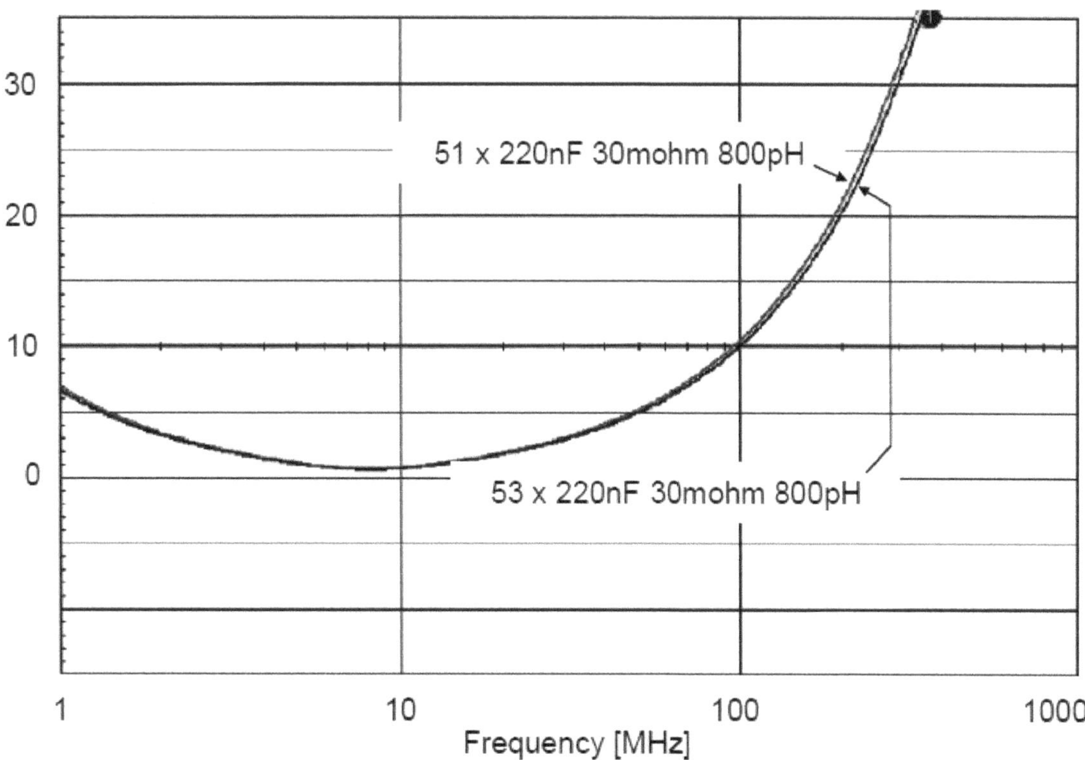

Figure 11.24: Big "V" Open Component Sensitivity

Network Synthesis

Big "V" network synthesis is straightforward and can be achieved by doing the following:

- Determining the high frequency cutoff frequency and impedance
- Translating this to an equivalent inductance
- Finding the device count by dividing the target inductance into the mounted inductance of a single capacitor
- Mapping the parallel capacitance response against the VRM/bulk capacitor response and determining if adequate phase margin is available to prevent excessive peaking
- Inserting an additional intermediate capacitor value with SRF approximately at the impedance crossover point only if peaking is excessive
- Scaling the quantity until the impedance profile is satisfied

X^N Synthesis is not much more difficult and is carried out by doing the following:

- Selecting a transition frequency from the VRM/bulk capacitors that affords at least 45-degree phase margin
- Determining the high frequency cutoff and target impedance
- Estimating the number of capacitor values NMAX as:

$$N_{VAL} = Round\_up\left(ln\left(\frac{F_{HF\_cut\_off}}{F_{LF\_cross}}\right) / ln(X)\right)$$

- For each N, from 0 to N_{VAL-1}, determining an initial capacitor count based on K2 = -0.5:

$$Count\_initial_0 \approx K4 \frac{ESR_0}{|Z_{target}|}$$

$$Count\_initial_N \approx K4 \frac{ESR_N}{|Z_{target}|}$$

Note that K4 decreases with decreasing X and increases with increasing mounted capacitor inductance.

- For each N, from 0 to N_{VAL-2}, correcting the capacitor count to reflect K2 > -0.5:

$$K_{COUNT\_COMP(N)} \approx K5 - \frac{ln\left(\frac{ESR_{CAP(N+1)}}{ESR_{CAP(N)}}\right)}{\ln(X)}$$

$$Count_N \approx Round\_up\left(Count\_initial_N^{K_{Count\_comp\_\exp}(N)}\right)$$

- Verifying the synthesized network performance in SPICE. Adjust as necessary. Note that for small values of X, sensitivity to count quantization can be quite high.

Summary

Bypass filter design is as much about insuring adequate phase margin at each frequency transition as any other task. The multipole approach developed by Sun researchers maintains phase margin not only at the network extremes, but also across the entire filter band as well. This provides the greatest advantage of lowest amplitude impedance at parallel resonance with the PCB plane cavity, but it is likely inadequate for many I/O rails.

Device ESR greatly affects phase looking up in frequency and transitions between adjacent MLCC networks, along with the transition from the discrete capacitor networks to the PCB plane cavities. However, MLCC ESR is of little concern down in frequency at the transition from the VRM/bulk capacitors. At the bulk transition, we need sufficient capacitance from the higher frequency network so that the transition occurs at a low enough frequency that the VRM/bulk capacitor network phase is well below 90 degrees. This is usually a significant advantage for big "V" networks.

Total component count and component cost advantage swings between the multi/many pole approaches, and the single capacitor value approach depending on the combined circumstances of the PWB target impedance and VRM/bulk capacitor configuration.

Viable bypass capacitor networks may be synthesized by any of the three popular techniques using straightforward spreadsheet equations backed by SPICE simulations. 1D network design fails to account for important spatial effects. 2D simulations should be used to verify and adjust any design. No matter which technique is used, low mounted inductance reduces required component account across the frequency spectrum, not just at mid- to high frequencies.

Of the three basic approaches, the most appropriate method depends on how one weighs various design and manufacturing criteria. Table 11.2 summarizes a number of these criteria and serves to illustrate that each method has specific strengths and limitations relative to the other two.

Criteria	1.5^N	3.2^N	"V"	Comments
Plane cavity parallel resonance with discrete capacitors	▲	▲-	▼	Matters when the application uses bypass capacitors for signal returns
Total parts count, optimized	?	?	?	Depends on VRM / Bulk cap configuration, target impedance, bandwidth and spatial power delivery requirements
Total parts cost, optimized	?	?	?	
Transient pulse, initial peak	▼	▲	—	Rings at discrete capacitor to plane resonance
Transient single pulse, p-p	—	▲	▼	All methods require damping to support wide bandwidths
Transient pulse decay time	▲	—	▼	
Response close to HF cut-off	▲-	▲	▼	3.2^N capable of deep SRF closest to cut-off
Low and mid frequency response	▼	—	▲	Is the noise profile flat?
Manufacturing complexity	▼	▼+	▲	Typical twelve values versus three versus one X^N boards S/B swept with SA or VNA after assy' to insure proper stuffing
Immunity to ESR variation	▲-	▼	▲	Continuous supply monitoring critical for 1.5^N
Immunity to stuff errors	▼	▲-	▲	1.5^N requires removal and replacement of many components in the event of one or a few mis-stuffs. Errors in larger capacitor values cause more severe impedance impacts.
Immunity to open components	▼	▲-	▲	

Table 11.2: Summary of Bypass Method Trade-Offs

Conclusion

Each of the three common bypass capacitor methods—many-pole, capacitors on decade values, or single bypass capacitor value—have strengths and weaknesses compared to the alternative methods.

When properly applied, each method can yield well-performing networks. A dispassionate review of design and manufacturing criteria will aid the astute designer in the selection of a preferred method and the wisdom to apply either of the alternatives under the appropriate circumstances.

Fine multipole can reduce required parts counts where either the PCB, and/or the IC package, use very thin plane cavities. Fine multipole also yields the lowest impedance peaks at the capacitor to board PRF. However, these are typically still too severe to support I/O rails where the ICs lack internal capacitors. Fine multipole has better resilience to ESR variation than capacitors on decades spacing, but correct population particularly of the larger capacitor values is critical.

Capacitors on decade values can provide some impedance advantages over fine multipole at the cost of somewhat higher peak impedance at PRF. They do not improve component count over fine multipole at the high-frequency end, fail to provide the bulk capacitor transition advantages of big "V" at the low end. They also suffer the highest sensitivity to ESR variations. For equal capacitor counts, they can selectively stretch bandwidth compared to big "V."

Big "V" offers great simplicity and manufacturing tolerance. In many cases, big "V" delivers high-frequency IC power equally as either X^N method using the same number of components. Due to its wider bandwidth, big "V" will often result in fewer total components and lower cost when the transition to the bulk capacitor network is taken into account. Big "V," however, exhibits a much higher board PRF than an equivalent performance X^N network.

This can actually be a significant factor depending on the application.

References

[1] Larry D. Smith et al., *Power Distribution System Design Methodology and Capacitor Selection for Modern CMOS Technology*, IEEE Transactions on Advanced Packaging, Vol. 22, No. 3, August 1999, pp. 284-291.

[2] Howard Johnson and Martin Graham, *High-Speed Digital Design: A Handbook of Black Magic*, Prentice Hall, 1993, pp. 281-293.

[3] William J. Dally and John W. Poulton, *Digital Systems Engineering*, Cambridge University Press, 1998, pp. 247-256.

[4] Brian Young, *Digital Signal Integrity*, Prentice Hall, 2001, pp 412-420.

[5] Tanmoy Roy, Larry Smith, and John Prymak, *ESR and ESL of Ceramic Capacitor Applied to Decoupling Applications*, EPEP Proceedings 1998.

[6] Larry D. Smith and Jeffery Lee, *Power Distribution System for JEDEC DDR2 Memory DIMM*, EPEP Proceedings, 2003.

[7] Steve Weir and Scott McMorrow, *High Performance FPGA Bypass Networks*, DesignCon Proceedings 2005.

Chapter 12—Comparison of Power Distribution Network Design Methods: Bypass Capacitor Selection Based on Time Domain and Frequency Domain Performances

István Novák, Senior Signal Integrity Engineer

This chapter is a revised version of the paper presented in TF-MP3 *Comparison of Power Distribution Network Methods: Bypass Capacitor Selection Based on Time Domain and Frequency Domain Performances*, DesignCon 2006, Santa Clara, California, February 6-9, 2006.

POWER DISTRIBUTION NETWORK (PDN) design of high-speed and high-power systems appears to be black magic: there are contradictory design philosophies, component selection, and layout rules. Suboptimum or not properly designed power distribution increases the supply-rail noise, but this impacts system performance in a statistical manner, convoluted with many other variables.

As a result, different PDN designs may appear to perform equally well over a wide range, thus giving rise to several misconceptions. This paper compares some of the popular board-level PDN design methods by their impedance profiles and worst-case transient noise. It is shown that smooth R-L type self-impedance profiles result in the lowest worst-case transient noise. Component placement is also analyzed, and it is concluded that bypass capacitor location matters mostly in those PDNs, where the impedance of the power/ground planes does not match the target impedance. Finally, measured impedance profiles are shown for some of the popular SUN servers, illustrating some of the possible PDN design philosophies.

Introduction

With the growing number of different supply rails, continuously increasing currents, shrinking supply voltages, and supply-rail noise limits, the PDN has become one of the most challenging subsystems in today's high-end systems. Figure 12.1 shows three illustrations from SUN servers.

The top photo shows the central processing unit (CPU) module from a SUN Enterprise 450 server. The 4" x 6" board, designed in the mid-1990s, has one CPU, two major supply rails, and a total of 351 bypass capacitors. The middle photo shows a V890 server CPU module. The 10" x 20" board, designed in the early

2000s, has two quad-core CPUs, eight major supply rails, and a total of 1,907 bypass capacitors. The bottom photo shows a T2000 server, with a recently designed 8" x 10" CPU board. It has one UltraSparc T1 CPU, six major supply rails, and a total of 596 bypass components.

Part of the challenge is that PDNs possibly have to serve a set of diverse functions in the systems [1], including the following:

- Provide clean power to the active devices
- Serve as return path for signals
- Ensure that PDN radiation does not violate legal limits

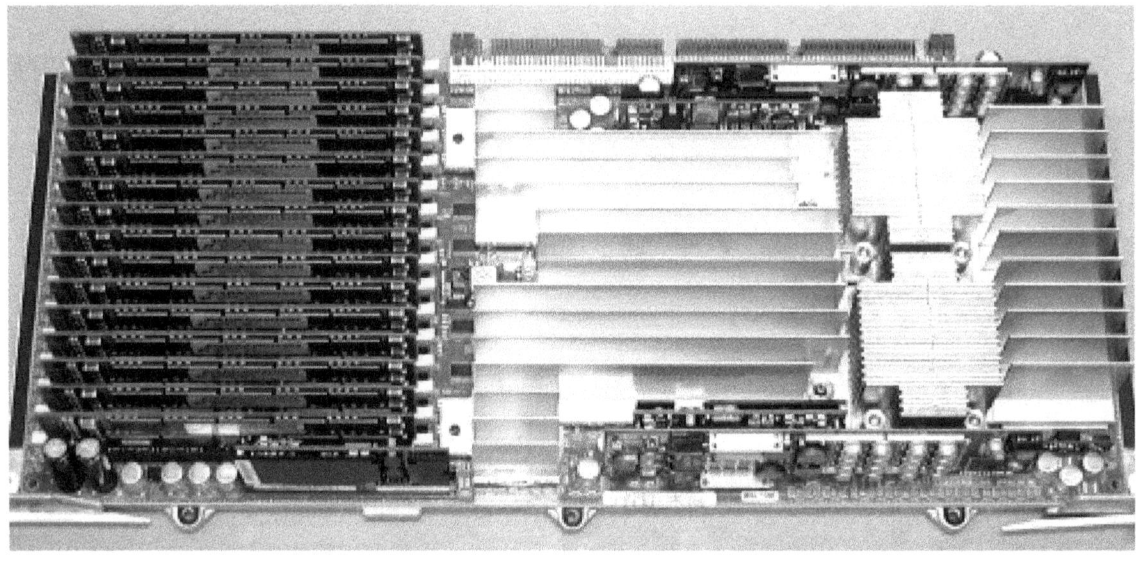

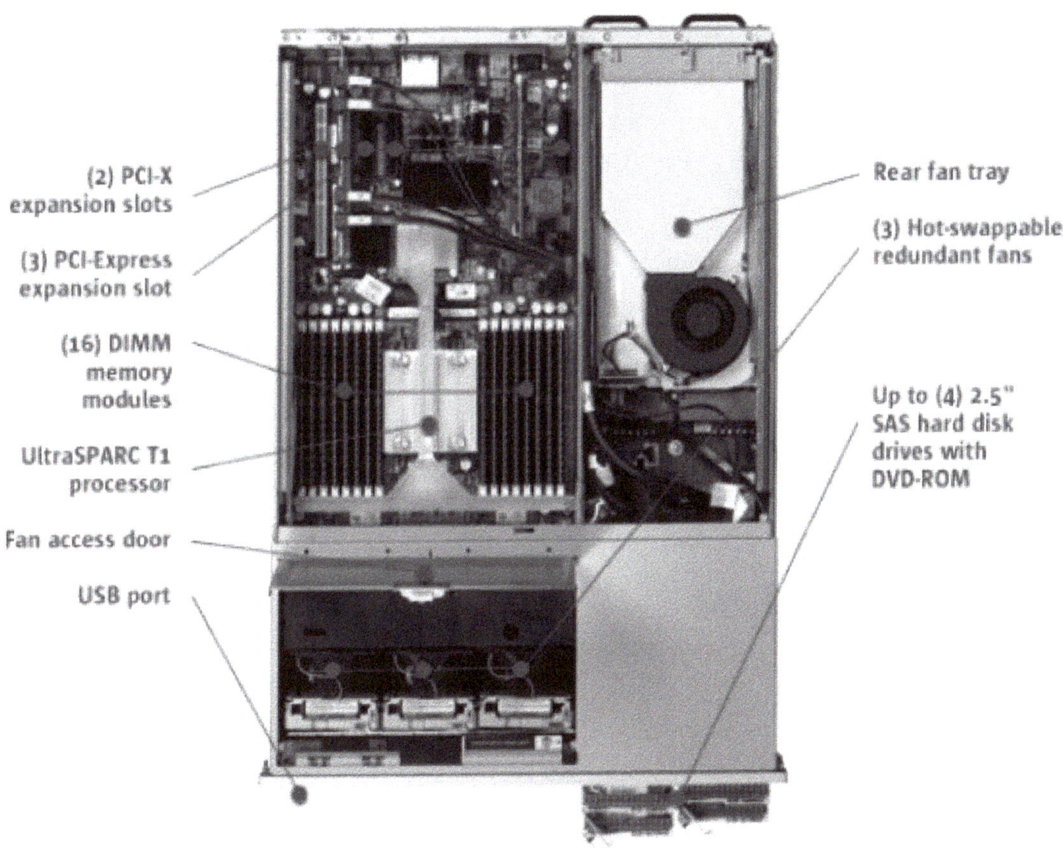

Figure 12.1: Three Illustrations from SUN Servers. Top: E450 CPU Module, Approximately 4" x 6" In Size, Designed in the mid-1990s. Middle: V890 CPU Module, Approximately 10" x 20" in Size, Designed in the Early 2000s. Bottom: T2000 2U Server with 8" x 10" CPU Module, Recent Design. Note That the Scales Are Different for the Photos

The simple sketch in Figure 12.2 illustrates a board-level PDN with one supply rail. In reality, most of the time, we have multiple supply rails in a system, each having different requirements, but being interrelated by shared planes and cross couplings among elements. In a complex system, these three major functions can also be interrelated and, dependent on the nature of the PDN, one or two of the major functions may not be present or may be irrelevant for a particular rail.

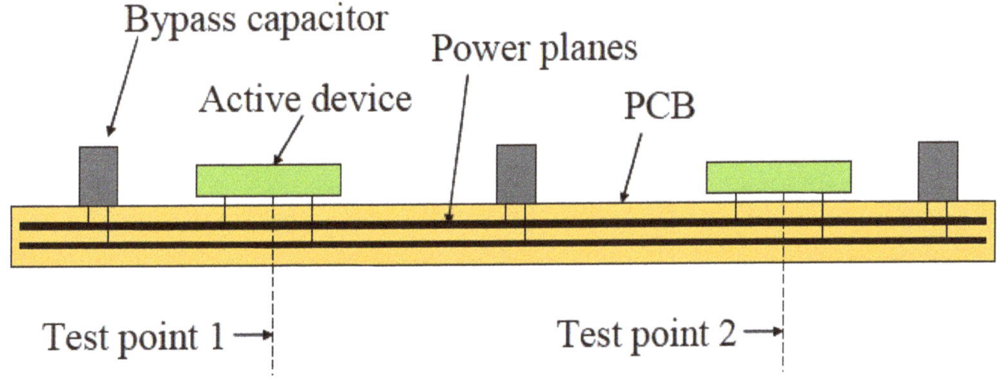

Figure 12.2: Sketch of Board-Level PDN and Its Major Components

In fact, [1] outlines two major classes of PDNs: core PDN and I/O PDN. In its simplest form, a core PDN feeds only one active device with a dedicated network. If, moreover, in the implementation the core plane is sandwiched between neighboring ground planes, the core plane does not serve as a signal return path, and being between ground planes, electromagnetic interface (EMI) radiation may also be less of a concern. The push for more cost-effective designs, however, many times results in shared PDNs: a PDN, with its primary function of feeding one or more cores or distributed I/O networks, may also serve as reference plane for slow or high-speed signals. In such situations, the functions have to be looked at separately, and the most restrictive will have to drive the design. For instance, a core may be able to live with more noise on the rail than what is being allowed for some high-speed reference planes. Similar considerations apply to the EMI limits, for instance, since core distributions through packages behave like low-pass filters, resonances of board PDN at several hundred MHz or above will have very little influence on the power delivery to the core, but it may be detrimental for EMI radiation and/or signal-return functions.

The requirements to supply clean power to the chips and the function of reference path for high-speed signals can both be captured by the maximum transient noise as a result of worst-case activity of the system. With the exception of low-end systems, where occasional failure may be acceptable, in synchronous interconnects, we usually have to keep the worst-case peak-to-peak transient noise under control so that we can maintain the proper operating conditions of the chips and limit simultaneous switching noise, jitter, and inter-symbol interference introduced by the rail noise through the return path. Since most PDN components can be fairly accurately described by linear models, the worst-case peak-to- peak transient noise can be obtained from the impedance matrix of the PDN. Assuming an arbitrary sequence of step-like excitation current with a fixed rise time, the worst-case excitation pattern and peak-to-peak transient noise can be obtained by processing the network functions of the PDN [2].

CPU clock frequencies and signaling rates used today push more features on the board into the red zone, such that their relative dimensions with respect to the wavelength of clock signals approach or exceed the quarter-wave limit. Quarter-wave or longer structures may become effective primary or secondary radiators; therefore the PDN has become one of the primary EMI risk factors. It is generally understood that the best way to avoid EMI radiation problems from PDNs is to ensure their resonance-free impedance profile. The resonance-free impedance profile also helps to minimize simultaneous sequence noise (SSN) and jitter of high-speed signals when the PDN serves as signal reference.

In a good industrial design, we want to ensure that the system and its PDN will properly function over the entire foreseeable range of parameter variations such as component tolerances and aging. It is an added bonus if the design is less sensitive to missing and/or broken components. In short, low sensitivity is preferred.

To improve long-term reliability of the PDN, we also need to ensure that stress distribution among the PDN components is as uniform as possible.

The constant push for more cost-effective solutions results in optimizing the margins and removing the extra padding of performance in interrelated PDN subsystems. This may result in designs, where changing one dimension or component value may have a ripple effect and a whole chain of components and dimensions have to be realigned and redesigned. Therefore, portability of the design is becoming increasingly valuable and important.

In addition to these electrical requirements, we may have a set of important non-electrical parameters and targets to meet, including cost, size (e.g., area, height), weight, component connection style (through hole or surface mount), placement restrictions, and simplicity of bill of material.

When we design the board PDN, we have to select the direct current (DC) sources, bypass capacitors, printed circuit board (PCB) planes, and stackup. More important, we need to find the proper location for all of the parts.

The primary questions about the bypass capacitors of a board-level PDN include the following:

- What values of bypass capacitors should we use and how many?
- Where should we place the components?

One should note that selecting the proper DC sources (DC-DC converters) is becoming a challenge by itself, but we do not address that question in this chapter.

So, What Is the Metric?

Since we usually do not accurately know the transient noise current excitation, it is rather customary to design the PDN to meet a required impedance profile [3]. For single-load and small-size core PDNs, not serving the additional task of reference path for high-speed signals, the requirements can be eventually distilled to the self-impedance profile provided to the silicon core. In this case, the transfer functions among various points of the PDN are less important. For large, distributed core and I/O PDNs, possibly with multiple noise sources, the description has to be done by full impedance matrices: both self- and transfer-impedance terms at the key locations are important.

Since we can derive the worst-case transient noise from the impedance matrix, it may appear that the impedance profile of PDN is the key parameter. While the impedance profile is very important, we should not lose sight of the original requirement, which was the worst-case transient noise. As we will show later, impedance profiles that may look better than others may in fact be worse and generate more noise.

In this paper, we will analyze some of the popular PDN design methods by first comparing their performances based on lumped self-impedance profiles. Later, we look at some of the important distributed aspects such as component placement and modal resonances. The analyzed approaches will be illustrated by simulated and measured performance data.

Comparison of Popular Methods Based on Lumped Self-Impedance

In the early days of PDN design, the focus was on the number of 0.1 μF capacitors to be placed close to the active devices' leads. This approach was sufficient as long as the active devices did not generate significant transient noise spectrum above the series resonance frequency (SRF) of the bypass capacitors. Higher system speeds later created the need to systematically design the PDN to meet certain impedance requirements. Here, we compare the following four possible ways of synthesizing a lumped self-impedance profile:

- Multipole (MP) [3]
- Capacitors-by-the-decade (CBD) [4]
- Big "V" [4]
- Distributed matched bypassing (DMB) [5]

One should note that in terms of impedance profile, CBD and big "V" are variations of MP and DMB. There is no clear boundary between these design approaches, as we can gradually transform one into the other by varying parameters, and there are an infinite number of variants in between. The illustrations in this section depict simulated scenarios with hypothetical component values. Implementation details will be given later with measured illustrations.

We start with DMB, which calls for a flat impedance profile from DC up to a certain corner frequency, beyond which frequency the impedance may rise linearly, following the reactance of the PDN inductance. For the sake of simplicity, let us assume that we want to synthesize 10 mOhms of impedance, and the corner frequency is 10 MHz, corresponding to 150 pH of PDN inductance. The impedance profile and the corresponding step response are shown in Figure 12.3.

The horizontal scales both in the frequency and time domains are logarithmic. On a logarithmic frequency scale, we cannot show the DC value, but we assume that it is the same 10 mOhm value. The finite DC output resistance is generated by adaptive voltage positioning (AVP) [6] in the DC source. We further assume that we generate the step response with a 1A 10 ns rise-time step excitation after a 10 MHz single-pole low-pass filter. The 10 MHz high-pass and 10 MHz low-pass filter functions cancel each other, which restores the 10 ns piece-wise-linear response. This is what we see on the bottom graph in Figure 12.3.

This graph shows the positive and negative step responses. Let us note again that the horizontal scale is logarithmic, which is used here and on subsequent step-response graphs to enable us to see details close to the initial edge as well as changes much later in the step response. The step-response is monotonic, there is no overshoot/undershoot or ringing, therefore the worst-case transient noise for any arbitrary sequence of step excitation with 30 ns or longer rise-time excitation will be 10 mVpp for each ampere of current-step magnitude. This 10 mVpp/A number serves as a baseline in comparing other designs.

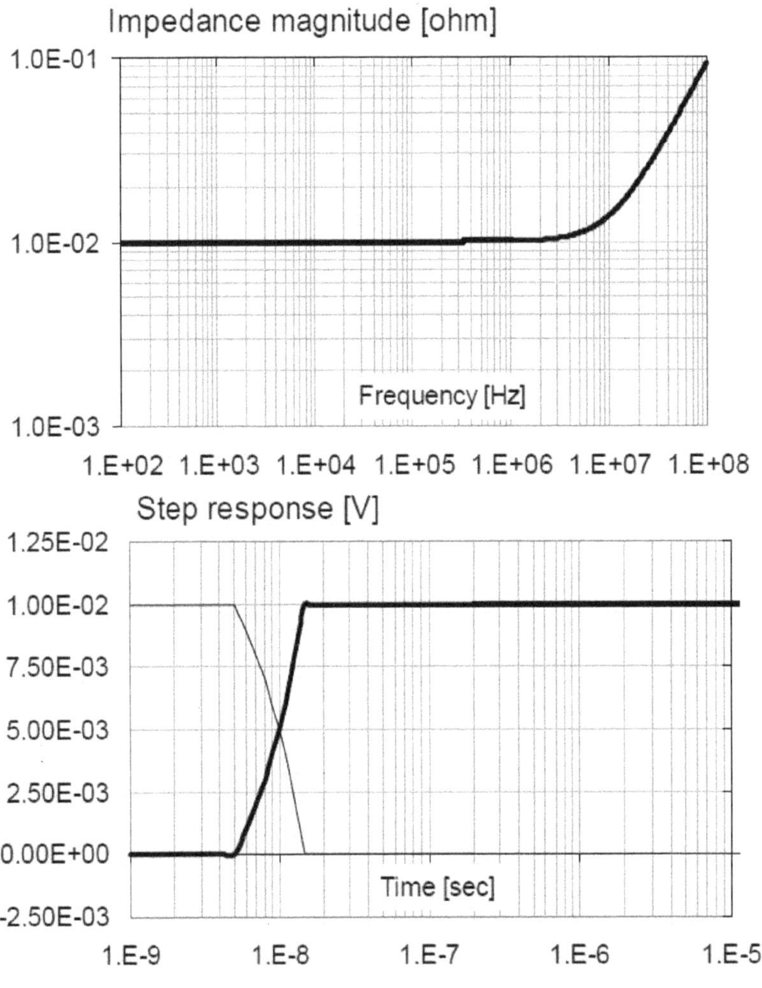

Figure 12.3: Impedance Profile (Top) and Step Response (Bottom) of a 10 mOhm PDN Based on Distributed Matched Bypassing with AVP

In case we decide to use high DC gain in the DC source instead of AVP, the DC output resistance will be low. Figure 12.4 shows the impedance profile with 1 mOhm DC resistance and the step response with the same filtered 10 ns step excitation.

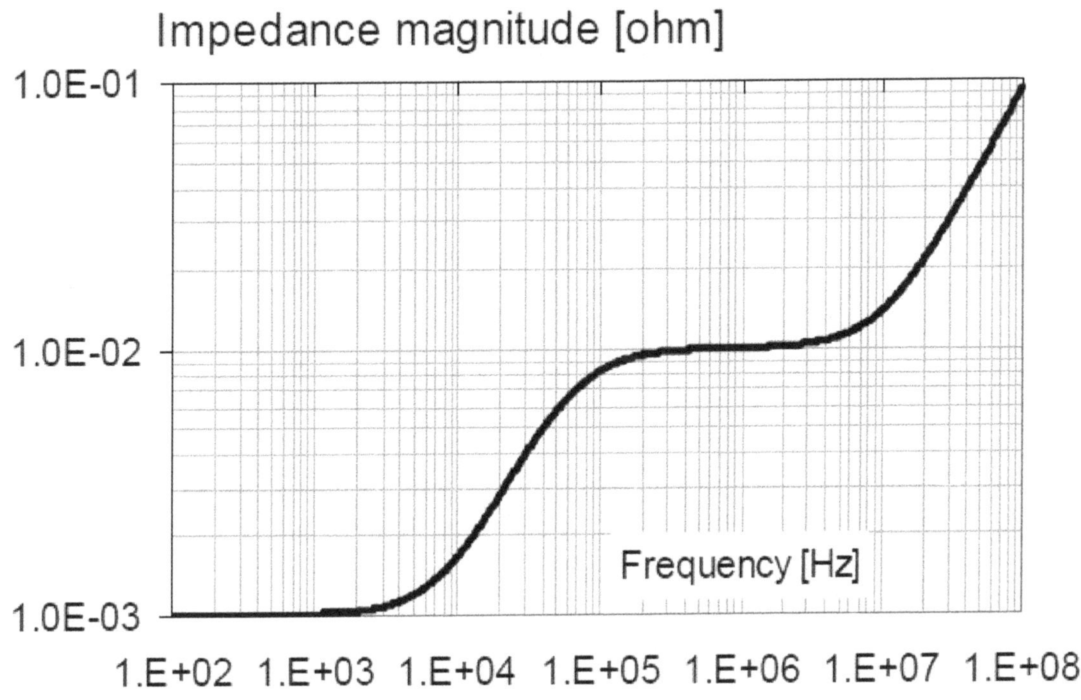

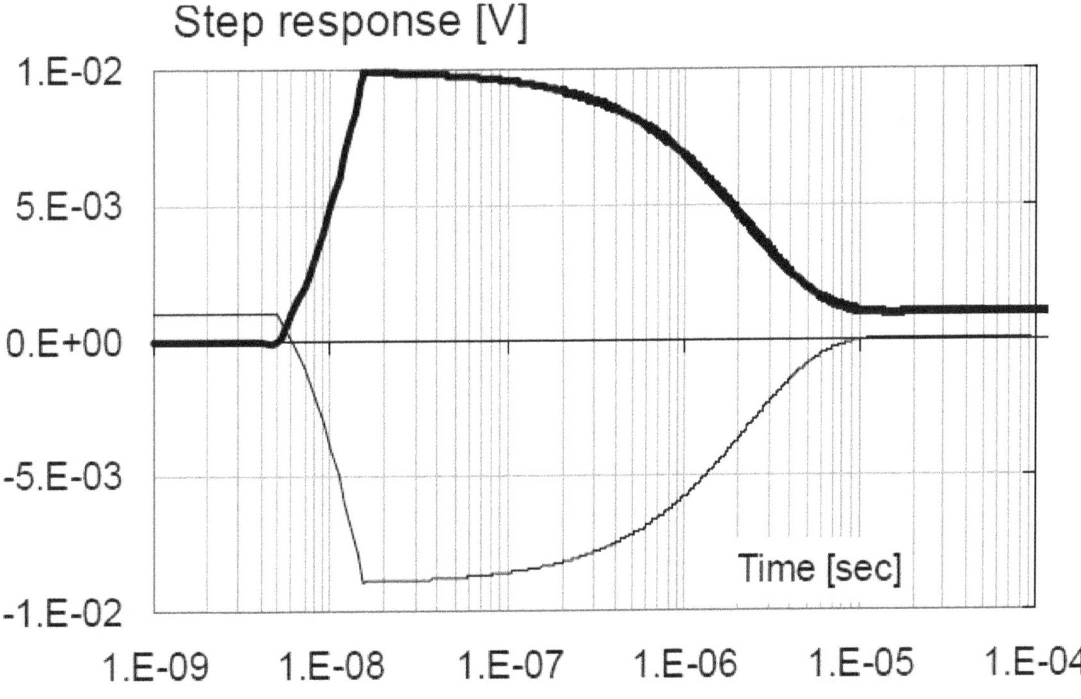

Figure 12.4: Impedance Profile (Top) and Step Response (Bottom) of a 10 mOhm PDN Based on Distributed Matched Bypassing with no AVP

The step response is still smooth, with no ringing, but with respect to its steady state, it has a 9 mV overshoot. The worst-case peak-to-peak transient noise thus becomes two times the overshoot plus the steady-state swing: 2 * 9 + 1 = 19 mVpp.

Next, we look at an MP design with AVP. We assume five capacitor values, which cover one decade of frequency in the range of 1 to 10 MHz. We have four capacitance values per decade, and each part is assumed to have a Q of approximately 3. As shown in Figure 12.5, the impedance and step responses in this case have multiple peaks and ringing. We need to follow the procedure in [2] to get the worst-case peak-to-peak transient noise, which comes out as 15.7 mVpp/A. The assumed component values are shown in Figure 12.6.

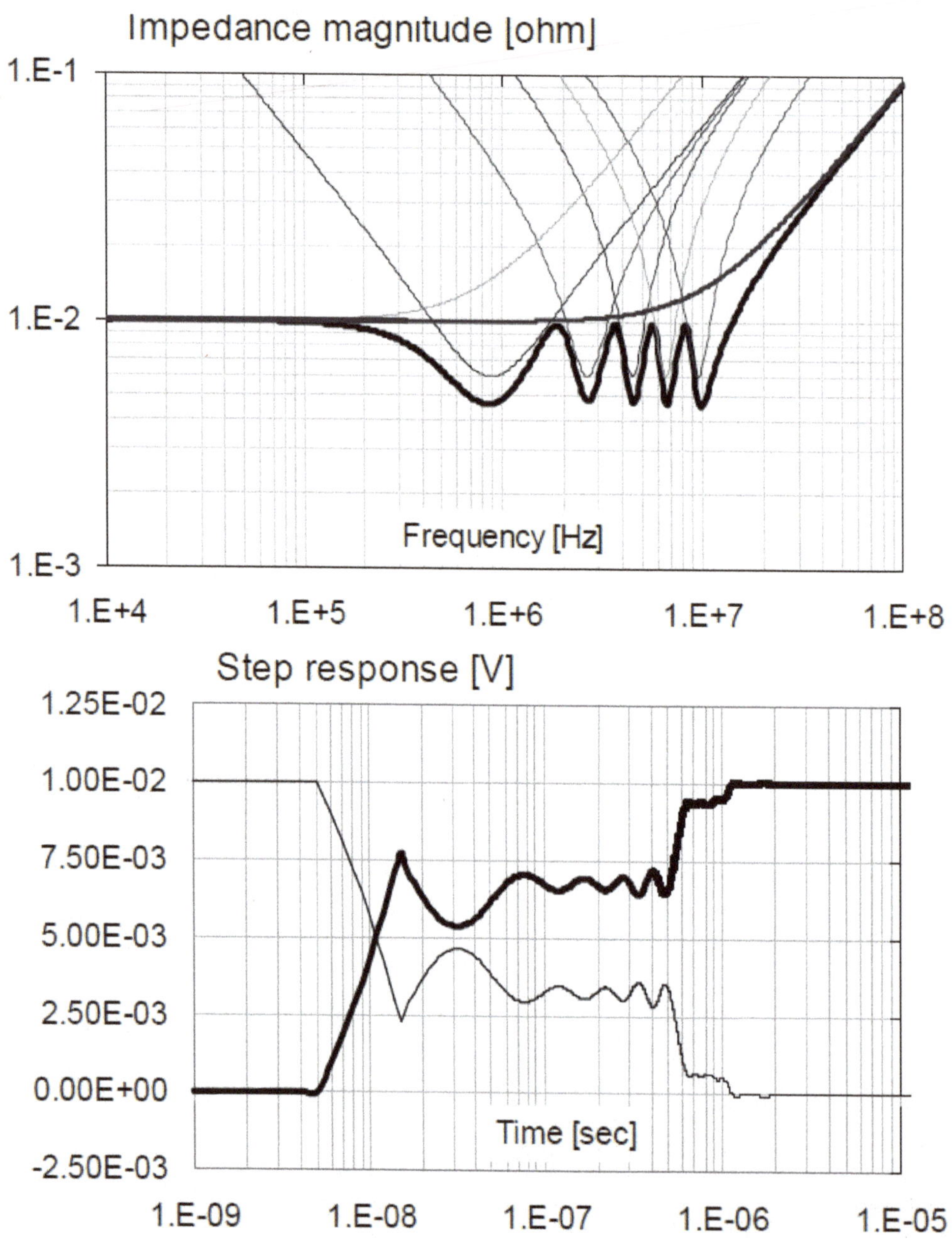

Figure 12.5: Impedance Profile (Top) and Step Response (Bottom) of a 10 mOhm Multipole PDN with AVP

Figure 12.7 shows the impedance profile and step response for the same multipole PDN design with high DC gain in the DC source. The worst-case peak-to-peak transient noise increases to 25 mVpp/A. Figure 12.8 shows the component values and the detailed impedance profile. By comparing the component values in Figure 12.6 and Figure 12.8, we can see that only the DC source (R-L) and the bulk capacitor (C1) changed; the other component values are identical in the two designs.

	C [uF]	ESR [ohm]	L [nH]
R-L	-	0.01	2
C1	33	0.006	1
C2	3.6	0.006	1
C3	1.3	0.006	1
C4	0.75	0.006	0.8
C5	0.56	0.006	0.5

Figure 12.6: Component Values for the PDN Shown in Figure 12.5

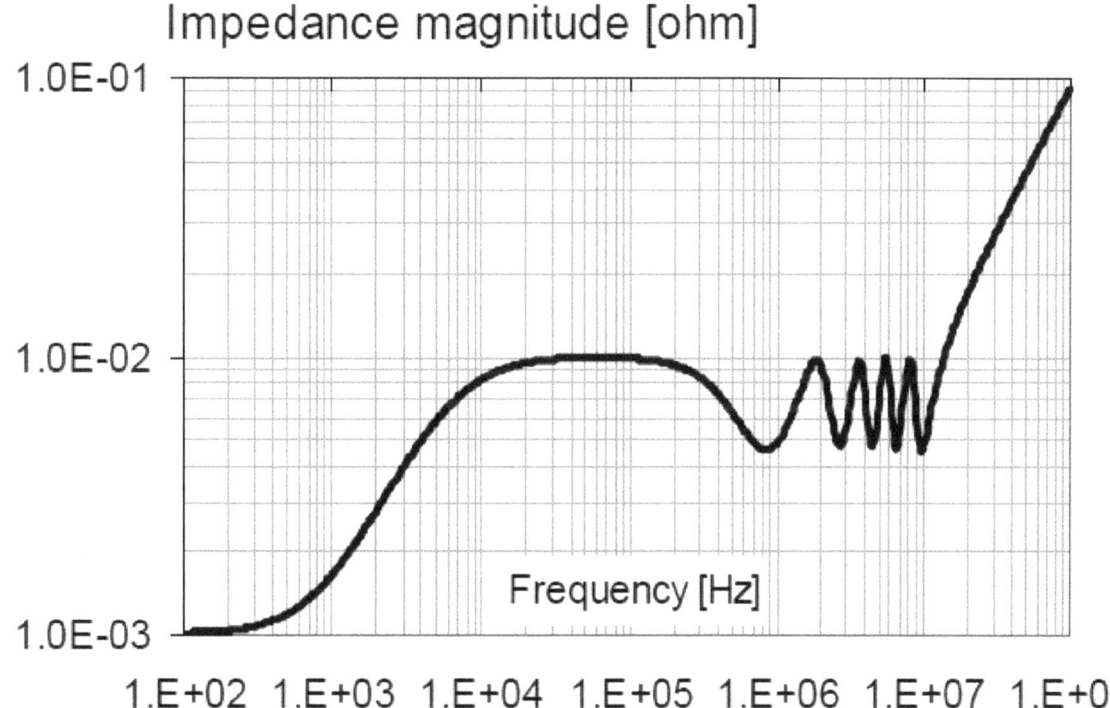

Figure 12.7a: Impedance Profile (Top) and Step Response (Bottom) of a 10 mOhm Multipole PDN with High DC Gain

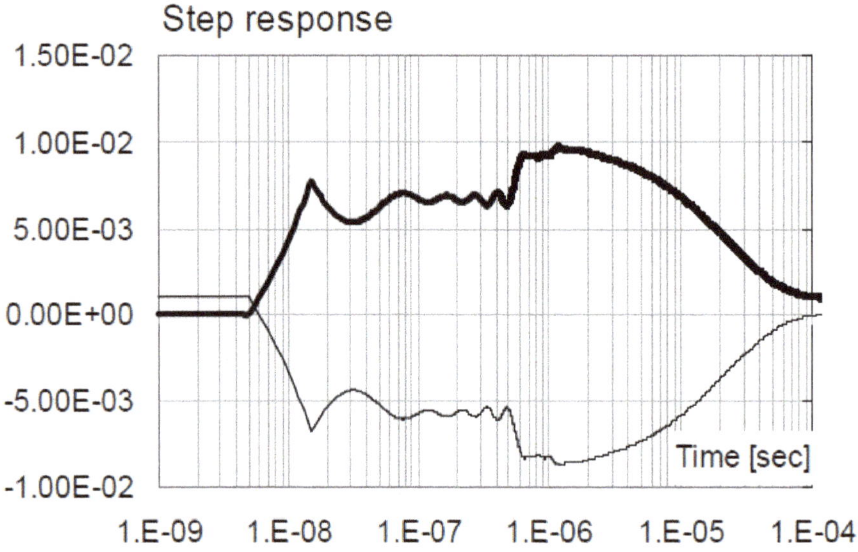

Figure 12.7b: Impedance Profile (Top) and Step Response (Bottom) of a 10 mOhm Multipole PDN with High DC Gain

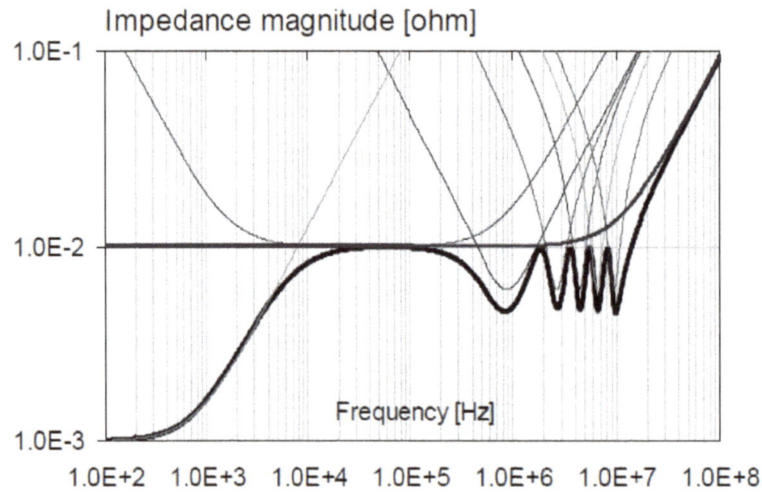

	C [uF]	ESR [ohm]	L [nH]
R-L	-	0.001	20
C0	1000	0.01	2
C1	33	0.006	1
C2	3.6	0.006	1
C3	1.3	0.006	1
C4	0.75	0.006	0.8
C5	0.56	0.006	0.5

Figure 12.8: Component Values and Detailed Impedance Profile for the PDN Shown in Figure 12.7

We skip the CBD since it is a variant of the MP design approach and continue with a Big "V" illustration. It will be illustrated with its measured performance. Figure 12.9 shows the impedance profile and step response of a 10 mOhm Big "V" design with AVP. This impedance profile assumes a 10 mOhm 1.5 nH DC source, in parallel to a multitude of bypass capacitors, cumulatively generating a C = 30 μF, ESR = 1E-3 Ohm, ESL = 165 pH impedance. The step response has one pronounced dip and one peak, resulting in a worst-case transient peak-to-peak noise of 21.4 mVpp/A.

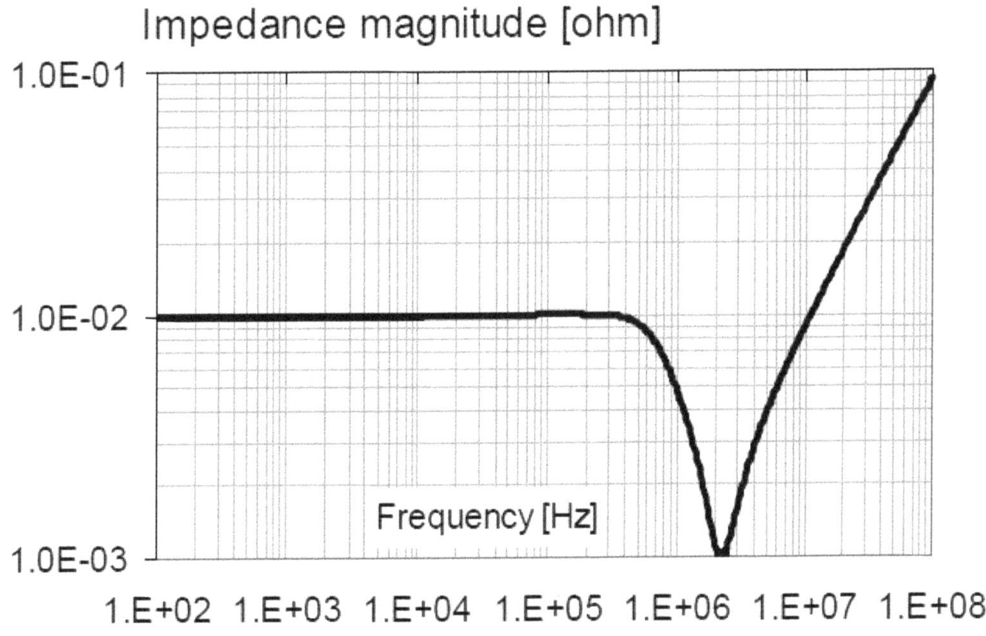

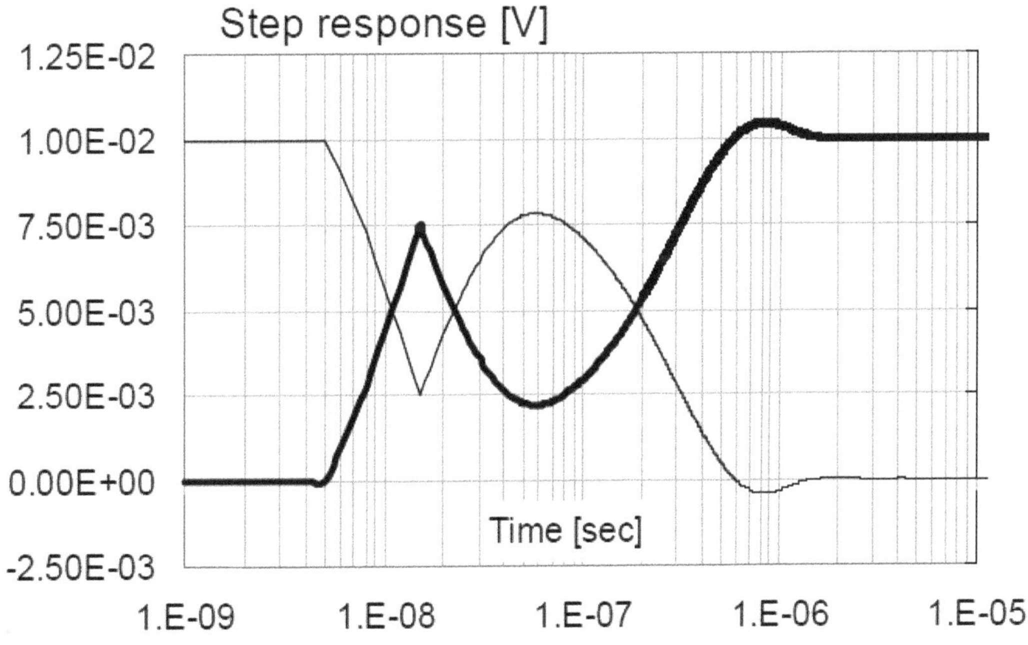

Figure 12.9: Impedance Profile (Top) and Step Response (Bottom) of a 10 mOhm Big "V" PDN with AVP

If we use high DC gain with the big "V" impedance profile instead of AVP, we get the impedance and step responses shown in Figure 12.10. This impedance profile can be obtained with the same set of bypass capacitors by replacing the 10 mOhm 1.5 nH DC source with the combination of a 1 mOhm 100 nH DC source and C = 10,000 μF, ESR = 1E-2 Ohm, ESL = 1.5 nH bulk capacitor. This impedance profile results in a 27.5 mVpp/A worst-case transient noise.

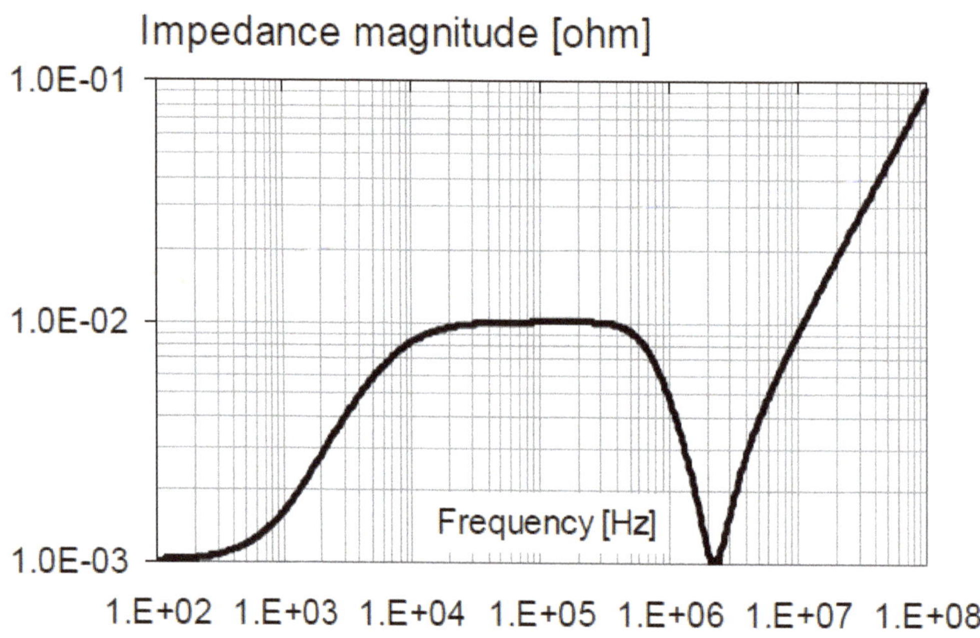

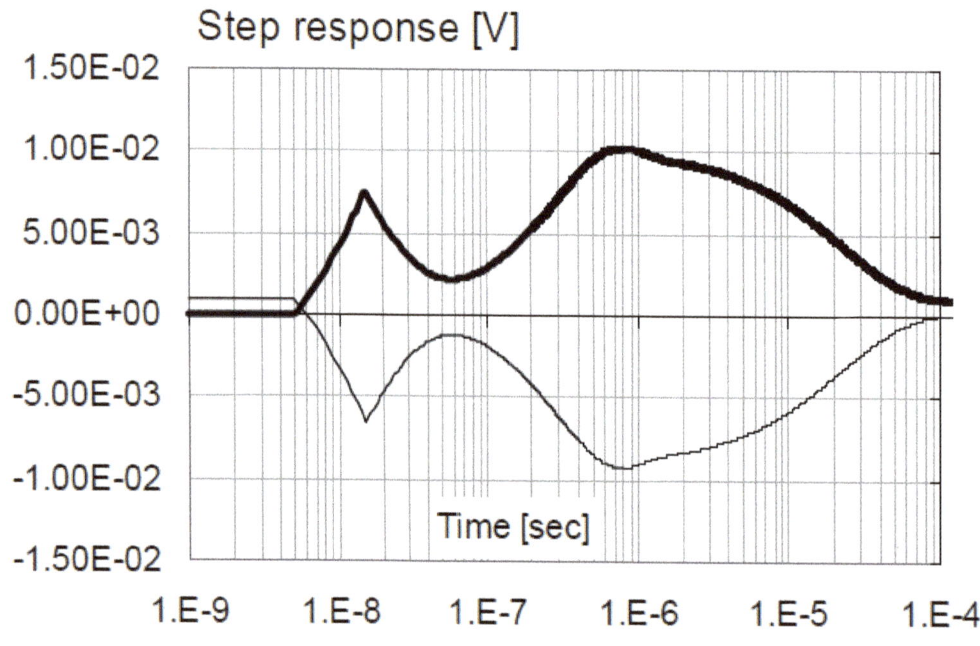

Figure 12.10: Impedance Profile (Top) and Step Response (Bottom) of a 10 mOhm Big "V" PDN with No AVP

When we compare the impedance and step responses with the worst-case noise performance numbers, we notice that the 10 mOhm 150 pH DMB impedance profile of Figure 12.3 resulted in the lowest noise number, even though this impedance response is an upper bound to all consecutive impedance profiles. This suggests that making the impedance profile "better" by driving dips to make the impedance profile meet an impedance target over a wider frequency range is counter-productive, as it creates more noise. We can clearly show this by comparing the impedance and transient response profiles on one graph. Figure 12.11 compares the MP, big "V," and DMB designs with AVP.

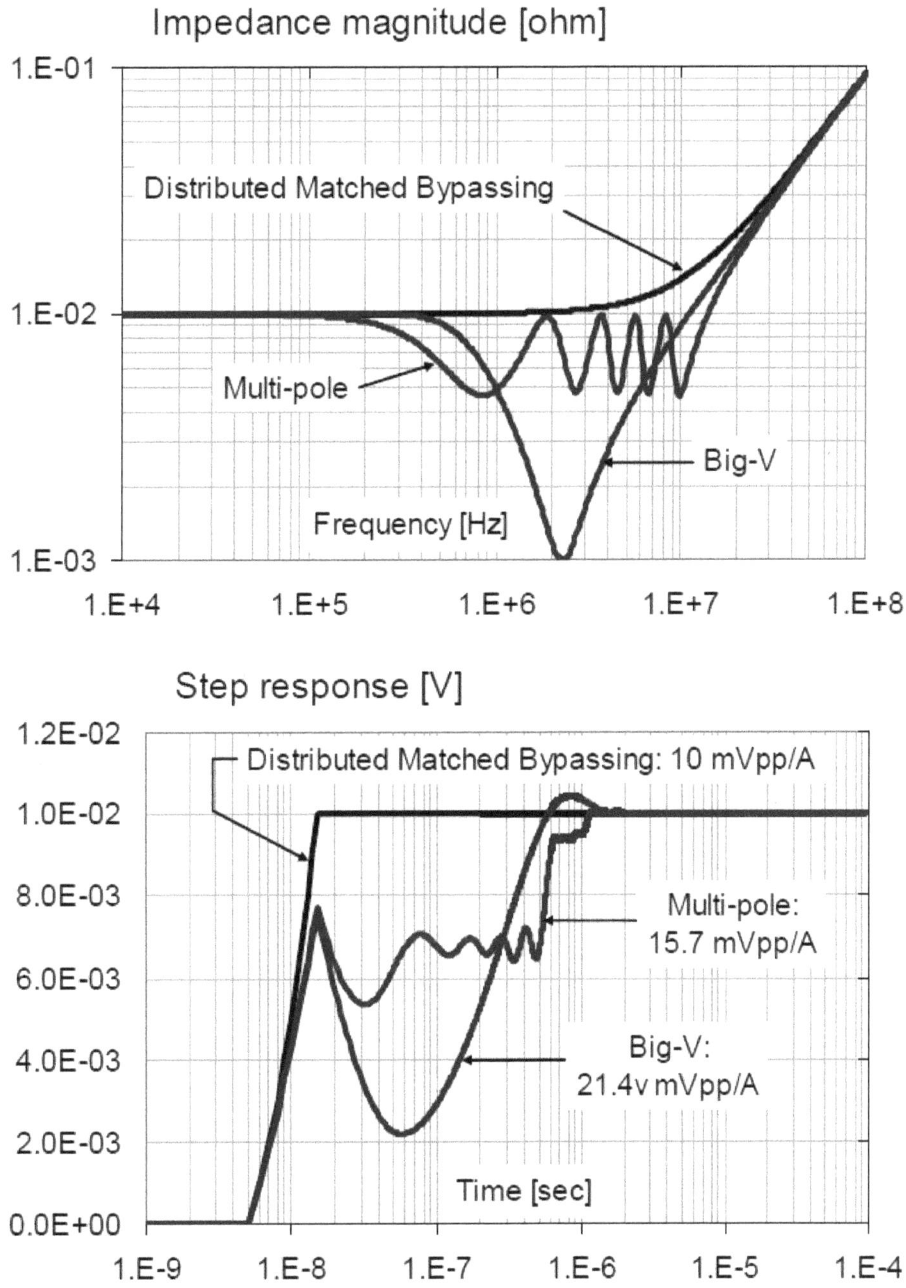

Figure 12.11: Comparison of Impedance Profiles and Step Responses of Multipole, Big "V," and Distributed Matched Bypassing Design Approaches with AVP

Figure 12.12 first shows the result of partial AVP, where we change the DC resistance of the source. As the DC resistance approaches the mid-frequency AC impedance, the overshoot in the transient step response becomes smaller.

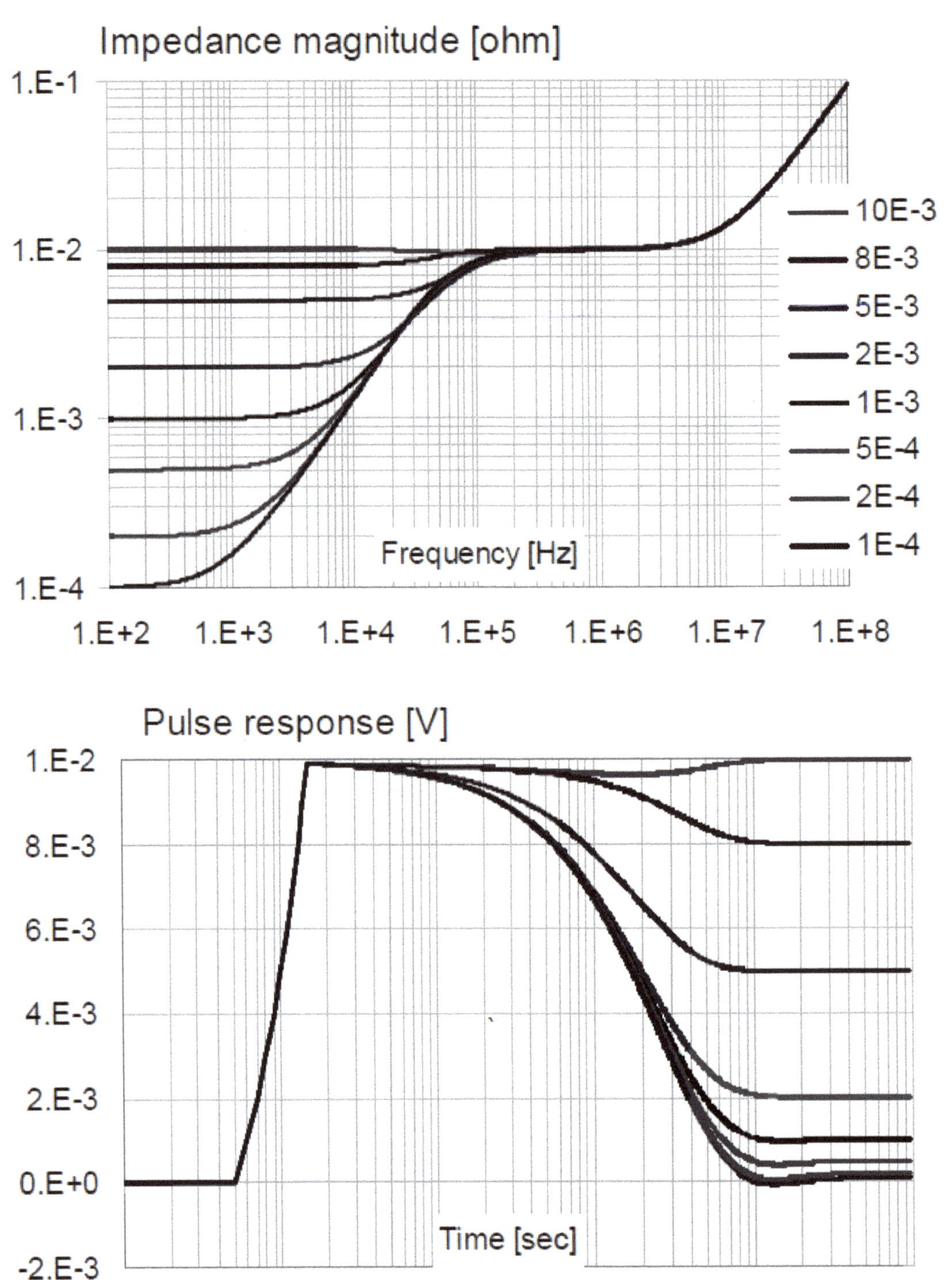

Figure 12.12: DMB Design with Different DC Resistances. The Scale on the Impedance Plot Refers to Both Plots and Indicates the Sequence of Traces from Top to Bottom

From the comparison in Figure 12.11, we may be inclined to conclude that the big "V" design is generally worse than MP. This generalization, however, is again not correct; eventually, the worstcase transient noise depends on the number and depth of dips and peaks in the impedance profile. For the comparison of Figure 12.11, the MP design performs better than the big "V," because we assumed a 2:1 ripple magnitude in the MP impedance profile (5 mOhm minima and 10 mOhm peaks) as opposed to a 10:1 ratio with the big "V" design. If we can manage to reduce the impedance variation of the big "V" design, the worst-case transient noise will improve.

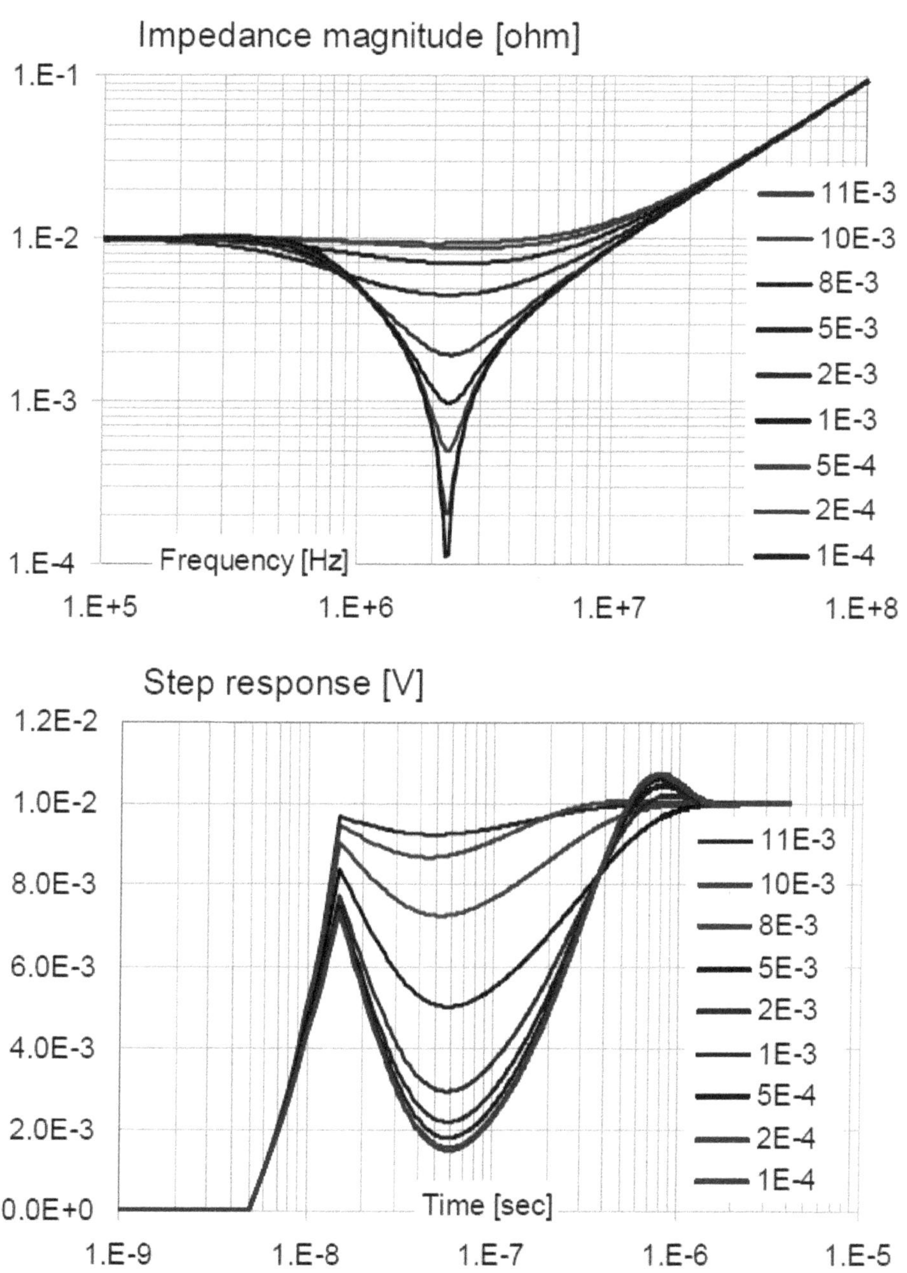

Figure 12.13: Big "V" Design of Figure 12.9, Except ESR of the Capacitor Bank Generating the Impedance Dip Varies between 0.1 and 11 mOhm. The Scale on Each Plot Shows the Sequence of Traces from Top to Bottom

We can also look at impact of changing the value of the impedance minimum in the big "V" design. The impedance and step responses of Figure 12.13 span the minimum impedance in a 100:1 ratio. Let us note that the deeper the minimum, the more fluctuation we get in the step response; hence, we get larger worstcase transient noise.

We can ask similar questions about the MP design.

How can we control the fluctuations in the step response—by increasing the number of capacitance values per decade (N) or by reducing Q of each capacitor bank? The correct answer is not one or the other, but a combination of the two, because eventually what matters is the peak/valley ratio in the impedance profile. We can use lower N with lower-Q parts, or we can afford higher-Q parts if N is higher.

This is illustrated in Figure 12.14.

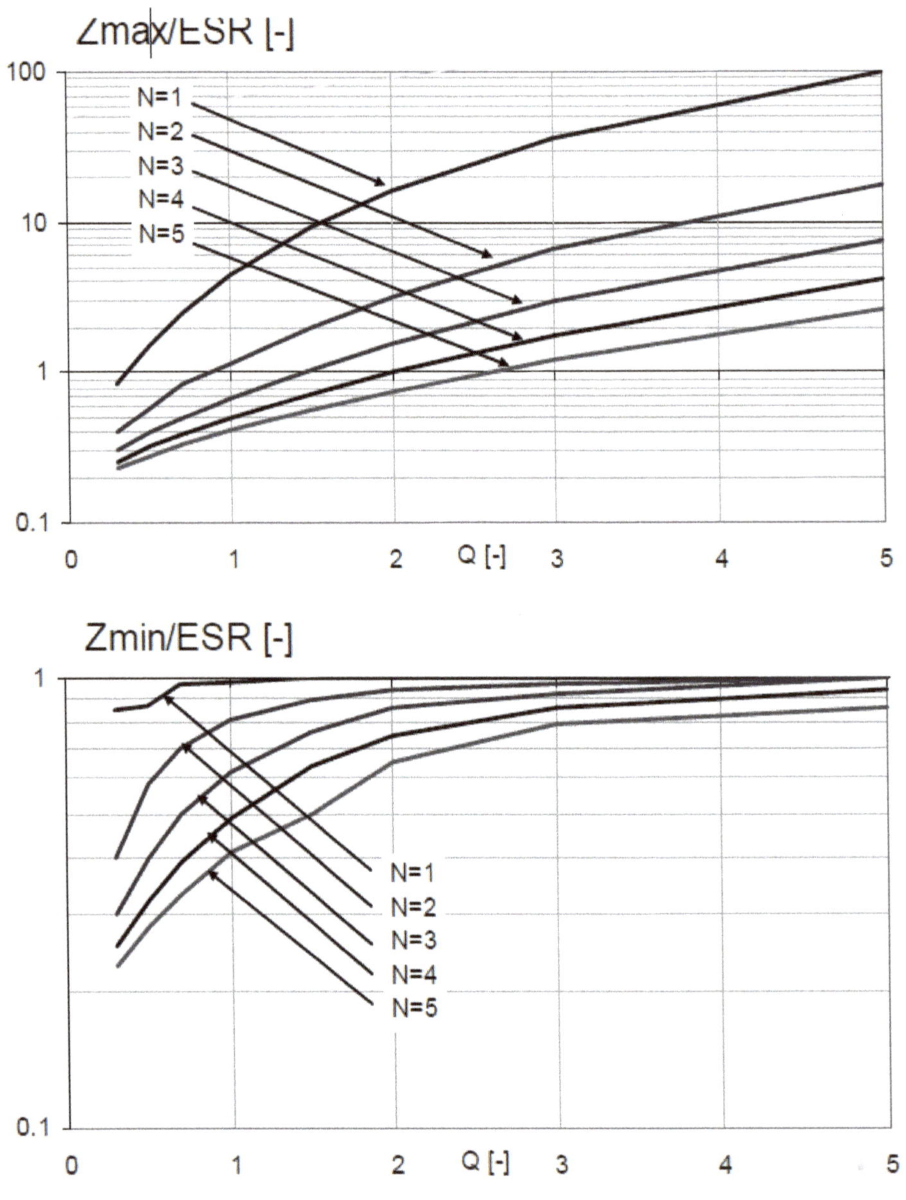

Figure 12.14: Normalized Impedance Maximum (Top) and Minimum (Bottom) Values of the Multipole Design as a Function of the Number of Capacitance Values per Decade (N) and Q of Capacitors

The charts assume the circuit in Figure 12.6, with five capacitor banks. ESR in each bank was held constant, and the capacitance and inductance values were varied to get different Q and N values.

The series resonance frequencies of capacitor banks were logarithmically allocated on the frequency scale; this, for a given set of N and Q, results in equal-height peaks and dips in the impedance response. The top plot in Figure 12.14 shows the impedance peak values normalized to the common ESR as a function of Q and N. The plot on the bottom shows the impedance dip values normalized to the common ESR. One should note that the values on the bottom graph are less than one, indicating that the impedance minimum of parallel-connected capacitors with closely spaced SRF is below their respective ESR values. We can see from these plots that for a given ratio of impedance maximum and minimum we can trade N for Q and vice versa.

Component Placement

The previous simulated examples assumed lumped component connections: delays and impedance among the components were neglected. In addition to the proper selection of bypass capacitor values, another hotly debated question is where to place them.

Most application notes recommend placing bypass capacitors close to the active device's supply pins. With high-power area connection packages (pin grid array [PGA], ball grid array [BGA], land grid array [LGA]), implementing this advice is becoming very hard. So, the designer faces the question: how close should we put the bypass capacitors? Conventional wisdom argues that we need to place bypass capacitors close to the active device in order to have charge available quickly when the device needs it for fast current transients [7]. This argument, however, ignores the fact that charge can be stored also by inductors; moreover, charge can also be supplied by transmission lines. Ultimately, what matters most is the connection of the active device to an impedance low enough that the current transient will generate noise voltages within our specified limits. Therefore, if we assume a matched transmission line with suitable characteristic impedance feeding the active device, what matters here is that this transmission line should reach the active device, in which case, the locations of charge-storage elements do not matter.

In PDNs, the power planes and/or puddles behave like two-dimensional transmission lines. The approximate impedance of a rectangular plane pair is [5]

$$Z_p = \frac{532}{\sqrt{\varepsilon_r}} \frac{h}{P}$$

...where Zp is the approximate plane impedance in Ohms, h and P are the plane separation and plane periphery, respectively, in the same but otherwise arbitrary units. For instance, a 10" x 5" plane pair with a 12 mil plane separation yields approximately 0.1 Ohm impedance. The impedance can be set for instance by 60 pieces of RC components placed around the plane periphery, one at every half-inch interval. If we select R = 5.6 Ohm and C = 0.1 µF, the 60 pieces in parallel will give a flat impedance profile with about 0.1 Ohm. The 12-mil plane separation is very easy to ensure in a stackup; in fact we have room also to place one or maybe even two signal layers between these power/ground planes. If we need lower than 0.1 Ohm impedance, we can place the planes closer. Without extra cost we can get down to 3 mil plane separation, yielding approximately 25 mOhm impedance on the 10" x 5" planes. For added cost we can get plane separations down to 1 mil or 0.5 mils ([8–10]), resulting in a plane impedance of 12.5 and 6 mOhm, respectively.

But what happens if we need a 1 mOhm supply-rail impedance? To get it with matched planes, the plane separation would need to shrink to thin-film dimensions, which is not practical for large rigid boards today. Or, what happens if we need just 10 mOhm, but we want to maintain for instance 10 mil plane separation (for cost or maybe other reasons)? In these cases, we can still create the necessary low- and mid-frequency impedance with bypass capacitors, but if these bypass capacitors are connected to the active device with higher-impedance planes, now the capacitor placement does matter.

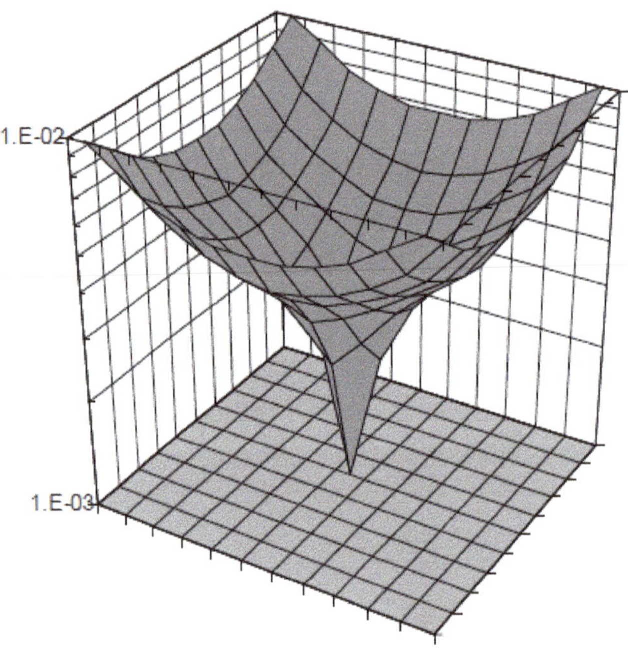

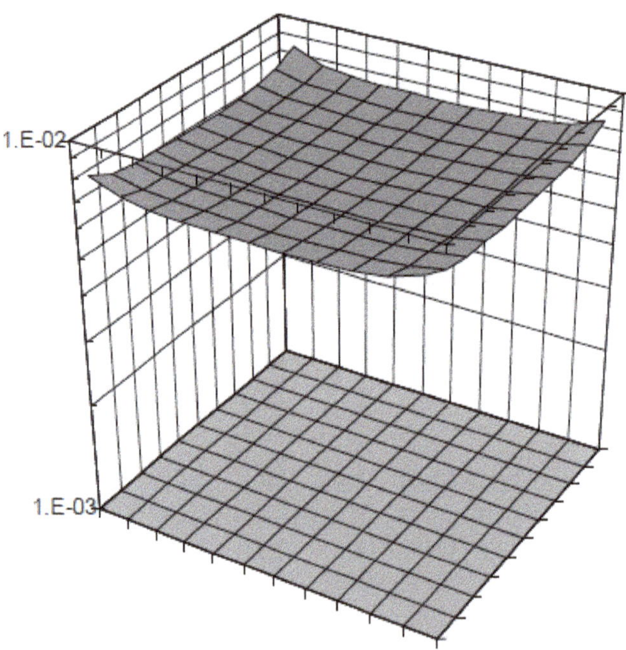

Figure 12.15: Self-Impedance at 0.5 MHz on a 2"x 2" Plane Pair with One Capacitor Placed in the Middle. Top: 50 mils Dielectric Separation, with a 100 μF, 1 mOhm, 1 nH Capacitor. Bottom: 2 mils Dielectric Separation, with 100 μF, 7 mOhm, 1 nH Capacitor

As shown in [11], if we have terminated planes, the impedance on the planes will not strongly depend on where we are on the planes, therefore component placement is largely irrelevant. If, however, we do not or cannot match the characteristic impedance of planes and real part of bypass capacitor impedance, component placement does matter.

These points are illustrated by impedance plots generated by a spreadsheet illustration tool [12]. We assume one piece of capacitor located in the middle of a pair of 2" x 2" planes. For this example, we use C = 100 μF, ESR = 0.001 Ohm, ESL = 1 nH. The SRF of this part is 0.5 MHz. The surface plots of Figure 11.15 show the variation of self-impedance magnitude over the plane at 0.5 MHz. The gray bottom area of the graph represents the top view of the planes. The grid on the bottom area shows the locations where the impedance was calculated: the granularity was 0.2". The logarithmic vertical scale shows the impedance magnitude between 1 and 10 mOhm. The impedance surface on the top has a sharp minimum in the middle; here the capacitor forces its ESR value over the plane impedance. However, as we move away from the capacitor, the impedance rises very sharply. At 0.2" away, the impedance is approximately 50 percent higher; 0.4" away, the impedance magnitude doubles. At the corners of the 2" x 2" plane shape, the impedance magnitude is almost 5 mOhm. When changing either the plane impedance or the ESR of capacitor so that their values are closer, the variation of impedance over the plane shape gets smaller. The plot on the bottom shows the impedance surface of the same plane shape and same capacitor in the middle, except we increased its ESR from 1 to 7 mOhm and decreased the plane separation from 50 to 2 mils. Now the impedance surface at SRF varies only about 10 percent over the plane area.

Implementation Examples

The implementation examples are taken from the server modules shown in Figure 12.1. The MP PDN design is illustrated in Figure 12.16. On the top graph, the percentage usage versus capacitance values shows an approximately uniform distribution, characteristic of MP PDNs. The measured impedance profile on one of the supply rails is shown on the bottom graph. The finite granularity of available capacitance and ESR and ESL values results in unequal peaks and valleys. ESR of MLCC parts with several μF and higher capacitance is approaching a couple of milliOhms, making the synthesis of 10 mOhms and higher target impedances difficult. In addition, having just one or two capacitors covering each narrow frequency range comes with higher sensitivity to component tolerances and missing or broken components and results in high possible component dissipation.

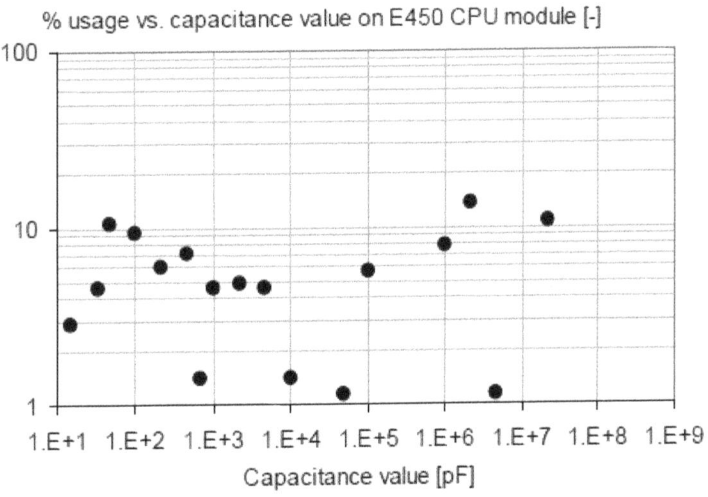

Figure 12.16a: Distribution of Capacitance Values on the E450 CPU Module

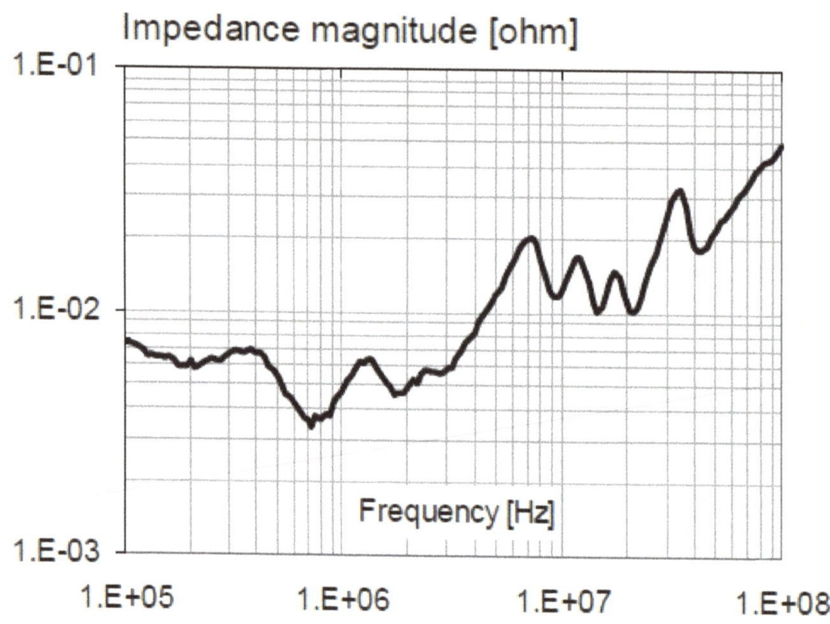

Figure 12.16b: Measured Self-Impedance Profile on the E450 CPU Module

Figure 12.17 is an illustration of the CBD PDN design from one of the V890 supply rails. The top graph shows the capacitance distribution over the entire CPU module. The impedance plot on the bottom is the measured self-impedance of a high-current rail. Low-Q bulk capacitors were complemented by a set of 10 μF and 1 μF MLCC parts. The choice of ceramic capacitors was dictated by the required low cumulative inductance requirement and the restrictions on component placement. Let us note that the minima created by the MLCCs do not go deep; this minimizes the transient noise.

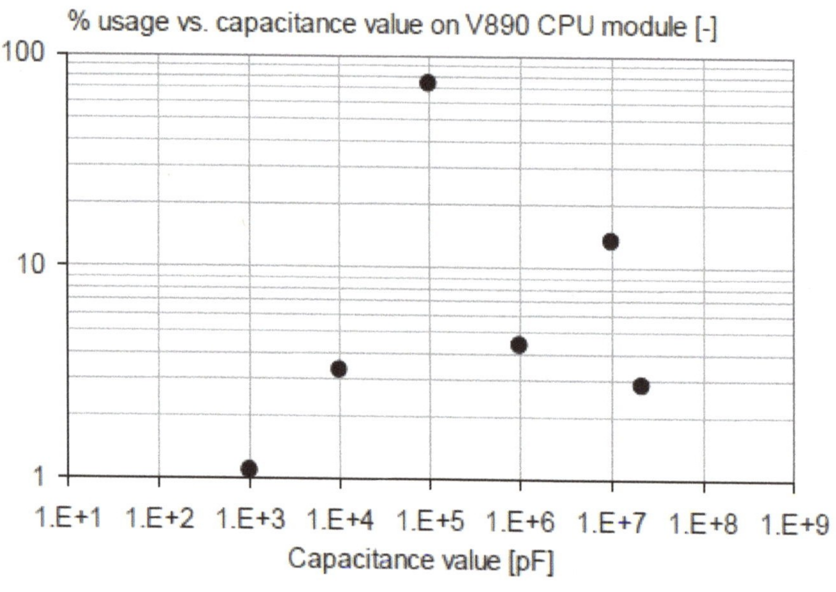

Figure 12.17a: Distribution of Capacitance Values on the V890 CPU Module

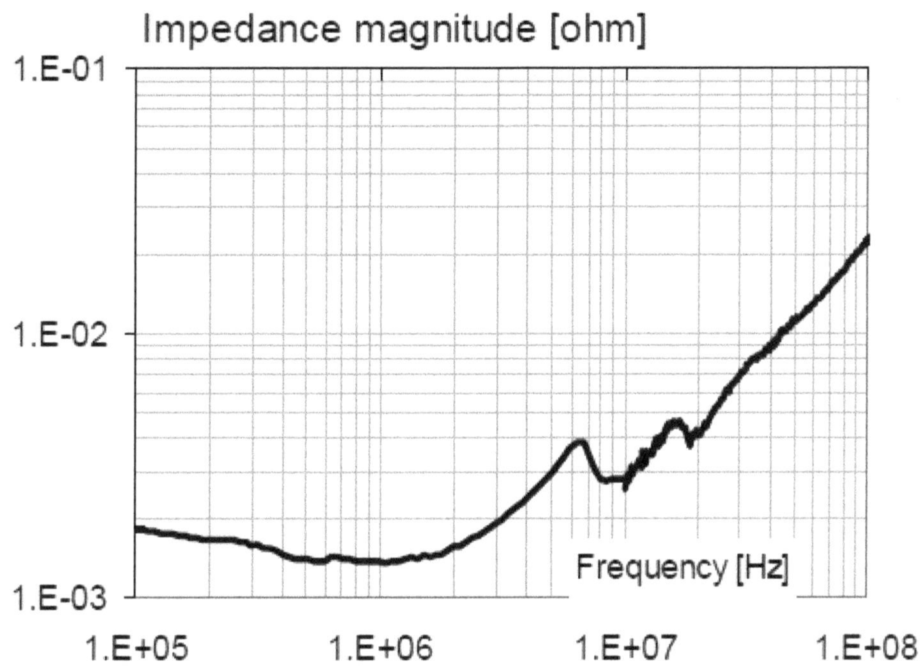

Figure 12.17b: Measured Self-Impedance Profile on the V890 CPU Module

Figure 12.18 and Figure 12.19 are illustrations from the T2000 CPU module. The top graph of Figure 12.18 shows the capacitance distribution over the entire module. The lowest value is 0.1 μF MLCC used for plane termination. Otherwise, each supply rail has just one MLCC value. The impedance plot on the bottom of Figure 12.18 is an illustration of the big "V" PDN design for a medium current rail.

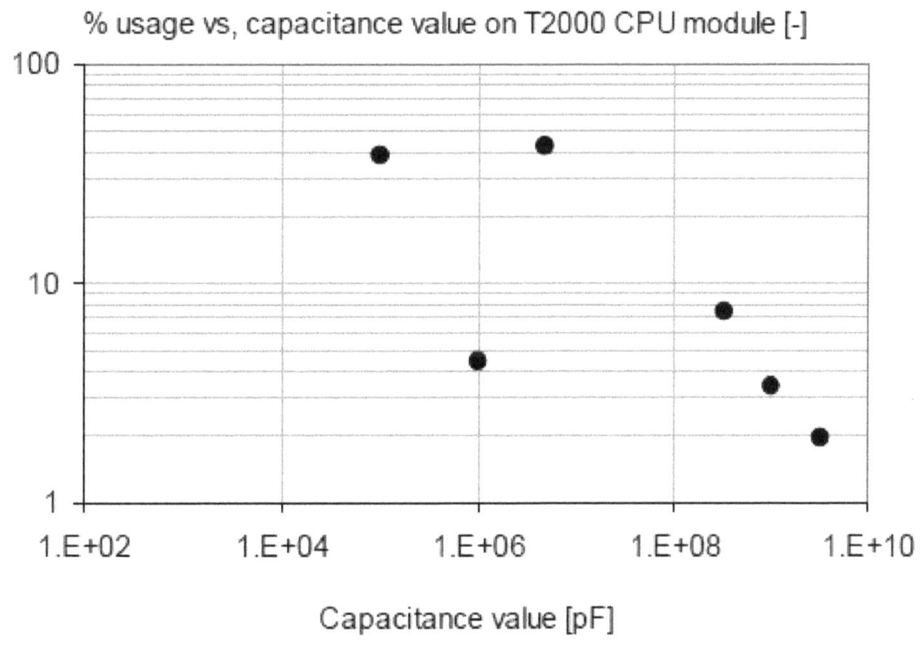

Figure 12.18a: Distribution of Capacitance Values on the T2000 CPU Module, Low-Current Rail with Big "V" Design

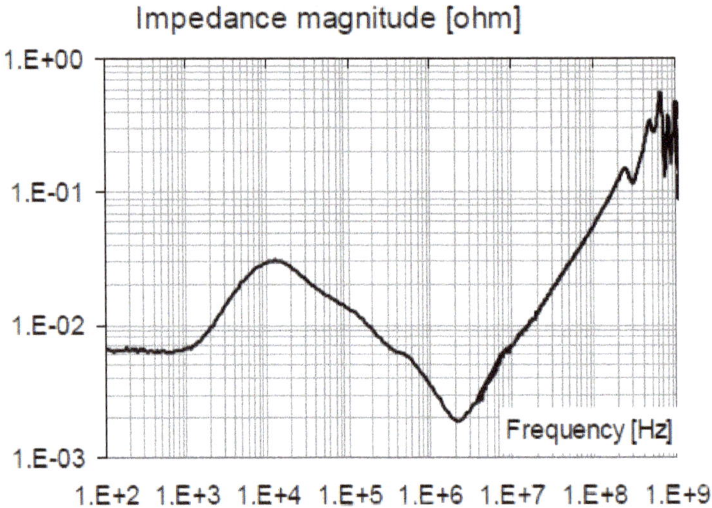

Figure 12.18b: Measured Self-Impedance Profile on the T2000 CPU Module, Low-Current Rail with Big "V" Design

Figure 12.19 is an illustration of a high-current rail. In addition to low-Q bulk capacitors, only one value of MLCC was used, formally creating a big "V" design. The high-current rail had multiple power planes in the stackup. A combination of horizontal plane resistance, component placement, and plane allocation made it possible to almost completely eliminate the impedance minimum at the SRF of the MLCC parts, effectively creating a flat R-L type response characteristic of DMB. (With zero horizontal plane resistance between capacitors, the simulated impedance minimum would be 0.06 mOhm.)

This supply rail was also a signal reference, the modal plane resonances were suppressed by RC termination along the plane edge. Figure 12.20 compares the measured impedance profiles with and without dissipative edge termination (DET) components [13]. One should note the significant reduction of impedance peaks despite the fact that the big "V" design results in a very low cumulative ESR.

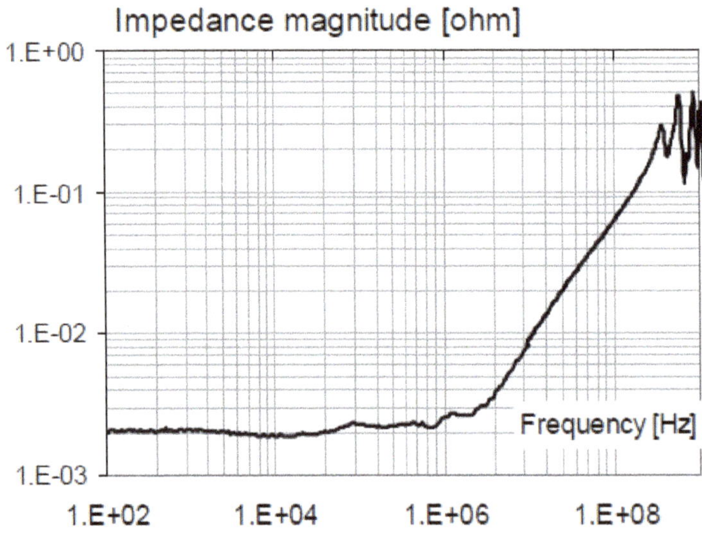

Figure 12.19: Measured Self-Impedance Profile on the T2000 CPU Module, High-Current Rail with Big "V" Design

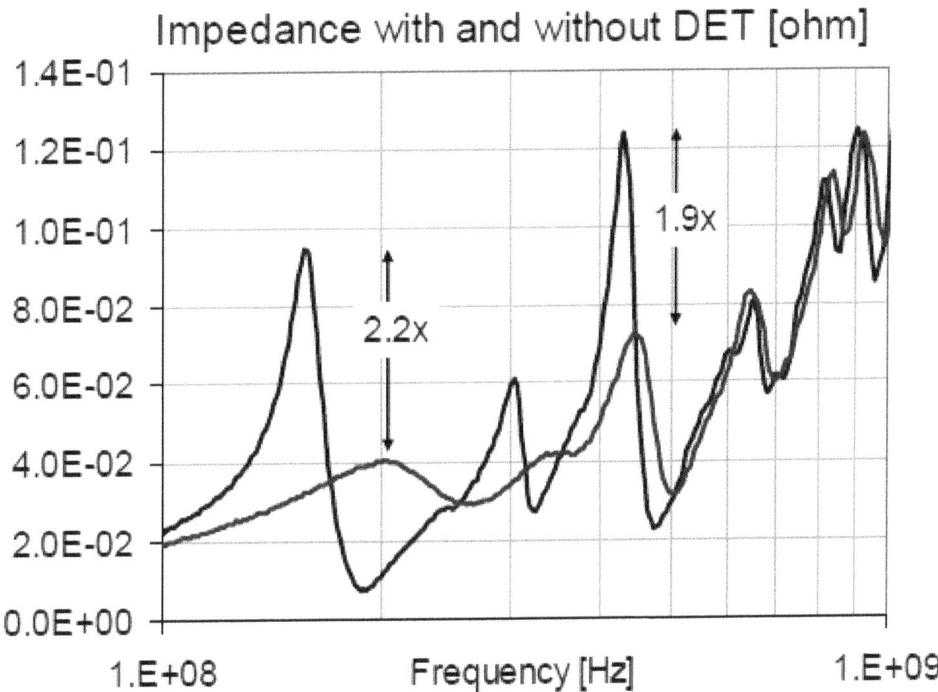

Figure 12.20: High-Frequency Impedance Profile of the Same Rail with and without Dissipative Edge Termination

Conclusion

It has been demonstrated that, if properly implemented, each of the popular PDN design methodologies can be made to work. The multipole approach has high flexibility in terms of component selection but results in higher sensitivity and higher transient noise. The big "V" approach has simple bill of materials (BOM) and low sensitivity but also has higher transient noise.

The DMB then results in very low sensitivity, simple BOM, and the lowest possible noise.

References

[1] Ravi Kaw et al., *Towards Developing a Standard for Data Input/Output Format for PDN Modeling & Simulation Tools*, IEEE EMC 2005 Symposium, August 8-12, 2005, Chicago, Illinois.

[2] Drabkin et al., *Aperiodic Resonant Excitation of Microprocessor Power Distribution Systems and the Reverse Pulse Technique*, Proceedings of EPEP 2002, p. 175.

[3] Larry D. Smith et al., *Power Distribution System Design Methodology and Capacitor Selection for Modern CMOS Technology*, IEEE Tr. AdvP, Vol. 22, No. 3, August 1999, pp. 284-291.

[4] Steve Weir, *Bypass Filter Design Considerations for Modern Digital Systems, A Comparative Evaluation of the Big 'V,' Multi-pole, and Many Pole Bypass Strategies*, DesignCon East 2005.

[5] István Novák, Leesa M. Noujeim, Valerie St. Cyr, Nick Biunno, Atul Patel, George Korony, and Andy Ritter, *Distributed Matched Bypassing for Board-Level Power Distribution Networks*, IEEE Transactions on Advanced Packaging, Vol. 25, No. 2, May 2002, pp. 230-243.

[6] Alex Waizman and Chee-Yee Chung, *Extended Adaptive Voltage Positioning (EAVP)*, Proc. of EPEP conference, October 23-25, 2000, Scottsdale, Arizona.

[7] Bruce Archambeault, *The Effect of Decoupling Capacitor Distance on Printed Circuit Boards Using Both Frequency and Time Domain Analysis*, Proc. of IEEE EMC Symposium, 2005.
[8] www.mcm.dupont.com/MCM/en_US/Products/ embedded_passives/emb_pass_pcl.html.
[9] www.oakmitsui.com/pages/advancedTechnology/faradFlex.asp
[10] ww.3m.com/us/electronics_mfg/microelectronic_packaging/materials
[11] István Novák, *Tolerance Calculations in Power Distribution Network*, XCell Summer 2004, Xilinx.
[12] http://home.att.net/~istvan.novak/tools/Caprange_rev10.xls
[13] *Reducing Simultaneous Switching Noise and EMI on Ground/Power Planes by Dissipative Edge Termination*, IEEE Transactions of CPMT, ITAPFZ, Vol. 22, No. 3, August 1999, pp. 274-283.

Chapter 13—PDN Design Strategies: Ceramic SMT Decoupling Capacitors—What Values Should I Choose?

James L. Knighten, Senior Staff Engineer, NCR Corp.
Bruce Archambeault, Distinguished Engineer, IBM
Jun Fan, Senior Hardware Engineer, NCR Corp.
Giuseppe Selli, PhD, University of Missouri-Rolla
Samuel Connor, Senior Engineer, IBM
James L. Drewniak, Professor, University of Missouri-Rolla

Introduction

THIS CHAPTER IS the first in a series on strategies for the design of direct current (DC) power distribution networks (PDNs) on modern digital printed circuit boards (PCBs). In high-speed digital circuit designs, the PDN associated with the PCB plays a vital role in maintaining signal integrity (SI) (e.g., necessary fidelity of signal and clock wave shapes) and minimizing electromagnetic noise generation. However, the design of the power distribution system presents an increasingly difficult challenge for digital circuits employing active devices. As integrated circuit (IC) technology is scaled downward to yield smaller and faster transistors, the power supply voltage must decrease. As clock rates rise and more functions are integrated into microprocessors and application-specific ICs (ASICs), the power consumed must increase, meaning that current levels (e.g., the movement of electrical charge) must also increase [1, 2].

One design engineer who confronts this design challenge is the SI engineer, whose goal is to ensure adequate fidelity of the individual signal and clock wave shapes on the PCB [2-4]. Another engineer who faces similar design challenges is the electromagnetic interference/compatibility (EMI/EMC) engineer, whose goal is to minimize electrical noise generated by the circuitry to prevent interference with other systems and within the same system [5-8].

While both engineers wrestle with the same physics of the PDN on a digital PCB, practitioners of different design disciplines may view the same physical phenomena differently. For instance, the SI engineer may be more familiar with circuit behavior and analysis expressed in the time domain than with the behavior of electromagnetic waves and analysis expressed in the frequency domain. The EMI engineer's experience is likely just the reverse.

Therefore, these engineers may employ different methodologies and approaches to PCB design. These different design methodologies may sometimes seem contradictory and/or incompatible, but both engineers have similar goals of assuring adequate charge transfer between active devices and the PDN with minimum noise generation. This series of articles is intended to review the state of knowledge of DC power distribution design, offer practical design advice, and address schools of design that appear to offer conflicting advice.

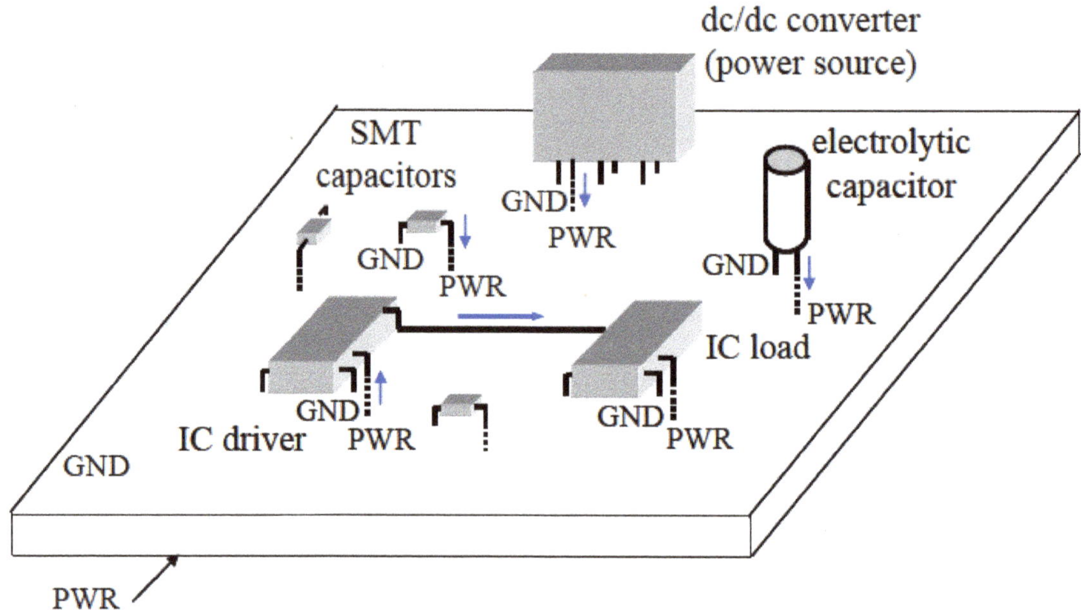

Figure 13.1: A Power Bus (Power and Ground Plane Pair) on a Typical Digital PCB Will Be Populated with Active IC Devices along with Power Delivery Devices Such as DC-DC Converters (e. g., VRMs) and Capacitors of Various Types

The PDN for modern medium- to high-speed digital PCBs is usually formed from one or more pair of conducting planes used as power and ground (power return). The PDN for digital circuitry has evolved over time, as signal and clock speeds have increased, from discrete power supply wires to discrete traces to area fills and ground islands on single/two-layer slow-speed boards to the planar power bus structure used extensively in today's multilayer high-speed PCBs. The low inductance associated with charge delivery from the plane to circuit element allows for the storage of relatively easy-to-deliver charge available all over the board. Often, the term power bus is used to identify an individual plane pair, whereas the term PDN is used for the entire system of supplying power to circuits placed on the PCB. Figure 13.1 illustrates a typical power bus populated with an assortment of devices often found on digital PCBs, e.g., DC-DC converters, ICs, and capacitors. As speeds of active devices have increased, digital data rates have escalated and signal rise and fall times have dropped so that the frequency regime of operation on the PCB has risen into the GHz band. Operation at high frequencies can blur the boundaries between circuit behavior and electromagnetic behavior.

Noise is generated in the power bus when a digital active device (integrated circuit or transistor) switches between its high and low logical states (switching noise) [5], or it can be coupled to the power bus when a high-speed signal transits through the power bus by signal vias (transition noise) [9, 10]. Noise generated in the power bus can be easily propagated throughout the board.

Propagated noise can affect the operation of other active devices (SI) as well as radiate from the PCB (e.g., EMI). At the PCB level, there is no way to eliminate the production of noise by IC devices.

However, a good PCB design can ensure that the generated noise be constrained to a level that permits successful circuit operation and the resulting low levels of radiation produced do not violate regulatory requirements. The use of decoupling capacitors are one of the key elements in achieving this goal, along with the board stack up design, power/ground plane pair, usage of losses, power islands, board edge termination, etc.

The Power Bus Function

The PDN has two primary purposes. The first purpose of the PDN is functionality. The PDN is a charge storage and delivery system that supplies charge (current) when an IC switches state and requires additional current. As seen in Figure 13.2, the voltage at the device varies with current draw from the PDN because the PDN impedance is non-zero. The lower the value of the PDN impedance, the lower its effect on the device voltage will be. If sufficient current is not provided, the IC may experience a functional failure. The voltage (V_{device}) at the IC power pin fluctuates with charge demand to manifest an AC voltage ripple that is added to the DC level. The magnitude of this ripple is related to the magnitude of the current and the PDN impedance, which is discussed in succeeding sections.

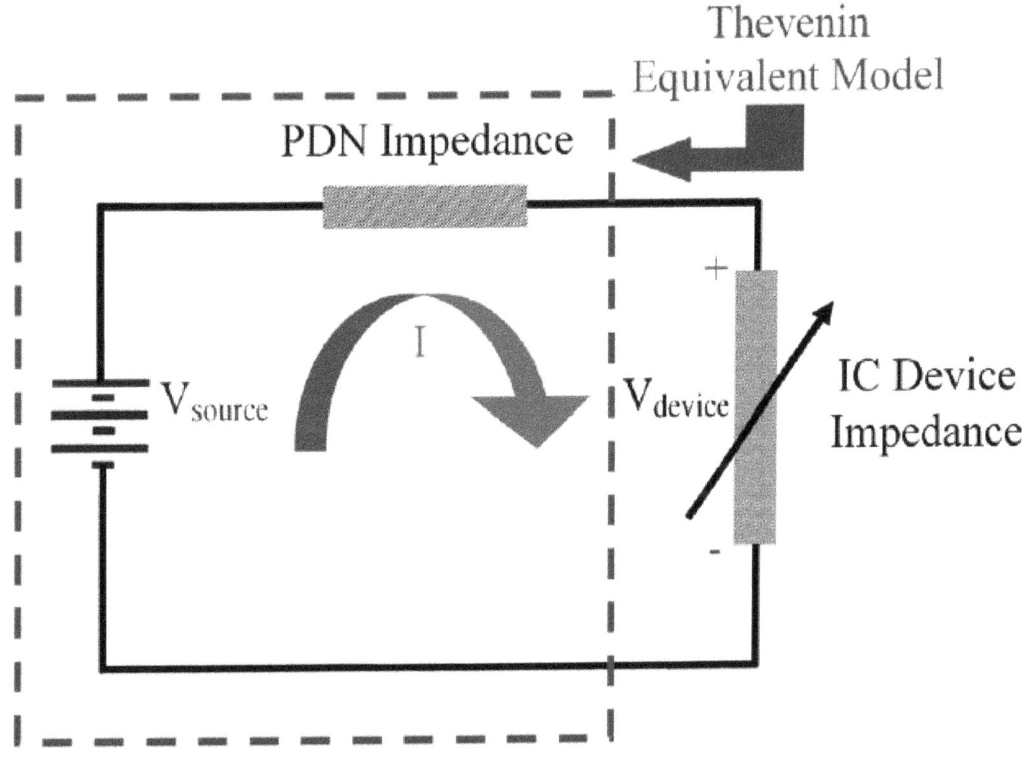

Figure 13.2: A Digital IC Is Powered by a PDN. Changes between Low and High States Cause a Changing Current Demand from the IC, which Causes the DC Voltage across the IC to Fluctuate. This Voltage Fluctuation Is Inevitable and Represents an AC Voltage Ripple on the DC Voltage Level, which Is a Source of AC Noise on the PDN.

The second purpose of the PDN is to reduce or minimize the noise injected into the power and ground-reference plane pair and thus reduce the potential of noise propagation in the board and EMI emissions from the circuit board. There are several mechanisms for EMI emissions; for instance, the edge of a board may be near the seams of a metal enclosure or an air vent area, allowing this noise to escape the enclosure. Alternatively, PDN noise may couple onto input/output (I/O) connector pins or onto a grounded cable shield and be directly coupled out of the metal enclosure through any of the cables. There are a variety of coupling mechanisms that are possible once this noise is created. To avoid undesirable consequences from noise on the PDN, the impedance of the PDN should be low over a wide frequency range that includes the spectrum of the critical signals and their harmonics.

The Decoupling Capacitor

A PDN is comprised of several elements, including the voltage regulator module (VRM), bulk capacitors, surface mount technology (SMT) decoupling capacitors, and power/ground plane pairs (power bus). The effectiveness of each element in delivering sufficient charge with adequate speed is not uniform.

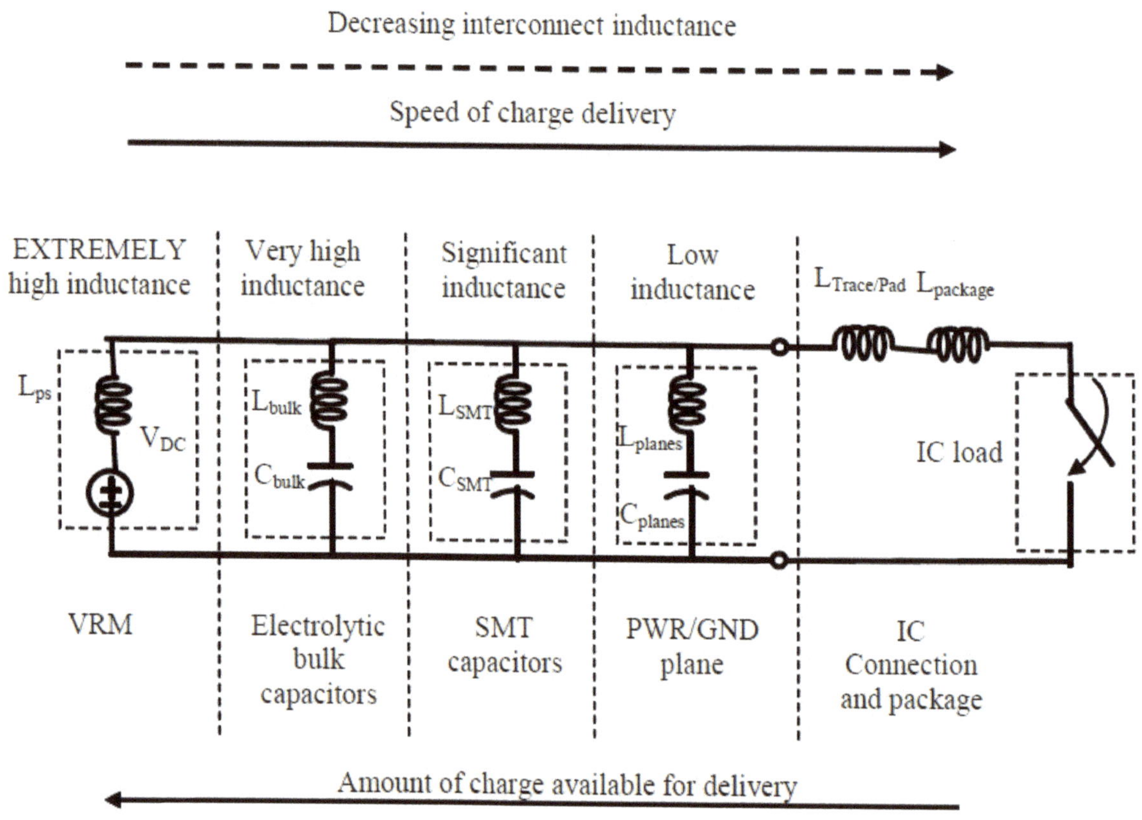

Figure 13.3: A Hierarchal Design of Capacitive Decoupling Provides Both the Needed Quantity of Available Charge and the Required Rapidity of Delivery

A charging hierarchy, as seen in Figure 13.3, exists based on the rate of charge delivery (usually impeded by distance and inductance) and charge storage capacity [2].

The VRM (e.g., DC-DC converter), the largest source of charge, is able to store and release a lot of charge, but it cannot rapidly deliver charge due to the large inductance connecting it to the PDN. It cannot keep up with charge demands that vary or oscillate with rapidity greater than a megaHertz. Hence, it cannot deliver charge in a timely manner when the circuits demanding charge have time constants that are much shorter than one microsecond. Bulk capacitors constitute the second largest source of charges in this hierarchy and are typically capacitors with values that range from a few hundred microFarads to a few milliFarads. These components are able to supply charge with sufficient speed to meet the demands by systems characterized by time constants as low as a few hundred nanoseconds and even shorter.

Decoupling capacitors, sometimes referred to as high-frequency ceramic capacitors, are the second to last category of components in this charging hierarchy [2]. Decoupling capacitors usually exhibit capacitance values from a few tens of nanoFarads to as high as a few microFarads. These capacitors can usually support charge demand from circuits with time constants as low as a few tens of nanoseconds.

The power (PWR)/ground (GND) planes form the last component in the charging hierarchy and can usually deliver charge to circuits whose time constants are shorter than a few tens of nanoseconds, e.g., a charge demand frequency above several hundreds of MHz. The VRM and the bulk capacitors are usually few in number and are located in specific areas of the PDN due to their dimensions and other constraints. High-frequency decoupling capacitors are usually large in number and are typically easily located with a greater flexibility. A subsequent article will address where to place high-frequency capacitors for effective decoupling. Figure 13.4 shows a typical impedance profile of a decoupled PDN.

Impedance plots (driving point, or Z11, and transfer impedance, or Z21) versus frequency are often used to evaluate the behavior of a power bus. In general, Zij is defined as:

$$Z_{ij} = \frac{V_i}{I_j} | I_k = 0 \; for \; k \neq j$$

...where V_i is the voltage at a location on the power bus, labeled, port i and I_j is the current at a location on the power bus, labeled, port j, and all other ports are open circuited, i.e., $I_k = 0$ for $k \neq j$.

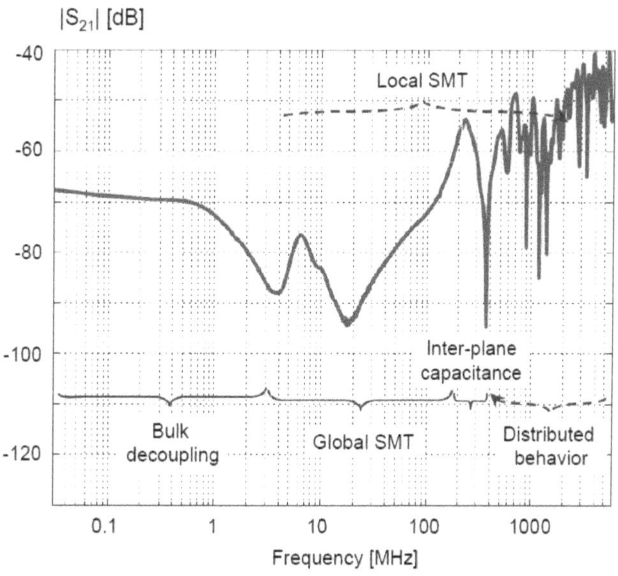

Figure 13.4: An |S21| Profile of a Decoupled PDN Shows Typical Ranges of Effectiveness of Contributors to the Hierarchal Decoupling Strategy. S21 Is Related to the Transfer Impedance between Two Points on a PDN. Much of This Article Refers to "Global" SMT Decoupling. Local SMT Refers to Location Dependant Decoupling That Is Addressed in Future Articles.

Therefore, Z_{11} provides an indication of the voltage created by the injection of noise current. Z_{21} indicates the noise transmission from noise source to anywhere on the board. Z_{21} is very useful for circuit susceptibility and radiated emission studies. Z_{ij} is a vector quantity in that it has both magnitude and phase. For these types of studies, in fact, often solely the magnitude is examined.

Interconnect Inductance

The decoupling capacitor exhibits parasitic inductance and resistance in addition to its capacitance. The parasitic inductance consists of an inductance associated with the capacitor itself (equivalent series inductance [ESL]) and inductance associated with the means of connecting the capacitor between power and ground planes (inductances associated with the solder pads used to secure the capacitor to the PCB and any traces and/or vias used to make the electrical connections). This is illustrated in Figure 13.3 as L_{SMT} for high-frequency SMT capacitors and L_{BULK} for bulk capacitors. The parasitic inductance impedes changes in the current; hence, it impedes the prompt availability of charge. The parasitic inductance and resistance when combined with the device's capacitance form a series resonant circuit with an impedance that dips to a minimum at the frequency where the inductive and capacitive reactances cancel, as shown in Figure 13.5.

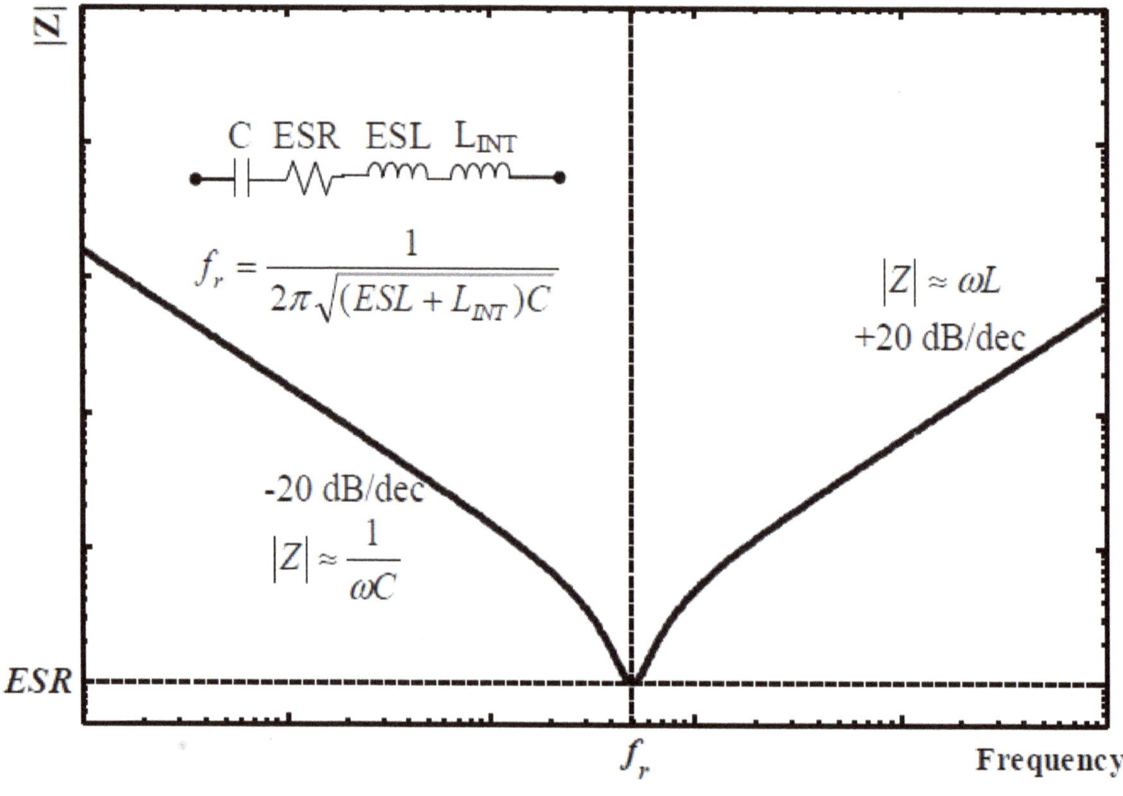

Figure 13.5: Parasitic Elements of the Capacitor (Resistance and Inductance), Along with Inductance of the Traces, Pads, and Vias That Connect the Capacitor to the Planes Form a Series Resonant Circuit That Exhibits an Impedance Minimum at the Frequency Where the Reactances Cancel. Lowering Inductance Moves This Resonance Higher in Frequency, Causing the Capacitor to Function as a Capacitor at Higher Frequencies, Demonstrating Why Inductance from Traces, Pads, and Vias Can Be Detrimental to Effective Decoupling

At higher frequencies, the capacitor behaves inductively and is ineffective in decoupling [2, 5]. If the resonant frequency is shifted higher in frequency by lowering the parasitic inductance, decoupling can be made more effective at higher frequencies. Since the capacitance is fixed as the value of the component, design attention is usually paid to the inductance associated with the capacitor and its interconnect. Low values in interconnect inductance can often be achieved by careful attention to the design of solder pads lands with low inductance properties along with having no traces in the ideal case, or in the realistic case, very short traces connecting them to the planes [4].

The design of the land patterns for the decoupling capacitors has evolved during the years [2]. The parasitic inductance of the interconnects has been lowered from several nH to less than 1 nH. In brief, a good PDN design is characterized by a low interconnect inductance between each decoupling capacitor and the PDN itself. Figure 13.6 illustrates the connection of an SMT capacitor to a power bus, which provides a parasitic inductive component from the current path above the plane (Loop 2) and a parasitic inductive component associated with current flow between the planes, first in the via and then returning as displacement current (Loop 1). As stated earlier, the parasitic inductance associated with current flow above the plane also includes effects from the solder pads that connect to the capacitor and any traces used to connect solder pads to the vias.

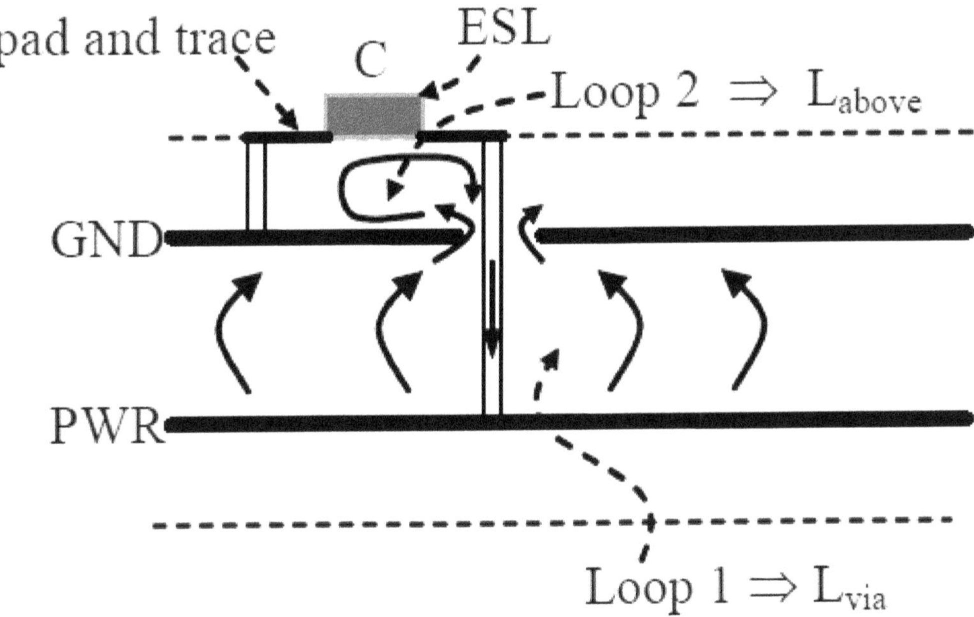

Figure 13.6: The Connection of an SMT Decoupling Capacitor to the Surface of a PCB Produces Parasitic Inductance Associated with Current Flow in the Loop above the Plane and Parasitic Inductance Due to Current Flow between the Planes

The decoupling capacitor's capacitance, along with the overall parasitic inductance and its ESR, forms a resonant resistor-inductor-capacitor (RLC) circuit, with a time constant or envelope proportional to the overall parasitic inductance. The time constant of a series RLC resonant circuit is given by:

$$\tau = \frac{2L}{R}$$

The lower the value of the inductance, the faster the capacitor can supply and store charge and the more importance that is attached to the distance of the decoupling capacitor from an IC in achieving effective decoupling capacitor behavior.

The ESL of the decoupling capacitors is a function of the length, width, and height of the capacitor itself. Due to improvements in the material selection and manufactory technology, the size of SMT decoupling capacitors have shrunk from the early 1206 package size (120 mils length x 60 mils width) down to the more recent 0201 (20 mils length x 10 mils width) package size, allowing a significant reduction of the equivalent series inductance, which is always less than 1 nH.

The ESL, as well as the ESR, is usually measured by employing impedance analyzers and/or network analyzers. In both cases, special fixtures are utilized along with calibration procedures and measurement techniques in order to minimize the parasitic elements associated with the measurement setup itself [11]. Values reported by capacitor manufacturers are influenced by the specific measurement techniques employed and should be viewed critically when the use of the specific values of parasitic elements is desired.

A very large selection of decoupling capacitors is available to designers. Given the wide range of package sizes, materials (electrolytic, tantalum, or ceramic X7R, X5R, Y5V, etc.), and manufacturing technologies (MLC, low-inductance chip capacitor [LICC], interdigitated configuration [IDC], or low-inductance chip array [LICA]), a good PDN design might be carried out in various ways. Ceramic capacitors are characterized by lower ESL than electrolytic and tantalum capacitors, even if the latter are usually available in the same package size. In fact, the multilayer configuration of the ceramic capacitors (e.g., MLC) allows controlling the values of the equivalent series inductance by adjusting the height, width, length, and number of pad connections accordingly [2]. Several fabrication technologies are available for ceramic capacitors, i.e., the reverse geometry or LICC configurations, where the current flows into the decoupling capacitor from the wide sides, the IDC, where multiple connections are employed, and the LICA configuration, where the decoupling capacitor is mounted as a flip-chip component [12].

Determining Individual Decoupling Capacitor Values—Differing Approaches

Approach A: The SI Community

Two general approaches have developed in the design community on how to deploy decoupling (high-frequency ceramic) capacitors in order to reduce the impedance of the PDN between frequencies in the range of approximately 1 MHz to a few hundred MHz. A prominent approach, referred to here as Approach A, is used in the SI community and has developed out of the experience of server motherboard design and other high-performance digital PCBs [2], [3]. This approach uses an array of values of decoupling capacitors. This technique generally uses three capacitor values per decade to achieve the flattest PDN impedance versus frequency profile to maintain an upper-bound “target impedance” to provide an upper bound on the AC ripple voltage on the PDN [4, 13].

In Approach A, the capacitor values are typically chosen so that they are logarithmically spaced (e.g., 10, 22, 47, 100 nF). The effectiveness of this approach is somewhat dependant on the value of ESR of the capacitors and the resulting series/parallel resonant (resonant/anti-resonant) frequencies of the decoupling capacitors to maintain the impedance to be below the desired target impedance over the frequency range of interest.

The use of two capacitor values (e.g., 10, 33, 100 nF) or a single capacitor value (e.g., 10, 100, 1000 nF) per decade can be employed, but it does not usually provide a PDN impedance profile that is sufficiently flat [4]. An example of Approach A with differing densities of arrays of capacitance values is shown in Figure 13.7. The target impedance is chosen to be -20 dBOhm (0.10 Ohm) between 10 and 100 MHz. Three decoupling capacitor arrays are considered: a three capacitor/decade array of 100, 47, 22, and 10 nF; a two capacitor/decade array of 100, 33, and 10 nF; and a single capacitor/decade array of 10 and 100 nF. A series inductance of 1 nH and resistance of 50 mOhm are considered in series with each capacitor. The target impedance is exceeded at some frequencies due to the parallel resonances.

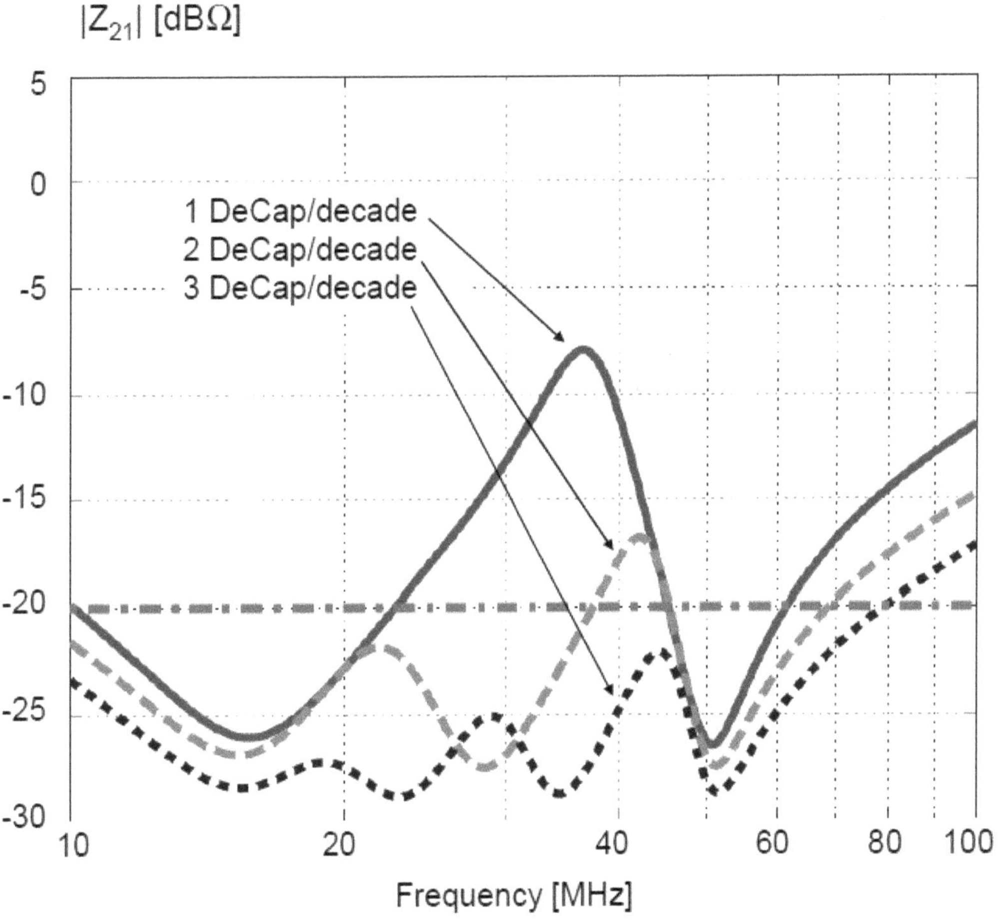

Figure 13.7: Approach A Is Shown with Three Different Densities of Capacitor Values. Three Capacitor Values per Decade Is a Commonly Used Array Density That Produces an Impedance Profile That Is Reasonably Flat

Approach B: The EMI Community

On the other hand, a prominent view in the EMI community for PDN design for high-speed digital PCBs is that the specific values of decoupling capacitors need not be as carefully chosen as in the previous approach [5]. This design methodology, Approach B, addresses the high-frequency ceramic decoupling specifically and employs the largest value of capacitance available in the specific SMT package size to yield a PDN impedance profile that is acceptably flat. For the same number of high-frequency ceramic decoupling capacitors, more total capacitance is often achieved in Approach B than with Approach A.

A comparison of the two approaches in the frequency range between 1 kHz and 1 GHz using a 2D cavity model method, allowing parallel plane characteristics to be included is shown in Figure 13.8 [14, 15].

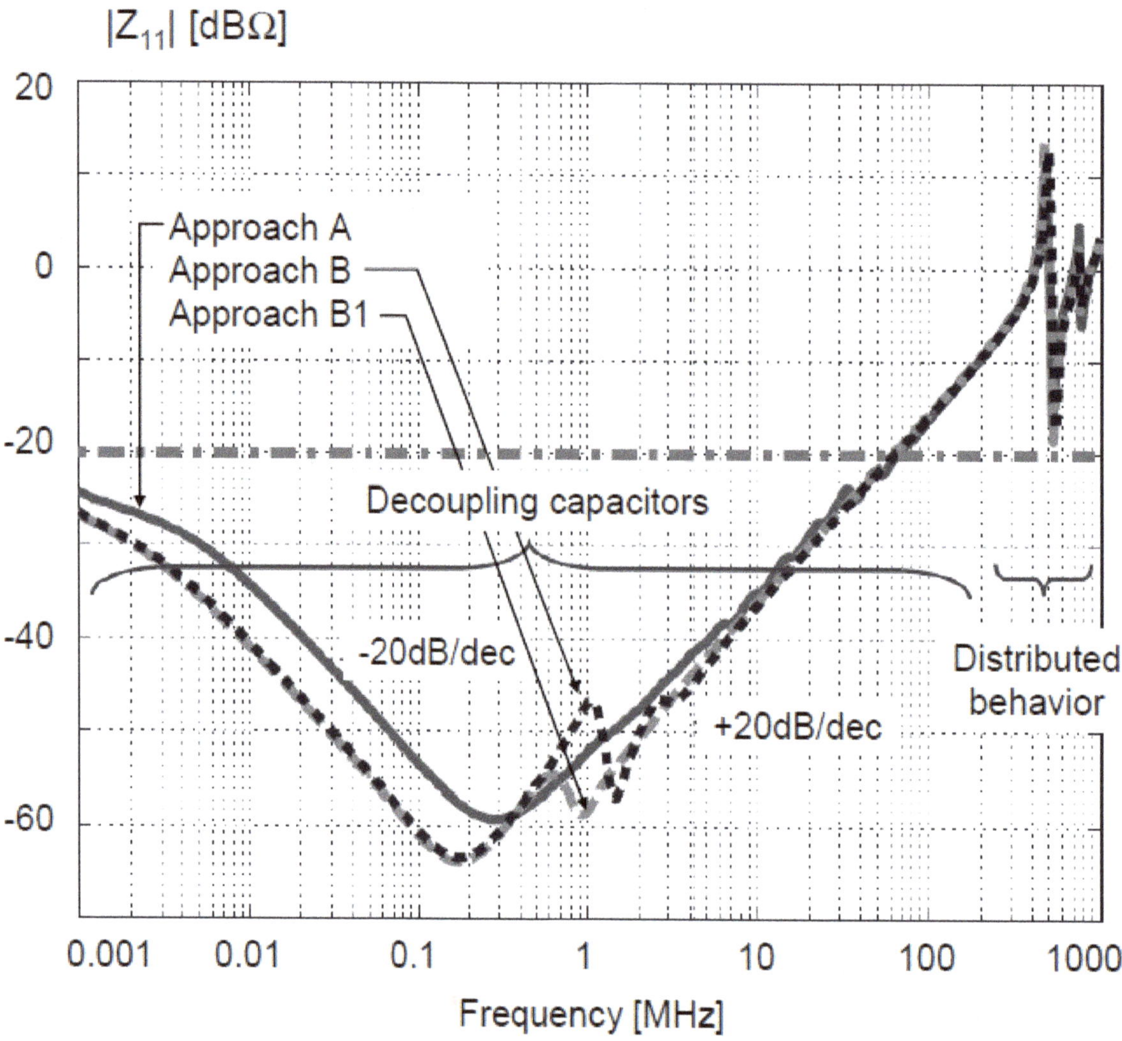

Figure 13.8: The Driving Point Impedance, Z11, of a Power Bus Showing a Comparison of Design Approaches A and B for Choosing the Values of Individual Decoupling Capacitors in the Frequency Range from 1 kHz to 1 GHz. Both Approaches Meet the Desired Target Impedance Requirements in the Frequency Range of Interest. The Impedance in Approaches B and B1 Reach an Initial Minimum at a Lower Frequency than Approach A Because the Overall Capacitance in These Approaches Is Higher. The Impedance Minimum for Approach A Is Not as Low as the Minimum for the Other Approaches because the Net ESR of the Capacitors in Approach A is Higher than the Net ESR in Either Approach B or B1

The PDN dimensions correspond to a PCB that is 6" x 9" with a single power/ground plane pair power bus of thickness 10 mils. The PCB material is chosen to exhibit a dielectric constant of 4.5 and a loss tangent of 0.02; a relative permeability of unity; and a plane capacitance of 2.426 nF. For each example with this PCB, one bulk decoupling capacitor and 60 ceramic decoupling capacitors were chosen. In addition, it was assumed that the power bus was located at the center of the 62 mil PCB stackup and that all decoupling capacitors were placed on the board's surface, allowing for inclusion of the via interconnect inductance in the simulation.

As in Figure 13.7, a target impedance of -20 dBOhm was chosen. Table 13.1 shows the specific values of capacitors chosen. For each capacitor type, typical values of ESR and ESL were selected from typical values from a specific vendor's catalog for X7R MLC capacitors [16].

A third approach, Approach B1, a subset of Approach B, was included to investigate the effects of making all of the smaller decoupling capacitors in the 0402 package size instead of dividing them between the 0603 (60 x 30 mils) and the 0402 (40 x 20 mils) package sizes. Figure 13.8 shows the driving point impedance, $|Z_{11}|$ of the PDN and Figure 13.9 shows the transfer impedance, $|Z_{21}|$.

Capacitor Description					# of Caps		
Value (nF)	ESR (mΩ)	ESL (nH)	Inter-connect (nH)	Type	A	B	B1
3.30E+06	60	15	2	E-lytic	1	1	1
100000	11	1.4	2	1812	4	16	16
47000	12	1.4	2	1812	4		
22000	14	1.4	2	1812	4		
10000	16	1.4	2	1812	4		
4700	16	0.5	1.6	0603	4	24	
2200	19	0.5	1.6	0603	4		
1000	23	0.5	1.6	0603	4		
470	29	0.5	1.6	0603	4		
220	23	0.5	1.6	0603	4		
100	30	0.5	1.6	0603	4		
47	40	0.4	1.35	0402	4	20	44
22	55	0.4	1.35	0402	4		
10	75	0.4	1.35	0402	4		
4.7	104	0.4	1.35	0402	4		
2.2	211	0.4	1.35	0402	4		
Total # of Decoupling Capacitors =					61	61	61
Total Capacitance (milliF) =					4.05	5.01	4.90

Table 13.1: The Details of the Numbers of Capacitors and Their Values Used in the Simulation Comparisons of Approaches A, B, and B1.

While there are differences between the three approaches shown in the figures, all three provide an impedance well below the target impedance up to frequencies in the range of 100 MHz. At low frequencies, Approaches B and B1 provide a lower impedance, which is a manifestation of the higher capacitance used. It is also interesting to note that there is very little difference between Approaches B and B1, except near 1 MHz, where the impedance is already very low compared to the target impedance. Above a few MHz, when the impedance rises are proportional to frequency, i.e., at a rate of 20 dB/decade, there is virtually no difference between any of the methods. In this example, above 100 MHz, the discrete decoupling capacitors do not do a good job of maintaining a low PDN impedance, regardless of the design strategy.

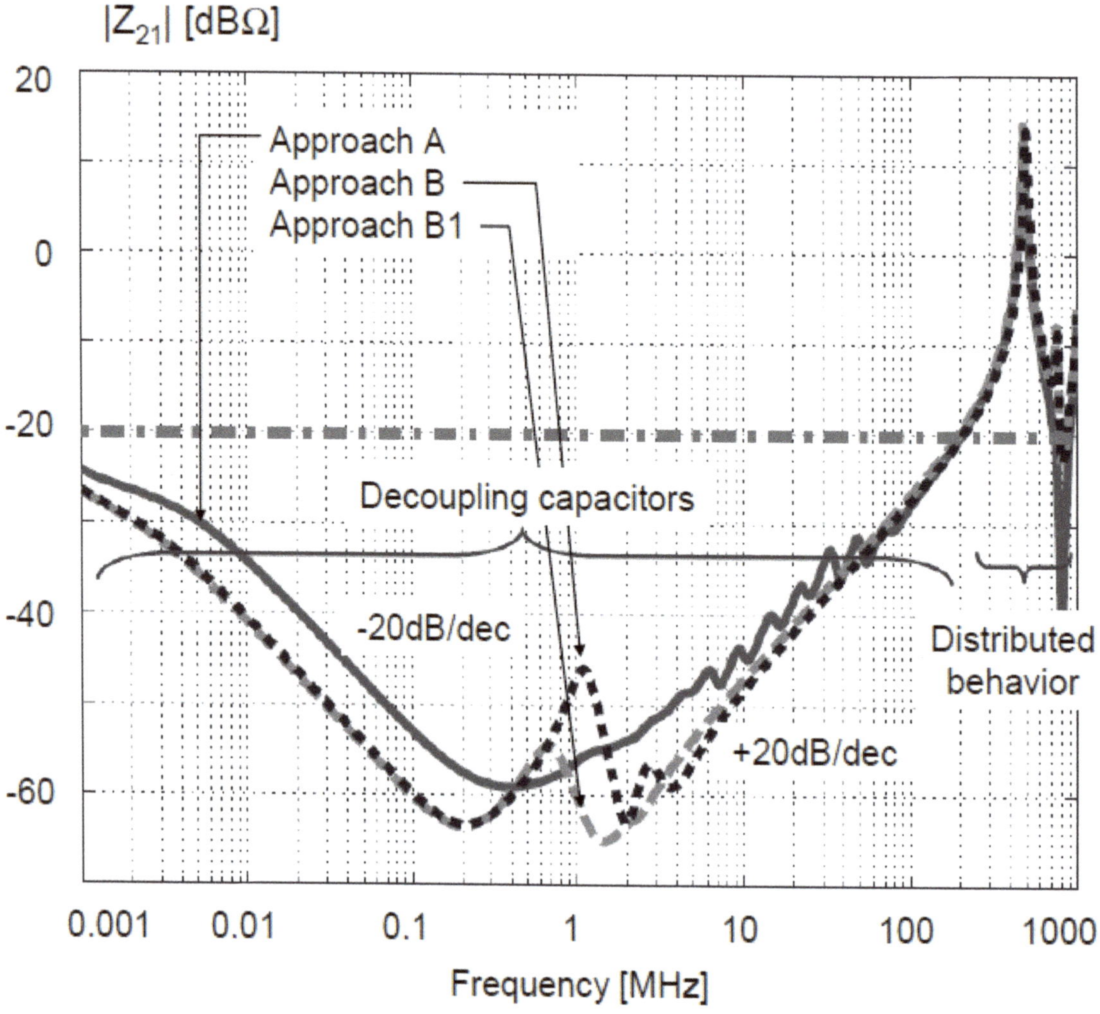

Figure 13.9: The Transfer Impedance Magnitude, $|Z_{21}|$, between Two Points on a Power Bus Showing a Comparison of Design Approaches A and B for Choosing the Values of Individual Decoupling Capacitors in the Frequency Range from 1 kHz to 1 GHz. Both Design Approaches Meet the Desired Target Impedance Requirements in the Frequency Range of Interest

In these examples, it is clear that either approach can achieve the design goal on PDN transfer impedance and have nearly identical performance above frequencies of a few hundred MHz. The use of a single value of capacitance in the largest value in the package size may provide the benefit of simplicity of design and manufacture [7]. Changing the design parameters (PCB characteristics, power bus characteristics, capacitor characteristics, etc.) will alter the impedance curves regardless of the design approach used but will not change the overall conclusion that there is little difference in the PDN impedance profiles between Approaches A and B (and B1).

Although the values of the decoupling capacitors employed are different in the two strategies, the need for lowering the parasitic inductance associated with the decoupling capacitors is consistent [2, 3, 5, 8]. In fact, lowering this inductance shifts all the series/parallel resonant frequencies higher, particularly the last one, allowing the PDN to meet the design specifications on the target impedance in a broader frequency band.

Conclusion

A low-noise PDN on a digital PCB is important to reduce EMI emissions and provide supply voltages to ICs that are sufficiently noise-free to prevent data errors. Decoupling capacitors are usually placed between the power and ground-reference planes to help maintain this voltage. There are two main purposes for decoupling capacitors: to provide sufficient charge/current to the IC as it switches states and to reduce noise between the planes that may result in noise propagation and EMI emissions.

The overall decoupling strategy usually employs a hierarchy of decoupling capacitors. Large capacitors (bulk) are used as high volume charge/current storage reservoirs, but they deliver current slowly due to the relatively high inductance associated with the capacitor and its circuit. Smaller capacitors usually have less inductance, so they can provide current faster. Charge can be provided even faster from the stored charge between the planes in the vicinity of the IC. (Decoupling capacitors in the IC package or on the actual die itself provide the fastest charge.) Typically, the capacitors that deliver charge the fastest are the smallest capacitors, although they have limited amounts of charge they can deliver. Hence, the hierarchical design of decoupling provides some capacitors that yield a limited amount of charge very quickly, and others deliver more charge, but more slowly.

The overall limiting factor of the speed of charge delivery is inductance. Inductance is dominated by the current loop area. Therefore, the spacing between the vias connecting the capacitors to the planes, the spacing between the planes, and the distance between the capacitor and the IC influence the inductance associated with the decoupling capacitor. Often, the only impact a physical designer of a PCB can have is to keep the connection inductance of the capacitors as small as possible. This means keeping the vias that connect the capacitor and the planes close to the capacitor body (avoid traces between pad and via) and using solder pads that minimize the interconnect inductance.

This chapter compares the effectiveness of two common design approaches to maintain a PDN impedance below a desired target impedance over a wide frequency range. Approach A, often used in the signal integrity design community, relies on an array of decoupling capacitor values (the prevalent array being a logarithmic array of three capacitor values per decade of value).

Approach B, more prevalent in the EMI design community, is less structured but demonstrates that use of the largest capacitor value in a package size is sufficient to meet a low target-impedance requirement. Simulations illustrate that either approach can be effective in maintaining a low target impedance. Designers using Approach A often regard the ESR capacitor as an important factor in managing the amplitudes of the alternating resonances and antiresonances that the array of capacitor values produce. Designers using Approach B are often less concerned with ESR. A primary benefit of Approach B resides in its simplicity. Decisions on which approach is better for a particular design should depend on design factors other than the ability to maintain a low PDN impedance, since both approaches can be effective.

General design implications may be drawn based on this discussion of the methods of properly choosing high-frequency ceramic decoupling capacitor values in a modern digital PCB.

Design Implications

- Decoupling capacitors serve the following two purposes:
  - Help meet demands for charge by switching ICs.
  - Reduce noise in power/ground-reference planes.
- Design for a low PDN impedance for good signal and EMI behavior.
- Use a hierarchical array of capacitors to satisfy charge delivery demands. Large (bulk) capacitors provide large amounts of charge slowly. Small capacitors provide lesser amounts of charge rapidly.
- Inductance of the decoupling capacitor circuit determines its effectiveness. ESL, current path loop area, and interconnect detail determine this inductance.
  - Avoid the use of traces to connect solder pad to via.
  - Use low-inductance solder pads for SMT decoupling capacitors.
  - Low-ESL capacitors may enhance decoupling effectiveness at high frequencies.
- Either a logarithmic array of ceramic decoupling capacitor values or just a few values (largest value in a package size) can keep the PDN impedance satisfactorily low over a wide frequency range.

References

[1] M. Swaminathan, K. Joungho, I. Novak, and J. P. Libous, *Power Distribution Networks for System-on-Package: Status and Challenges*, IEEE Transactions on Advanced Packaging, Vol. 27, No. 2, May 2004, pp. 286-300.

[2] L. Smith, R. E. Anderson, D. W. Forehand, T. J. Pelc, and T. Roy, *Power Distribution System Design Methodology and Capacitor Selection for Modern CMOS Technology*, IEEE Transaction on Advanced Packaging, Vol. 22, No. 3, August 1999, pp. 284-291.

[3] L. Smith, *Decoupling Capacitor Calculations for CMOS Circuits, Electrical Performance of Electronic Packages (EPEP)*, Monterey, California, November 1994.

[4] L. Smith and J. Lee, *Power Distribution System for JEDEC DDR2 Memory DIMM, Electrical Performance of Electronic Packages (EPEP)*, Princeton, New Jersey, October 2003, pp. 121-124.

[5] T. Hubing, J. Drewniak, T. Van Doren, and D. Hockanson, *Power Bus Decoupling on Multilayer Printed Circuit Boards*, IEEE Transactions on Electromagnetic Compatibility, Vol. 37, No. 2, May 1995, pp. 155-166.

[6] T. Hubing, T. Van Doren, F. Sha, J. Drewniak, and M. Wilhelm, *An Experimental Investigation of 4-Layer Printed Circuit Board Decoupling*, IEEE International Symposium on Electromagnetic Compatibility, August 1995, pp. 308-312.

[7] Jun Fan, James L. Knighten, Lin Zhang, Giuseppe Selli, Jingkun Mao, Bruce Archambeault, Richard E. DuBroff, and James L. Drewniak, *An Investigation of the Importance of Decoupling Capacitor Values in High- Speed Digital PCBs*, IMAPS Advanced Technology Workshop on High-Speed Interconnect, EMC and Power Aspects of System Packaging for High-Performance Computing, Telecom and Semiconductor Capital Equipment, Palo Alto, California, October 2003.

[8] N. Na, J. Choi, S. Chun, M. Swaminatham, and J. Srinivasan, *Modeling and Transient Simulation of Planes in Electronic Packages*, IEEE Transaction on Advanced Packaging, Vol. 23, No. 3, August 2000, pp. 340-352.

[9] W. Cui, X. Ye, B. Archambeault, D. White, M. Li, and J. L. Drewniak, *Modeling EMI Resulting from a Signal via Transition through Power/Ground Layers*, Proceedings of the 16th Annual Review of Progress in Applied Computational Electromagnetics, Monterey, California, March 2000, pp. 436-443.

[10] M. Tanaka, Y. Ding, J. L. Drewniak, and H. Inoue, *Diagnosing EMI Resulting from High-Speed Routing*

between Power and Ground Planes, IEICE Transactions on Communication, vol. E84-B, no. 7, July 2001, pp. 1,970-1,972.

[11] www.agilent.com, in particular LCR Meters, Impedance Analyzers

[12] J. Galvani, *Low Inductance Capacitors for Digital Circuits*, AVX Technical Information. www.avxcorp.com/docs/ techinfo/li_ti.pdf.

[13] Smith, Larry, RE: Decoupling Capacitor, SI-LIST Subscriber E-Mail Mailing List Thread, June 06, 2002, silist@ freelists.org, www.freelists.org/webpage/si-list.

[14] T. Okoshi, *Planar Circuits for Microwaves and Lightwaves*, Springer-Verlag Berlin Heidelberg, 1985.

[15] Y. Lo, D. Solomon, W. Richards, Theory and Experiment on Microstrip Antennas, IEEE Transactions on Antennas and Propagation, vol. 27, March 1979, pp. 137-145.

[16] AVX SpiCap3: www.avxcorpo.com

Chapter 14—Power Integrity Analysis of DDR2 Memory Systems during Simultaneous Switching Events

Ralf Schmitt, Signal Integrity Engineer, Rambus, Inc.
Joong-Ho Kim, Signal Integrity Engineer, Rambus, Inc.
Chuck Yuan, Signal Integrity Engineer, Rambus, Inc.
June Feng, Signal Integrity Engineer, Rambus, Inc.
Woopoung Kim, Signal Integity Engineer, Rambus, Inc.
Dan Oh, Signal Integrity Engineer, Rambus Inc.

This paper was presented at DesignCon 2006, Santa Clara, California, February 6-9, 2006.

SIMULTANEOUS SWITCHING NOISE (SSN) in systems using single-ended drivers poses significant design challenges as data rates continue to increase. In this paper, we analyze the impact of SSN on a double data rate (DDR2) memory system using a wire-bond package for the controller and operating at 667 MHz. We not only focus on the supply rail where the output driver is located, but also on the other ones where sensitive circuits are. We demonstrate that the noise coupled into these sensitive supply rails through either other supply rails or signals can be significant. In addition, we present a methodology to find data patterns that cause the worst-case supply noise on each supply rail. Finally, the simulated supply noise is correlated with hardware measurements to validate the modeling approach.

Introduction

Today's computing systems require ever-increasing bandwidth even for cost-sensitive consumer applications such as high-definition TV (HDTV). The bandwidth requirements in these systems are reaching levels previously associated with high-performance workstations, servers, and routers.

At the same time, cost is a major design factor for these systems and, as a result, there is a great interest to implement high-speed interfaces using low-cost technologies such as wire-bond packages. The data rate of DDR2 systems is expected to move from the current 400 MHz to 667 MHz and 800 MHz in the near future. At these high data rates, jitter introduced by supply noise can significantly reduce the timing margin in the system. As typical for interface systems using single-ended signaling, DDR2 memory systems can generate substantial current peaks when output drivers are switching simultaneously. These current peaks in return generate simultaneous switching noise (SSN) in the system if the impedance of the power distribution system is not sufficiently low. In particular, many systems using DDR2 memory interfaces are implemented using limited silicon area for on-chip bypass capacitors and packages using wire-bond technology.

Wire-bond packages typically add a substantial amount of inductance to the power distribution network (PDN), resulting in a large impedance of the power distribution system around the package/chip resonance frequency. As a result, there is a significant amount of supply noise in the system by design, especially on the power rails used for output buffers. A way to address this problem is by moving noise sensitive circuits away from the noisy buffer supply.

Additional supply rails are introduced for supply noise sensitive circuits such as phase-locked loops (PLLs) and clock drivers in an effort to minimize the jitter caused by supply noise. During SSN events, the current dissipation on these rails usually does not change directly. However, noise can be generated on these rails by coupling bond wires of different rails, signal wires and supply wires, and ground bounce on the shared ground rail. This coupled noise often has significant impact on the device jitter during SSN events in addition to the noise on the output driver supply rail itself. Therefore, predicting the SSN leading to jitter in the interface requires a power distribution model that includes these sensitive supply rails in the model and accounts for all coupling mechanisms into these rails.

In this paper, we present a methodology to accurately model the supply noise of a 667 MHz DDR2 during SSN events. First, a model for the power distribution system of the interface is presented, which addresses self-induced noise as well as coupled noise on all supply rails. Next, the dependency of each noise component to the data pattern transmitted in the interface is discussed and worst-case access profiles are derived. Finally, simulation results of supply noise and jitter profile on a test system are compared to measurements, showing correlations with good accuracy. Amplitude as well as waveform of supply noise simulated on different rails and for different access patterns in the interface are confirmed by measurements with an accuracy of few percent.

Measurements of jitter sensitivity to local bonding implementations are correlated to noise coupling simulation results, confirming the accuracy of the noise coupling model and emphasizing the importance to include noise coupling into the SSN analysis.

Supply Noise Modeling Methodology for Interface Systems

General System-Level Power Distribution Models

High-speed interface systems require high-quality power supply systems. Power and ground bounce on the supply rails causes signal distortion, affects timing and noise margin, and introduces jitter. As the data rate of interfaces increases, the acceptable timing loss due to supply distortion decreases.

Self-induced supply noise is generated when the current dissipated in the system changes over time. Interface systems, in particular input/output (I/O) systems using single-ended signaling schemes, are causing current changes at every signal transition. A major challenge for the power distribution system is simultaneous switching events, when many interface signals are switching at the same time, causing severe current changes in the system. In order to limit the supply voltage fluctuations caused by these current changes, the impedance of the PDN seen by the interface circuits has to be low over a wide frequency range.

It was shown in [1] that modeling the impedance of the PDN over the entire frequency range of interest requires a model covering several levels of the design hierarchy. Figure 14.1 shows the schematic of a power distribution model for the supply rails used for internal circuits. It models the impedance Z_{PDN} of the power distribution system at low and medium frequency, which is the frequency range most interesting for simultaneous switching events.

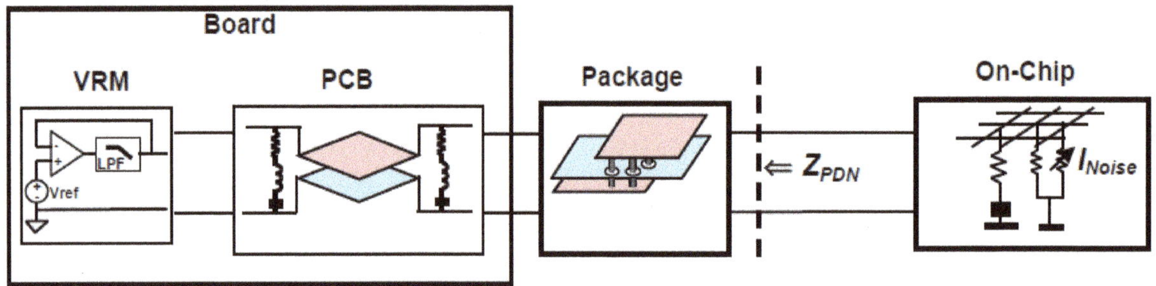

Figure 14.1: Schematic of PDN for Internal Supply Rails at Low and Medium Frequency

The impedance Z_{PDN} seen between the supply voltage and ground by the circuits on the silicon is dominated by components on different system hierarchy levels at different frequencies. At high frequencies, the on-chip distribution system and the package are dominating, while at very low frequencies the voltage regulator module and decoupling capacitors on the printed circuit board (PCB) are determining Z_{PDN}. The profile of Z_{PDN} over the frequency range of interest is used as a figure of merit for the power distribution system.

Power Distribution Model for Interface Systems

A concept widely used in the design of power distribution systems is the target impedance concept. The design methodology in this concept is to keep the power distribution impedance Z_{PDN} below a target impedance Z_{target} over the entire frequency range of interest. The target impedance Z_{target} is calculated as:

$$Z_{PDN}(f) \leq Z_{target}(f) = \frac{V_{Noise}(f)}{I_{Noise}(f)}$$

...where V_{Noise} is the allowed voltage noise and I_{Noise} is the supply current change causing the noise.

In this concept, it is commonly assumed that the supply noise current I_{Noise} is flowing between the power supply node and the ground node.

Therefore, the impedance Z_{PDN} is defined between the supply nodes at the interface between package and chip, as shown in Figure 14.1. For supply rails providing power to internal circuits this is a reasonable assumption for the average currents of larger circuit blocks, even if it is incorrect for an individual logic gate. In complementary metal oxide semiconductor (CMOS) technology, e.g., the current is flowing during pull-up from the positive supply (VDD) to an internal capacitance at the output node, and at a later time during pull-down from the internal capacitance to the negative (VSS) node. Strictly speaking, therefore, the gate current is not flowing directly between VDD and VSS. However, for a larger circuit block the superposition of pull-up and pull-down currents of many gates appears as a direct current between the supply rails VDD and VSS.

For interface systems, the assumption that the current flow is mainly between the supply rail (e.g., VDD) and ground is not true anymore. As an example, Figure 14.2 shows the schematic of an interface using voltage-mode signaling. In such an interface, the main current components will often still be flowing directly between the supply rail and ground, since both pull-up and pulldown path are conducting typically at least temporarily during transitions. Additional current components, however, will charge the external load capacitance from the supply voltage VDD during pull-up and discharge this capacitance to ground at a later time during pull-down.

Even if both charging and discharging current are occurring with the same frequency, the response of the power distribution system to these two excitations cannot be described by the single PDN impedance Z_{PDN} defined in Figure 14.1, mostly due to the phase shift between charging and discharging currents in the package and PCB supply. In order to accurately model the current flow during switching operations, the signal traces and the load capacitances have to be added to the PDN model of the transmitter.

Neglecting these elements in an SSN analysis introduces errors into the analysis, overestimating the supply noise during SSN events. However, as long as the direct current between the supply rails of the transmitter dominates the total switching current, the resulting accuracy is often sufficient.

This is not true any more, however, for modern high-speed interfaces like DDR2, which are using current mode signaling. Figure 14.3 shows an example of an interface using current-mode signaling.

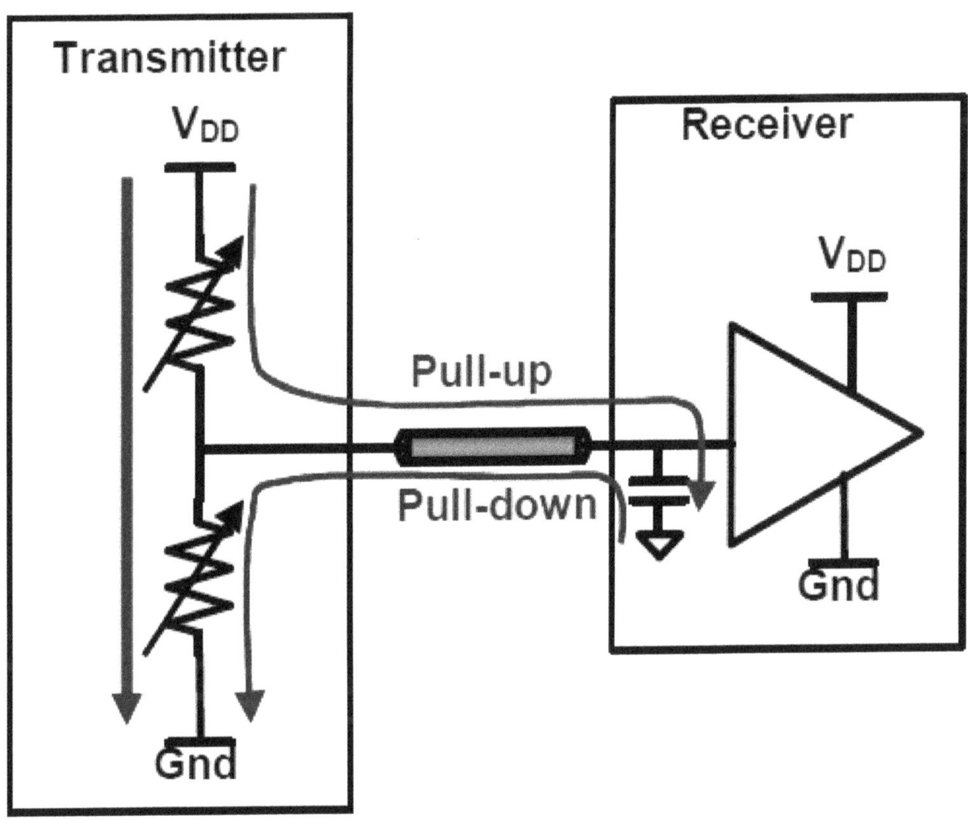

Figure 14.2: Current Components in Interface Using Voltage-Mode Signaling

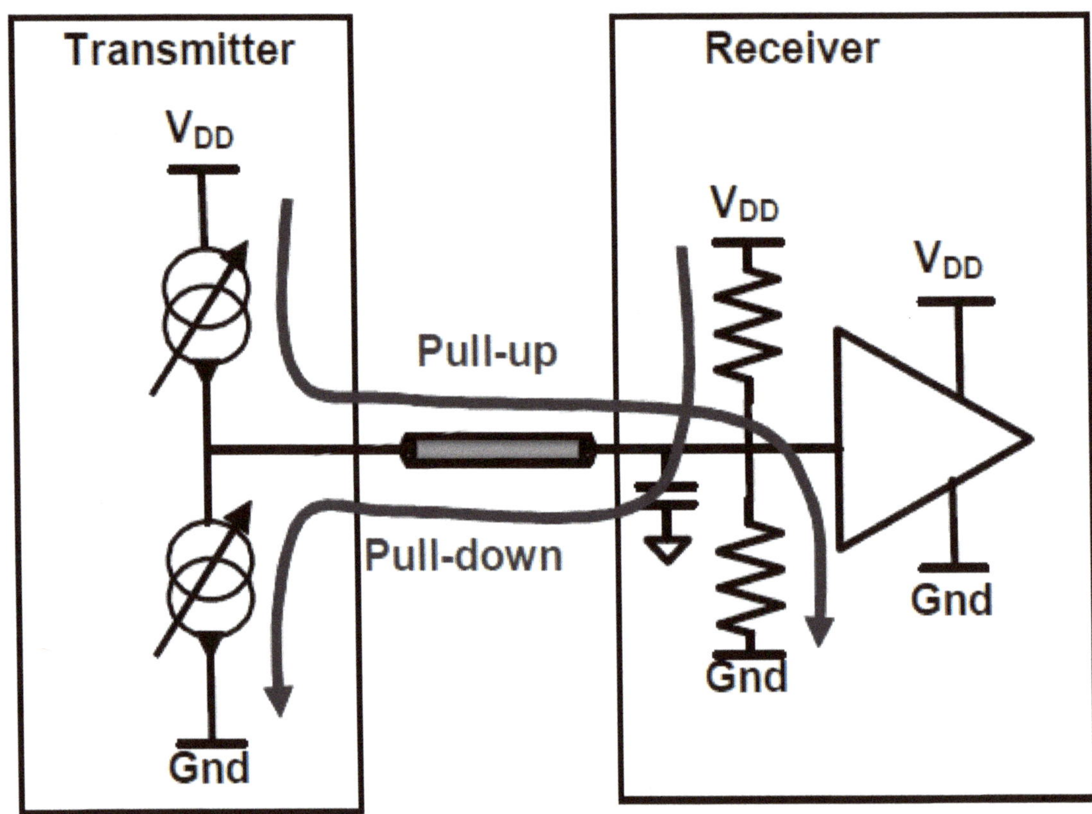

Figure 14.3: *Current Components in Interface Using Current-Mode Signaling and On-Die Termination*

In current-mode signaling, the major current components are flowing between one supply rail of the transmitter and load resistors at the end of the signal channel. These load resistors can be either on the receiver chip or on the PCB. In both cases, in order to model the current flow during switching operations, it is necessary to add the signal channel and the load resistor to the power distribution model.

Thus, in order to model SSN noise in a high-speed interface, it is necessary to combine the model of the PDN of both the transmitter and receiver with a model of the signal traces in the system. This has the additional advantage that noise coupling between signal traces and the supply network can also be modeled in the same model. A simple model of the power distribution of, for example, the transmitter alone is not sufficient, since it does not model the current distribution in the system correctly during switching operation.

Since there are multiple current loops in the system contributing to supply noise in the system, it is also not possible anymore to define a single power supply impedance Z_{PDN}. Separate target impedances have to be at least defined for each current loop in the system.

There is a further limitation of the target impedance concept that limits its usability for the design of interface systems. The target impedance concept assumes that the worst-case current excitation can be modeled as a single-frequency excitation. In this case, the supply noise amplitude can be calculated as product of noise current amplitude and power distribution impedance:

$$V_{noise} = Z_{PDN}\ x\ I_{Noise}$$

If, for a given frequency range, the acceptable voltage noise amplitude V_{noise} and a maximum, single-frequency alternate current (AC) excitation I_{noise} are known, then a single value Z_{target} can be calculated for this frequency range. Z_{target} can be a function of frequency, as the acceptable voltage noise amplitude changes over frequency or if the maximum possible (single frequency) current excitation varies with frequency.

However, the target impedance concept is not well suited for cases with largely distributed I_{noise} spectrum, where noise contributions at different frequencies are superimposing. This is the typical case for interface systems, where current component contributions at different frequencies and phase relationships are superimposing. In these cases, it would be possible to calculate the supply noise contribution of each of these spectral components and add them together. However, this is very pessimistic, since it ignores the phase relationship between the different components. Therefore, Z_{PDN} can be used for these cases as a relative figure of merit to compare implementations and identify critical frequencies. For final supply noise verification, however, a simulation in time domain is required, which naturally considers the phase relationship of the noise contributions.

Self-Induced and Coupled Noise in Interface Systems

The goal of our SSN analysis for interface systems is to predict the margin loss in the system due to the simultaneous switching of output drivers. Often, SSN analysis focuses on the self-induced supply noise on the output driver supply rails, e.g., the noise generated by the output drivers on their supply rails during switching. The sensitivity of most single-ended drivers to noise on their supply rails, however, is usually not severe. Thus, this self-induced noise component often has only limited impact on the timing and voltage margin of the interface system. In contrast, noise on supply rails not used by the output drivers, but used for sensitive circuits such as PLLs and clock drivers, can have a much larger impact on system jitter. Therefore, supply noise generated on these sensitive supply rails during simultaneous switching events has to be included in the SSN analysis in order to predict the system margin loss during these events.

There are three major sources for noise coupling into internal supply rails other than the buffer supplies during simultaneous switching events. The first source is signal-to-supply coupling in package and PCB, which can easily occur in wire-bond packages.

Modeling this noise coupling requires a package model that reflects the coupling between individual bond wires. It also requires that all signal lines be excited with worst-case pattern to excite the largest noise amplitude in the sensitive supply rail.

The second source of coupled noise into internal supply rails is supply-to-supply coupling from the output driver rails in package and PCB. Similar to the signal-to-supply coupling, wire-bond packages can easily couple noise between the driver supply rails, which experience large current surges during simultaneous switching events, and bond wires of sensitive supply rails. In addition, supply noise can be coupled between the supply rails on the PCB, especially if both supply rails are sharing the same regulator on the PCB. In this case, the PCB supply and any filter between the different supply rails have to be included in the SSN simulation model.

The third source of coupled noise into internal supply rails is ground bounce on shared VSS nodes inside the chip. The inductance of the ground distribution, especially in the package, causes ground bounce during simultaneous switching events that change the voltage difference between the internal supply rail and the shared ground rail. Another source of ground bounce is the voltage drop on the highly resistive on-chip power distribution (IR drop). As the activity in the circuits on the chip is changing, the distribution of IR drop over the chip is changing as well. Therefore, simulating the noise coupled into internal supply rails, other than the buffer supply during simultaneous switching events, requires a complete model of the driver supply, the internal supply, and the entire signal channel all at the same time.

Driver Model for SSN Analysis

The simulation of supply noise under simultaneous switching conditions requires an accurate model of the PDN under switching conditions as well as accurate current profiles that excite this network.

Ideally, accurate transistor-level transmitter and receiver models could be used for the SSN simulation. These transmitter and receiver models naturally provide accurate current profiles on the supply rails, including feedback effects due to rail collapse. They also provide accurate waveforms on the signal lines modeled in the PDN system, and thus allow modeling the coupling between signal and supply traces. Furthermore, combining the power distribution model with the channel model while using the full transistor-level driver and receiver models all allow the impact of supply noise to be simulated on the signal timing. The disadvantage of full transistor driver and receiver models is the large complexity of these models. Common analog simulation engines such as SPICE are only able to simulate interfaces with a small number of signal lines using transistor-level transmitter and receiver models. Larger interfaces require simplified transmitter and receiver models that can be simulated more efficiently.

IBIS models are simple receiver models commonly used for signal integrity simulations. These models are very efficient to simulate, thus making analyzing a large interface is easily possible. However, common IBIS models do not provide accurate supply current waveforms and do not include the impact of SSN noise on the internal power rail [2]. There are currently activities to improve the current profile accuracy of IBIS models (e.g., [3]), but even with these extensions only some of the interactions between supply noise and current profile are modeled and accuracy of the resulting current profiles is still a concern.

An elegant way to achieve transistor-level accuracy with limited complexity is by using current controlled current sources (CCCS) to duplicate the current profiles of one or a few accurate drivers. In this approach, a single driver is implemented using a transistor-level accurate model. The current profiles at all supply and signal ports of this master driver are measured and duplicated to the corresponding position of other drivers. This creates correct current profiles on the supply net, identical to the current profile expected if all drivers were implemented using transistor-level models, and also accurate waveforms on the signal lines. This approach also preserves the feedback of rail collapse on the drivers, as the master driver experiences the full supply noise caused by all drivers at the same time, reflecting the effect of rail collapse in its port currents. It is also very computationally effective, since CCCS elements require very little simulation effort for an analog simulator such as SPICE.

Duplicating port currents with CCCS circuits requires that the port impedances of the master driver be identical with the port impedances of the duplicated drivers. For the supply ports, this is usually the case, since all drivers share the same power distribution environment. For signal lines, the reflection behavior of signal lines with CCCS-based duplicated drivers should be identical to the reflection behavior seen by the "master" driver. In systems with well-terminated signal buses, this is easily achieved.

If different signal lines are operating at varied frequencies or with different data patterns, separate master drivers are needed for each frequency or pattern. Even in this case, large interface systems can be easily modeled with a small number of transistor-level models, limiting the complexity of the interface model.

SSN Model for DDR2 Test System

Figure 14.4 shows a picture of the DDR2 test system analyzed in this paper. A detailed description of the system can be found in [4].

The system consists of one DDR2 controller and a single rank of two x16 DDR2 devices, operating at 667 MHz. The DDR2 interface consists of 32 data signals operating at a data rate of 667 MHz and 21 address and control signals operating at half the data signal rate.

Figure 14.4: DDR2 Test System Board

The system is implemented on a six-layer PCB. The controller uses a wire-bond package showing that DDR2 interfaces of this data rate can be implemented in a low-cost package system. The interface in the controller chip uses power supply rails that are separated from the power supply of the application-specific integrated circuit (ASIC) core. All power supply rails use individual voltage regulators to prevent noise coupling on the PCB board.

The controller DDR2 interface uses the supply rail VddIO of 1.8V for output drivers, and the supply rail Vddr of 1.2V for other circuits in the interface. Both supply rails share a common ground node VSS. Figure 14.5 shows the power delivery model of the data bus of the interface system. A similar model was added for the address/control bus of the system.

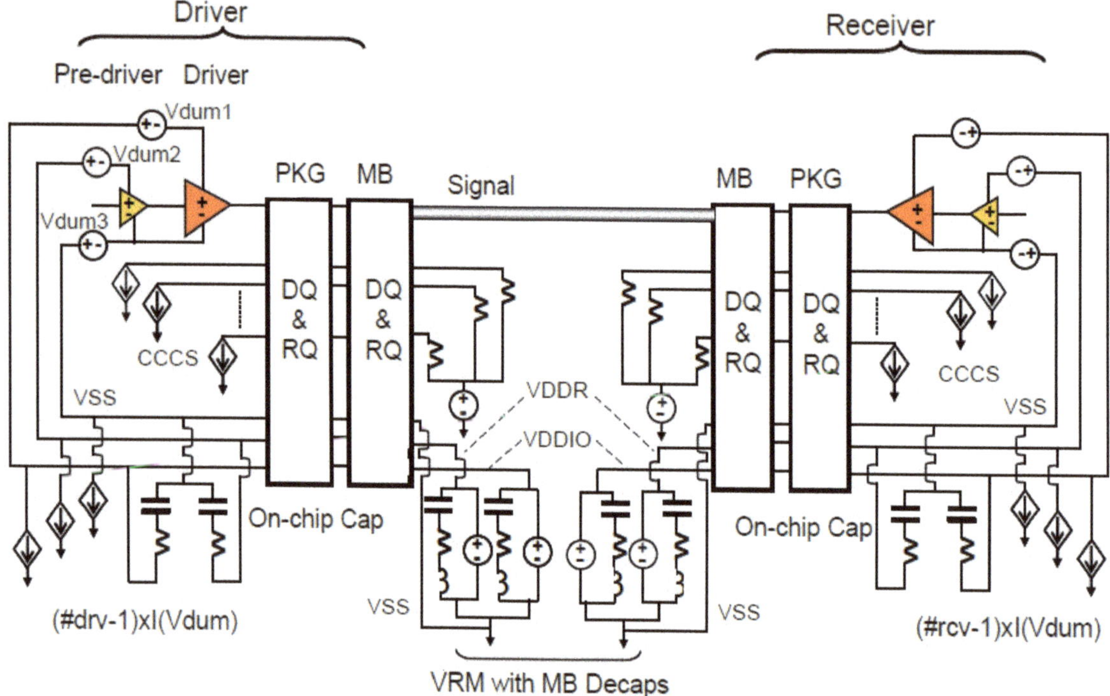

Figure 14.5: Power Supply Model of DDR2 Data Bus for SSN Analysis. CCCS = Dependent Current-Controlled Current Source

The SSN model shown in Figure 14.5 allows simulating supply noise inside the controller as well as inside the DDR2 device during "read" and "write" operations. In the scope of this work, we will focus on the supply noise inside the controller as we verify the controller design. Few details were available about the package and on-chip supply network of the DDR2 device. Therefore, simplified models of the package and the on-chip power distribution of the DDR2 devices were used in this analysis. This has little impact on the noise simulations for supply rails inside the controller, but it makes noise simulations inside the DDR2 devices less accurate.

Special attention was given to the package model of the controller.

Wire-bond packages add a substantial amount of inductance to the supply network. In many cases, the impedance of the supply loops is dominated by the inductance of the bond wires in the package.

Furthermore, coupling between bond wires are a major source of signal-to-supply and supply-to-supply coupling in the system.

Therefore, a 3D field solver was used to extract a package model from a 3D picture of the entire wire-bond section of the package based on the partial inductance concept. This model preserves the inductive coupling between all bond wires. Additional traces and planes in the package can be added using conventional modeling methodologies such as those presented in [1].

Figure 15.5 shows a single transistor-level driver and receiver circuit for the data bus of the system. The remaining 31 data lines are excited using CCCS current mirrors. This configuration requires that all data signals are transmitting the same data pattern. If groups of signal lines are required at different frequency or data patterns, additional transistor-level driver and receiver models are necessary. For SSN analysis, this is typically not the case, since the largest current excitation is achieved if all drivers are switching at the same time. The address/control bus of the system, which is not shown in Figure 14.5, operates at half the frequency of the data bus. The characteristic impedance and termination of address/control transmission lines is also different from data bus lines [4].

Therefore, implementing the address/control bus requires an additional transistor-level driver.

Determining Worst-Case Switching Profiles

Supply noise in a system is generated as a reaction of the supply network to current changes in the system. The amplitude of the supply noise generated by a (single frequency) current excitation is dependent on the impedance of the supply network at that frequency. Larger supply noise amplitudes are created when current changes excite the power delivery system at frequencies of high power supply impedance. A common way to find worst-case excitation pattern in a system is analyzing the power distribution impedance profile $Z_{PDN(f)}$ and identifying excitation patterns that cause large current spectral components at frequencies of high $Z_{PDN(f)}$ [5].

In interface system, finding the worst-case excitation to create the largest system jitter is more complicated, as the system is described by more than one power supply impedance, and each of these impedances can have different profiles over frequency.

Worst-Case Access Pattern for Self-Induced Noise

Figure 14.6 shows the current flow in the system during different stages of switching activity at the output drivers.

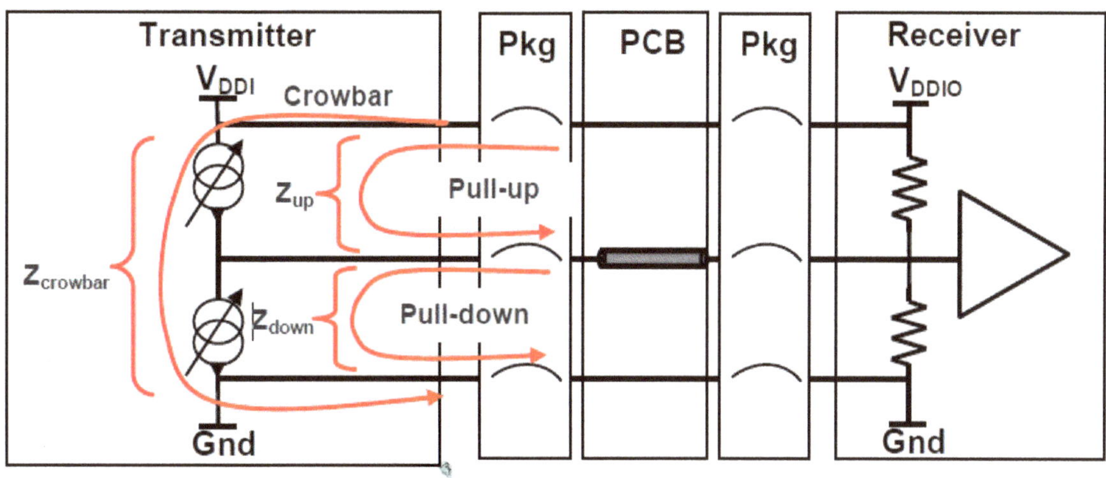

Figure 14.6: Current Flow during Different Stages of Output Driver Switching

During pull-up, the current is flowing through the transmitter from the V_{ddIO} supply rail to the signal lines. The impedance of this current loop, as seen by the transmitter circuits, is Z_{up}. Similarly, the impedance Z_{down} is the supply impedance seen by the transmitter during pull-down operation. For the crowbar current, which is flowing while both the pull-up and the pull-down path of the output driver are active, the supply impedance is $Z_{crowbar}$, which is the traditional Z_{PDN} between V_{ddIO} and Gnd. All three current loops contribute to the total supply noise in the system. In order to identify the worst-case activity pattern in the system, all three impedances have to be analyzed. Figure 14.7 shows the impedance profiles of Z_{up}, Z_{down}, and $Z_{crowbar}$ for the test system.

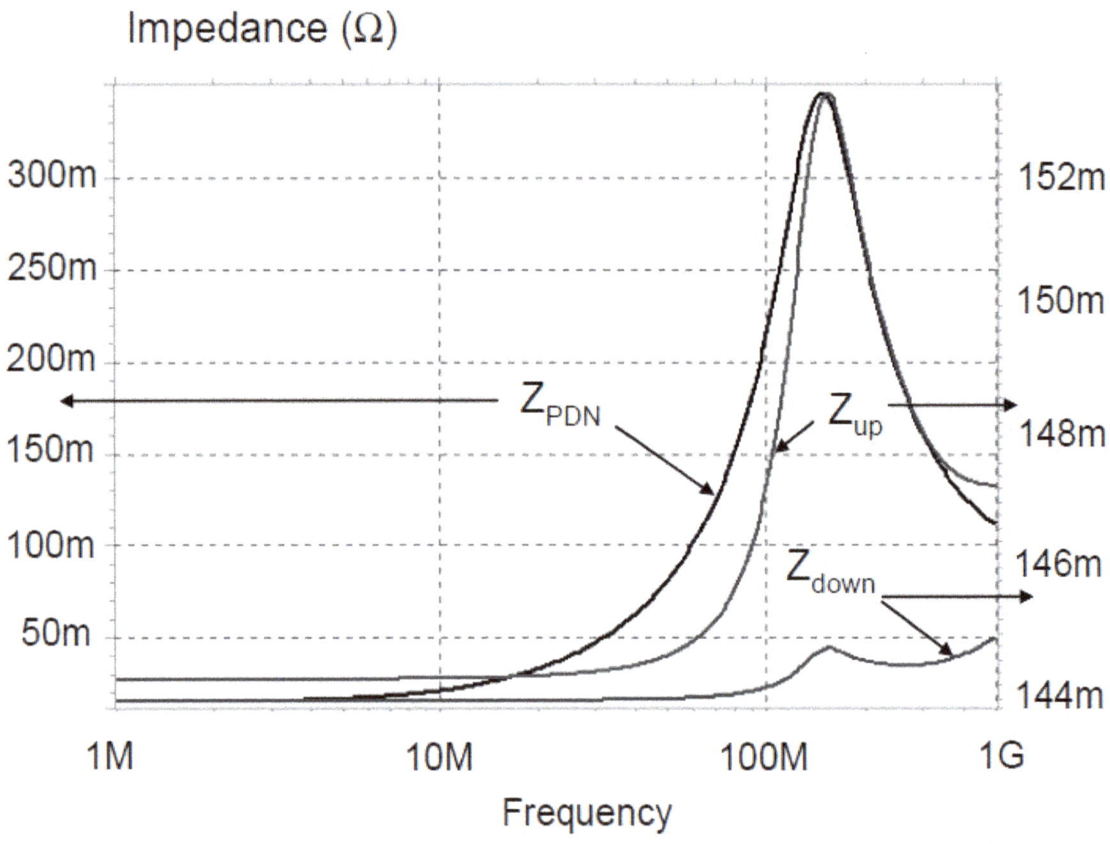

Figure 14.7: Impedance Profiles for Self-Induced Noise on Output Driver Supply Rail

Figure 14.7 shows that in our test system all three supply impedances have maxima at approximately the same frequency of 150 MHz. Assuming that all noise contributions superimpose with worst-case phase offset, maximum noise amplitude on (V_{ddIO}Vss) is expected for data pattern creating large current components at this frequency. Dependent on the phase offset between the different noise components, however, patterns at a slightly different frequency can result in even worse supply noise.

Therefore, several patterns with current frequencies close to the frequency of maximum supply impedance will have to be tested using time-domain simulations.

In a DDR2 system operating at a data rate of 667 MHz, a pattern of 1100 ... on the data lines will cause current excitations at 167 MHz, which is close the frequency of maximum supply impedances. The address/control signals, operating at half the data rate, will create current excitation at the same frequency when transmitting a pattern of 1010.

Worst-Case Access Pattern for Coupled Noise

To analyze the sensitivity of the system in order to couple noise into another supply rail, we again have to analyze the impedance between supply noise and the current exciting the noise. This time, however, this impedance is the transfer impedance, not the self-impedance, between the voltage noise created in the internal supply rail and the current excitation on another supply rail or signal line causing this noise. This makes the process to find the worst-case excitation for coupled noise very similar to the same process for self-induced noise.

It also creates the same difficulties in interface systems: there are several possible contributions causing noise coupling, and each of these contributions requires different worstcase access pattern in the system. On top of this, self-induced noise on these internal supply rails has to be considered as well, especially if it is correlated to the activity of the output drivers.

This is for example the case if a pre-driver is operating from the internal supply rail. Figure 14.8 shows the different transfer impedances for noise coupling into the Vddr supply rail as well as the impedance of Vddr-Vss for self-induced noise.

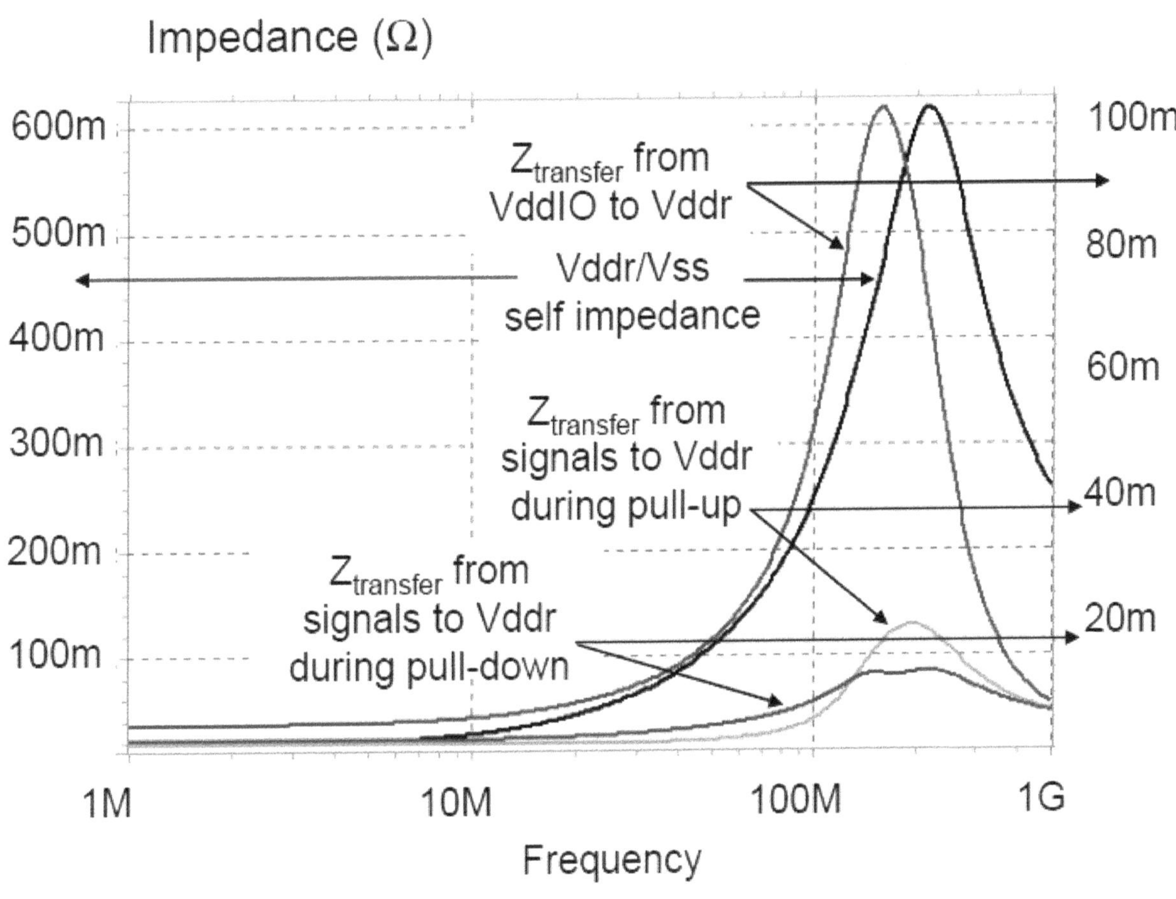

Figure 14.8: Self- and Transfer Impedance for Supply Noise Generation on Vddr

In order to maximize the contributions of all components, the system would have to be excited at different frequencies for pullup, pull-down, and crowbar operation.

This is not possible, however, since the cycle time is always the sum of high and low time of the data sequence, and as such it is the same for all noise contributions. The noise amplitude will depend on the amplitude of each contribution as well as on the phase relationship between the contributions. Thus, it is not possible to determine a single worst-case excitation sequence from the alternate current (AC) analysis of the supply impedances any more.

There are several patterns possible to be worst-case, and these patterns will have to be tested using a time-domain simulation. The worst-case excitation, however, is still expected at a frequency where all supply impedances have larger impedance values, i.e., at frequencies between 150 and 230 MHz. Thus, the AC analysis determines a frequency range where supply noise can easily be generated.

The final noise amplitude for pattern exciting this frequency range, however, has to be determined using time-domain simulations.

Correlating Supply Noise Parameters—Overview of Verification System

Simulations using the SSN analysis model were correlated with measurements on the test system shown in Figure 14.4. This test system was designed as a verification system. As such, it was designed to provide access to data and address/control lines for signal and power integrity measurements.

The ASIC core of the controller chip provided full control over all data patterns transmitted on either the data or the address/control signal lines. In this way, it was possible to send well-defined worstcase pattern over every signal lines, even if these patterns would not be possible in a real system due to the communication protocol on the control bus of a DDR2 system.

As a result, the measured supply noise is more pessimistic than can be expected at normal operation, since a real system would not be able to excite the worst pattern sequence continuously on all signal lines. The measured noise, however, is an upper limit of the supply noise possible in the system under any operating conditions.

Measuring Supply Noise on Driver Supply Voltage

The supply noise on the supply rails of the output drivers can be directly measured at the output of one driver. A detailed description of this measurement methodology can be found in [5].

During this measurement, one driver is transmitting a constant value, either 1 or 0, while all other drivers are transmitting the same worst-case data pattern. As long as the driver is transmitting a constant 1, the driver supply rail V_{ddIO} is connected to the output signal through the on-die termination resistor of this driver. Any noise on the V_{ddIO} supply rail is reflected as noise on the signal line. The on-die termination resistor of the driver used for noise sensing can be added to the SSN simulation and correlated with measurements at the same location. The model then allows estimating the supply noise signal at the supply rail itself.

Figure 14.9 shows the signal measured at a data line of an output driver that transmitted a long string of 0's followed by a long string of 1's while all other drivers were transmitting worst-case pattern.

The measurement shows the supply noise on the V_{ddIO} and Gnd rails separately.

During pull-up operation the output is connected to the V_{ddIO} rail, while during pull-down operation the signal is connected to the ground node. It can be seen that the nose on both rails is comparable in amplitude.

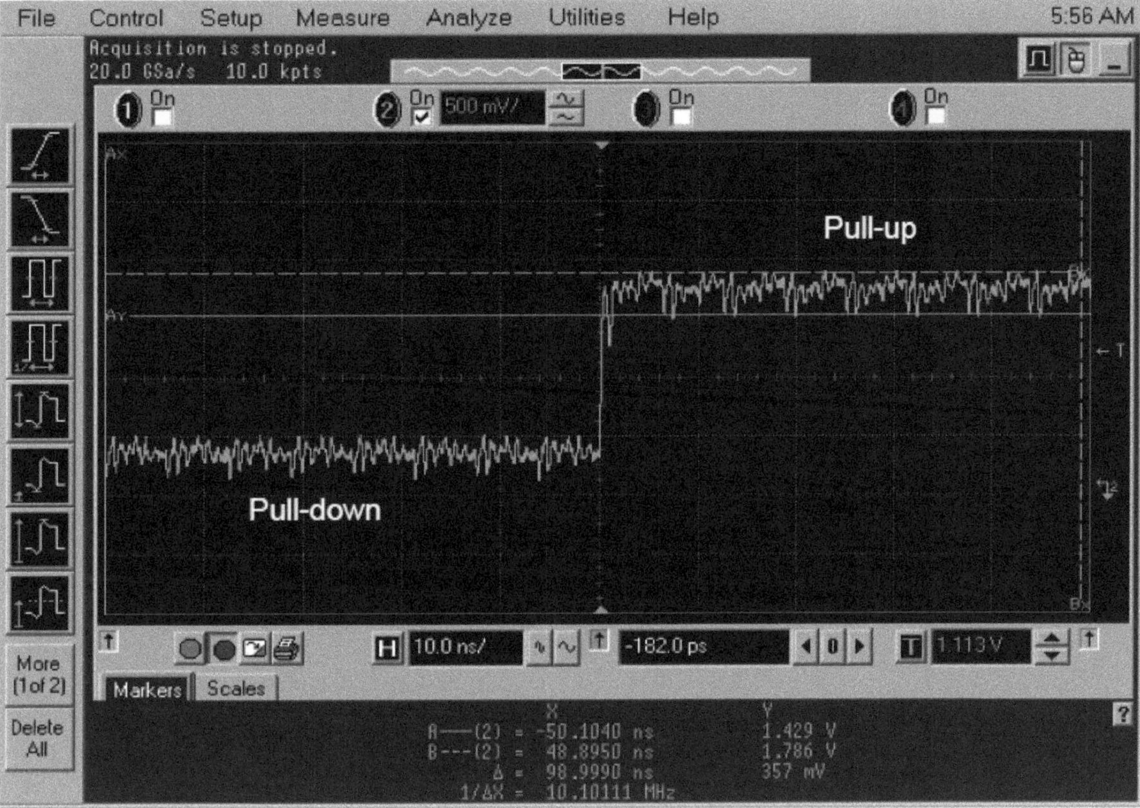

Figure 14.9: V_{ddIO} and Gnd Supply Noise Measurement at Driver Output

Table 14.1 lists two of the data sequences used for supply noise correlation. The table also lists the dominating current excitation frequency generated by these sequences for a data rate of 667 MHz. These sequences were chosen based on the supply impedance analysis described earlier.

The current excitation frequencies generated by these sequences are close to the frequencies where the different supply impedances approach their maximum values.

Name	Data bus signals		Address/control signals	
	Sequence	Frequency [MHz]	Sequence	Frequency [MHz]
Pattern 1	1100	167	1010	167
Pattern 2	110	222	1010	167

Table 14.1: Data Sequences for Supply Noise Correlation

Since the address/control bus is operating with half the data rate in a DDR2 design, the maximum frequency achievable on address/control lines is 167 MHz for a data rate of 667 MHz. The data bus signals, however, can achieve higher frequency.

Figure 14.10 shows the correlation between simulated and measured noise signals for Pattern 1.

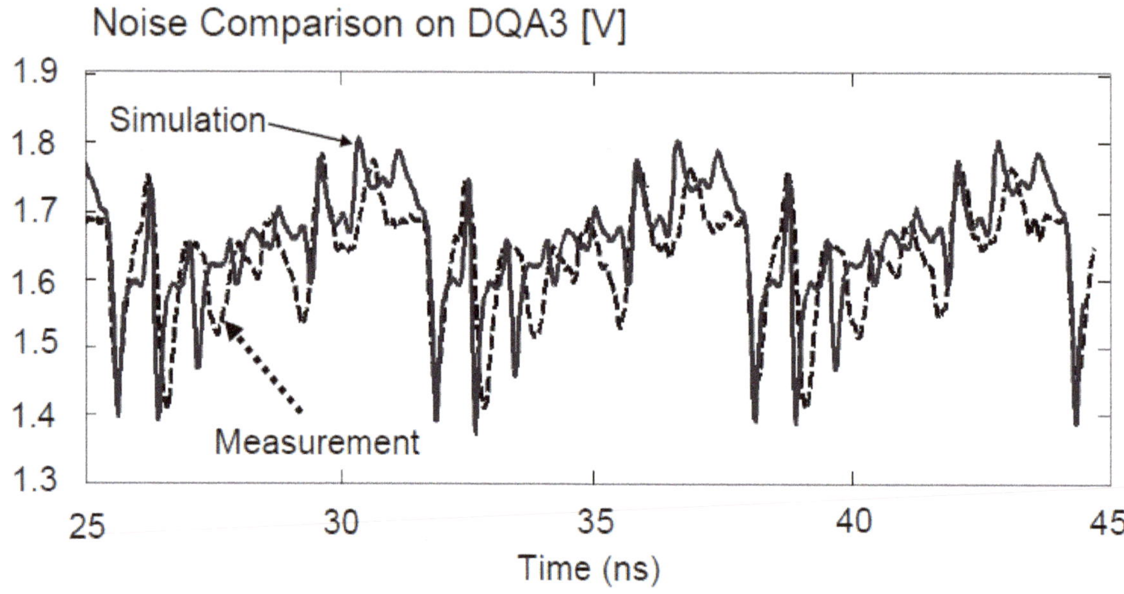

Figure 14.10: Correlation of V_{ddIO} Supply Noise Simulation and Measurement for Pattern 1

Figure 14.11 shows the correlation between simulation and measurement for Pattern 2.

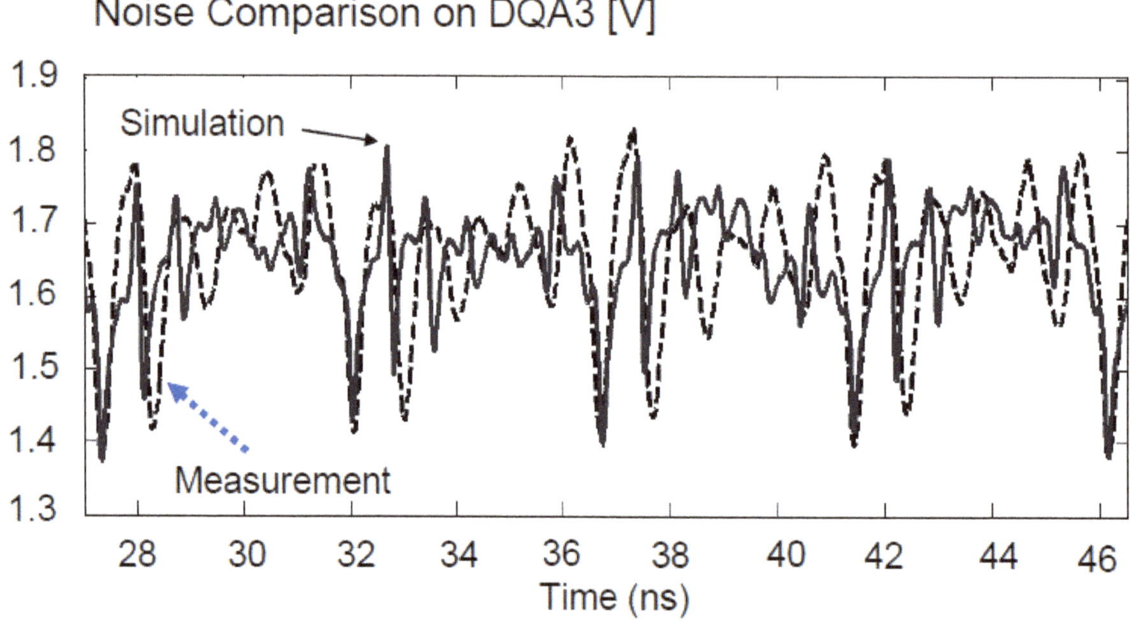

Figure 14.11: Correlation of V_{ddIO} Supply Noise Simulation and Measurement for Pattern 2

Very good correlation is achieved for both data patterns. The simulated and measured waveforms not only correlate very well in noise amplitude, but they also correlate reasonably well in noise waveform. This confirms the accuracy of the modeling methodology for SSN analysis of interface systems.

Table 14.2 summarizes the comparison of simulated and measured supply noise amplitude on the

VddIO supply during simultaneous switching of the different pattern:

Pattern Name	Data Signal Pattern	Address/control Signal Pattern	Measured Noise Amplitude	Simulated Noise Amplitude	Amplitude Error
Pattern 1	1100...	1010...	380 mV	410 mV	8%
Pattern 2	110...	1010...	447 mV	431 mV	2%

Table 14.2: Comparison of Simulated and Measured Supply Noise Amplitude

The accuracy of the simulation is noticeably better than 10 percent.

The table also shows that the noise amplitude for Pattern 2 is larger than for Pattern 1. Based on the AC analysis of supply impedances shown in Figure 14.7, the opposite was expected. This demonstrates the limitation of noise predictions based on the AC analysis of supply impedances. Since the AC analysis does not consider the phase offset between different noise contributions, time domain simulations are necessary to predict the accurate noise amplitudes of different excitation patterns.

Measuring Supply Noise on Internal Supply Voltage

Measuring supply noise on the internal Vddr rail is more difficult, as there is no direct way to observe this rail from outside the chip.

During the chip design, however, differential pairs of large probing pads were added to the topmost metal layer to allow measurement of Vddr and ground noise using a differential probe station.

Figure 14.12 shows the correlation of measurement and simulation of Vddr supply noise while the output drivers are transmitting Pattern 1 listed above.

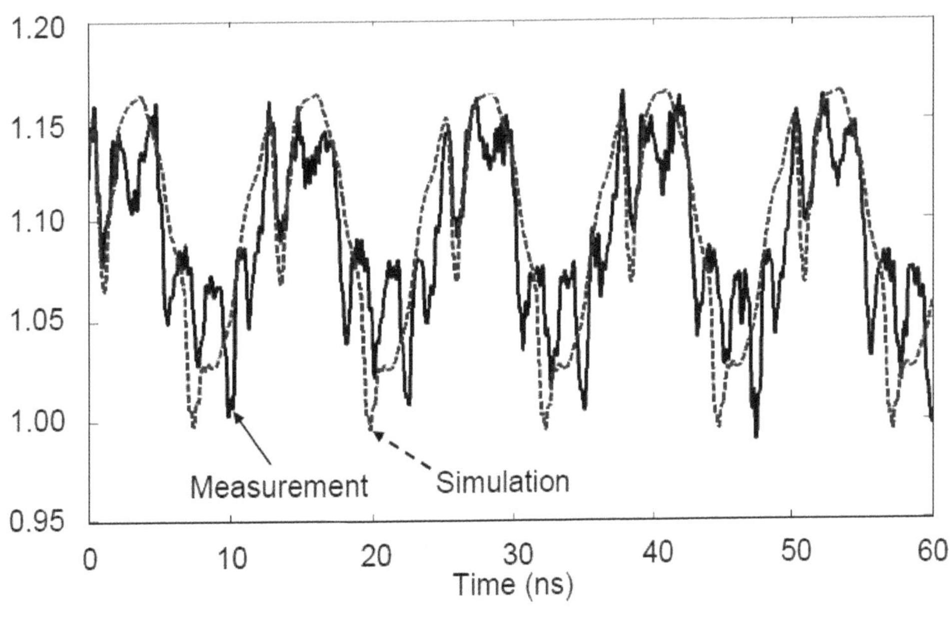

Figure 14.12: Correlation of Vddr Supply Noise Simulation and Measurement for Pattern 1

Figure 14.12 shows a good correlation of the supply noise amplitude for the base frequency component determined by the transmitted data pattern, but the measurement shows more high-frequency noise superimposing this base frequency. The main reason for this discrepancy is that the simulation only includes noise coupled into the Vddr rail or generated directly by the output drivers. It does not include additional self-induced noise generated by the circuits using Vddr as supply. These circuits are operating at a higher frequency and add additional self-induced high-frequency noise to the Vddr rail.

Table 14.3 summarizes the comparison of simulated and measured supply noise amplitude on the Vddr supply during simultaneous switching of the different pattern.

Pattern Name	Data Signal Pattern	Address/control Signal Pattern	Measured Noise Amplitude	Simulated Noise Amplitude	Amplitude Error
Pattern 1	1100...	1010...	179 mV	171 mV	4%
Pattern 2	110...	1010...	201 mV	196 mV	2%

Table 14.3: Comparison of Simulated and Measured Supply Noise Amplitude

Disregarding the high-frequency component not covered by the simulation model, the table shows very good correlation between measured and simulated noise amplitude on Vddr for different data patterns. This verifies the modeling strategy for noise coupling in our SSN model.

The dependency of Vddr supply noise on the data pattern transmitted by the output drivers is clear. The table also shows that significant noise is generated on this internal supply rail, which can significantly affect system jitter. Thus, it is mandatory to include this noise analysis in the SSN analysis of the interface system.

Measuring Jitter Sensitivity to Coupled Noise

In order to analyze the impact of signal-to-supply noise on system jitter in the DDR2 interface, the jitter introduced by a single toggling data line was measured and correlated with the supply noise caused by this data line. In this measurement, the data line BA3 was used to transmit a clock pattern. The jitter on this signal line was measured using an oscilloscope. From the remaining data lines, one data line was transmitting a toggle pattern, creating noise in the system, while the other data lines were kept quiet. The jitter caused by this additional noise was measured on data line BA3.

This measurement was repeated using the remaining data lines one by one as noise generators.

Figure 14.13 shows the jitter measured at BA3 for noise generated at the different signal lines.

The order of signal lines in Figure 14.13 reflects the neighborhood of wire bond pads in the design. It can be seen that the largest jitter is created by signal lines not in the direct vicinity of the signal BA3 used for jitter measurement. This means that the observed jitter is not simply caused by signal-to-signal coupling. In this case, we would have expected the jitter to be a clear function of proximity to the signal line BA3. The pattern in Figure 14.13 cannot be explained by the generation of self-induced noise on the V_{ddIO} supply rail either. Since all driver circuits are identical in this measurement, we would expect that each signal creates the same amount of noise on the V_{ddIO} rail inside the chip and therefore creates the same jitter on the BA3 line.

We used the SSN analysis model to simulate the supply noise coupled from each signal line into the Vddr supply rail. This noise amount is different for each signal line, as the bonding pattern and the vicinity to supply bonds acting as return paths varies throughout the design. Figure 14.14 shows the supply noise amplitude caused by each signal line when it is toggling. For comparison, the order of signal lines in Figure 14.14 is the same as the order in Figure 14.13.

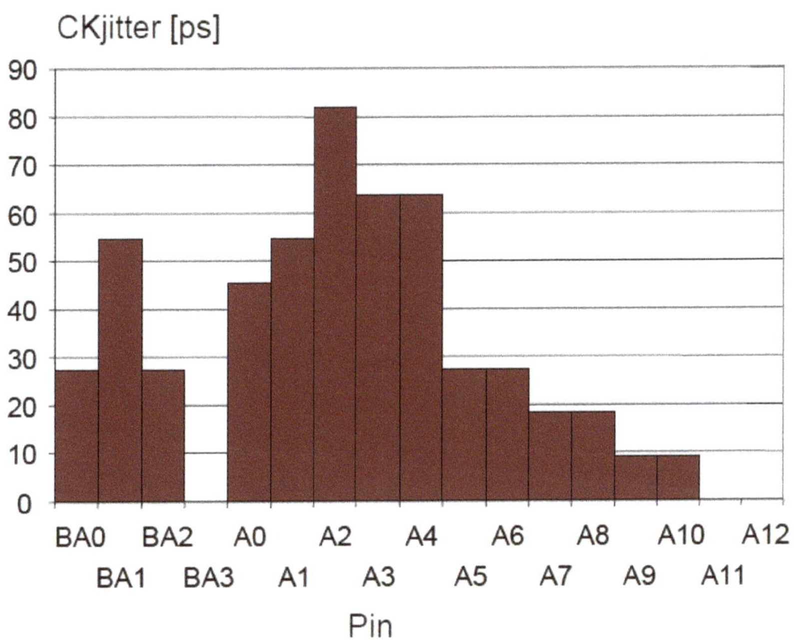

Figure 14.13: Jitter Caused by Toggle Pattern on Different Signal Lines

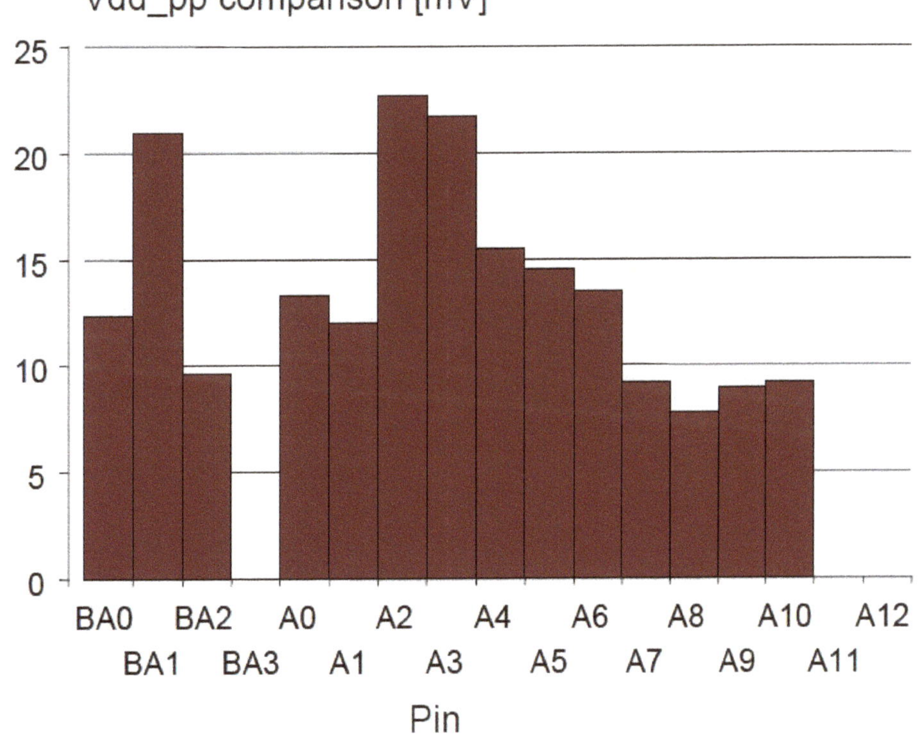

Figure 14.14: Vddr Supply Noise Amplitude Caused by Signal-to-Supply Coupling

The patterns in Figures 14.13 and 14.14 show a close correlation between the noise amplitude on Vddr and the jitter measured at the signal BA3. This demonstrates that signal-to-supply coupling in a wire-bond package can cause significant amount of noise and therefore add noticeable jitter to the system. The correlation in Figures 14.13 and 14.14 also verifies the accuracy of the noise coupling prediction by the SSN model, despite the use of CCCS-based current mirrors for almost all of these signal lines.

Summary

In this paper, we presented the analysis of supply noise for a 667 MHz DDR2 system in a low-cost wire-bond package. The high inductance of wire-bond packages poses a serious challenge for power integrity during simultaneous switching events. It is essential to achieve an accurate understanding and modeling of the worstcase supply noise in such a system to obtain the desired data rate.

We presented a modeling methodology for interface systems that considers the supply noise on the output driver supply and on other supply rails where noise generated during simultaneous switching events can contribute to system jitter. The supply noise analysis model is so compact that a large interface with more than 50 lines switching all at once can be simulated in a short time.

Next, a methodology was presented to help find an activity pattern generating worst-case supply noise in the system. We addressed the limitations of conventional AC supply impedance concepts for interfaces and demonstrated the need for time-domain simulations for supply noise prediction.

Finally, we presented correlations between measurements and simulations of supply noise waveforms and system jitter profiles.

These correlations demonstrate the accuracy of the analysis model. They also confirm the importance of considering noise coupling as a serious source of system jitter in these interface systems.

References

[1] R. Schmitt, X. Huang, L. Yang, C. Yuan, *System-Level Power Integrity Analysis and Correlation for Multi-Gigabit Designs*, DesignCon 2004, Santa Clara, California, February 2004.

[2] A. Varma, M. Steer, P. Franzon, *SSN Issues with IBIS Models*, IEEE 13th Topical Meeting on Electrical Performance of Electronic Packaging, pp. 25-27, Portland, Oregon, October 2004.

[3] A. Varma, M. Steer, P. Franzon, *A New Method to Achieve Improved Accuracy with IBIS Models*, IEEE 14th Topical Meeting on Electrical Performance of Electronic Packaging, pp. 119-122, Austin, Texas, October 2005.

[4] D. Oh, W. Kim, B. Stott, L. Yang, C. Yuan, *Channel Timing Error Analysis for DDR2 Memory Systems*, IEEE 14th Topical Meeting on Electrical Performance of Electronic Packaging, pp. 3-6, Austin, Texas, October 2005.

[5] R. Schmitt, C. Yuan, W. Kim, *Modeling and Correlation of Supply Noise for a 3.2GHz Bidirectional Differential Memory Bus*, DesignCon 2005, Santa Clara, California, February 2005.

Chapter 15—Analysis of Supply Noise-Induced Jitter in Gigabit I/O Interfaces

Ralf Schmitt, Signal Integrity Engineer, Rambus, Inc.
Hai Lan, Signal Integrity Engineer, Rambus, Inc.
Chris Madden, Signal Integrity Engineer, Rambus, Inc.
Chuck Yuan, Signal Integrity Engineer, Rambus Inc.

This paper was presented at DesignCon 2007, Santa Clara, California, January 29-31, 2007.

SUPPLY NOISE-INDUCED JITTER limits the performance of gigabit input/output (I/O) systems. Analyzing this jitter component requires the analysis of two parameters. First, we have to analyze the spectral contents of supply noise generated in the system. Second, the sensitivity of the interface circuits to supply noise at different frequencies has to be determined. We will present measurement and simulation results from a serial link operating at 6.4 Gbps, analyzing the spectral content of the supply noise and the noise sensitivity of the interface circuits over frequency. By combining this information, we can predict worst-case supply induced jitter and guide the design optimization.

Introduction

I/O interfaces operating in the gigabit range have to achieve very low jitter in order to meet their timing budget. One major source of jitter in these systems is power supply noise. Supply noise affects the jitter of the phase-locked loop (PLL) controlling the internal timing in the interface. Additionally, it affects the timing of other circuits such as output drivers and clock distribution. Many of these circuits are more sensitive to supply noise in some frequency ranges and are less sensitive to supply noise in other frequency ranges. Therefore, understanding supply noise-induced jitter requires the analysis of two system characteristics.

First, it is necessary to analyze the supply noise generated in the system. It is necessary to understand the spectral content of the supply noise as well as the dependency of this spectrum on activity pattern in the interface and package design.

Second, it is necessary to determine the circuit sensitivity to supply noise at different frequencies. By combining this information, we can predict worst-case supply noise-induced jitter for the interface.

In this paper, we will present results from the analysis of supply noise-induced jitter in a serial link operating at 6.4 Gbps. In this design, we implemented on-chip noise monitors on all power supply rails of the interface circuit. These noise monitors were used to measure the power spectrum density of supply noise on the supply rails over a wide frequency range, from the lower MHz range up to several harmonics of the data rate. The noise monitors were also used to reconstruct time-domain supply noise waveforms similar to a sampling oscilloscope. In this paper, we also compare supply noise measurements in time-domain and frequency-domain with noise simulation results obtained from circuit simulations.

Measurements and simulations show a good correlation for the major noise contributions, confirming in general the simulation methodology for supply noise.

After, we present measurement results for the sensitivity of the interface circuits to noise at different frequencies and on different supply rails. By using noise generators implemented on each supply rail, we generate noise at a fixed frequency, determine the noise amplitude using the noise monitors, and measure the jitter caused by this noise at the signal output of the interface. By sweeping the operating frequency of the noise generators, we can determine the noise-sensitivity profiles over frequency for each supply rail.

Combining the supply noise spectrum with the noise-sensitivity profiles allows us to predict the supply noise-induced jitter in the interface for typical activity and allows us to guide the circuit and power distribution design to reduce supply noise-induced jitter.

Finally, we present supply noise results for activity patterns in the interface. We demonstrate the dependency of the supply-noise spectrum seen on the different power rails on data patterns in the interface and power distribution design in the package. By using the profile of the power distribution impedance seen by the system as a function of frequency, we can derive worst-case data pattern for the system and compare this worst-case noise with supply noise measured during typical data activities.

Power Supply Design Environment Requirements for Gigabit I/O Interfaces

High-speed interface systems require very low random and systematic jitter in order to meet their timing budget. As the data rate increases, the acceptable timing loss due to supply distortion decreases. In general, decreasing timing jitter can be achieved in two ways: either the circuit sensitivity to supply noise can be reduced or the amplitude of supply noise can be reduced.

A common way to address this problem is to scale the noise budget of the supply rails in the systems so it is inversely proportional to the data rate in the system. Assuming a constant sensitivity of the circuits to supply noise, scaling the supply noise in this way assures that the relative contribution of supply noise-induced jitter to the total bit time (e.g., UI) stays constant. However, this approach makes the design of the power distribution systems increasingly difficult. For gigabit I/O systems, a pure linear scaling of the supply noise budget with the data rate would easily lead to unrealistic supply budgets that cannot be achieved in a system with acceptable package and decoupling resources.

In high-speed interface systems, this problem is commonly addressed by introducing several independent power supply rails with different noise budgets. Circuits that are controlling the internal timing of the interface—including PLLs and data-link layers (DLLs)—are supplied by a separate power rail—often called analog supply (e.g., VDDA)—due to the analog nature of these circuits, with a very tight noise budget. This tight noise budget reflects the high jitter sensitivity of these circuits to supply noise.

Output drivers and on-chip terminations of the signaling bus, on the other hand, are often supplied by an extra supply rail (called VddIO or, as in the case of our test system, VTT). This extra output circuit supply is often required because different (and often higher) voltage supply levels are required for the channel signaling than what is available for the internal circuits. It also helps in mitigating the supply noise problem in the system. The supply noise requirement for this supply rail is usually much more relaxed, since the jitter sensitivity of these circuits to supply noise is typically small. Finally, the remaining circuits in the interface are supplied by a third supply rail (called VDD) with a moderate supply noise budget.

This separation of the interface circuits into circuit groups supplied by independent supplies has two major advantages. The first advantage is that the resources of the power distribution system can now be focused on the supply rails with the tighter supply noise budget. A larger amount of on-chip decoupling capacitors can be assigned to, e.g., the supply VDDA to achieve a low supply noise level on this rail. The second advantage of the separation into different supply rails is that this separation also removes the largest noise excitations from the more sensitive supply rails. Self-induced supply noise is generated on a supply rail when the current drawn from this rail changes over time. Circuits with data pattern-dependent current dissipation—particularly large output driver circuits driving large off-chip loads like the signaling channel—are able to cause large current changes at various fractions of the data rate frequency, which can easily excite significant noise on the supply rail of these circuits. Separating these current changes from the sensitive supply rails removes a major source of supply noise from them, thus making it much easier to achieve low supply-noise levels on the supply rails of sensitive circuits.

Separating the power supply system of the interface into several separate supply rails with different noise sensitivities and different noise current profiles now generates two important questions that have to be addressed by the power distribution design. The first question is, what target budget for supply noise-induced jitter should be defined for each of the supply rails, and what is the supply noise budget for each supply rail to achieve this jitter budget? The second question is, once a supply noise budget is defined as an answer to the first question, how can this supply noise budget be achieved with reasonable power supply resources on silicon, package, and printed circuit boards (PCBs)?

In the past, many authors (e.g., [1, 2]) have addressed the second question, namely, deriving power delivery design requirements to achieve a predefined supply noise budget. These activities have provided many insights into the generation of supply noise as well as the optimization of the power delivery system in the frequency domain. However, these analyses primarily focus on the supply noise in the system, not on the jitter generated by this supply noise.

Little information is given about the exact process used to derive the supply noise budget. In many designs, the supply noise budget is derived from a negotiation between the circuit team and the system designer responsible for the power delivery design, defining supply noise levels that are both achievable with reasonable system resources and acceptable as a basis for the design and optimization of the circuits. The supply noise budget in these cases is usually either constant over the entire operating frequency range of the system, from direct current (DC) to several harmonics of the data rate, or the entire frequency range is separated into a small number of smaller ranges with different supply noise target, based on the knowledge of the circuit designer about frequency regions where high or low noise sensitivity is expected from the circuits. By doing so, this concept assumes a nearly constant jitter sensitivity to supply noise in the different frequency ranges, and the circuit design assumes that noise can occur at any frequency with amplitude defined by the supply noise budget.

For gigabit I/O systems, the requirements on supply-induced jitter are so demanding that a more detailed understanding of the jitter generation in the system is necessary. Supply-induced jitter is the result of the interaction of two separate and largely independent parameters—the supply noise spectrum generated on each supply rail and the sensitivity of the circuits supplied by these supply rails to noise at different frequencies. Together, these two parameters define the total supply-induced jitter impact to the system due to supply noise on each power rail. Understanding both parameters independently as well as the combination of these parameters (or the total jitter impact) provides us with insight that can be used to optimize the jitter performance of the system. Comparing the total jitter impact created by supply noise on different rails helps identify which supply rails are contributing most to the system jitter and should be addressed in further supply delivery optimizations. A supply rail with smaller noise sensitivity but larger supply noise levels can contribute as much or even more jitter to the system than a supply rail with a much larger noise sensitivity but very small supply noise levels.

Analyzing both the noise sensitivity profile and the supply noise spectrum also gives insight into the interaction between both parameters and directs the optimization of these parameters for optimized system performance. For example, the power delivery system design can provide larger attenuation for noise in a frequency range where the noise sensitivity is known to be large, or the circuit design can be optimized with the knowledge that larger supply noise contributions have to be expected in a particular frequency range.

Overview of Gigabit I/O Interface Test System

In order to analyze the jitter impact due to supply noise, we used the test system of a gigabit I/O interface system with the block diagram shown in Figure 15.1.

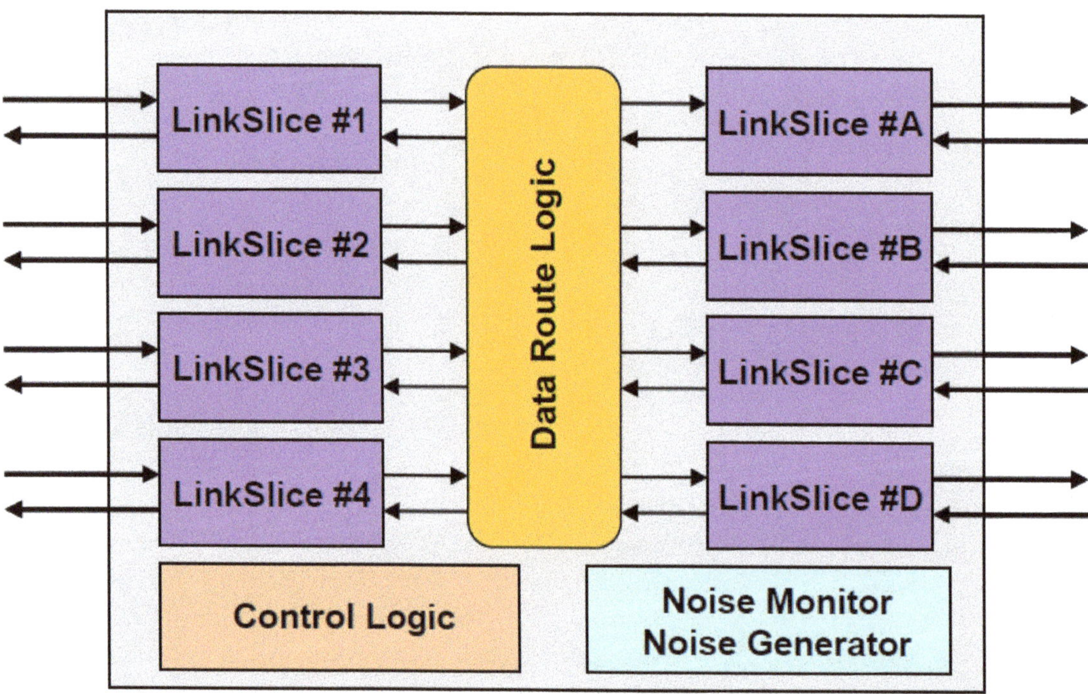

Figure 15.1: Block Diagram of Gigabit I/O Interface Test System

The test interface system consists of eight serial links, each operating at a data rate of up to 6.4 GHz. Further, it contains a data route logic to control the data flow between the different links and control logic containing general logic for link initialization as well as logic used for characterization such as pattern generators.

The test system uses the following four separate supply rails for different parts of the circuits:

- The supply VDDA for sensitive timing controlling circuits such as PLLs

- The supply VTT for channel termination resistors
- The supply VDD for all circuits inside the link slices not supplied by VDDA or VTT
- The supply VDD core for the control logic and the data route logic of the test system

An on-chip supply noise measurement system was implemented on the test chip. For each of these four power supplies, one noise generator and two noise monitors are dedicated to generate and measure noise in that supply domain. The noise generator is used to intentionally generate the noise current with controllable strength at a desired frequency, which is injected into the supply rail to excite the supply noise. The two noise monitors are used to capture the supply noise dynamics and convert it to digital outputs for post-processing.

Noise Generator Circuit

Noise generator circuits are implemented attached to each supply rail to generate additional supply noise in the system.

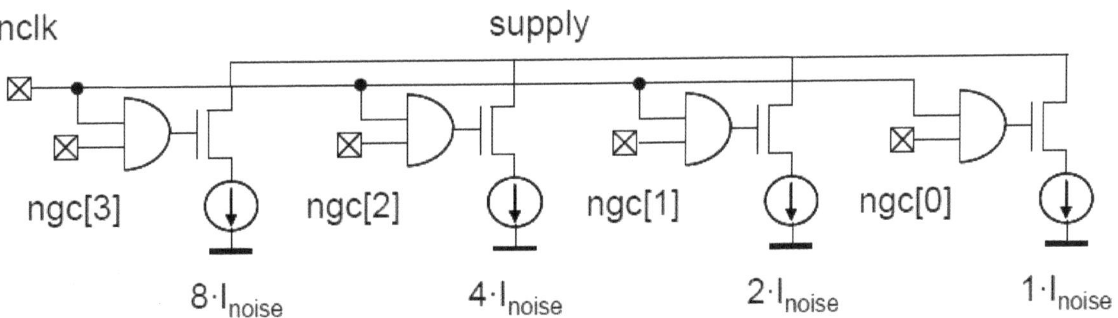

Figure 15.2: Block Diagram of Noise Generator Circuits

The noise generator clock signal nclk connects current sources to the supply rail, creating a clock signal–shaped current as noise source to the supply rail. The fundamental frequency of this clock-shaped noise signal is the frequency of the noise generator clock nclk. The amplitude of the noise waveform is adjusted by a 4-bit control register (ngc[3] … ngc[0]). This allows the noise amplitude to be adjusted to the impedance of the power distribution network (PDN) at any frequency. For fixed noise current amplitude, the resulting supply voltage noise would be large at frequencies where the supply impedance of the PDN is high and small at frequencies where the supply impedance is small. Through the 4-bit noise generator control register, the noise current amplitude can be adjusted to achieve a reasonable supply voltage noise at any frequency.

Noise Monitor Circuit

The block diagram of the noise monitor system is depicted in Figure 15.3. Two samplers are enabled by an external clock generator.

Only one of the two samplers is required to collect the distribution of the noise by recording the sampling instants as a real-time "oscilloscope." Moreover, two samplers are needed in order to characterize the statistical properties of the noise, e.g., its autocorrelation. The noise power spectrum density can be derived from the auto-correlation measured by using two samplers with precise sampling instants and controlled delay between the two clocks, but at low sampling rates. To realize a sub-mV-level resolution, an

ADC is needed to convert the sampled analog voltages to digital outputs, allowing more flexibility in post-processing and analyzing the sampled supply noise with higher precision. Since this monitor system should be easily integrated at different locations on-chip, relatively simple circuitry is preferred to implement the ADC part so that the whole monitor system can be realized in a compact form with low power consumption.

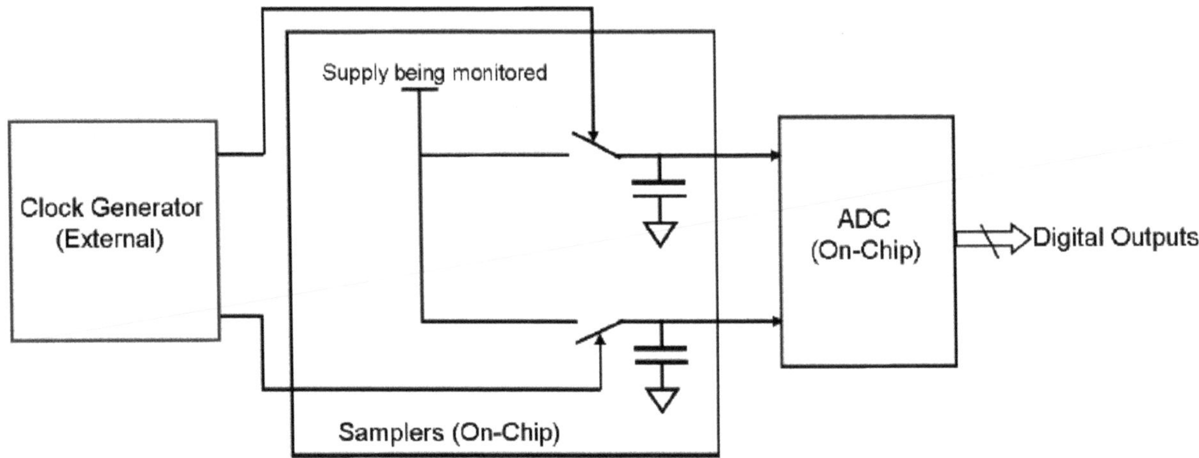

Figure 15.3: Block Diagram of Noise Monitor System

For our analysis, we need to measure supply noise on the internal power rails for frequencies up to a multiple of the fastest toggle frequency. For a system with a data rate of 6.4 Gbps, e.g., a fastest data toggle rate of 3.2 GHz, noise has to be measured for frequencies up to 10 GHz or more. Measuring supply noise on an internal supply rail at these high frequencies is a challenging task, and it requires special noise monitor circuits. We are primarily interested in the frequency domain spectrum of the supply noise.

For correlation with simulated noise waveforms, however, we would also like to have the ability to measure characteristic supply noise waveforms in time domain. Thus, we need a noise monitor with sufficiently high bandwidth suitable for frequencies beyond 10 GHz, which not only can provide measurement results for the frequency domain noise spectrum, but can also be used to construct characteristic time domain noise waveforms.

Figure 15.4 shows the circuit schematic of the noise monitor implemented along with the test gigabit I/O system.

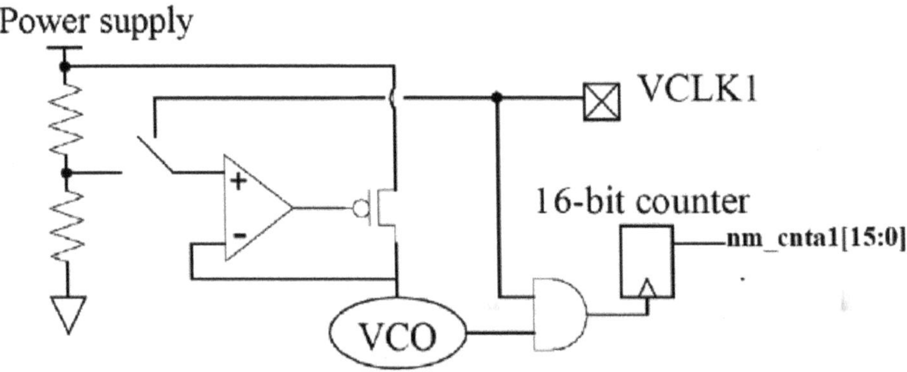

Figure 15.4: Circuit Implementation of Noise Monitor

The supply voltage being monitored is first sensed by the sample-and-hold circuit.

The sampling circuit is implemented using positive-channel metal oxide semiconductor (PMOS) switches so that it is easier to measure the VDD supply. Moreover, a simple PMOS switch can achieve very high bandwidth, which is necessary for the monitor system to adequately capture the dynamic behavior of the supply noise up to several harmonics of the data rate.

During the hold mode, the buffered sampled voltage is used as the control voltage to set the frequency of a voltage-controlled oscillator (VCO). The VCO is implemented as a ring oscillator to obtain high frequency since ring-type oscillators can easily achieve several times fanout-of-four cycle time. As shown in Figure 15.4, the output of the VCO is the input to a 16-bit counter. The counter is enabled during the same specified time window as the hold-time window of the hold circuit to digitally estimate the VCO frequency by counting the number of clock edges of the VCO output. The resulting digital count is proportional to the VCO control voltage, which is the sampled supply voltage.

After obtaining the VCO calibration curve, the original supply voltage can be easily derived from the stored digital counts. The counting process actually ensures that the averaged VCO frequency gets measured. Hence, it essentially filters out the high-frequency noise, typically coupled from other sources other than the supply voltage of interest. The achievable resolution by the VCO analog/digital (A/D) converter is determined by the VCO gain in Hz/V and the conversion time.

Each of the power supplies has two samplers to monitor the voltage on it. Both samplers are attached to individual VCO counters. As shown in Figure 15.3, the two samplers are enabled by two externally supplied clocks, which are at the same frequency but have controllable relative delay between them. By controlling the clock duration and relative phase of the two clocks, the autocorrelation of the two samples can be collected by comparing the values obtained from each of the VCO converters.

Typically, supply noise can be characterized as a cyclostationary random process [4]. Hence, the obtained auto-correlation can be used to describe many important statistical properties of supply noise, in time domain and in frequency domain. The Nyquist frequency of the measurement is determined by the minimum adjustable delay between the two clocks and not by the repetition rate of the sampling itself. This significantly reduces the usually stringent requirement on the throughput of the high speed sampling circuits.

Before using the noise monitor to capture the alternating current (AC) supply noise, it is necessary to obtain the calibration curves for each monitor, e.g., the mapping relationship between the digital code and the sampled voltage value.

The test chip is set in a power-down state and thus no data gets transmitted through the I/O circuit. Only the noise monitor blocks are powered up. The supply voltage is swept over the range of interest, typically a few hundred millivolts, using an external power supply source. At each swept voltage point, the digital code is repeatedly recorded several times.

The averaged count is then used in the final calibration process. Figure 15.5 shows typical DC calibration curves of two noise monitors with individual VCO converters for the same power supply. The figure illustrates that, for this particular case, both monitors achieve a nominal LSB of about 1mV for given conversion time window.

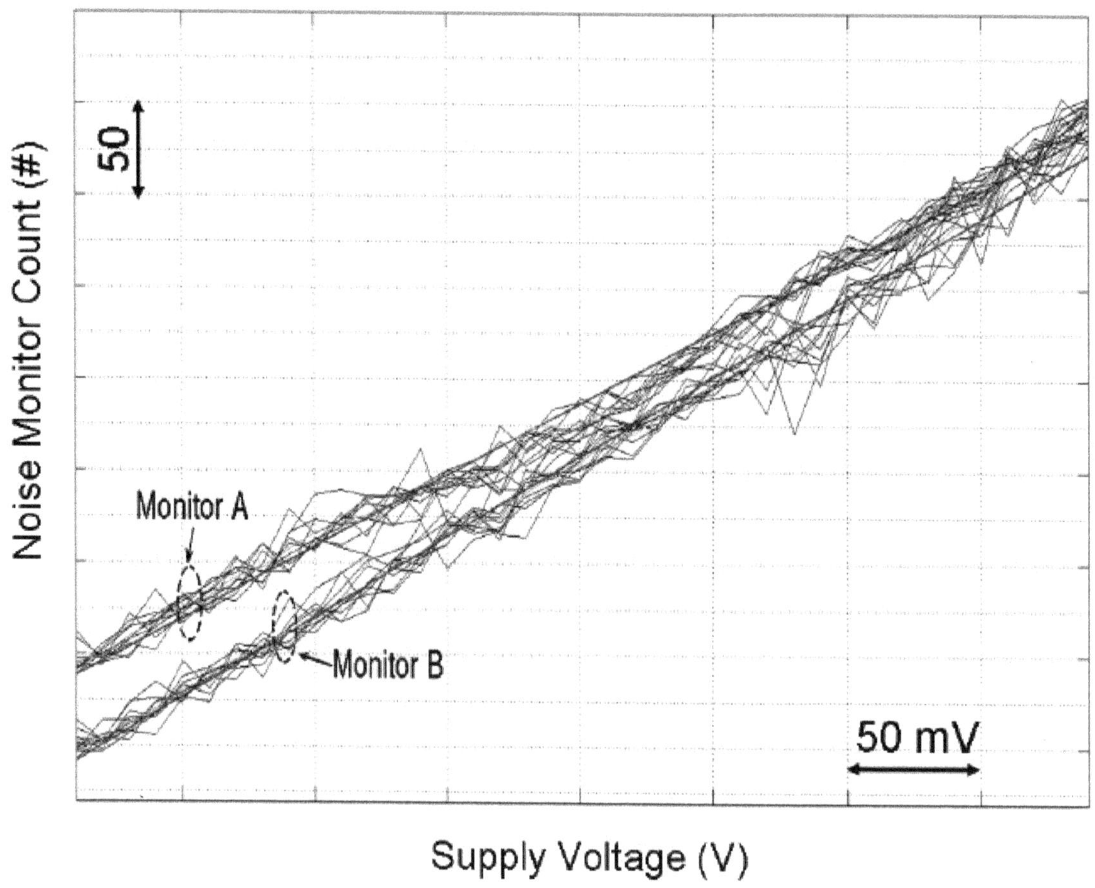

Figure 15.5: Typical DC Calibration Curves for Noise Monitors for Given Conversion Time

As previously discussed, the noise monitor system can be used in different ways to capture different aspects of power supply noise. By using only one monitor as a sampling oscilloscope, one can collect the distribution of the supply noise at each point in time.

Deterministic noise waveform can be extracted by taking the mean of all the sampled voltages at each point in time. Meanwhile, the extent of the variation around the deterministic, repetitive noise waveform indicates how much random noise exists on the supply rail. Typically, one would expect the noise on VDD exhibits much larger random variation than that on VDDA.

When both noise monitors are used with controlled clocks by varying the relative phase (delay) between them, one can collect the auto-correlation data and therefore derive the cyclostationary, or time-averaged power spectrum density (PSD), of the supply noise.

Figure 15.6 shows a representative measured supply noise data from two noise monitors, where monitor A is enabled by clk1 with zero delay with respect to the global clock and monitor B is enabled by clk2 with varying delay with respect to clk1.

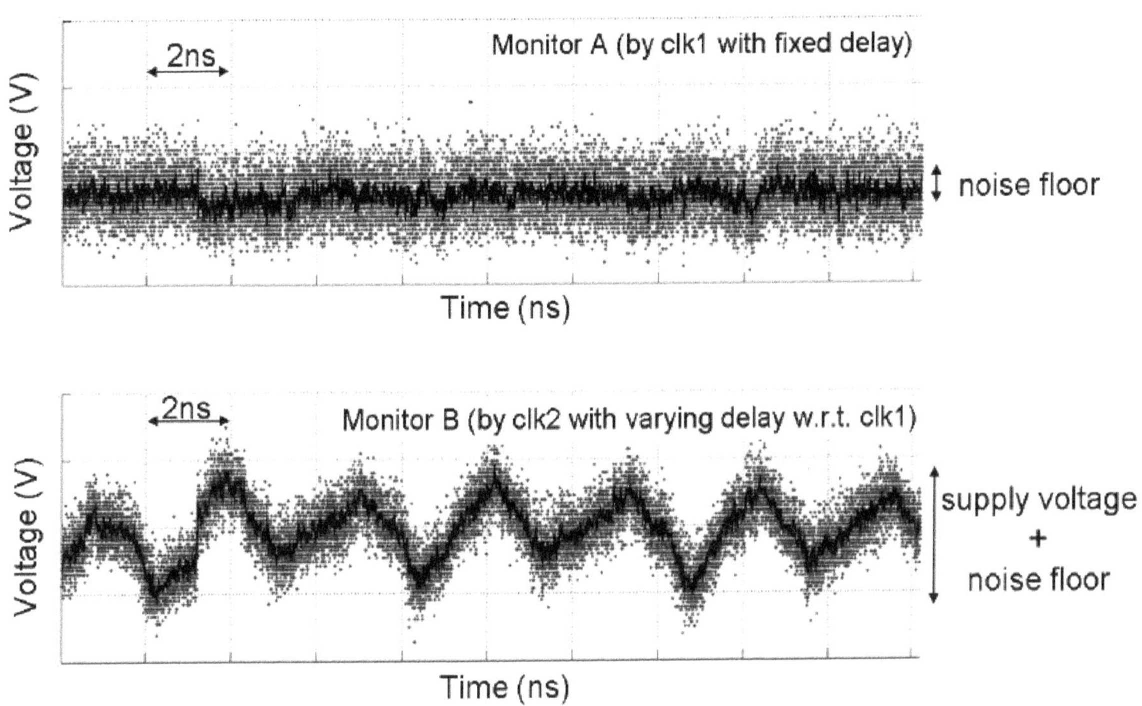

Figure 15.6: Example of Sampler Outputs

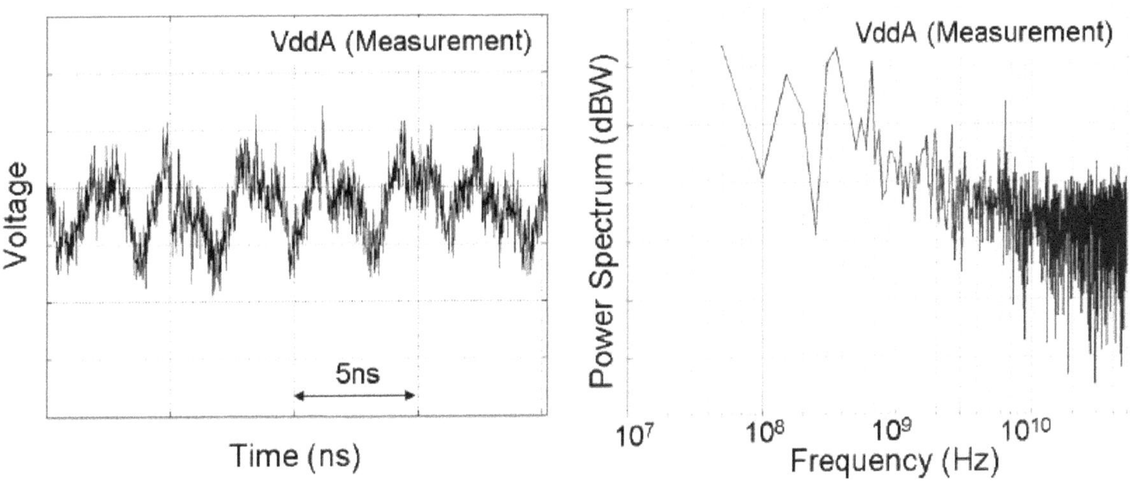

Figure 15.7: Extracted Supply Noise Waveform in Time Domain and Its Corresponding PSD in Frequency Domain

Examining the data from any one of the monitors gives the sampled noise distribution, from which the deterministic and random variation can be characterized. Examining the outputs from both monitors, the auto-correlation function can be computed. Consequently, the noise PSD can be obtained, as shown in the right side of Figure 15.7. The time domain waveform can also be constructed from the sample outputs, as shown in the left side of Figure 15.7.

Measuring Jitter Sensitivity Profiles

We define the jitter sensitivity of a circuit as the ratio between the jitter generated by a single-frequency supply noise signal divided by the amplitude of this noise signal. The typical unit of jitter sensitivity is [ps/mV] and it is a function of the noise frequency.

Figure 15.8 shows the measurement setup for measuring the supply noise sensitivity in our test system.

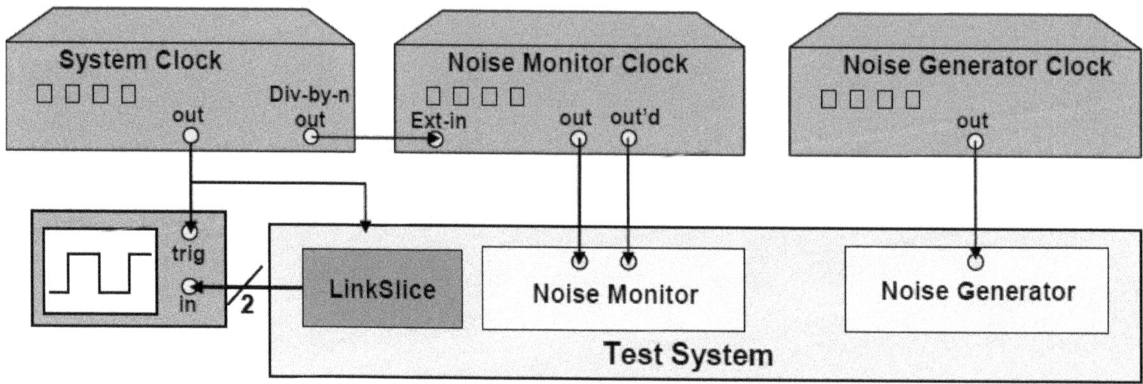

Figure 15.8: Measurement Setup Used to Measure Jitter Sensitivity

Supply noise is generated in the system through the noise generator circuits, while a link of the circuit transmits a clock-like signal. The amplitude of the supply noise at the frequency of interest is measured using the noise monitor circuits. The system jitter is measured at the clock-like signal transmitted by the link using a real time scope. Dividing the measured jitter by the measured supply noise amplitude provides the jitter sensitivity at the frequency of the supply noise:

$$Sensitivity = \frac{Jitter(f_{sample})}{Noise(f_{sample})} \left[\frac{ps}{mV}\right]$$

Repeating these measurements while sweeping the noise generator frequency provides jitter sensitivity values for the entire frequency range of interest.

In a real system environment, the system will generate supply noise contributions at many frequencies at the same time. Extracting a single noise frequency together with the jitter component caused by this single noise frequency is difficult in this case. In our test system, we can generate a dominating supply noise signal at a predefined frequency using the noise monitors implemented on the system. Adjusting the strength of these noise generators we can generate a dominant supply noise spectral component that is significantly stronger than any other natural supply noise component in the system during the measurement. Figure 15.9 shows the power spectrum of the VDD supply noise, measured with the noise monitor circuits, while generating additional supply noise with the noise generators at 400 MHz.

As can be seen in this figure, the intentional supply noise power is at least 5 dB, in most cases more than 10 dB larger than the natural noise power contributions in the system.

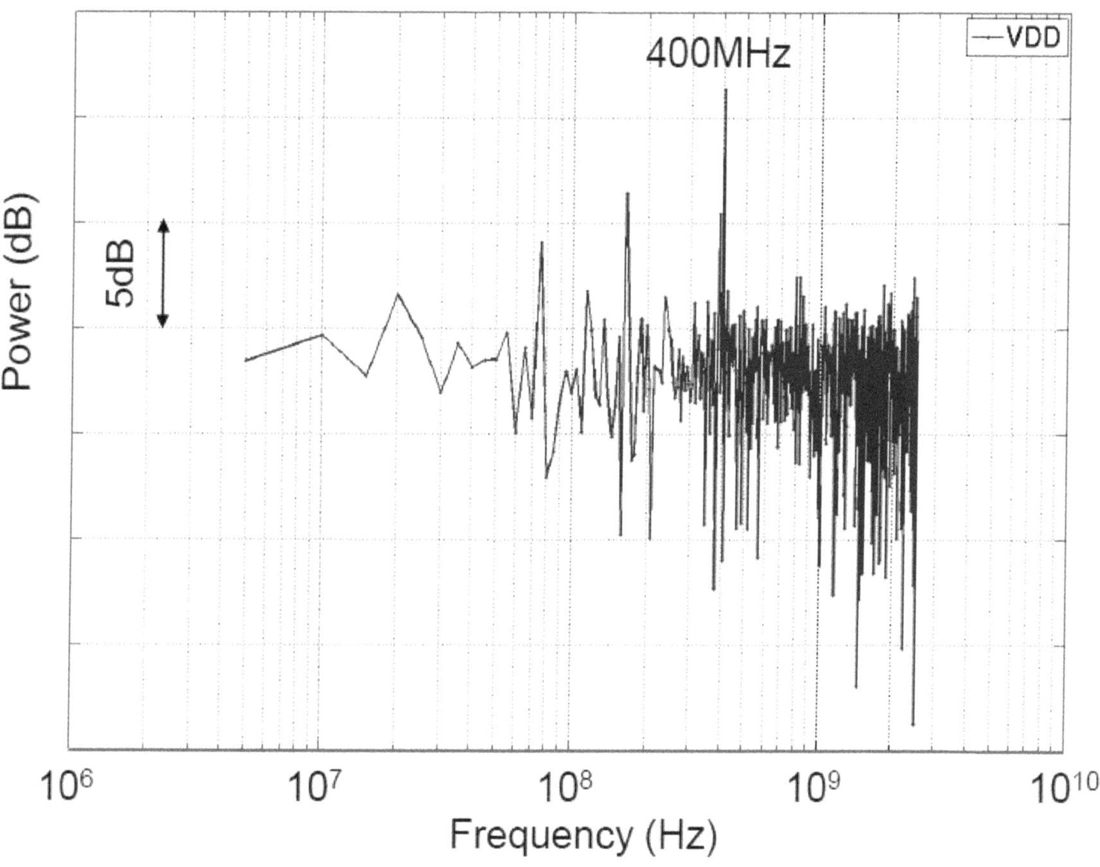

Figure 15.9: Example of Supply Noise Power Spectrum with Intentional Modulation Noise at 400 MHz

The impact of unwanted natural noise frequency contributions can be further reduced by extracting only the jitter at the target frequency. Jitter analysis software is used to analyze the jitter spectrum of the signal measured at the real-time scope, and only the jitter contribution at the target frequency of this measurement is used for the calculation of the jitter sensitivity.

Measurement of Supply Noise-Induced Jitter in a Gigabit I/O Interface

The noise generator and noise monitor circuits previously described were used to characterize the supply noise-induced jitter in the high-speed serial data links of the test system. In this analysis, we focused our attention on the two power rails VDDA and VDD of the supply of the serial link. These two supply rails are expected to dominate the contribution of supply-induced jitter to the total link jitter.

The supply rail VTT, which is used for the termination resistors of the serial link, was excluded from this analysis due to measurement problems with the noise monitors connected to this rail. The supply rail VDDcore was excluded because it is exclusively used for control logic outside the high-speed links and does not participate in the data flow of the link itself. Therefore, supply noise on this power rail does not generate link jitter. This assumption was verified with a small number of jitter measurements for different supply noise levels on VDDcore.

Supply Noise Spectrum under Typical Operating Conditions

In the first analysis step, the supply noise spectrum on VDDA and VDD was measured during typical operating conditions by using the noise monitor circuits on the system. These measurements were compared to the results of noise spectrum simulations used for the design of the PDN of this system. This comparison should verify that the current noise modeling methodology used during the design stage correctly predicts the supply noise spectrum in the system. This prediction capability is essential to direct the optimization of the PDN during the design stage of the system.

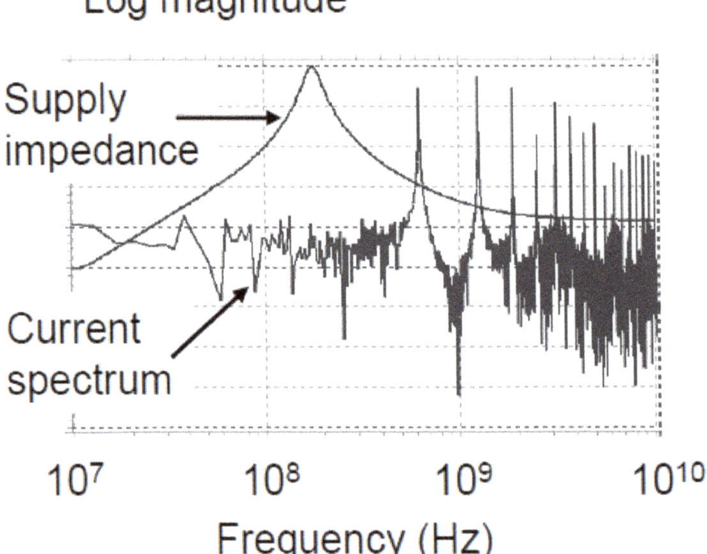

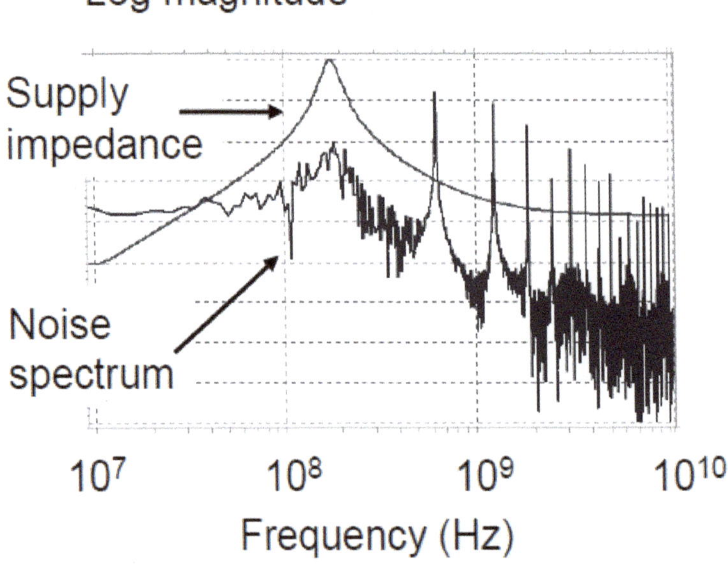

Figure 15.10: Simulated Current (Top) and Supply Noise (Bottom) Spectrum of VDDA Supply Rail for PRBS7 Data Pattern

In order to predict the supply noise spectrum, the system's supply current is simulated using device-level simulations during different operating conditions. These current profiles are used to excite a power distribution model of the system, predicting waveform as well as the spectrum of the noise on different supply rails. Figure 15.10 shows an example of the current profile for the supply VDDA during typical operating conditions (PRBS7) as well as the impedance profile of the power distribution system for this supply rail.

The VDDA current spectrum on the left side of this figure shows primarily high-frequency components at multiples of the different clock rates used in the design. There are no spectral spikes visible at low or medium frequency. This result is not surprising, since the circuits supplied by VDDA, e.g., PLLs, are operating at constant clock frequency, providing the timing reference for the link. The current dissipation of these circuits is expected to be nearly constant (e.g., DC) with noise components only at multiples of the constant clock frequencies these circuits are operating at. Since these clocks are operating at high frequencies, the impedance of the power distribution can be kept small in this frequency range by adding on-chip decoupling capacitors, achieving small noise levels on this supply rail.

Figure 15.10 also shows the profile of the power distribution impedance $Z_{PDN}(f)$ over frequency. As typical for power distribution impedances, the impedance achieves small amplitudes at low and high frequencies, but shows a large amplitude peak at medium frequencies, where the inductance of the package system resonates with the on-chip capacitors (e.g., [3]). The resulting noise on the power supply rail of the system can be calculated as the product of the current spectrum $I_{noise}(f)$ and the impedance of the power supply network $Z_{PDN}(f)$:

$$V_{noise}(f) = Z_{PDN}(f)\ x\ I_{noise}(f)$$

Using the simulated current profile and the supply impedance profile from the power delivery model of the test system, the resulting supply noise spectrum for the VDDA supply rail of the test system is shown on the right side of the figure. As can be seen in this figure, the supply noise spectrum shows the same spectral peaks as the original current spectrum on this rail, but the "background" level of the supply noise spectrum follows the profile of the supply impedance $Z_{PDN}(f)$. This effect emphasizes any noise contributions at medium frequency, where the supply impedance $Z_{PDN}(f)$ shows a large amplitude due to package-chip resonance.

We used the supply noise monitors to measure the power spectrum of the noise on the supply rails VDDA and VDD of the test system during the transmission of a PRBS7 data pattern. The PRBS7 data pattern represents a typical data pattern with random data contents. Figure 15.11 shows the power spectrum of the supply noise on both supply rails measured at the test system during this operation.

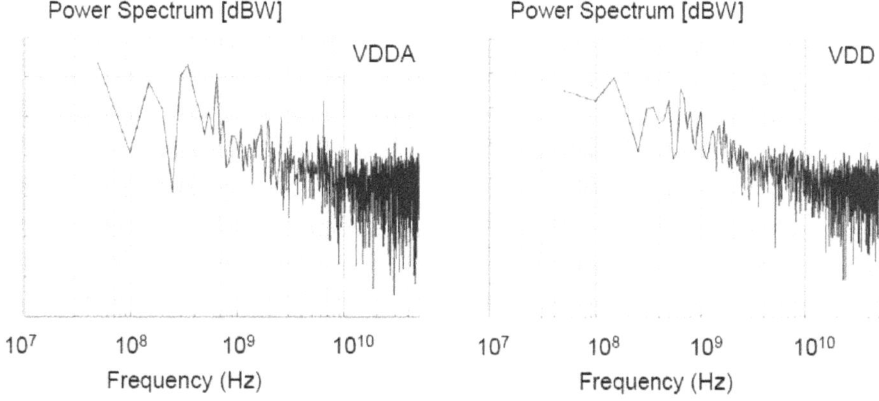

Figure 15.11: Power Spectrum Density on VDDA during PRBS7 Data Transmission, Measured Using Noise Monitor Circuits

The measurements show a noise power profile over frequency similar in shape to the noise (amplitude) spectrum estimated using simulated current waveforms and supply impedance models. The measured noise power spectrum shows an emphasis of noise at medium frequencies, in the frequency range where the power distribution system encounters package/chip resonance. At higher frequencies, the noise background decreases and, especially for VDDA, single noise peaks are visible. At lower frequencies, the resolution of this measurement is limited due to the limited range of delay times between both sampler clocks. Therefore, it is difficult to determine the shape of the supply noise power spectrum in the low-frequency range.

The measurements clearly show that supply noise at medium frequency contributes significantly to the total supply noise in the system. Even for the VDDA supply, which shows little noise excitation at medium frequency, the supply noise power at medium frequency dominates the total supply noise power in the system.

This emphasis of medium-frequency noise is caused by the shape of the power supply impedance ZPDN(f) with its characteristic peak at medium frequencies due to package/chip resonance. The frequency of this resonance peak is determined primarily by the packaging technology available today and the silicon space reasonably available for in-chip decoupling capacitors. As a result, the resonance frequency typically falls into a quite narrow frequency range between e.g., 10 MHz and 400 MHz, which is a much smaller frequency than the data rate of high-speed interfaces and also smaller than most internal clock frequencies inside these interfaces. This means that, independent of the data rate of a highspeed interface, a major noise contribution has to be expected at a medium frequency range, at the frequency of the package-chip resonance, even if there is little current excitation at this frequency range.

Correlation of Supply Noise Simulations and Measurements

The noise monitor measurement results were used to construct time domain supply noise waveforms. These waveforms were compared to simulated noise waveforms for the same data activity.

Figure 15.12 shows the simulated and measured waveforms for the VDDA supply rail.

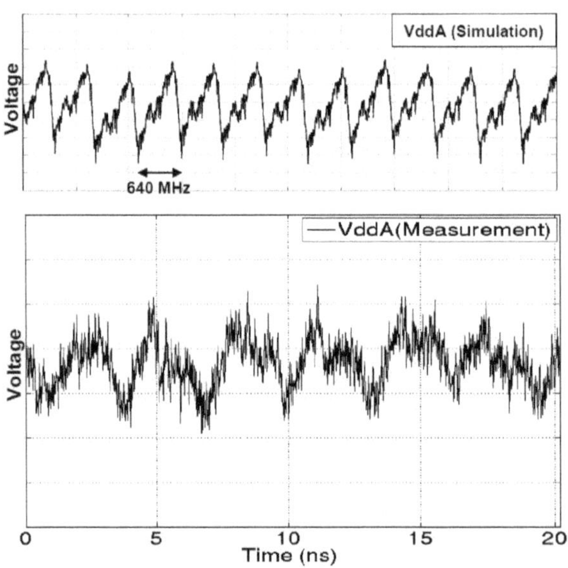

Figure 15.12: Simulated and Measured Supply Noise Waveforms for VDDA Supply

The peak-to-peak amplitude, as well as the general waveform of the supply noise simulation, correlates well to the measurements on the test system. The measurement, however, shows an additional 320 MHz component of the noise waveform that is not predicted by the simulation. The corresponding current component at 320 MHz was not predicted by the circuit simulations used to extract the current profile of the circuits supplied by this rail. This can be caused by the ideal assumptions used commonly for the device-level simulations.

Examples of these ideal assumptions are an ideal external input clock and PLL filter capacitors without leakage. The peak-peak noise prediction of the simulations, however, is still quite accurate. Similarly, the noise waveform measured on the VDD supply of the test system was compared to the simulated waveforms for the same data activity. Figure 15.13 shows the measured and simulated waveforms.

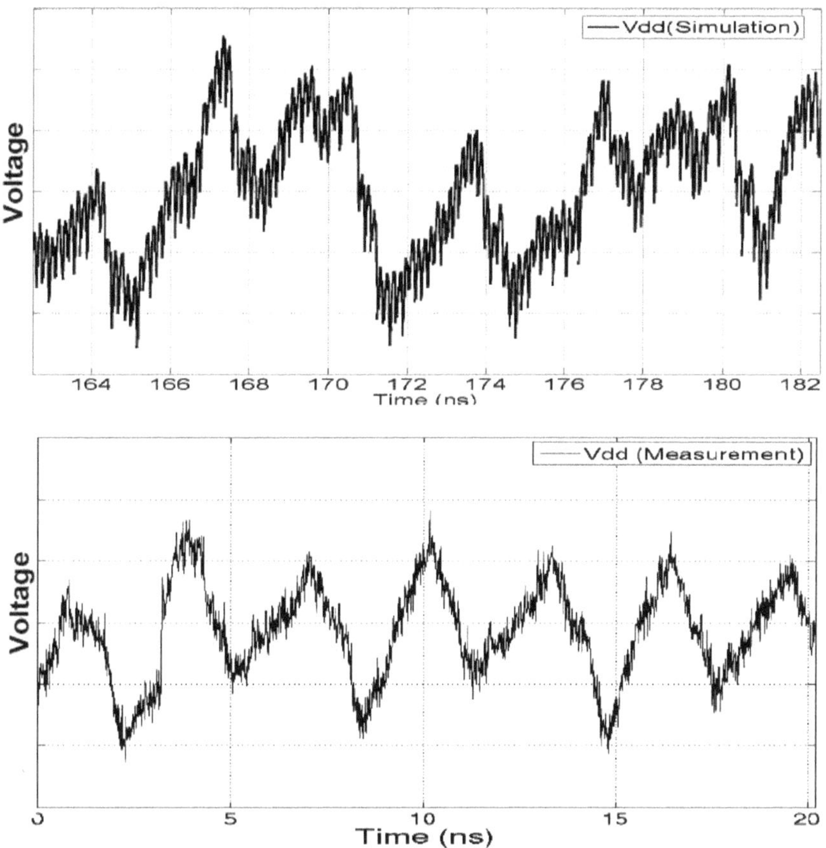

Figure 15.13: Simulated and Measured Supply Noise Waveforms for VDD Supply

Again, the peak-peak supply noise, as well as the general waveform shape, is correlating well between simulation and measurement. The exact waveform shape on VDD is changing depending on which fraction of the data pattern (PRBS7) is observed. Since we cannot easily recognize the fraction of the data pattern observed by the measurement, the simulation will show the noise at a different fraction of the PRBS7 pattern, which explains the difference in the detail of the measured and simulated waveforms.

For both supply rails, noise at medium frequencies dominates the total supply noise amplitude. This is consistent with our expectations based on the frequency domain analysis of the supply noise spectrum presented earlier and shows the importance to address medium-frequency supply noise in the jitter analysis even for high-speed interfaces with data rates much higher than the package/chip resonance frequency of the system.

Jitter Sensitivity Profiles

The supply-noise monitors and supply-noise generators were used to measure the sensitivity of system jitter as a function of supply noise at different noise frequencies.

Figure 15.14 shows sensitivity profiles for the VDDA and the VDD supply of the test system.

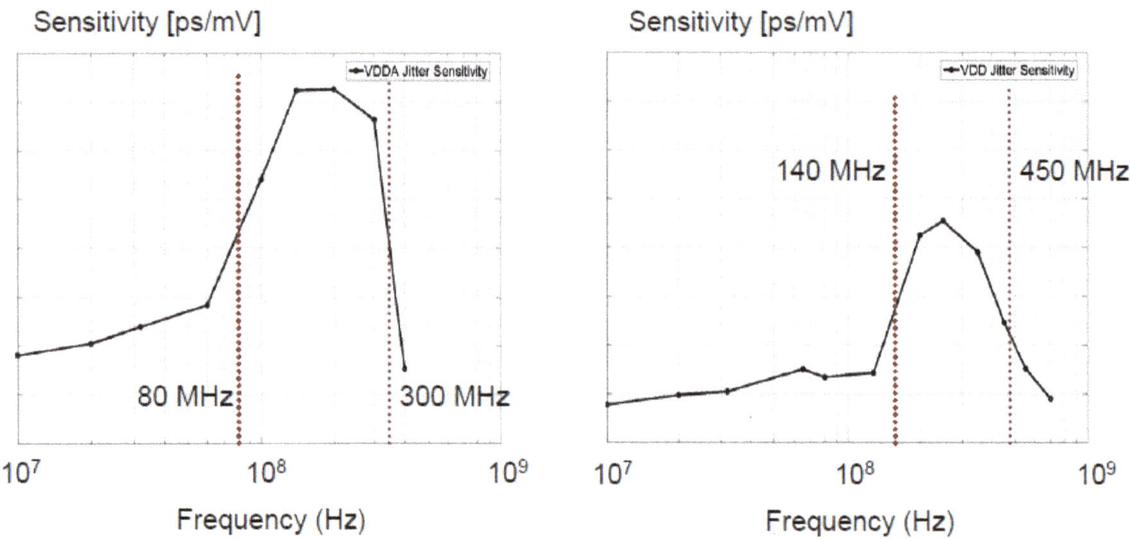

Figure 15.14: Jitter Sensitivity Profiles for VDDA and VDD Supply of the Test System

The sensitivity profiles for VDDA and VDD show similar shapes. The sensitivity is very small at low and high frequencies, but it reaches a maximum at medium frequencies. This means that low-frequency and high-frequency supply noise have little impact on the total jitter in the system. Supply noise at medium frequency, however, has the largest impact on system jitter and should be tightly controlled.

Unfortunately, following our discussion of supply noise spectrum, we expect a major contribution of supply noise at medium frequencies, close to the package/chip resonance frequency.

The package/chip resonance frequencies of common power distribution implementations easily fall into the frequency ranges where our measurements show circuits' large jitter sensitivity to supply noise. Thus, we have the unfortunate position that we expect significant noise at medium frequencies due to the package/chip resonance, and at the same time the circuits show large sensitivity to noise in exactly this frequency range.

This means that for gigabit I/O interface systems, the power distribution system and the circuits determining the sensitivity profile of the system should be designed together in a co-design process.

The goal of this co-design has to be to separate the frequency range of the largest noise generation from the frequency range of the largest noise sensitivity.

A second interesting result of the sensitivity measurements on the test system is the fact that, as expected, the jitter sensitivity of circuits on the VDDA supply is larger than for circuits on the VDD supply, but the difference is not as large as expected. The jitter sensitivity on VDD still reaches approximately 50 percent of the peak sensitivity observed on the VDDA supply.

Since the supply noise on VDDA is expected to be significantly smaller than on VDD, the supply noise on VDD can be still a significant source for system jitter compared to the contribution of supply noise on VDDA.

System Jitter Impact Due to Supply Noise

Combining the supply noise spectrum generated by the system on each rail with the jitter sensitivity profile of the system to supply noise on these rails, we can estimate the total jitter impact of supply noise on each power rail. For this we multiply the spectrum of the supply noise measured in the system with the jitter sensitivity of the system at this frequency.

$$(jitter\ impact)(f) = sensitivity(f)\ x\ V_{noise}(f)$$

Figure 15.15 shows the jitter impact profiles for VDDA and VDD of the test system, using the measured supply noise spectra measured for typical (PRBS7) data activity.

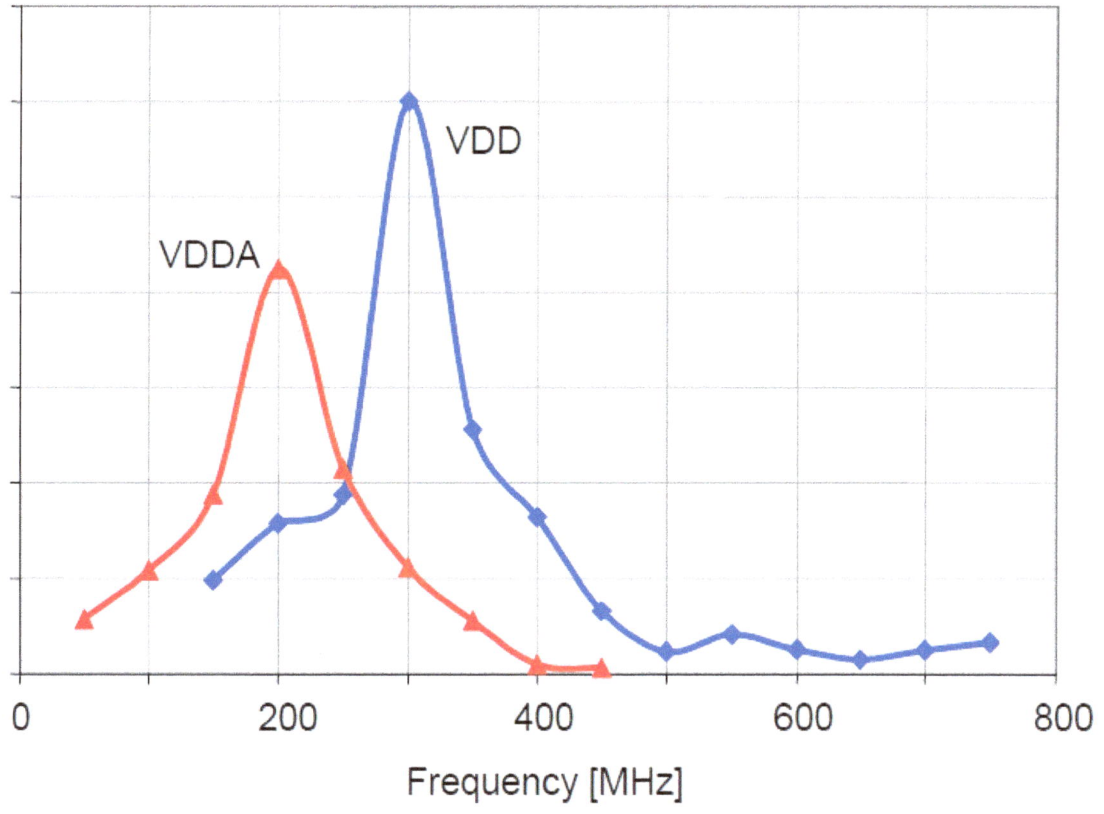

Figure 15.15: Jitter Impact Profiles for VDDA and VDD during PRBS7 Data Pattern

Figure 15.15 shows that supply noise predominantly contributes jitter in the medium-frequency range, as we expected from our previous discussion. This means that a supply noise budget for gigabit designs similar to the test case has to pay special attention to the amplitude as well as the exact frequency of supply noise in this frequency range.

The figure also shows that the total jitter contribution from noise on VDD can exceed the jitter due to supply noise on VDDA. As already speculated earlier, despite the smaller jitter sensitivity, VDD contributes significantly to the total jitter impact due to the larger noise level on this supply rail. Based on these results, a further optimization of this system's PDN should address the medium frequency noise on VDD.

Pattern Dependency of Supply—Noise Profiles

In a final analysis, we investigate the dependency of supply noise on the data pattern transmitted by the interface link. Until now, our analysis has been focused on the supply noise during the transmission of a typical random data pattern, e.g., using a PRBS7 pattern. Next, we will investigate whether there are pathological data patterns that could generate significantly higher supply noise levels or jitter contributions. Figure 15.16 shows the time domain noise waveforms on VDDA for three data patterns transmitted by the link.

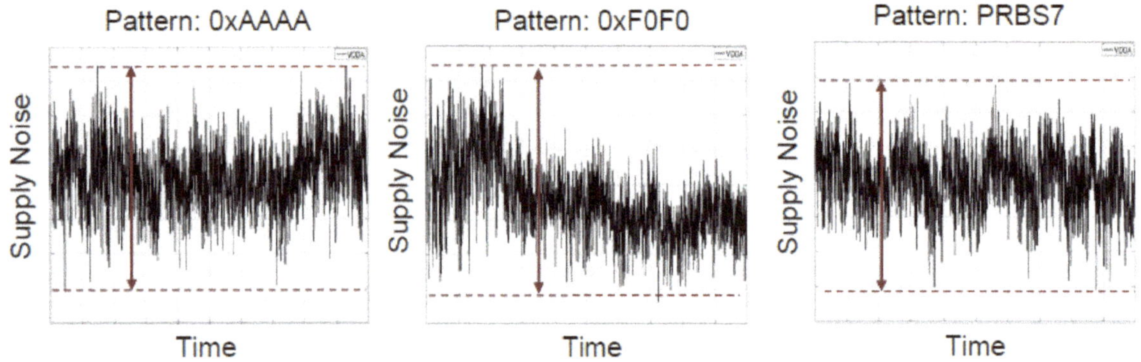

Figure 15.16: VDDA Supply Noise Waveform for Three Data Patterns

The VDDA waveforms show only very minor differences in peak-to-peak noise amplitude, which could be easily caused by small changes of the noise waveform over a longer pattern. This observation confirms the expectation of supply noise on VDDA.

Since the circuits supplied from VDDA are not part of the data path in the link, but instead are operating at constant frequencies to generate timing signals for the link, the supply noise on this rail should not show a dependency on the data pattern in the link.

Therefore, the jitter impact of supply noise on VDDA can be analyzed for any arbitrary data pattern in the system. Next, we analyzed the data pattern dependency of supply noise on the VDD rail. Figure 15.17 shows the time domain waveforms on VDD for the same three data pattern transmitted by the link.

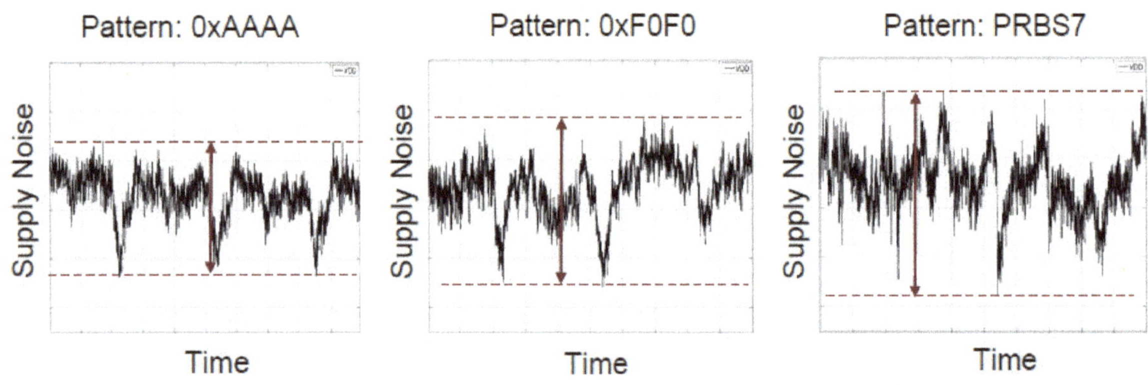

Figure 15.17: VDD Supply Noise Waveform for Different Data Pattern

The VDD waveforms show a significant dependency on the data pattern transmitted. In this small collection of data patterns, the PRBS7 pattern, the "typical" random data pattern used for most of our analysis, generates the largest peak-peak supply noise. The noise peak-peak amplitude while transmitting a PRBS7 pattern is twice as large as the peak-peak amplitude during the transmission of a 0xAAAA pattern. The major reason for the smaller noise during 0xAAAA and 0xF0F0 pattern is the very short pattern length for these patterns, which limits the noise current spectrum to higher frequencies.

This means that the excitation in the medium-frequency range, where the supply impedance has the highest magnitude, is very small for these patterns. For a PRBS7 pattern, on the other side, the random pattern is generating some noise current excitation in the medium frequency range.

Based on this observation, we would expect even larger peak-peak supply noise for data pattern that have more medium-frequency noise current components than a PRBS7 pattern. One such pathological pattern would be a load change pattern where the pattern generates a change in current dissipation at a frequency close to the package/chip resonance frequency.

Unfortunately, it was not possible to generate long arbitrary data patterns in the test system due to limitations in the pattern buffer implementation. We can, however, simulate the current profile for a load change pattern in a link and simulate the resulting supply noise using the power supply model. From the correlation results presented earlier, we know that these simulations are correlating very well for PRBS7 patterns with the measurements on the test system. Figure 15.18 shows a comparison for the VDD supply noise waveform in frequency and time domain, simulated for a PRBS7 pattern and for a load switch pattern that targets the package/chip resonance frequency of this system.

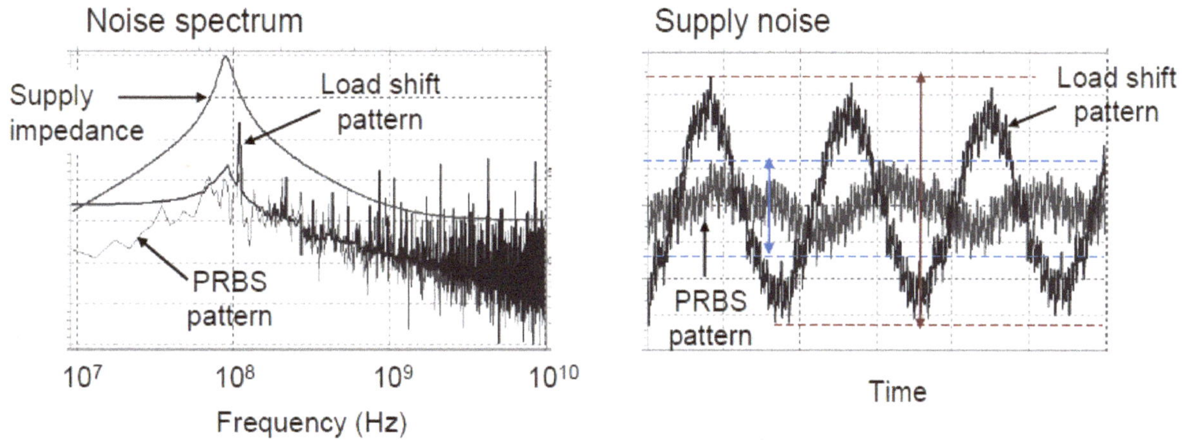

Figure 15.18: Supply Noise on VDD Supply Simulated for PRBS and for Load Shift Pattern

The left side of Figure 15.8 shows the noise spectrum on VDD for both patterns as well as the profile of the supply impedance. The load shift pattern creates a clearly visible noise spike close to the frequency of the package/chip resonance. It was not possible to address the resonance frequency exactly due to limitations on the transmitted pattern due to the data coding in the link, but the noise spike of the load shift pattern clearly occurs at a frequency of higher supply impedance.

The right side of Figure 15.8 shows the time domain supply noise waveforms for both patterns. It can be seen that the peak-peak noise generated by the load shift pattern is twice the size of the peak-peak noise caused by the PRBS7 pattern. It has to be noted that the impact on supply-induced jitter does not necessarily scale with the peak-peak noise amplitudes, since both patterns will result in different noise spectra and the interaction of these spectra with the jitter sensitivity profiles of the system has to be analyzed to predict the jitter impact. The results suggest, however, that it is essential to separate the frequency of package/chip resonance from the frequency of large jitter sensitivity in the system to avoid the coincident of large supply noise with high jitter sensitivity.

Summary

In this chapter, we presented the analysis of supply noise on link jitter in a serial I/O interface operating in the gigabit range. As the data rate in high-speed interfaces is scaled to ever-higher rates, it is essential to scale the link jitter with the decreasing unit time UI of the link. Supply noise is one significant source of link jitter, and a thorough understanding of the jitter generation, due to supply noise, is necessary to reduce this jitter contribution for future datarate scaling.

Supply noise-induced jitter is dependent on the interaction of two system characteristics. The total jitter contribution due to supply noise depends on the magnitude and spectrum of noise on the different supply rails and the sensitivity of the circuits to noise on their supply rail. We analyzed both of these parameters as well as the final jitter impact generated by the interaction of both parameters.

First, we analyzed the noise generated on the different supply rails of the data link. We discussed the dependency of the noise spectrum on the shape of the power supply impedance and showed that noise in the medium-frequency range, around the frequency of the package/chip resonance, dominates the total supply noise spectrum in many cases. We also presented measurements in frequency and time domain on a test system confirming this prediction for different supply rails.

Next, we analyzed the sensitivity of circuits to supply noise at a different frequency. The jitter generated by supply noise is the parameter determining link performance in the system, while supply noise magnitude and spectrum are merely inputs to this parameter. Analyzing the jitter sensitivity of the circuits to supply noise provides the connection between supply noise and link jitter and allows defining a supply noise budget based on the link jitter requirements.

We presented a methodology to measure the jitter sensitivity of the link to noise on different supply rails. This methodology requires dedicated noise monitor and noise generator circuits that were implemented on our test system. With these noise monitor and generator circuits, the sensitivity profile for different supply rails was measured on the test system. The results showed that the jitter sensitivity is a strong function of the frequency of the noise signal. Additionally, the measurements on the test system showed large jitter sensitivity to supply noise at medium frequency, the same frequency range where package/chip resonance often generates major supply noise contributions. Separating the frequency range of large noise generation from the frequency range of high jitter sensitivity is essential for the optimization of supply-induced jitter and requires an understanding of the design parameters determining these two frequency ranges in the system.

After this, we calculated the total jitter impact of noise on different supply rails in the system. Jitter analysis often focuses on the noise on the supply rail with the highest jitter sensitivity. It has been demonstrated that supply rails with low jitter sensitivity and large noise magnitudes can significantly contribute to the total link jitter.

Calculating the total jitter impact for different supply rails can be used to direct the design effort and optimize all supply noise contributions that have a significant impact on the link jitter. Finally, we investigated the dependency of the supply noise on the data pattern in the system. We showed that the noise on some supply rails is strongly dependent on the data pattern transmitted by the link. This dependency has to be taken into consideration when estimating the worst-case jitter contribution of supply noise on this rail.

References

[1] O. Mandhana, *Optimizing the Output Impedance of a Power Delivery Network for Microprocessor Systems*, Proceedings 54th Electronic Components & Technology Conference 2004, Las Vegas, Nevada, June 2004.

[2] R. Schmitt, C. Yuan, W. Kim, *Modeling and Correlation of Supply Noise for a 3.2GHz Bidirectional Differential Memory Bus*, DesignCon 2005, Santa Clara, California, February 2005.

[3] R. Schmitt, X. Huang, L. Yang, C. Yuan, *System Level Power Integrity Analysis and Correlation for Multi-Gigabit Designs*, DesignCon 2004, Santa Clara, California, Feb. 6-9, 2004.

[4] E. Alon, V. Stojanovic, and M. Horowitz, *Circuits and Techniques for High-Resolution Measurement of On-Chip Power Supply Noise*, IEEE J. Solid-State Circuits, vol. 40, no. 4, pp. 820-828, April 2005.

Chapter 16—PCB Design Methods for Optimum FPGA SerDes Jitter Performance

Steve Weir, Consultant, Teraspeed Consulting Group
Scott McMorrow, President, Teraspeed Consulting Group
Al Neves, Consultant, Teraspeed Consulting Group
Tom Dagostino, Vice President, Teraspeed Consulting Group
Brian Vicich, Signal Integrity Engineer, Samtec, Inc.

This paper was presented at DesignCon 2007, Santa Clara, California, January 29-31, 2007.

Introduction

DESIGNING A PRINTED circuit board (PCB) to obtain minimum serializer/deserializer (SerDes) jitter is critical to maximize highspeed serial link bandwidth and reliability. The PCB power distribution methodology has a strong impact on SerDes performance. However, optimization for one impairment often comes at the cost of another. Considerable debate remains over which effects dominate and how to reliably yield quality designs.

Using carefully designed test vehicles, we characterize Virtex 4™ SerDes sensitivity to specific PCB impairments and physical PCB design tradeoffs. We reconcile current debate on best practices and offer a detailed design methodology.

FPGA SerDes Design Challenge

For several years field-programmable gate arrays (FPGAs) have had SerDes devices supporting at least 3.125 Gbps operation and sometimes beyond. Current families include the following:

- Altera Stratix II GX up to 6.375Gbps -3 speed grade [1]
- Xilinx Virtex 4™ FX up to 6.5Gpbs -12 speed grade [2]

Next-generation devices from Altera (Stratix III GX), Lattice (SC) and Xilinx (Virtex 5 FX) are other recent entries on the market.

SerDes imposes new design challenges that are often foreign to digital engineers. Many of these challenges occur in the channel design. Gbps serial channel design is similar for application-specific integrated circuits (ASICs) and FPGAs alike, and many excellent works have been published on the subject.

However, power delivery is peculiar to individual IC implementations. Whereas for an entirely digital device about 5 percent and often 10 percent rail noise over a broad spectrum does not present a serious problem, such noise levels can readily cripple SerDes operation. For this reason, FPGA manufacturers have been careful to publish power design guidelines, some in the form of rigid recipes.

The recipes are incomplete and contain elements that are subject to considerable debate. Some industry experts have assailed some of these guidelines as either unnecessary or, even worse, counterproductive. The current information gaps and requirements uncertainty serve neither users nor FPGA manufacturers. In 2006, Samtec, Inc. and Teraspeed Consulting Group undertook to research actual device requirements. This paper reports our findings as they relate to transmit jitter in Xilinx Virtex 4™ FX parts.

Based on our research, we believe that these findings will prove consistent on Virtex 5, as well as the Altera Stratix II and Stratix III parts scheduled for testing in late 2007.

Virtex 4™ Rocket I/O Transmitter

In the Virtex 4™ product line, Xilinx alternately refers to their SerDes offerings as RocketI/Os™ or multi-gigabit transceivers (MGTs). These devices support up to 3.125 Gbps operation in all speed grades and up to 6.5 Gbps operation in the fastest speed grade –12 [2]. A single FPGA may contain eight to 24 transceivers arranged in four to 12 tiles of two transceivers each. Each tile includes eight separate power supply inputs.

Supply	Function	Nominal Voltage	Max Current @ 3.125Gbps
AVCCAUXMGT	PLLs	2.5V	5 mA
AVCCAUXRXA	Tile A side receiver	1.2V	191 mA
AVCCAUXRXB	Tile A side receiver	1.2V	191 mA
AVCCAUXTX	Tile transmitters	1.2V	307 mA
VTRXA	Tile A side receive termination	1.5V	24 mA
VTRXB	Tile B side receive termination	1.5V	24 mA
VTTXA	Tile A side transmit termination	1.5V	105 mA
VTTXB	Tile B side transmit termination	1.5V	105 mA

Table 16.1: MGT Supply Voltages

As an example of the importance of clean SerDes power, Xilinx bluntly states that to receive factory support, customer designs must conform to each aspect of the stated power design requirements [3] contained in publication Virtex-4 RocketIO Multi-Gigabit Transceiver User Guide UG076 3.0, May 23, 2006, Chapter 6.

- Linear voltage regulators for each of the at least three MGT voltages:
  - AVCCAUXMGT 2.5V nominal
  - AVCCAUXRX/TX 1.2V nominal

- VTRX/TX 1.5V nominal
- Separate series decoupling filters for each MGT power pin:
    - Series ferrite bead TDK MPZ1608S221A
    - Shunt bypass capacitor generic 0402 220nF
- A specific capacitor bank at both the input and output of each linear regulator:
    - 10 µF capacitor input
    - 1ea 330 µF, 8ea 1 µF capacitors output

None of the five analog rails per MGT tile may be shared with digital devices.

The stated goal behind for all of these measures is to supply pristine power to the sensitive analog power supplies that support the SerDes.

MGT PDN PCB Reference Model

We examine a model of the resulting power distribution network (PDN) to determine how it may perform against the stated goals:

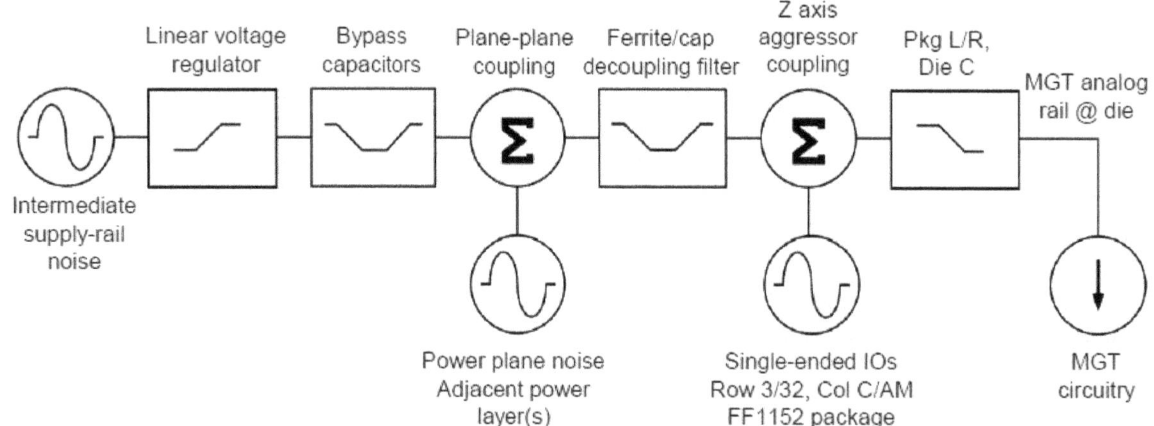

Figure 16.1: Model of Xilinx MGT Analog Power Distribution

The model includes three noise sources:

- Intermediate supply rail noise—this is typically the output of a switching power supply providing 3.3-5V input to a low drop-out linear regulator.
- Plane-to-plane noise coupling—when the stackup does not insert a shield layer between fill/plane used by the MGT power distribution, noise from aggressor signals and/or other power planes couples to the MGT power fill/plane.
- Z-axis coupling—single-ended signals adjacent to MGT pads couple through adjacent vias/balls in the PCB and the FPGA package.

Working against these noise sources are the linear voltage regulator, bypass capacitors, series decoupling

filters, and the low pass filter in the MGT power node of the FPGA.

Linear Regulator Ripple Rejection

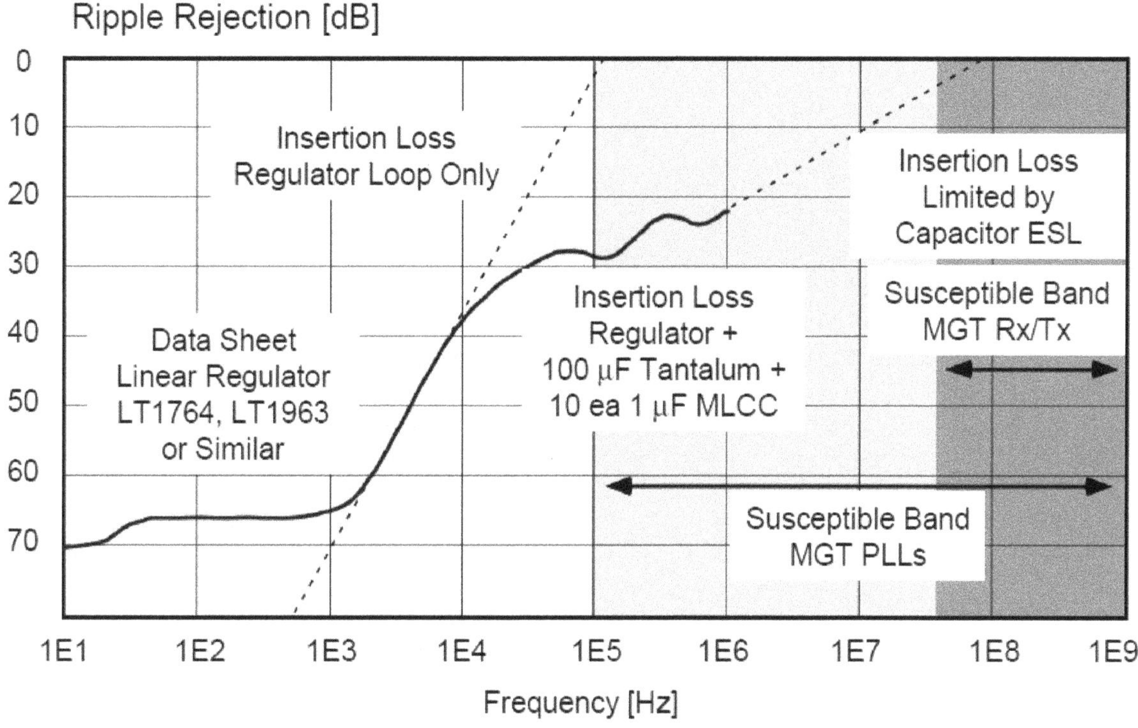

Figure 16.2: Ripple Rejection of MGT Linear Regulator with No Ferrite

Even high-quality modern linear regulators such as Linear Technology's (LT's) 1764xx, LT1963xx, or similar ones exhibit very limited ripple rejection above 100 kHz by themselves. Figure 16.2 replots data sheet ripple rejection as insertion loss.

In Figure 16.2, we can see that noise rejection from the regulator feedback loop drops off rapidly from 2 kHz to a projected 0 dB intercept near 100 kHz. Beyond about 50 kHz the regulator series parasitics loaded by the output bypass capacitor network provide most of the attenuation. Noise feed-through above 1 MHz depends on layout and bypass network equivalent series inductance (ESI).

Series Ferrite Decoupling Networks

Series ferrite decoupling networks can provide additional loss up to 1 GHz. However, unless applied carefully, ferrite beads can offer unwelcome surprises. At low frequencies, ferrites are low-loss magnetic materials, and they make excellent high Q inductors.

When mated with high Q capacitors such as ordinary ceramic bypass capacitors, the combination creates a low pass filter with high peaking at $F_{CUT\text{-}OFF}$. If we wish to use avoid amplifying noise voltage at resonance, we have the following three choices:

- Set the filter cut-off frequency to occur at or above the frequency at which the bead Q falls to approximately 1, (25 MHz for the MPZ1608221A).

- Add shunt or series damping to drop overall filter Q.
- Add enough series loss by other means to compensate resonant peaking.

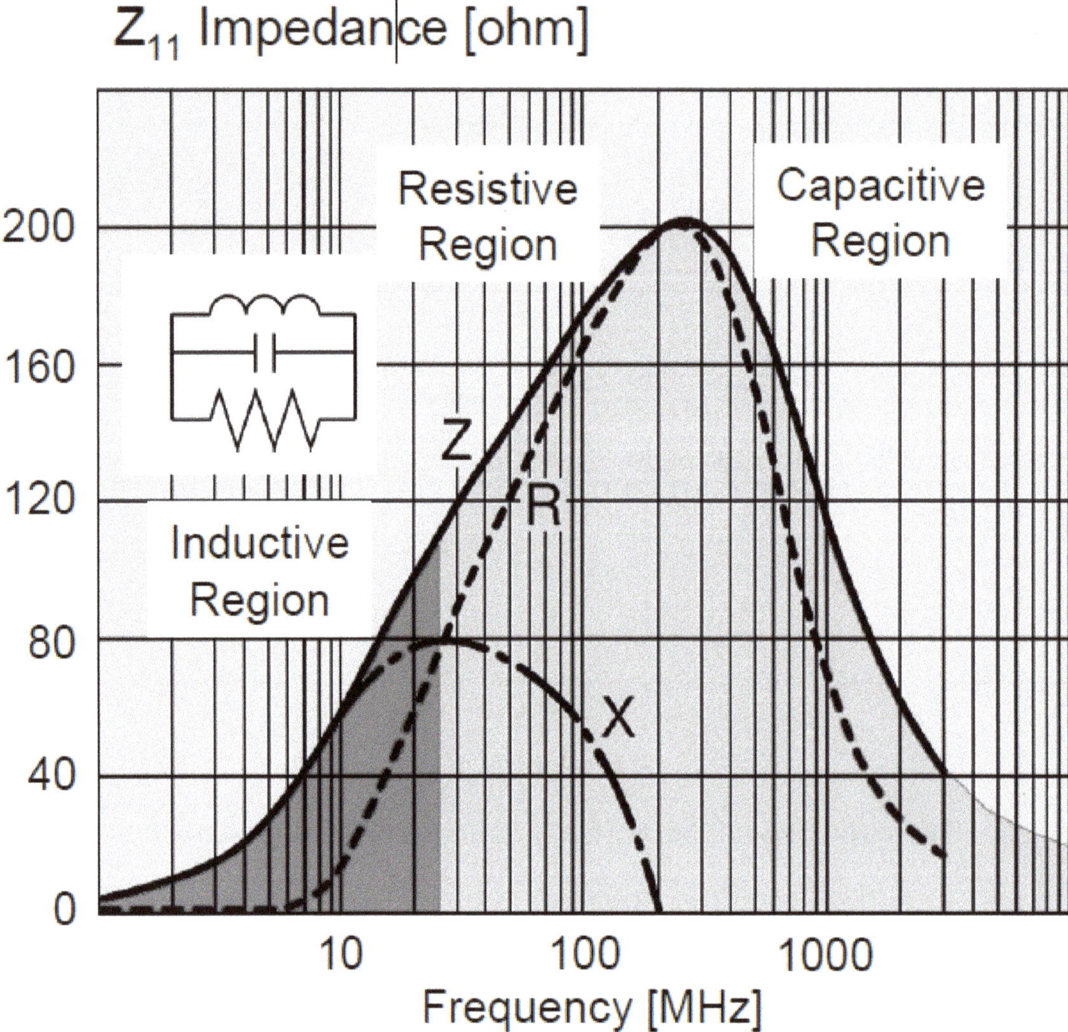

Figure 16.3: Z11 Impedance of TDK MPZ1608221A Ferrite [3]

$F_{CUT\text{-}OFF}$ of the low pass network stipulated by Xilinx: MPZ1608221A with a 220 nF capacitor occurs close to 300 kHz.

300 kHz is a very popular operating frequency for current generation SMPS modules. The following simulation demonstrates response of the specified filter network to a 300 kHz 60 mVp-p sawtooth at the filter input and a 500 Ohm load comparable to the AVCCAUXMGT load of a single MGT tile.

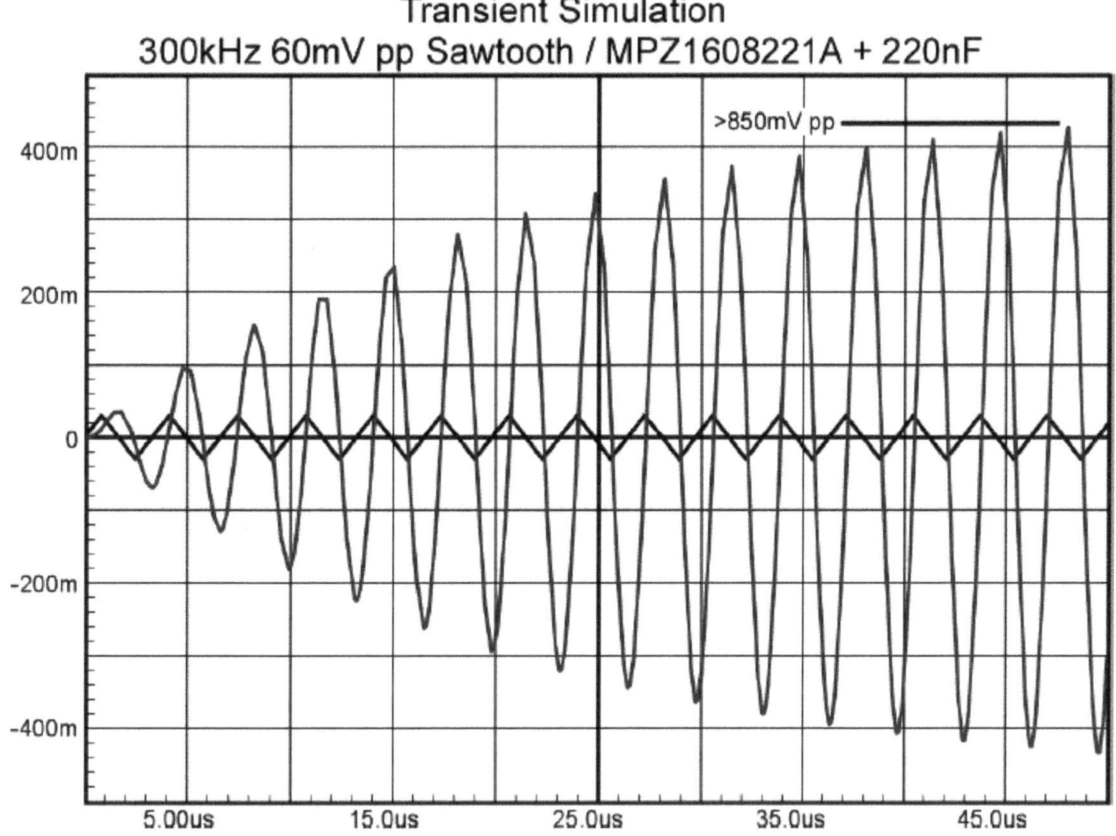

Figure 16.4: Transient Simulation, 300 kHz SMPS, MPZ1608221A + 220 nF

A high-quality linear regulator does not damp the filter resonance. It adds enough insertion loss so that the net insertion loss at F_{CUTOFF} is positive. Figure 16.5 illustrates composite frequency response of the linear regulator-based PDN recommended by Xilinx.

Beyond the sharp resonance at 300 kHz, insertion loss recovers quickly, strongly rejecting noise from less than 1 MHz to well over 100 MHz. The filter can be effective against the following two noise sources:

- High frequency SMPS ringing
- When present noise coupled from another adjacent copper fill layer, or adjacent aggressor signals

Properly chosen ferrite beads really shine suppressing high frequency switched mode power supply (SMPS) ringing. It is easy to obtain more than 30 dB insertion loss out to 250 MHz or more, reducing even 100 mV high frequency ringing to 3 mVpp or less.

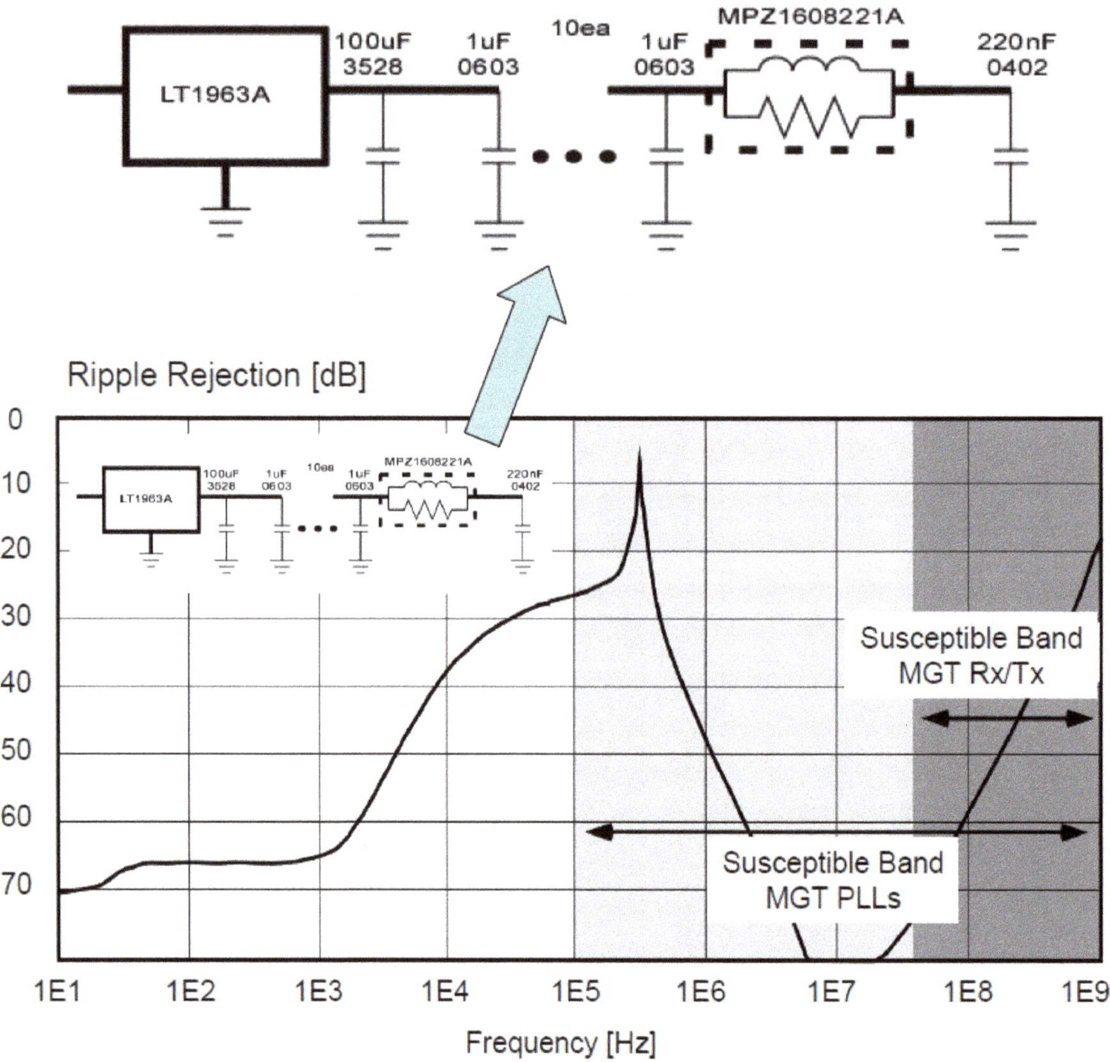

Figure 16.5: Ripple Rejection, MGT Linear Supply Plus MPZ1608221A and 220 nF

Digital noise coupling depends on the PCB stackup and routingcstrategy. Each MGT requires a minimum of three voltages:

- AVCCMGTAUX, 2.5V
- VTRX, typically 1.5V
- AVCCAUXRX, AVCCAUXTX 1.2V

For FX60 and larger devices, all three voltages encircle the package, and this dictates use of at least two PCB layers for MGT power. See Figure 16.7.

A strategy that minimizes bypass capacitor attachment inductance places MGT power on outer layers. The MGT power may be shielded by assigning global information grid (GIG) network defense (GND) to the next layer in on the PCB.

The downside of this arrangement is that either signal routing references VCCINT/VCCAUX on one layer creating a return path continuity issue, or additional GND layers must be added to the stackup.

Without GND shield layers, the adjacent power layers capacitively couple directly into the MGT power, and ferrite beads can attenuate this noise.

An alternate strategy is to place MGT power distribution on inner layers. Absent sandwiching each MGT plane shape between GND shields, noise for any other adjacent layers couples into the MGT power. Here again, ferrite beads may be used to suppress this noise. We can determine the amount of noise coupled from tools such as Si-Wave™. What we do not have is a specification of tolerable noise from Xilinx.

Figure 16.6 depicts three alternative stackups each with MGT power on the outer layers. Only the first configuration benefits from a series-decoupling filter. In the latter two configurations, a GND layer shields MGT plane shapes from digital noise pickup.

Each of these stackups suffers drawbacks. The 20-layer design adds the expense of two additional layers. The first configuration couples noise from both the core and VCCIO into MGTs. Several hundred pF coupling capacitance is easily obtained. The center configuration protects MGT power at the expense of complicating digital signal return paths. Aside from board level decoupling, VCCINT and VCCAUX have little to do with input/output (I/O) signal transmission.

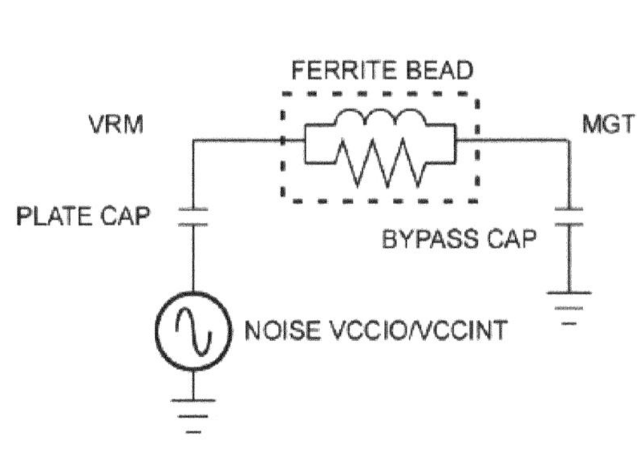

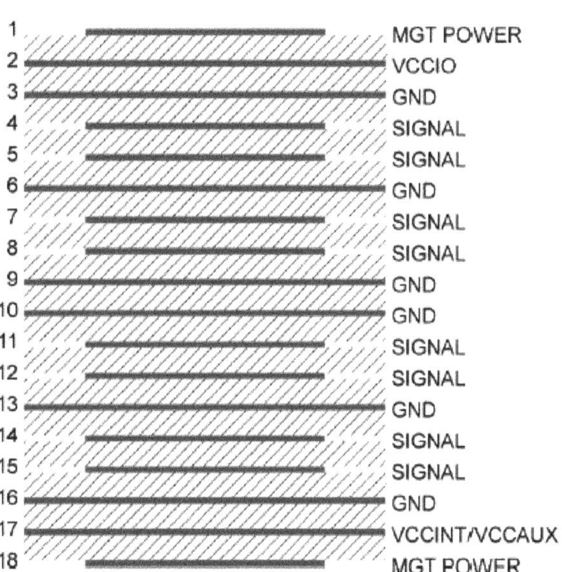

Figure 16.6.a: First Stackup Alternative with Eight Signal Layers

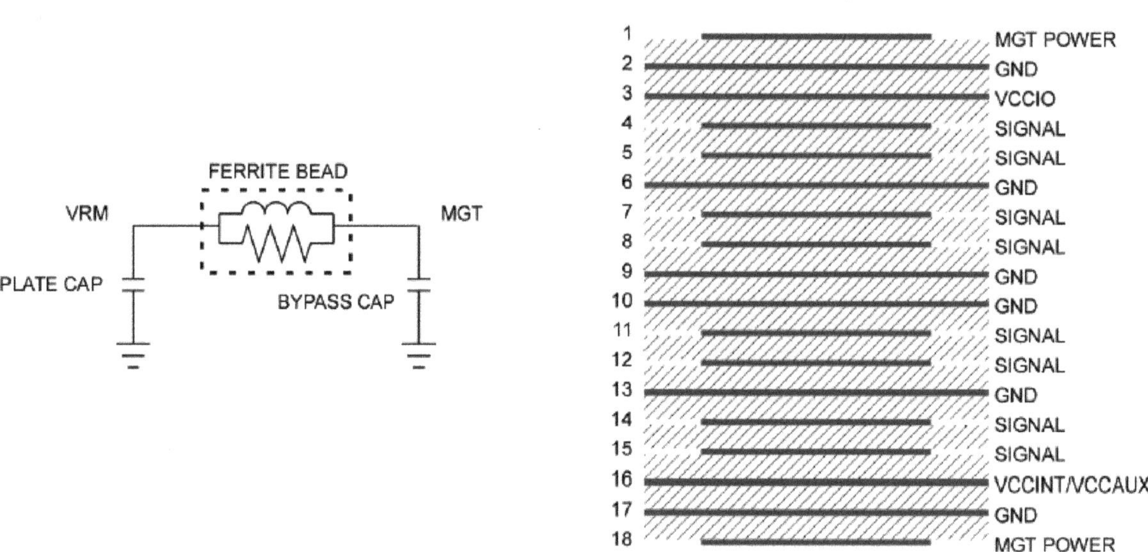

Figure 16.6.b: Second Stackup Alternative with Eight Signal Layers

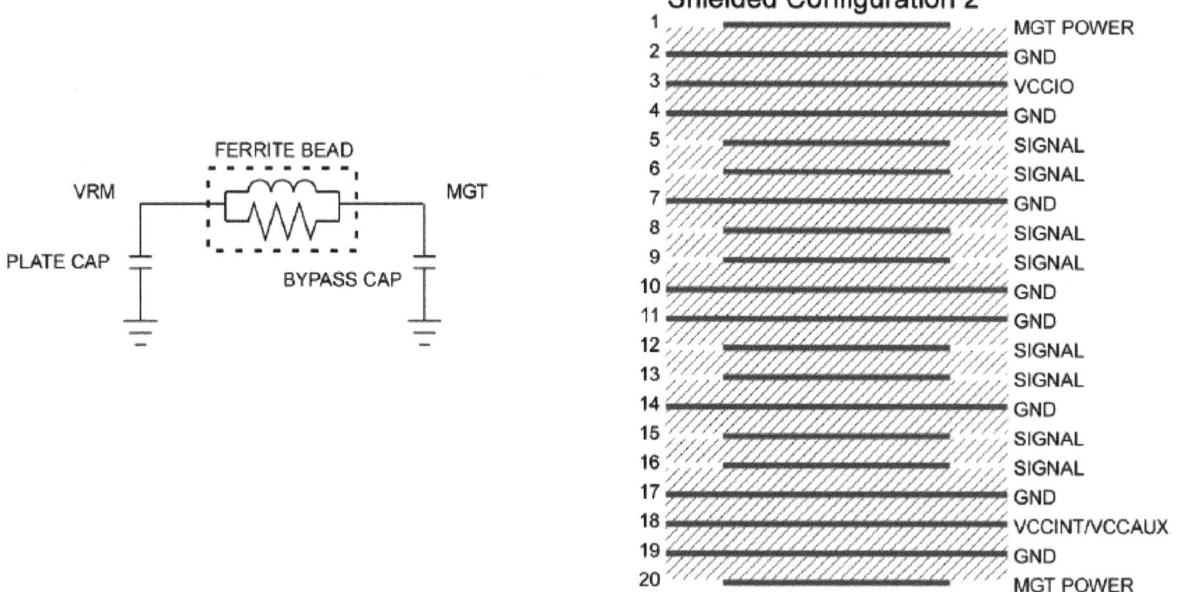

Figure 16.6.c: Third Stackup Alternative with Eight Signal Layers

Z Axis Crosstalk

The ferrite series-decoupling filter is to the left of the third noise source: crosstalk from digital I/O aggressors that couple through the PCB and device package Z axis. Refer to Figure 16.7.

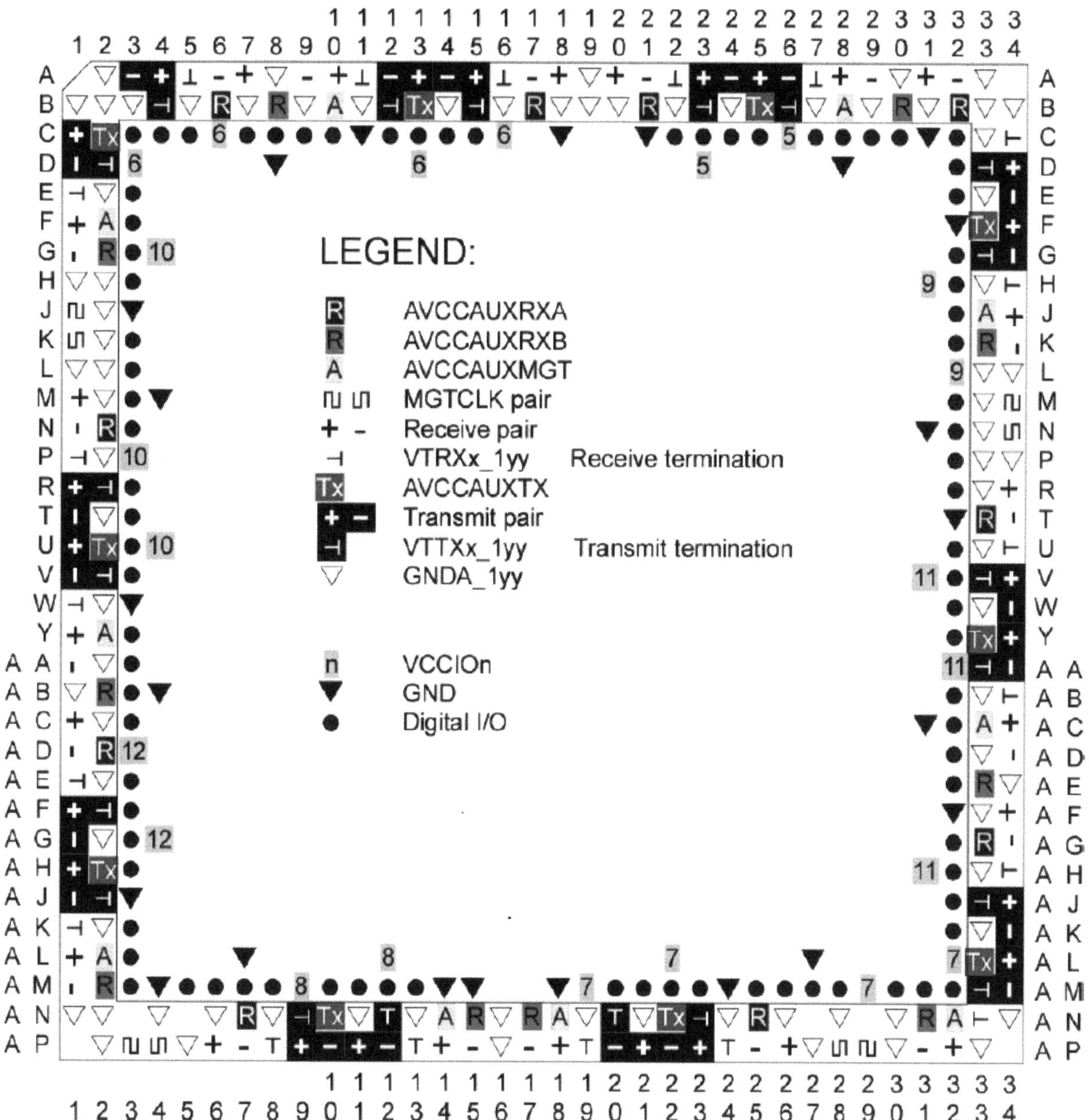

Figure 16.7: MGT I/O Map, Xilinx Virtex 4™ FX 1152 Package

The total Z axis path through a typical telecom board and the package is over 150 mils. This proves long enough to couple enough signal energy to affect MGT jitter performance even with the modest rise and fall times of Virtex 4™ digital I/Os.

Jitter Evaluation Test Vehicle

We evaluated PCB related jitter sources with a test vehicle that facilitates comparison of the following design choices and signal effects:

- Analog Vcc power supply architecture
- Aggressor signal coupling
- IC package ground bounce

Each vehicle consists of two similarly designed halves designated A and B.

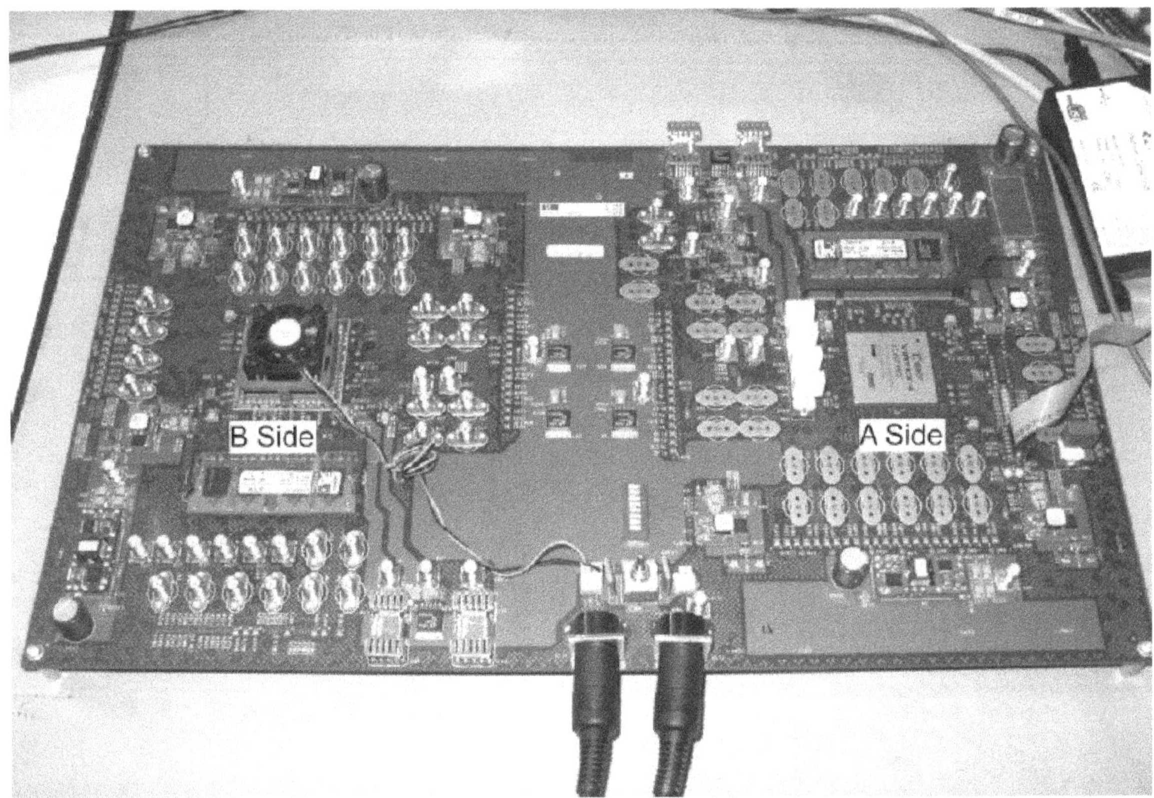

Figure 16.8: Jitter Evaluation Test Vehicle

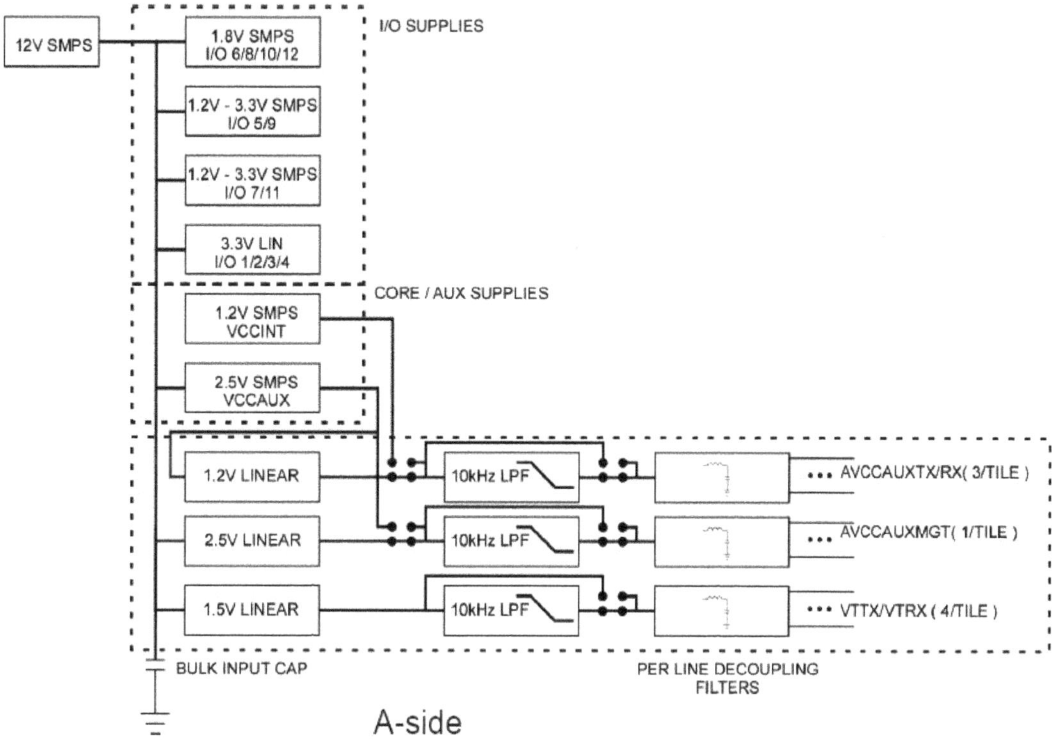

Figure 16.9.a: Structure, PDN, Jitter Evaluation Test Vehicle, A-Side

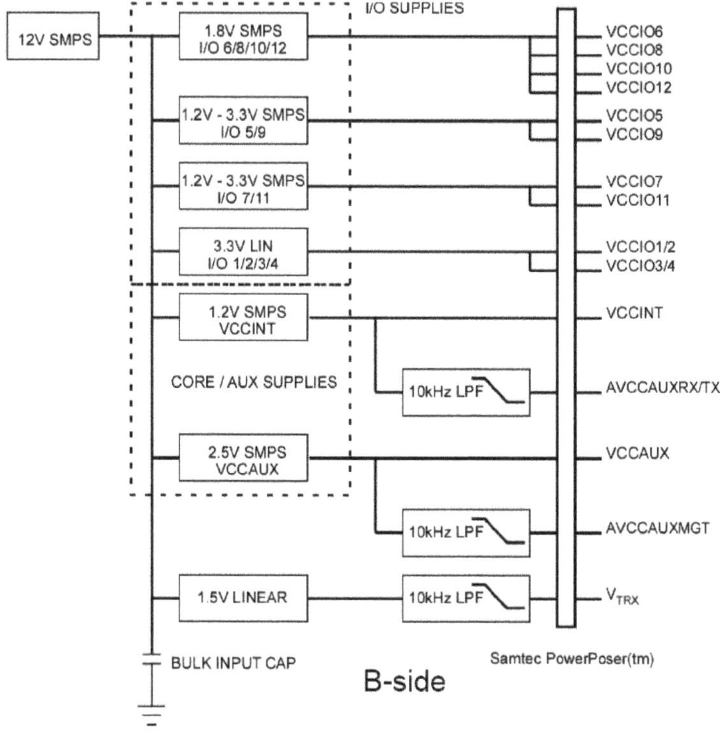

Figure 16.9.b: Structure, PDN, Jitter Evaluation Test Vehicle, B-Side

A Side

The A side serves primarily as a reference for the mandated Xilinx networks. The A side fully implements the Xilinx MGT power guidelines. The MGT 2.5V and 1.2V rails may source from linear regulators or from VCCAUX, and VCCINT respectively.

Additionally, overdamped 10 kHz low pass filters composed of a single ferrite bead and 3528-size tantalum capacitor may be switched into each MGT analog supply. The layout connects AGND BGA pads to PCB GND on all layers.

B Side

The B side eliminates the per power pin series ferrites in favor of direct plane connections. Instead of one capacitor per MGT power pin, one bypass capacitor on each side of the device bypasses each of the three analog MGT rails. Finally, the B side mounts the FPGA on a prototype power interposer assembly from Samtec. As with the A side, the B side includes provisions for powering MGT circuitry from either linear regulators or SMPS regulators that supply the FPGA digital sections. Here we report results using the SMPS regulators for the 1.2V and 2.5V rails.

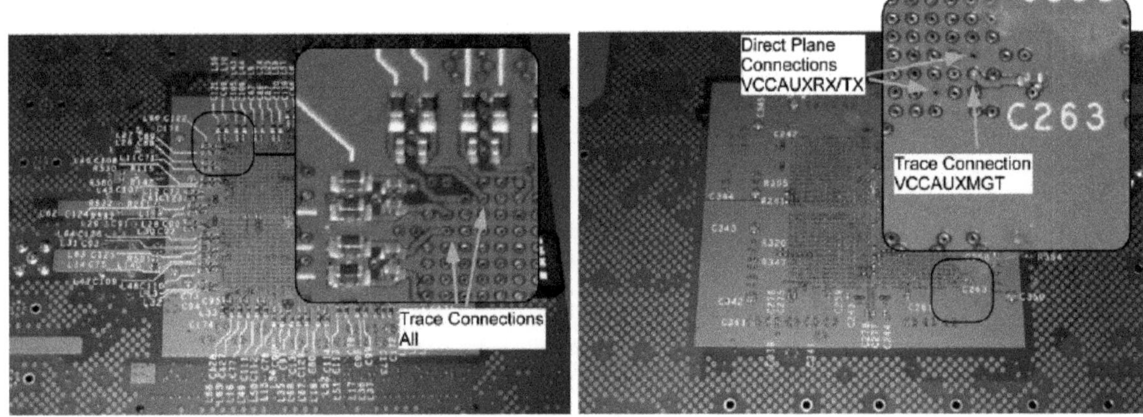

Figure 16.10: Jitter Evaluation PCB, FPGA Backside Detail Left: A Side; Right: B Side

Jitter Evaluation Tests

We conducted jitter evaluation tests by configuring both transceivers on one MGT tile for operation at 3.125 Gbps using 8b/10b encoding. Our test pattern consisted of PRBS9 for 254 symbol periods followed by a comma pattern for two symbol periods. We configured digital I/Os operating at 533 Mbps driving a common PRBS5 pattern and all active I/Os as SSTLII-1.8V class 2 (DDR2).

Test Results of Xilinx Guidelines-Based Design

We obtained very good results at 3.125 Gbps by following the Xilinx guidelines, combined with high-quality linear regulators and a careful layout.

The left side of Figure 16.11 shows the transmit data eye with no digital I/Os enabled. Jitter for this case is just 37 ps, 11.6 percent of the 320 ps unit interval (UI). The right-hand capture shows results with all digital I/Os except those in rows C and AM, and columns 3 and 32 simultaneous switching. Despite a 16 App total I/O return current jitter increases by just 1 ps to 38 ps. The Virtex 4™ package does a good job of blocking digital return currents from traversing the MGT analog domains.

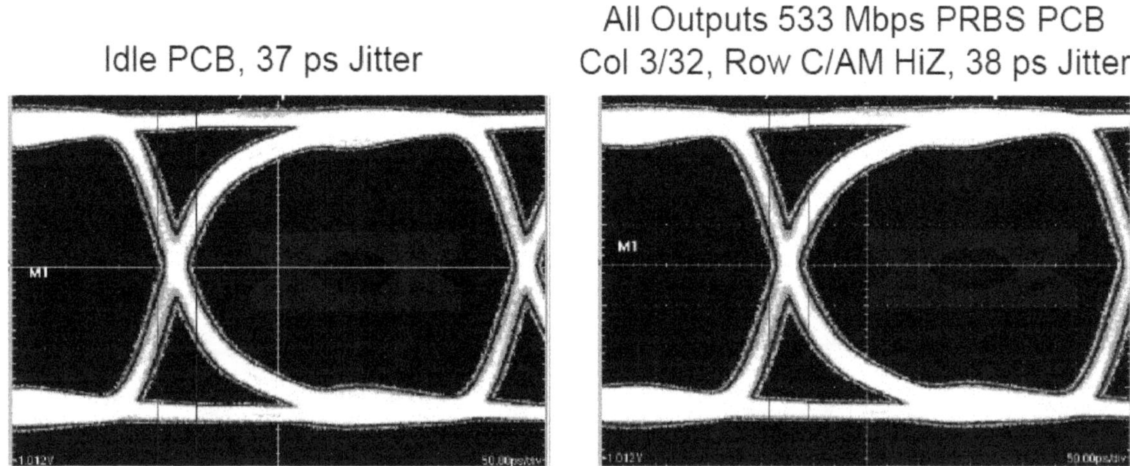

Figure 16.11: Jitter Xilinx Guidelines, Optimum PCB

We may reasonably conclude from this that package GND bounce and any related coupling between analog ground (AGND) and GND through the IC package is not a problem. Therefore, FPGA AGND pads should tie directly to PCB digital GND.

We evaluated Z axis crosstalk by separately enabling digital aggressors in columns 3/32, and rows C/AM as seen in Figure 16.12.

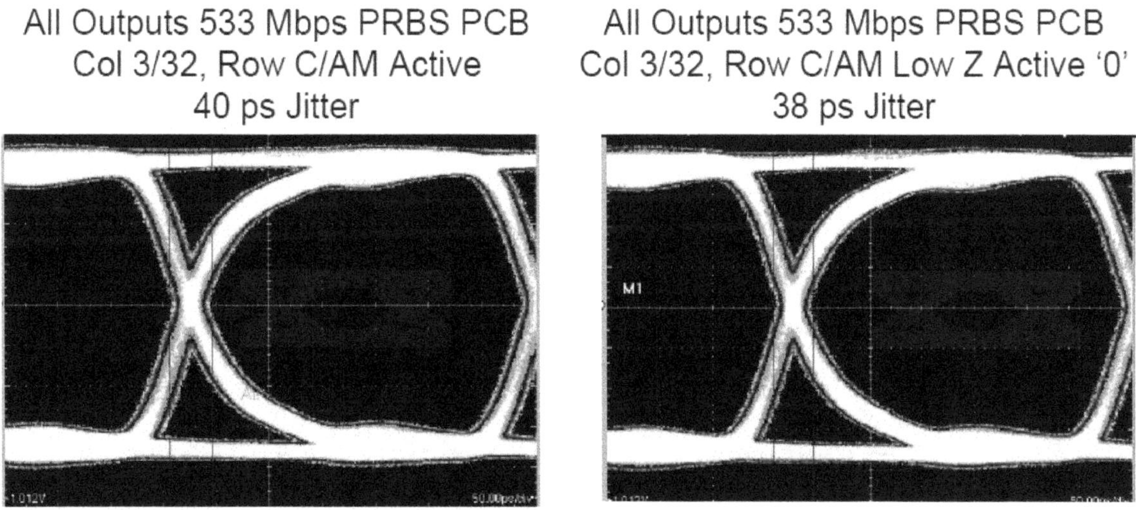

Figure 16.12: Jitter Evaluation Z Axis Crosstalk

Here, we see that jitter increases to 40ps p-p but, despite its small size, a minor caution is in order: these tests from the FPGA Virtex 4™ I/Os have relatively modest transition rates. Memory devices' transition times are four to six times faster. Commensurately higher crosstalk and jitter should be expected if memory-read data connects to FPGA pins adjacent to MGT pins than recorded here.

Powering SerDes Analog from SMPS

We configured the A side to power AVCCAUXRX and AVCCAUXTX supplies from VCCINT. We tested both with and without first passing VCCINT through an overdamped noise filter with FCO at 10 kHz. Transmit jitter performance degrades, but not critically, as seen in Figure 16.13.

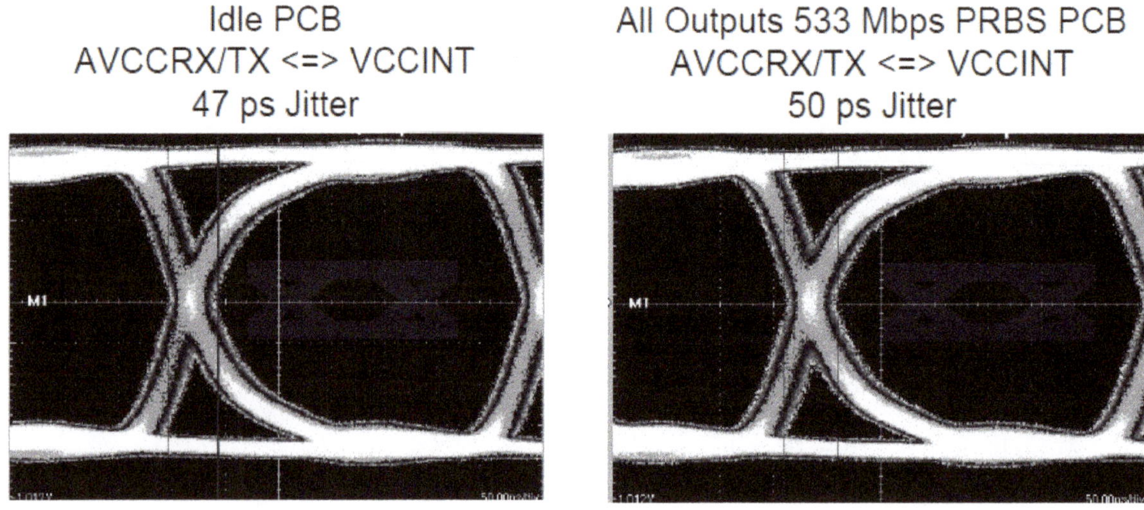

Figure 16.13: Ferrite Decoupling, AVCCAUXRX, AVCCAUXTX, 1.2V from VCCINT SMPS

Transmit jitter increases to 50 ps peak-to-peak with all digital I/Os switching, including those adjacent the MGT pins.

AVCCMGTAUX is an entirely different story. This node proves highly sensitive to noise over a wide spectrum. We first tested with just the original mandated networks. With this, configuration jitter increases to an unacceptable 110 ps peak-to-peak.

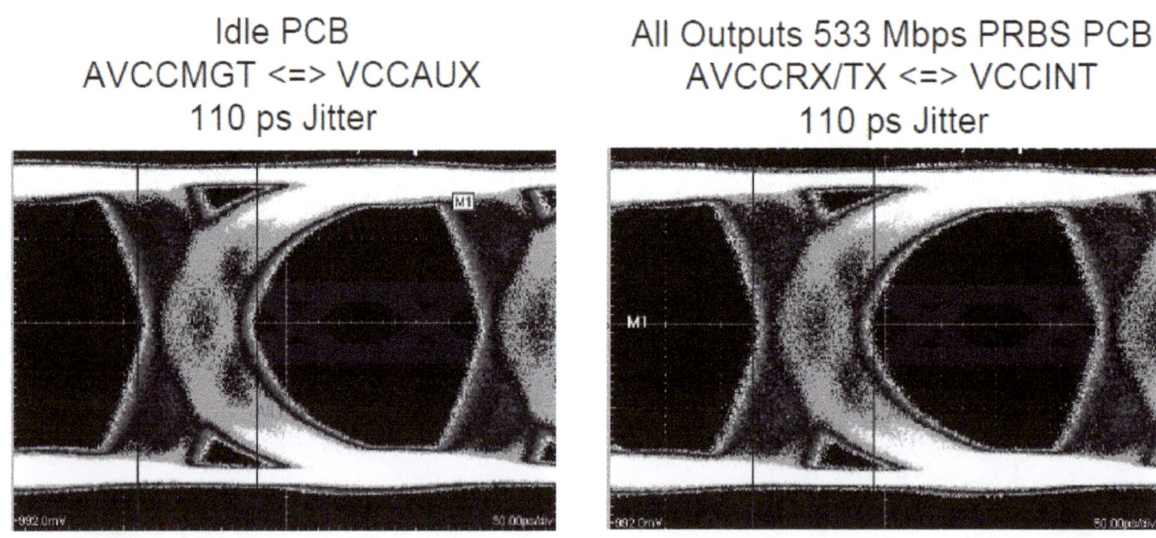

Figure 16.14: Transmit Jitter AVCCAUXRX/TX from AVCCINT, AVCCAUXMGT from VCCAUX

Results improve markedly when we pass VCCAUX through a 10 kHz FCO filter, as shown in Figure 16.15.

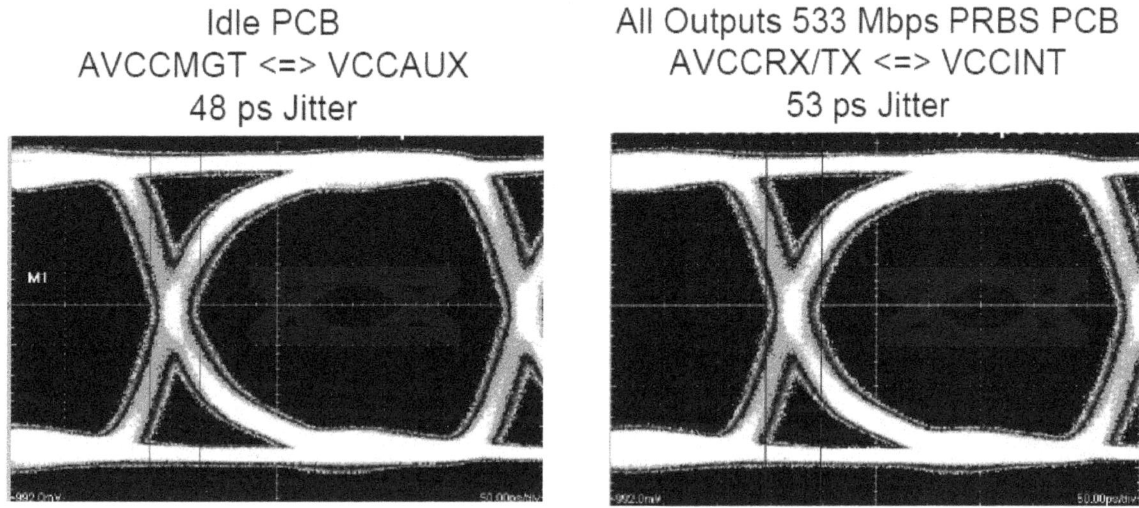

Figure 16.15: Transmit Jitter with 10 kHz LPF in AVCCAUXRX/TX, AVCCAUXMGT

Test Results B Side with Samtec PowerPoser™

The B side replaces 128 discrete components: 64 ferrites and 64 bypass capacitors with 12 0402 bypass capacitors. AVCCAUXRX and AVCCAUXTX source from the SMPS fed the 1.2V VCCINT core power supply, while AVCCAUXMGT obtains 2.5V from FPGA VCCAUX also through a 10 kHz low-pass filter.

Resulting jitter is very good. With no outputs switching or all outputs switching except for the Z axis aggressors, jitter is 46ps peak to peak. All digital outputs are switching, including the Z axis aggressors in rows C/AM, and columns 3/32 jitter is 50 ps peak to peak. A modest improvement in signal ripple is observable compared to the A side.

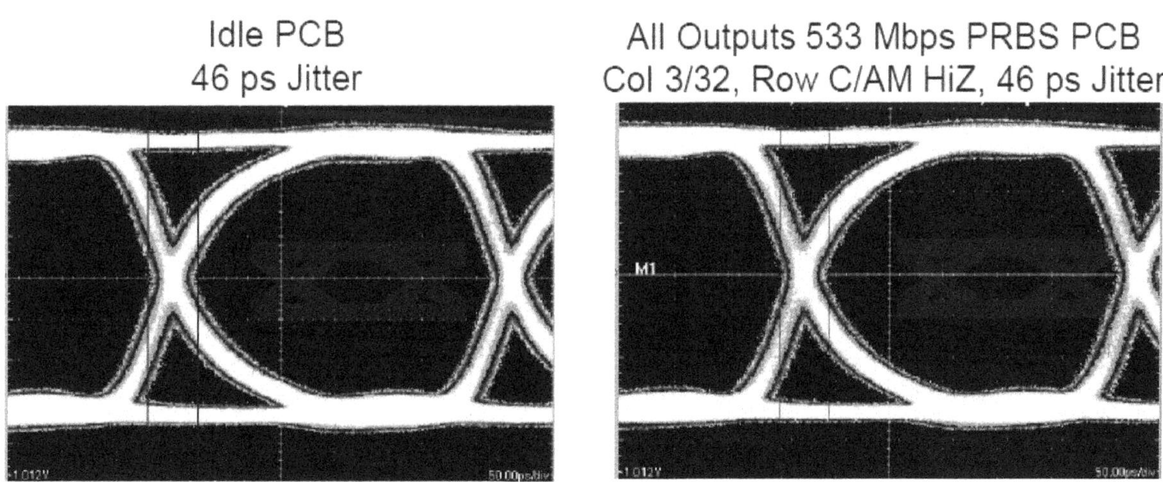

Figure 16.16: B Side Sourced from VCCINT, VCCAUX

All Outputs 533 Mbps PRBS PCB Col 3/32, Row C/AM Active, 50 ps Jitter

All Outputs 533 Mbps PRBS PCB Col 3/32, Row C/AM Low Z Active '0' 45 ps Jitter

Figure 16.17: B Side Sourced from VCCINT, VCCAUX

As with the A side, forcing all outputs adjacent to the MGT pins to drive output "0" has essentially the same effect as leaving those aggressors tri-stated.

Summary

While quantification of hard power-supply specifications remains elusive, given careful PDN design, PCB stackup and layout Virtex 4 FX SerDes are capable of low-jitter transmit operation. When implemented with care, the Xilinx design guidelines yield clean, low-jitter signaling. These guidelines include mandatory use of ferrites, which we show does help in some circumstances and is unnecessary in others.

The Virtex 4 package does a good job of isolating MGT circuitry from core and digital I/O ground bounce. From no digital I/Os switching to all I/Os switching (exclusive of Z axis aggressors), jitter increases only about 1 ps peak to peak. Z axis aggressors adjacent to MGT signal pins impart from 2 to 4 ps jitter when simultaneously driven by the FPGA. When driven by faster chips on the user PCB, we expect greater coupling efficiency and higher jitter.

By far, the greatest transmit jitter sensitivity in the Virtex 4 is predictably the AVCCAUXMGT power rail. This rail powers the PLLs and must be kept clean. We have demonstrated that with very simple 10 kHz board-level filters and the Samtec PowerPoser™, very good jitter performance is obtained even when sharing switching power supplies with the FPGA core and providing minimal application PCB decoupling.

Conclusion

We reached the following conclusions in our research:

- Xilinx Virtex 4 FX power delivery guidelines combined with best PCB practices yield low transmit jitter.
- The series decoupling network per pin requirement imposed by Xilinx benefits only certain circumstances.
- The Samtec PowerPoser™ relieves application PCB of best-practice requirements, enabling

elimination of more than 100 components and multiple linear regulators.

- While IC manufacturers work hard to provide usable design guidelines, users interested in optimum cost and performance need vendors to express power delivery requirements as tolerable external noise amplitude and PDN network impedance versus frequency.

References

[1] Stratix II GX Device Handbook, Volume 1, Altera Corp., February 2006.
[2] DS302, Virtex-4 Data Sheet: DC and Switching Characteristics, Xilinx, June 23, 2006.
[3] UG-076, Virtex-4 RocketIO MGT User Guide, Xilinx, May 23, 2006.
[4] E9415-MMZ, Chip Beads (SMD) for Signal Line, TDK Corp., June 6, 2006.

Chapter 17—Power Distribution System Design

Mark Alexander, Senior Staff Engineer, Xilinx Inc.

Excerpted from Chapter 2 of the Virtex-5 PCB Designer's Guide UG203 (V1.0) December 15, 2006, available at www.xilinx.com under Virtex-5 User Guides.

THIS CHAPTER DOCUMENTS the power distribution system (PDS) for Virtex™-5 field-programmable gate arrays (FPGAs), including decoupling capacitor selection, placement, and printed circuit board (PCB) geometries. A simple decoupling method is provided for each device in the Virtex-5 product family. Basic PDS design principles are covered, as well as simulation and analysis methods.

Required PCB Decoupling Capacitors—Recommended Networks per Device

Xilinx has determined a simple, PCB-decoupling network for each Virtex-5 device (refer to Table 17.1 through 17.19, many of which are not included in this version of the document). Many of the devices require few ceramic capacitors because high-frequency ceramic capacitors are present inside the device package (mounted on the substrate).

Thus, PCB-decoupling capacitor quantities can be significantly lower in Virtex-5 devices than in previous device families. Decoupling methods other than those presented in Table 17.1 through 17.19 of the Virtex-5 PCB Designer's Guide can be used, but the decoupling network should be designed to meet or exceed the performance of the simple decoupling network presented here.

The amount of in-package decoupling capacitance varies from package to package. The three package types are as follows:

- Packages with high-performance multi-terminal capacitors
- Packages with medium-performance capacitors
- Packages with no capacitors

Because CLB and input/output (I/O) utilization can vary the device capacitance requirements, PCB decoupling guidelines are provided on a per-device basis. Device performance at full utilization is equivalent across all devices when using these recommended networks.

Capacitor Specifications

In Table 17.1 through Table 17.19 of the Virtex-5 PCB Designer's Guide, tantalum capacitors are specified by recommended part number, whereas ceramic capacitors are specified by electrical characteristics.

Tantalum Capacitors

The tantalum capacitors specified in Table 17.1 through Table 17.3 are from Kemet, a capacitor manufacturer. These parts were selected for their value and controlled equivalent series resistance (ESR).

Parameter	Value	Instructions
Capacitance	330 µF	Parts substituted must have equal or greater capacitance.
Voltage Rating	2.5V	Parts substituted must be rated 2.5V minimum.
ESR	25 mΩ.	Parts substituted must have an ESR between 15 mΩ and 40 mΩ.
Case Size	V-case	Any case size is acceptable, if the ESL is less than 3 nH.
ESL	2.5 nH	

Table 17.1: T520V337M2R5ATE025 Tantalum Capacitor Substitution for V_{CCINT}

Parameter	Value	Instructions
Capacitance	33 µF	Parts substituted must have equal or greater capacitance.
Voltage Rating	6.3V	Parts substituted must be rated 6.3V minimum.
ESR	40 mΩ	Parts substituted must have an ESR between 30 mΩ and 60 mΩ.
Case Size	B-case	Any case size is acceptable, if the ESL is less than 2 nH.
ESL	2nH	

Table 17.2: T520B336M006ATE040 Tantalum Capacitor Substitution for V_{CCAUX}

Parameter	Value	Instructions
Capacitance	47 µF	Parts substituted must have equal or greater capacitance.
Voltage Rating	6.3V	Parts substituted must be rated 6.3V minimum.
ESR	70 mΩ	Parts substituted must have an ESR between 50 mΩ and 90 mΩ.
Case Size	B-case	Any case size is acceptable, if the ESL is less than 2 nH.
ESL	2 nH	

Table 17.3: T520B476M006ATE070 Tantalum Capacitor Substitution for VCCO

The purpose of the tantalum capacitors is to cover the low-frequency range between where the voltage regulator stops working and where the ceramic capacitors start working. If another manufacturer's tantalum capacitors or high-performance electrolytic capacitors meet the specifications listed in these tables, substitution is acceptable. Note that capacitor ESR must fall within a range of specified values. ESR greater than this range raises PDS impedance above the target impedance (no good). ESR lower than this range raises the network Q, increasing the severity of antiresonance impedance spikes.

Ceramic Capacitors

Two ceramic capacitor values appear in the tables: the 2.2 µF capacitor in an 0805 package and the 0.22 µF capacitor in an 0402 or 0204 package. For most characteristics, substitutions are permissible as long as they exceed a minimum value. Let us note that capacitor ESR must fall within a range of specified values.

ESR greater than this range raises PDS impedance above the target impedance (no good). ESR lower than this range raises the network Q, increasing the severity of anti-resonance impedance spikes.

Specifications are listed in Table 17.4 and Table 17.5.

Parameter	Value	Instructions
Capacitance	2.2 µF	Larger values are acceptable.
Voltage Rating	6.3V	Must be rated 6.3V minimum.
ESR	20 mΩ	Must have an ESR between 5 mΩ and 60 mΩ.
Case Size	0805	Larger body sizes are acceptable (1206 or 1812)

Table 17.4: 2.2 µF 0805 Ceramic Capacitor Requirements

Parameter	Value	Instructions
Capacitance	0.22 μF	Larger values are acceptable.
Voltage Rating	6.3V	Must be rated 6.3V minimum.
ESR	25 mΩ	Must have an ESR between 15 mΩ and 60 mΩ.
Case Size	0402 or 0204	A larger body size is NOT acceptable. A smaller body size or improved geometry (LICC, IDC, or X2Y) is acceptable, but only if it meets the ESR requirement.

Table 17.5: 0.22 μF 0402 or 0204 Ceramic Capacitor Requirements

Capacitor Placement and PCB Mounting Techniques

Placement and mounting restrictions presented in this section are different for each capacitor type, as listed in the section on capacitor specifications.

Tantalum Capacitors

Tantalum capacitors can be large and difficult to place very close to the FPGA. Fortunately, this is not a problem because the low-frequency energy covered by tantalum capacitors is not very sensitive to the capacitor location. Tantalum capacitors can be placed almost anywhere on the PCB, but the best placement is close as possible to the FPGA. Capacitor mounting should follow normal PCB layout practices.

0805 Ceramic Capacitor

The 2.2 μF 0805 capacitor covers the middle-frequency range, and placement has some impact on its performance. The capacitor should be placed as close as possible to the FPGA. Any placement within two inches of the device's outer edge is acceptable, and the capacitor mounting (e.g., solder lands, traces, vias) should be optimized for low inductance. Vias should be butted directly against the pads, and they can be located at the ends of the pads (see Figure 17.1B) or, optimally, located at the sides of the pads (see Figure 17.1C). Via placement at the sides of the pads decreases the mounting's overall parasitic inductance by increasing the mutual inductive coupling of one via to the other. Dual vias can be placed on both sides of the pads (see Figure 17.1D), but with diminishing returns.

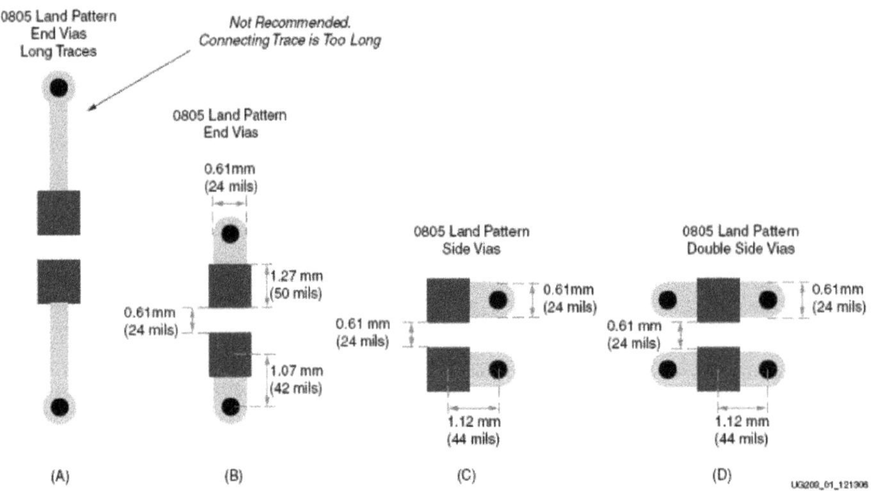

Figure 17.1: Example 0805 Capacitor Land and Mounting Geometries

0402 Ceramic Capacitor

The 0.22 μF 0402 capacitor covers the high-middle frequency range. Placement and mounting are critical for these capacitors.

The capacitor should be mounted as close to the FPGA as possible (so it achieves the least parasitic inductance possible). For PCBs with a total thickness of less than 1.575 mm (62 mils), the best placement location is on the PCB backside, within the device footprint (in the empty cross with an absence of vias). VCC and ground (GND) vias corresponding to the supply of interest should be identified in the via array. Where space is available, 0402 mounting pads should be added and connected to these vias.

For PCBs with a total thickness greater than 1.575 mm (62 mils), the best placement location may be on the PCB top surface. The depth of the VCC plane of interest in the PCB stackup is the key factor. If the VCC plane is in the PCB stackup's top half, capacitor placement on the top PCB surface is optimal. If the VCC plane is in the PCB stackup's bottom half, capacitor placement on the bottom PCB surface is optimal. Any 0402 capacitors placed outside the device footprint (whether on the top or bottom surface) should be within 0.5 inch of the device's outer edge.

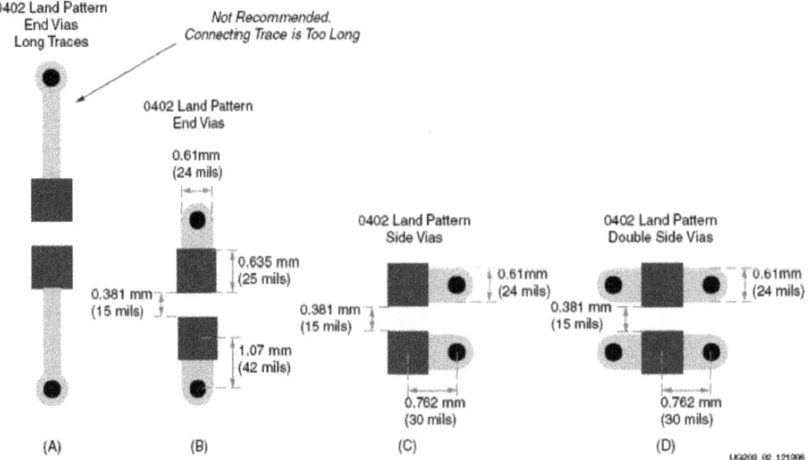

Figure 17.2: Example 0402 Capacitor Land and Mounting Geometries

The capacitor mounting (e.g., solder lands, traces, vias) must be optimized for low inductance. Vias should be butted against the pads with no trace length between. These vias should be at the sides of the pads if possible (see Figure 17.2C). Via placement at the sides of the pads decreases the mounting's parasitic inductance by increasing the mutual inductive coupling of one via to the other. Dual vias can be placed on both sides of the pads (see Figure 17.2D), but with diminishing returns.

It should be noted that many manufacturing rules prevent mounting any device within 0.1 inch of the FPGA on the PCB top surface. This decreases the available area for capacitor placement.

Basic Decoupling Network Principles

The purpose of the PDS and its components are discussed in this section, along with the important aspects of capacitor placement, capacitor mounting, PCB geometry, and PCB stackup recommendations.

Noise Limits

Each device in a system has operating power and cleanliness requirements for this power. Most digital devices, including all Virtex-5 FPGAs, require that VCC supplies not fluctuate more than around 5 percent of the nominal VCC value. In this document, the term VCC is used generically for all FPGA power supplies such as V_{CCINT}, V_{CCO}, V_{CCAUX}, and V_{REF}.

These requirements specify a maximum amount of noise present on the power supply, often referred to as ripple voltage (V_{RIPPLE}). The device requirements state that V_{CC} must be within 5 percent of the nominal voltage, meaning the peak-to-peak V_{RIPPLE} must be no more than 10 percent of the nominal V_{CC}. This assumes that nominal V_{CC} is exactly the nominal value provided in the data sheet. If not, then V_{RIPPLE} must be adjusted to a value correspondingly less than 10 percent. The power consumed by a digital device varies over time, and this variance occurs on all frequency scales, thus creating a need for a wideband PDS.

Low-frequency variance of power consumption is usually the result of devices, or large portions of devices, being enabled or disabled; this variance occurs in time frames from milliseconds to days.

High-frequency variance of power consumption is the result of individual switching events inside a device. This occurs on the scale of the clock frequency and the first few harmonics of the clock frequency.

Because the voltage level of VCC for a device is fixed, changing power demands are manifested as changing current demand. The PDS must accommodate these variances of current draw with as little change as possible in the power-supply voltage. When the current draw in a device changes, the PDS cannot instantaneously respond to that change. The voltage, at the device, changes in the short time before the PDS responds and the power-supply noise appears. Two main causes for this PDS lag correspond to the two major PDS components.

The first major component of the PDS is the voltage regulator, which observes its output voltage and adjusts the amount of current it is supplying to keep the output voltage constant. Most common voltage regulators make this adjustment in milliseconds to microseconds. Voltage regulators effectively maintain the output voltage for events at all frequencies from direct current (DC) to a few hundred kHz, depending on the regulator (some are effective at regulating in the low MHz). For transient events that occur at frequencies above this range, there is a time lag before the voltage regulator responds to the new current demand level.

For example, if the device's current demand increases in a few hundred picoseconds, the voltage at the device sags by some amount until the voltage regulator can adjust to the new, higher level of required current. This lag can last from microseconds to milliseconds. Decoupling capacitors replace the regulator during this time, preventing the voltage from sagging.

The second major PDS component is the decoupling capacitor (also known as a bypass capacitor). The decoupling capacitor works as the device's local energy storage. However, the capacitor cannot provide DC power because it stores only a small amount of energy (e.g., voltage regulator provides DC power). This local energy storage should respond very quickly to changing current demands.

The capacitors effectively maintain power-supply voltage at frequencies from hundreds of kHz to hundreds of MHz (in the milliseconds to nanoseconds range). Decoupling capacitors are not useful for events occurring above or below this range. For example, if current demand in the device increases in a few picoseconds, the voltage at the device sags by some amount until the capacitors can supply extra charge to the device. If current demand in the device maintains this new level for many milliseconds, the voltage-regulator circuit, operating in parallel with the decoupling capacitors, replaces the capacitors by changing its output to supply this new level of current.

Figure 17.3 shows the major PDS components: the power supply, the decoupling capacitors, and the active device being powered (in this case, the FPGA).

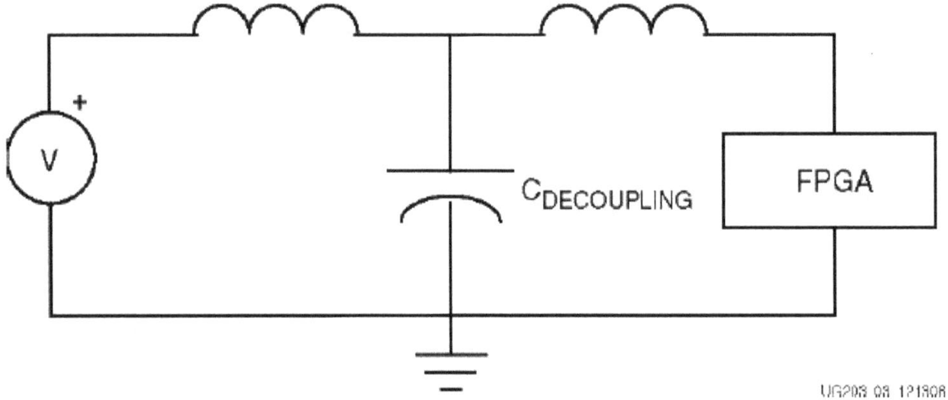

Figure 17.3: Simplified PDS Circuit

Figure 17.4 shows a simplified PDS circuit with all reactive components represented by a frequency-dependent resistor.

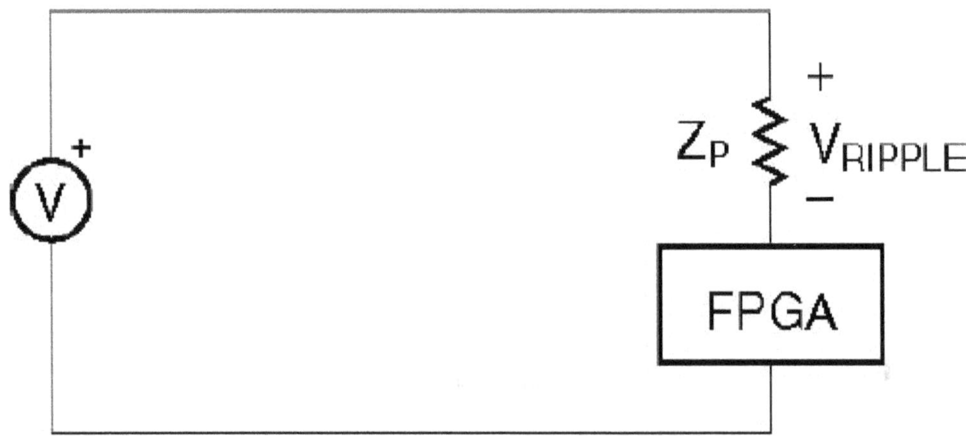

Figure 17.4: Further Simplified PDS Circuit

Role of Inductance

Inductance is the property of the capacitors and the PCB current paths that slows down changes in current flow. Inductance is the reason why capacitors cannot respond instantaneously to transient currents or to changes that occur at frequencies higher than their effective range.

Inductance can be thought of as the momentum of charge, especially as charge moving through a conductor represents some amount of current. If the level of current changes, the charge moves at a different rate. Since momentum (stored magnetic-field energy) is associated with this charge, some amount of time is required to slow down or speed up the charge flow. The greater the inductance, the greater the resistance to change and the longer the time required for the current level to change. A voltage develops across the inductance as this change occurs.

The PDS accommodates the device's current demand and responds to current transients as quickly as possible. When these current demands are not met, the voltage across the device's power supply changes. This is observed as noise. Inductance should be minimized because it retards the ability of decoupling capacitors to quickly respond to changing current demands.

Inductances occur between the FPGA device and capacitors and between the capacitors and the voltage regulator (see Figure 17.3). These inductances occur as parasitics in the capacitors and in all PCB current paths. It is important that each of these parasitics be minimized.

Capacitor Parasitic Inductance

The capacitance value is often considered the bypass capacitors' most important characteristic. In power system applications, the parasitic inductance (e.g., equivalent series inductance [ESL]) has the same or greater importance. Capacitor package dimensions (e.g., body size) determine the amount of parasitic inductance. Physically small capacitors usually have lower parasitic inductance than physically large capacitors.

Requirements for choosing decoupling capacitors include the following:

- For a specific capacitance value, the smallest package available
- For a specific package size (essentially a fixed inductance value), the highest capacitance value available in that package.

Surface-mount chip capacitors are the smallest capacitors available and are a good choice for discrete decoupling capacitors for the following reasons:

- For values from 10 μF to very small values such as 0.01 μF, X7R or X5R type capacitors are usually used. These capacitors have a low parasitic inductance and a low ESR, with an acceptable temperature characteristic.
- For larger values, such as 47 μF to 680 μF, tantalum capacitors are used. These capacitors have a low parasitic inductance and a medium ESR, giving them a low Q factor and consequently a very wide range of effective frequencies. High-performance electrolytic capacitors can also be used, provided they have comparable ESR and ESL values.

Surface-mount chip capacitors provide a comparatively high capacitance value in a small package size, reducing board real-estate costs. If tantalum capacitors are not available or cannot be used, low-ESR, low-inductance electrolytic capacitors are used. Other new technologies with similar characteristics are also available (OsCon, POSCAP, and polymer-electrolytic switch-mode transformer [SMT]).

A real capacitor of any type has not only capacitance characteristics, but also inductance and resistance characteristics.

Figure 17.5 shows the parasitic model of a real capacitor. A real capacitor should be treated as a radio link control (RLC) circuit, which consists of a resistor (R), an inductor (L), and a capacitor ([C], connected in series).

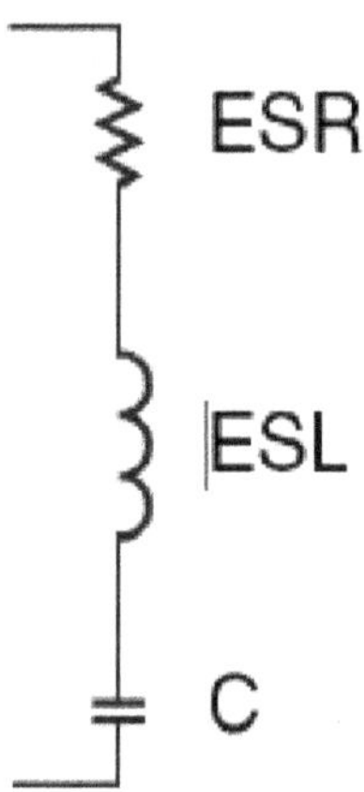

Figure 17.5: Parasitics of a Real, Non-Ideal Capacitor

Figure 17.6 shows a real capacitor's impedance characteristic. Overlaid on this plot are curves corresponding to the capacitor's capacitance and parasitic inductance (e.g., ESL). These two curves combine to form the RLC circuit's total impedance characteristic.

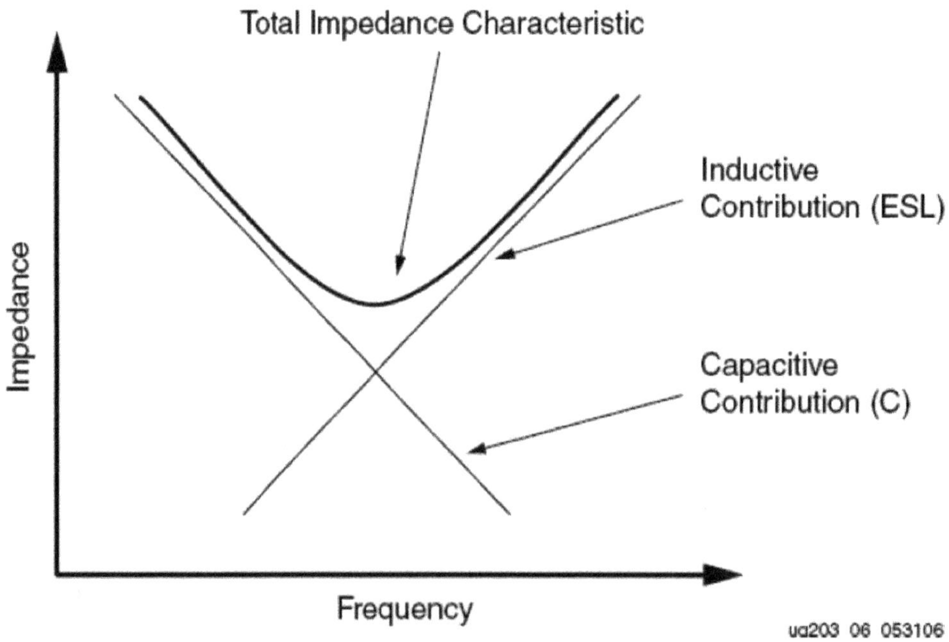

Figure 17.6: Contribution of Parasitics to Total Impedance Characteristics

As capacitive value is increased, the capacitive curve moves down and left. As parasitic inductance is

decreased, the inductive curve moves down and right. Because parasitic inductance for capacitors in a specific package is fixed, the inductance curve for capacitors in a specific package remains fixed.

As different capacitor values are selected in the same package, the capacitive curve moves up and down against the fixed inductance curve.

For a fixed capacitor package, the low-frequency impedance can be reduced by increasing the value of the capacitor; the high-frequency impedance can be reduced by decreasing the inductance of the capacitor. While it might be possible to specify a higher capacitance value in the fixed package, it is not possible to lower the inductance of the capacitor without putting more capacitors in parallel. Using multiple capacitors in parallel divides the parasitic inductance and, at the same time, multiplies the capacitance value. This lowers both the high- and low-frequency impedance at the same time.

PCB Current Path Inductance

The parasitic inductance of current paths in the PCB has the following three distinct sources:

- Capacitor mounting
- PCB power and ground planes
- FPGA mounting

Capacitor Mounting Inductance

Capacitor mounting refers to the capacitor's solder lands on the PCB, the trace (if any) between land and via, and via. The vias, traces, and capacitor mounting pads contribute inductance between 300 pH and 4 nH, depending on the specific geometry.

Because the current path's inductance is proportional to the loop area the current traverses, it is important to minimize this loop size. The loop consists of the path through one power plane, up through one via, through the connecting trace to the land, through the capacitor, through the other land and connecting trace, down through the other via, and into the other plane, as shown in Figure 17.7.

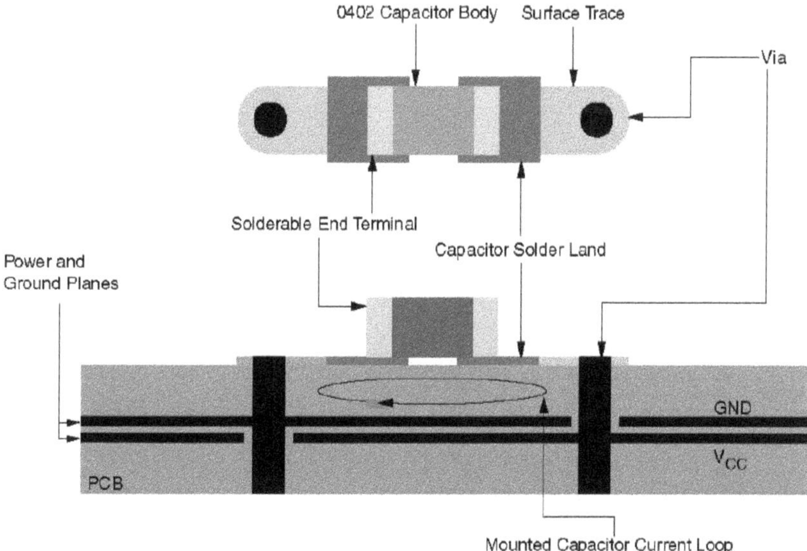

Figure 17.7: Example Cutaway View of PCB with Capacitor Mounting

A connecting trace length has a large impact on the mounting's parasitic inductance and, if used, should be as short and wide as possible. When possible, a connecting trace should not be used (Figure 17.1A) and the via should butt up against the land (Figure 17.1B). Placing vias to the side of the capacitor lands (Figure 17.1C) or doubling the number of vias (Figure 17.1D), further reduces the mounting's parasitic inductance.

Very few PCB manufacturing processes allow via-in-pad geometries, an option to reduce parasitic inductance. Using multiple vias per land is important with ultra-low inductance capacitors such as reverse aspect ratio capacitors (AVX's LICC).

PCB layout engineers often try to squeeze more parts into a small area by sharing vias among multiple capacitors. This technique should not be used under any circumstances. PDS improvement is very small when a second capacitor is connected to an existing capacitor's vias. For a larger improvement, reduce the total number of capacitors and maintain a one-to-one ratio of lands to vias.

The capacitor mounting (e.g., lands, traces, vias) typically contributes about the same amount or more inductance than the capacitor's own parasitic inductance.

Plane Inductance

Some inductance is associated with the PCB power and ground planes, which ultimately determine their own geometry. Current spreads out as it flows from one point to another, mostly due to a property similar to skin effect, in the power and ground planes. Inductance in planes can be described as spreading inductance, and it is specified in units of henries per square. The square is dimensionless, such that the shape of a section of a plane, not the size, determines the amount of inductance.

Spreading inductance acts like any other inductance and resists changes to the amount of current in a power plane (the conductor).

The inductance retards the capacitor's ability to respond to a device's transient currents and should be reduced as much as possible. Because the X-Y shape of the plane is typically something the designer has little control over, the only controllable factor is the spreading inductance value. This is primarily determined by the thickness of the dielectric separating a power plane from its associated ground plane.

For high-frequency power distribution systems, power and ground planes work in pairs, with their inductances coexisting dependently with each other. The spacing between the power and ground planes determines the pair's spreading inductance. The closer the spacing (the thinner the dielectric), the lower the spreading inductance. Approximate values of spreading inductance for different thicknesses of FR4 dielectric are shown in Table 17.6.

Dielectric Thickness		**Inductance**	**Capacitance**	
(micron)	**(mil)**	**(pH/square)**	**(pF/in2)**	**(pF/cm2)**
102	4	130	225	35
51	2	65	450	70
25	1	32	900	140

Table 17.6: Capacitance and Spreading Inductance Values for Different Thicknesses of FR4 Power-Ground Plane Sandwiches

Decreased spreading inductance corresponds to closer spacing of VCC and GND planes. When possible, the

VCC planes should be placed directly adjacent to the GND planes in the PCB stackup.

The facing VCC and GND planes are sometimes referred to as "sandwiches." While the use of VCC-GND sandwiches was not necessary in the past for previous technologies, the speeds involved and the sheer amount of power required for fast, dense devices demand it.

Besides offering a low-inductance current path, power-ground sandwiches also offer some high-frequency decoupling capacitance. As the plane area increases and as the separation between power and ground planes decreases, the value of this capacitance increases.

Capacitance per square inch is shown in Table 17.6.

FPGA Mounting Inductance

The PCB solder lands and vias that connect the FPGA power pins (V_{CC} and GND) contribute an amount of parasitic inductance to the overall power circuit. For existing PCB technology, the solder land geometry and the dogbone geometry are mostly fixed, and parasitic inductance of these geometries does not vary. Via parasitic inductance is a function of the via length and the proximity of the opposing current paths to one another.

The relevant via length is the portion of the via that carries transient current between the FPGA solder land and the associated VCC or GND plane. Any remaining via (between the power plane and the PCB backside) does not affect the parasitic inductance of the via (the shorter the via between the solder lands and the power plane, the smaller the parasitic inductance). Parasitic via inductance in the FPGA mounting is reduced by keeping the relevant VCC and GND planes as close to the FPGA as possible (close to the top of the PCB stackup).

Device pinout arrangement determines the proximity of opposing current paths to one another. Inductance is associated with any two opposing currents, for example, current flowing in a VCC and GND via pair. A high degree of mutual inductive coupling between the two opposing paths reduces the loop's total inductance. Therefore, VCC and GND vias should be as close together as possible.

The via field under an FPGA has many VCC and GND vias, and the total inductance is a function of the proximity of one via to another:

- For core VCC supplies (V_{CCINT} and V_{CCAUX}), the opposing current is between the VCC and GND pins.
- For I/O VCC supplies (V_{CCO}), the opposing current is between any I/O and its return current path, carried by a VCCO or GND pin.

To reduce parasitic inductance:

- Core VCC pins such as V_{CCINT} and V_{CCAUX} are placed in a checkerboard arrangement in the pinout.
- V_{CCO} and GND pins are distributed among the I/O pins.

Every I/O pin in the Virtex-5 FPGA pinout is adjacent to a return current pin. FPGA pinout arrangement determines the PCB via arrangement. The PCB designer cannot control the proximity of opposing current paths but has control over the tradeoffs between the capacitor's mounting inductance and FPGA's mounting inductance:

- Both mounting inductances are reduced by placing power planes close to the PCB stackup's top half and placing the capacitors on the top surface (reducing the capacitor's via length).

- If power planes are placed in the PCB stackup's bottom half, the capacitors must be mounted on the PCB backside. In this case, FPGA mounting vias are already long, and making the capacitor vias long as well is a bad practice. A better practice is to take advantage of the short distance between the underside of the PCB and the power plane of interest, mounting capacitors on the underside.

PCB Stackup and Layer Order

VCC and ground plane placement in the PCB stackup (the layer order) has a significant impact on the parasitic inductances of power current paths. Layer order must be considered early in the design process:

- High-priority supplies should be placed closer to the FPGA (in the PCB stackup's top half)
- Low-priority supplies should be placed farther from the FPGA (in the PCB stackup's bottom half)

Power supplies with high transient current should have the associated VCC planes close to the top surface (FPGA side) of the PCB stackup. This decreases the vertical distance (VCC and GND via length) that currents travel before reaching the associated VCC and GND planes. To reduce spreading inductance, every VCC plane should have an adjacent GND plane in the PCB stackup. The skin effect causes high-frequency currents to couple tightly and the GND plane adjacent to a specific VCC plane tends to carry the majority of the current complementary to that in the VCC plane.

Thus, adjacent VCC and GND planes are treated as a pair.

Not all VCC and GND plane pairs reside in the PCB stackup's top half because manufacturing constraints typically require a symmetrical PCB stackup around the center (with respect to dielectric thicknesses and etched copper areas). The PCB designer chooses the priority of the VCC and GND plane pairs: high-priority pairs carry high transient currents and are placed high in the stackup, while low priority pairs carry lower transient currents and are placed in the lower part of the stackup.

Capacitor Effective Frequency

Every capacitor has a narrow frequency band where it is most effective as a decoupling capacitor. This band is centered at the capacitor's self-resonant frequency FR_{SELF}. The effective frequency bands of some capacitors are wider than others. A capacitor's ESR determines the capacitor's quality (Q) factor, and the Q factor determines the width of the effective frequency band:

- Tantalum capacitors generally have a very wide effective band.
- X7R/X5R chip capacitors with a lower ESR generally have a very narrow effective frequency band.

An ideal capacitor only has a capacitive characteristic, whereas real non-ideal capacitors also have a parasitic inductance ESL and a parasitic resistance ESR. These parasitics work in series to form an RLC circuit (Figure 17.5). The RLC circuit's resonant frequency is the capacitor's self-resonant frequency.

To determine the RLC circuit's resonant frequency, use this equation:

$$F = \frac{1}{2\pi\sqrt{LC}}$$

Another way to determine the self-resonant frequency is to find the minimum point in the impedance curve

the FPGA. For any transient current demand in the FPGA, a round-trip delay occurs before any relief is seen at the FPGA.

The resonant frequency of the capacitor can be associated with these time delays to determine an adequate placement distance for each capacitor. This is done by considering the wavelength of the capacitor's resonant frequency. Due to the round-trip delay, negligible energy is transferred to the FPGA with placement distances greater than one quarter wavelength. Energy transferred to the FPGA increases from 0 percent at one quarter wavelength to 100 percent at zero distance. Therefore, efficient energy transfer occurs when the capacitor is placed within a fraction of a quarter wavelength of the FPGA power pins. This fraction should be small because the capacitor is also effective at some frequencies (shorter wavelengths) above its resonant frequency.

One-tenth of a quarter wavelength is a good target for most practical applications and leads to placing a capacitor within one-fortieth of a wavelength of the power pins it is decoupling. The wavelength corresponds to the capacitor's mounted resonant frequency, F_{RIS}. When using large numbers of external termination resistors or passive power filtering for transceivers, priority should be given to these over the decoupling capacitors. Moving away from the device in concentric rings, the termination resistors and discrete filtering should be closest to the device, followed by the smallest-value decoupling capacitors, then the larger-value decoupling capacitors.

V_{REF} Stabilization Capacitors

In V_{REF} supply stabilization, only one capacitor per pin is used. The capacitors used are in the 0.022-0.47 µF range. The V_{REF} capacitor's primary function is to reduce the V_{REF} node impedance, which, in turn, reduces crosstalk coupling. Little low-frequency energy is needed and, as a result, larger capacitors are not necessary.

Simulation Methods

Simulation methods exist to predict the PDS characteristics, ranging from very simple to very complex. An accurate simulation result is difficult to achieve without using a fairly sophisticated simulator and taking a significant amount of time.

Basic lumped RLC simulation is one of the simplest simulation methods, and although it does not account for the distributed behavior of a PDS, it is a useful tool to select and verify combinations of decoupling capacitor values.

Lumped RLC simulation is performed either in a version of SPICE or other circuit simulator, or by using a mathematical tool such as MathCAD or Microsoft Excel.

Table 17.7 also lists a few EDA tool vendors for PDS design and simulation. These tools span a wide range of sophistication levels.

Tool	Vendor	Website URL
ADS	Agilent	http://www.agilent.com
SIwave, HFSS	Ansoft	http://www.ansoft.com
Specctraquest Power Integrity	Cadence	http://www.cadence.com
Speed 2000, PowerSI	Sigrity	http://www.sigrity.com

Table 17.7: EDA Tools for PDS Design and Simulation

PDS Measurements

Measurements can be used to determine whether a PDS is adequate. PDS noise measurements are a unique task, and many specialized techniques have been developed. This section describes the noise magnitude and noise spectrum measurements.

Noise Magnitude Measurement

Noise measurement must be performed with a high-bandwidth oscilloscope (minimum 3 GHz oscilloscope and 1.5 GHz probe or direct coaxial connection) on a design running realistic test patterns. The measurement is taken at the device's power pins or at an unused I/O driven high or low (referred to as a *spyhole* measurement).

V_{CCINT} and V_{CCAUX} can only be measured at the PCB backside vias.

V_{CCO} can also be measured this way, but more accurate results are obtained by measuring static (fixed logic level) signals at unused I/Os in the bank of interest.

When making the noise measurement on the PCB backside, the via parasitics in the path between the measuring point and FPGA must be considered. Any voltage drop occurring in this path is not accounted for in the oscilloscope measurement.

PCB backside via measurements also have a potential problem: decoupling capacitors are often mounted directly underneath the device, meaning the capacitor lands connect directly to the VCC and GND vias with surface traces. These capacitors confuse the measurement by acting like a short circuit for the high-frequency AC current. To make sure the measurements are not shorted by the capacitors, remove the capacitor at the measurement site (keep all others to reflect the real system behavior).

When measuring V_{CCO} noise, the measurement can be taken at an I/O pin configured as a driver to logic 1 or logic 0. In most cases, the same I/O standard should be used for this "spyhole" as for the other signals in the bank. This measurement technique shows the crosstalk (via field, PCB routing, package routing) induced on the victim as well as the die-level power system noise.

Oscilloscope Measurement Methods

- Place the oscilloscope in infinite persistence mode to acquire all noise over a long time period (many seconds or minutes). If the design operates in many modes, using different resources in different amounts, these various conditions and modes should be in operation while the oscilloscope is acquiring the noise measurement.

- Place the oscilloscope in averaging mode and trigger on a known aggressor event. This can show the amount of noise correlated with the aggressor event (any events asynchronous to the aggressor are removed through averaging).

Noise measurements should be made at a few FPGA locations to ensure that any local noise phenomena are captured.

Figure 17.8 shows a single-capture noise measurement taken at the V_{CCINT} backside vias of a sample design.

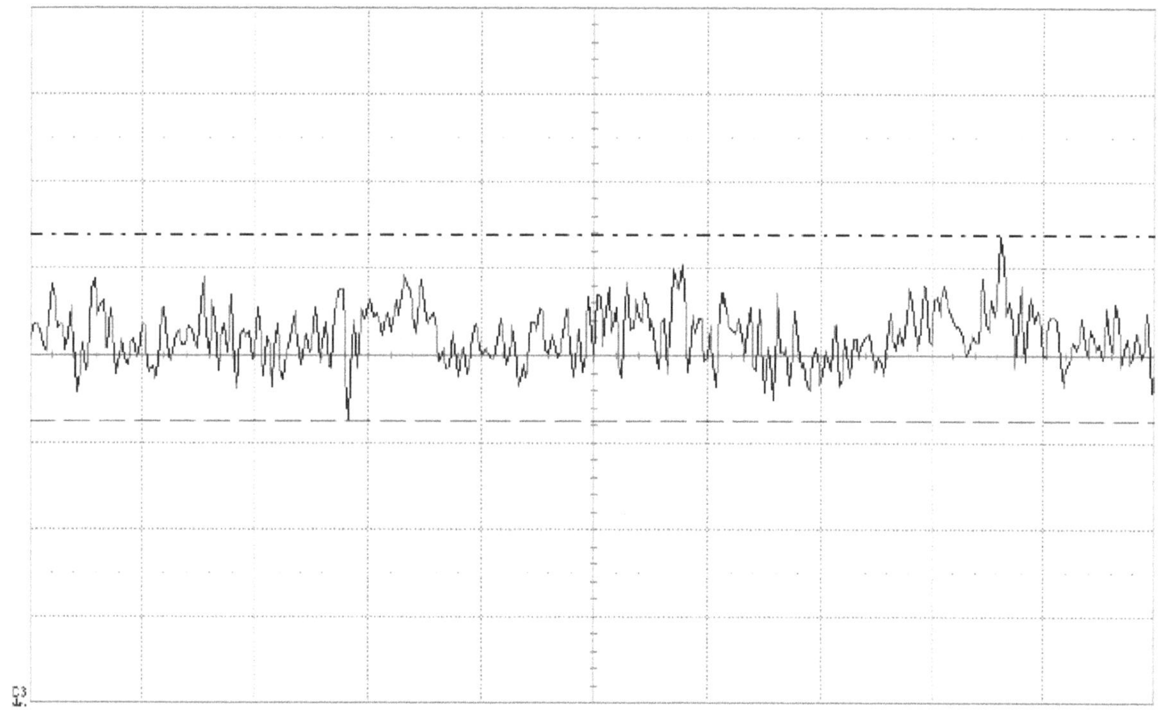

Figure 17.8: Single-Capture Measurement of V_{CCINT} Supply Horizontal Scale: 2 ns/div, Vertical Scale: 20 mV/div

Figure 17.9 shows an infinite persistence noise measurement of the same design. Because the infinite persistence measurement catches all noise events over a long period, both correlated and noncorrelated with the primary aggressor, all power system excursions are shown. The peak-to-peak noise after this long acquisition is 63.4 mV (it is necessary to measure at the highest excursion spots, which is where the measurement cursors are placed in Figure 17.9).

The measurements shown in Figure 17.8 and Figure 17.9 represent the peak-to-peak noise. If the peak-to-peak noise is outside the specified acceptable voltage range (data sheet value, VCC around 5 percent), the decoupling network is inadequate or a problem exists in the PCB layout.

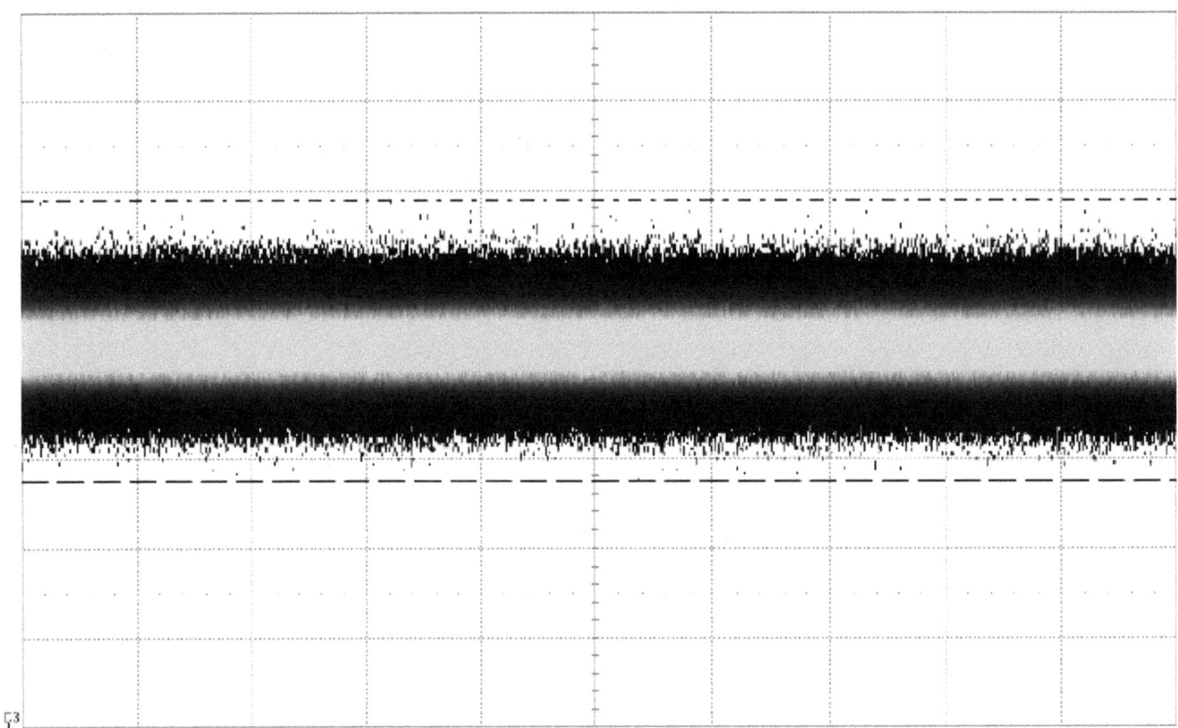

Figure 17.9: Infinite Persistence Measurement of Same Supply Horizontal Scale: 2 ns/div, Vertical Scale: 20 mV/div

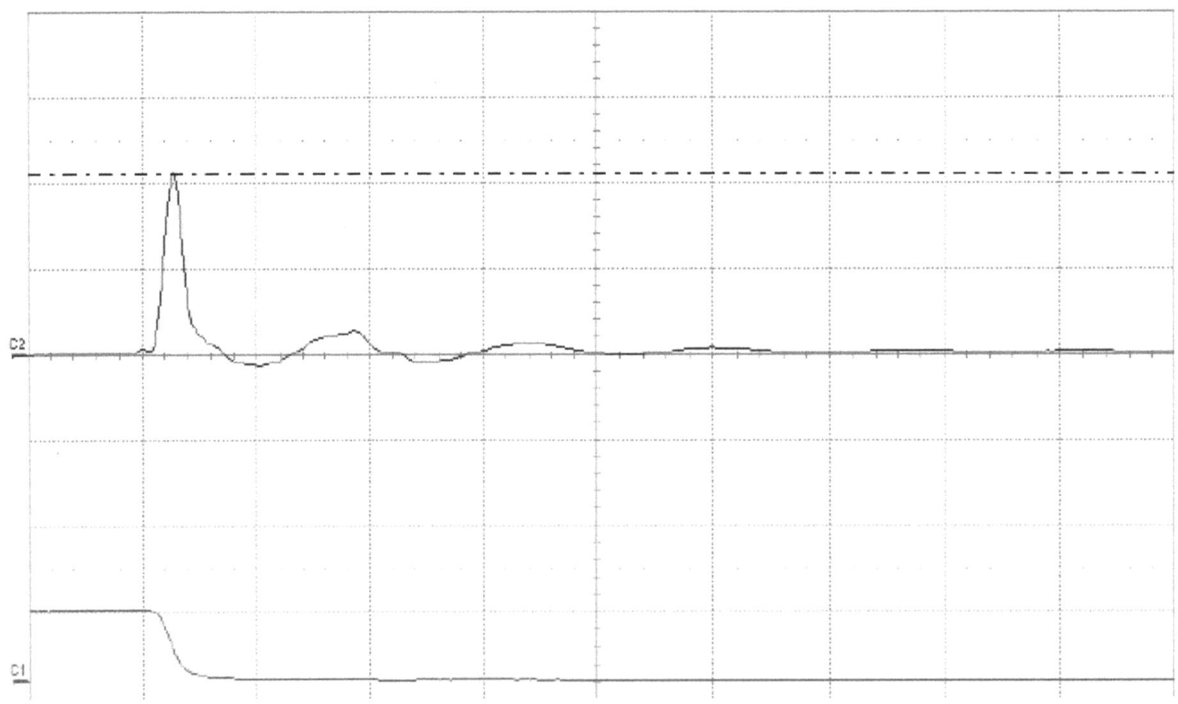

Figure 17.10: Averaged Measurement of Static-0 I/O Spyhole Horizontal Scale: 2 ns/div, Vertical Scale for the Top Trace: 20 mV/div

When looking for noise correlated to a particular aggressor event, it can be useful to trigger on that event and average out all other noise. This shows only the noise signature correlated with that aggressor event. Voltage regulator noise and non-correlated device switching noise is removed. Figure 17.10 shows an averaged noise waveform from an I/O pin driving a static-0, with the oscilloscope triggered by the switching of a group of nearby I/O pins. The static-0 I/O is in pink, and the falling aggressor (used as trigger) is in yellow. The positive excursion of the static-0 I/O is 41.8 mV, resulting from the group of nearby I/O switching from 1 to 0.

Noise Spectrum Measurements

Having the necessary information to improve the decoupling network requires additional measurements. Power spectrum measurement is necessary to determine the frequencies where the noise resides. A spectrum analyzer or a high-bandwidth oscilloscope with FFT math functionality work well.

Alternatively, a long sequence of time-domain data can be captured from an oscilloscope and converted to frequency domain through MATLAB or other post-processing software supporting FFT. The noise frequency content can be approximated by looking at the time-domain waveform and estimating the individual periodicities present in the noise.

A spectrum analyzer is a frequency-domain instrument, showing the frequency content of a voltage signal at its inputs. The user sees the exact frequencies where the PDS is inadequate when a spectrum analyzer is used.

Excessive noise at a certain frequency indicates a frequency where the PDS impedance is too high for the device's transient current demands. Using this information, the designer can modify the PDS to accommodate the transient current at the specific frequency.

This is accomplished by either adding capacitors with resonant frequencies close to the noise frequency or otherwise lowering the PDS impedance at the critical frequency.

The noise spectrum measurement should be taken at the same place as the peak-to-peak noise measurement, directly underneath the device, or at a static I/O driven high or low. A spectrum analyzer takes its measurements using a 50 Ohm cable instead of an active probe.

- A good method attaches the measurement cable through a coaxial connector tapped into the power and ground planes close to the device. This is not available in most cases.
- Another method attaches the measurement cable (of noise in the power planes) by removing a decoupling capacitor in the device vicinity and soldering the cable's center conductor and shield directly to the capacitor lands.

Alternatively, a probe station with 50 Ohm RF probes can be used.

A DC blocking capacitor or attenuator may be necessary to protect the spectrum analyzer from the device supply voltage.

Figure 17.11 shows an example of a noise spectrum measurement. It is a spectrum-analyzer measurement screenshot of the V_{CCO} power-supply noise, with multiple I/O sending patterns at 150 MHz.

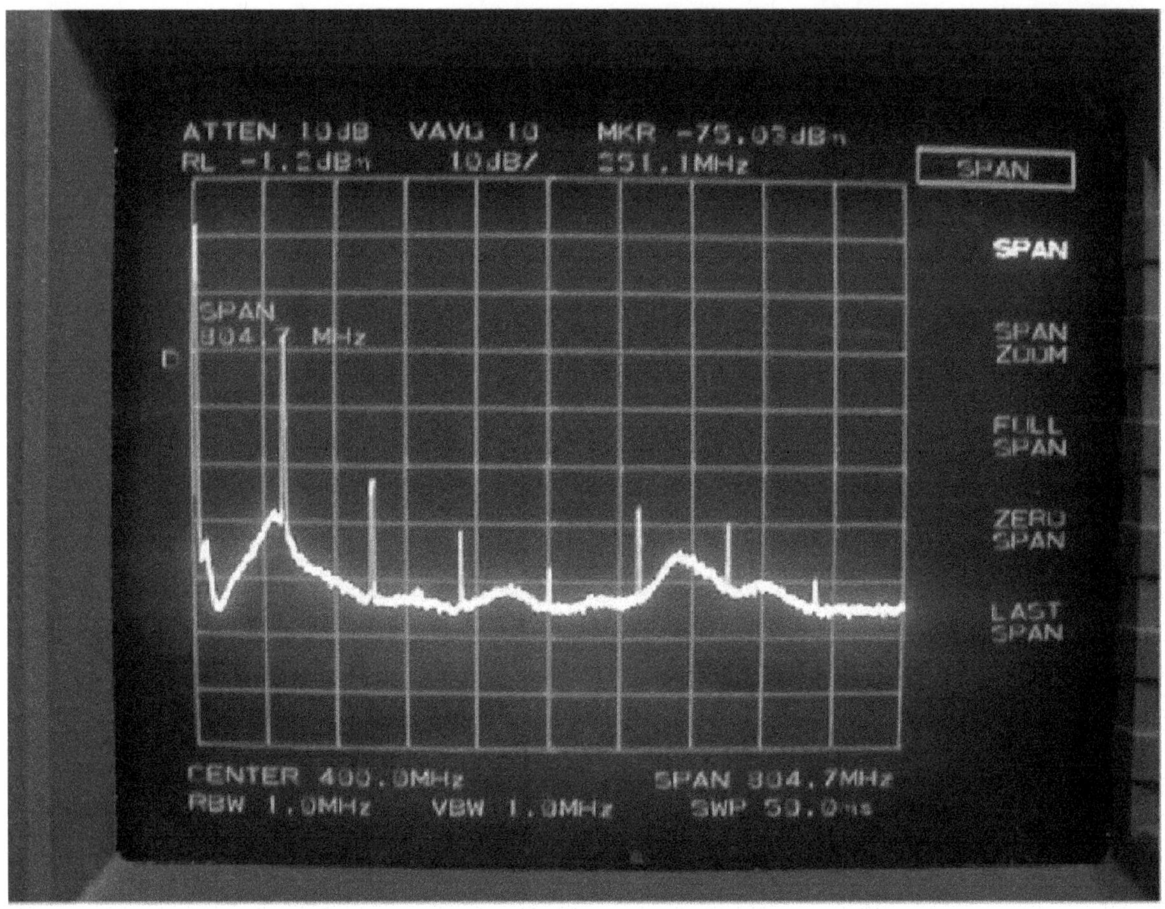

Figure 17.11: Screenshot of Spectrum Analyzer Measurement of V_{CCO}

Optical Decoupling Network Design

If a highly optimized PDS is needed, more measurements help to create a customized decoupling network design.

- A network analyzer measures the impedance profile of a prototype PDS. The network analyzer sweeps a stimulus across a range of frequencies and measures the PDS impedance at each frequency.
- A spectrum-analyzer measurement output is voltage as a function of frequency.

Using the previous two measurements, transient current as a function of frequency (Equation 4) is determined:

$$I(f) = \frac{V(f) From\ Spectrum\ Analyzer}{Z(f) From\ Network\ Analyzer}$$

Understanding the design's transient current requirements improves the designer's PDS choices. Through the data sheet's maximum voltage ripple value, the impedance value needed at all frequencies can be determined. This yields a target impedance as a function of frequency. A specially designed capacitor network can accommodate the specific design's transient current.

Other Concerns and Causes

Sometimes a design with the required noise specifications cannot be created by using the methods described in this document, and system aspects should be analyzed for possible changes.

Possibility 1: Excessive Noise from Other Devices on the PCB

Sometimes ground and/or power planes are shared among many devices, and noise from an inadequately decoupled device affects the PDS at other devices. Common causes of this noise are, as follows:

- RAM interfaces with inherently high-transient current demands because of temporary periodic contention and high-current drivers
- Large microprocessors

When unacceptable noise amounts are measured locally at these devices, the local PDS and the component decoupling networks should be analyzed.

Possibility 2: Parasitic Inductance of Planes, Vias, or Connecting Traces

Under this possibility, the decoupling network capacitance is adequate, but there is too much inductance in the path from the capacitors to the FPGA.

Possible causes are, as follows:

- Wrong connecting-trace geometry or solder-land geometry

- The path from the capacitors to the FPGA is too long

—and/or—

- A current path in the power vias traverses an exceptionally thick PCB stackup

For inadequate connecting trace geometry and capacitor land geometry, review the loop inductance of the current path. If the vias for a decoupling capacitor are spaced a few millimeters from the capacitor solder lands on the board, the current loop area is greater than necessary (see Figure 17.1A).

To reduce the current loop area, vias should be placed directly against capacitor solder lands (see Figure 17.1B). Never connect vias to the lands with a section of trace (see Figure 17.1A).

Other improvements of geometry are via-in-pad (via under the solder land), not shown, and via-beside-pad (vias straddle the lands instead of being placed at the ends of the lands), see Figure 17.1C.

Double vias also improve connecting trace geometry and capacitor land geometry (see Figure 17.1D). Exceptionally thick boards (> 2.3 mm or 90 mils) have vias with higher parasitic inductance.

To reduce the parasitic inductance, one should move both critical VCC/GND plane sandwiches and the highest frequency capacitors close to the top surface where the FPGA is located. Both of these changes reduce the parasitic inductance of the relevant current path.

Possibility 3: I/O Signals in PCB Are Stronger Than Necessary

If noise in the V_{CCO} PDS is still too high after refining the PDS, the I/O interface slew rate can be reduced. This likewise applies to outputs from and inputs to the FPGA. In severe cases, excessive overshoot on inputs to the FPGA can reverse-bias the IOB clamp diodes, injecting current into the VCCO PDS.

If large amounts of noise are present on VCCO, the drive strength of these interfaces should be decreased, or different termination should be used (on input and output paths).

Possibility 4: I/O Signal Return Current Traveling in Less Optimal Paths

I/O signal return currents can also cause excessive noise in the PDS. For every signal transmitted by a device into the PCB (and eventually into another device), there is an equal and opposite current flowing from the PCB into the device's power/ground system. If a low-impedance return current path is not available, a less optimal, higher impedance path is used. When I/O signal return currents flow over a less optimal path, voltage changes are induced in the PDS and the signal can be corrupted by crosstalk.

This can be improved by ensuring every signal has a closely spaced and fully intact return path.

Methods to correct a less optimal path include the following:

- Restrict signals to fewer available routing layers with verified continuous return current paths
- Provide low-impedance paths for AC currents to travel between reference planes (decoupling capacitors at specific PCB locations).

Chapter 18—Aperiodic Resonant Excitation of Microprocessor Power Distribution Systems and the Reverse Pulse Technique

Victor Drabkin, Senior Member of Technical Staff, Hewlett-Packard
Chris Houghton, Senior Member of Technical Staff, Hewlett-Packard
Isaac Kantorovich, Senior Member of Technical Staff, Hewlett-Packard
Michael Tsuk, Senior Member of Technical Staff, Hewlett-Packard

A shortened version of this chapter was published in the Proceedings of Electrical Performance of Electronic Packaging conference, 2002, pp. 175-178.

THIS CHAPTER DETAILS a method to determine the worst-case excursion of the voltage in a microprocessor power distribution system by considering architectural limitations on variations in required current.

Introduction

Because of the ever-increasing clock frequencies and power requirements of modern microprocessors (MPs), the design and modeling of the power distribution system has become an area of critical importance [1]. The purpose of the power distribution system is to provide the large current requirements of the microprocessor with a voltage fixed within certain tolerances. Since the amount of current the microprocessor requires changes in time, both within the period of the clock and over many clock cycles, the traditional method for analyzing the power distribution system has been to look at its impedance as a function of frequency [2].

Often, the criterion for acceptance of the power distribution system is that the impedance be less than some fixed value over a wide frequency range [3, 4].

However, a focus on the frequency domain behavior of the power distribution system does not take into account the information the designer may have about the actual current profile of the microprocessor. For example, the chip may have a very large difference between the minimum and maximum current it will draw, but if the architecture is such that the microprocessor is limited in how fast it can switch from minimum to maximum, the requirements on the impedance of the power distribution system at high frequencies are loosened.

This paper describes the reverse pulse technique, a method for using architectural information about the time-varying current drawn by a microprocessor to evaluate the worst-case voltage excursions at the chip. For clarity, the case shown is for a simple ramp of current from minimum to maximum, but the method can be easily generalized.

Definitions

We will consider the MP power distribution system, which is excited by the MP current. The system usually includes supply feeders, decoupling capacitors, voltage stabilizers, etc. The MP current I(t) is always bounded, $Imin \leq I(t) \leq Imax$. Let us consider current transitions between maximum and minimum bounds Imax and Imin with the highest speed possible in the system, depending on architectural properties.

We will name such current transitions step functions. V(t) is a system voltage step response when excited by one of the current transitions and $t0 < t1 < t2 < \ldots < tn , \ldots$ are the points of extreme values of the system voltage step response V(t): V0, V1, V2, V3, … , see an example in Figure 18.1.

Respectively, the negative step response,-V(t) is a voltage response of the system to the reverse transition of the current. We will see that such denotation does not cause misunderstanding. To simplify the first description of the method, we will assume that the system is linear time-invariant, that V(t) = -V(t), and V(0) = 0.

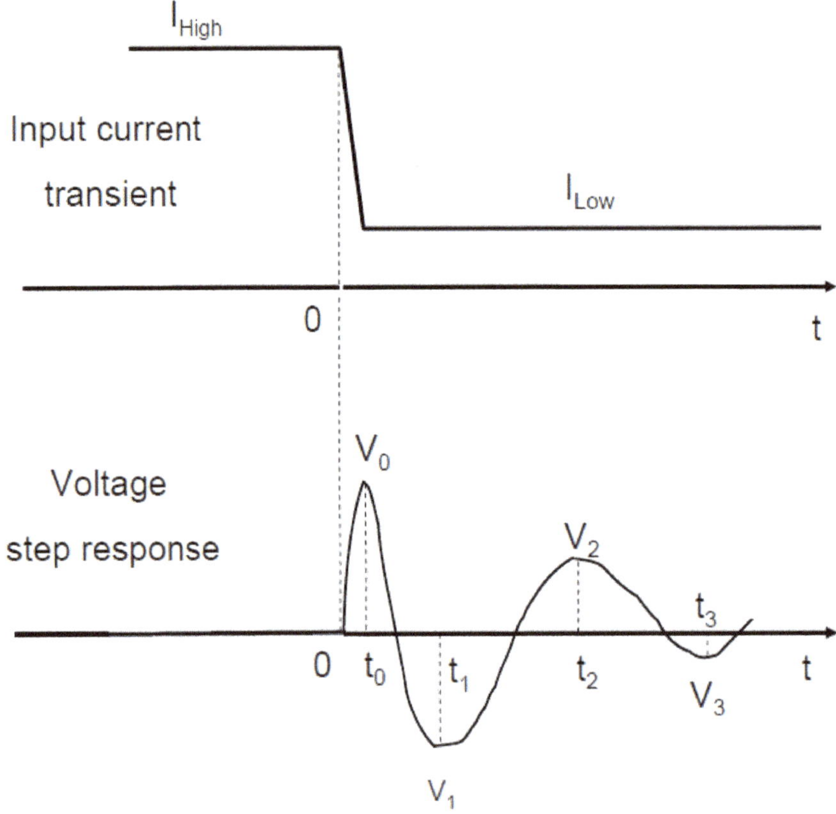

Figure 18.1: System Voltage Response

Reverse Pulse Technique—Plausible Reasoning

We will now derive a method for the finding the stimulus currents that provide the extreme values (max or min) of system voltage response. The idea is to gather all maxima or minima of the system voltage response V(t) and negative system response -V(t) at the chosen time. Figure 18.2 shows how to shift all maxima of V(t) and -V(t) to have them overlapped and added at the time t = 0.

Figures 18.3 to 18.6 show how V(t) and -V(t) should be shifted to achieve maximum system voltage excursion at t = 0. The current profile with respectively phased alternating positive and negative current transitions is represented in Figure 18.7. Figure 18.8 illustrates the result of four shifts: four voltage maxima V0, (-V1), V2, (-V3) have been put at the time point t = 0.

The resulting voltage is equal to:

$$V(0) = V_0 + (-V_1) + V_2 + (-V_3)$$

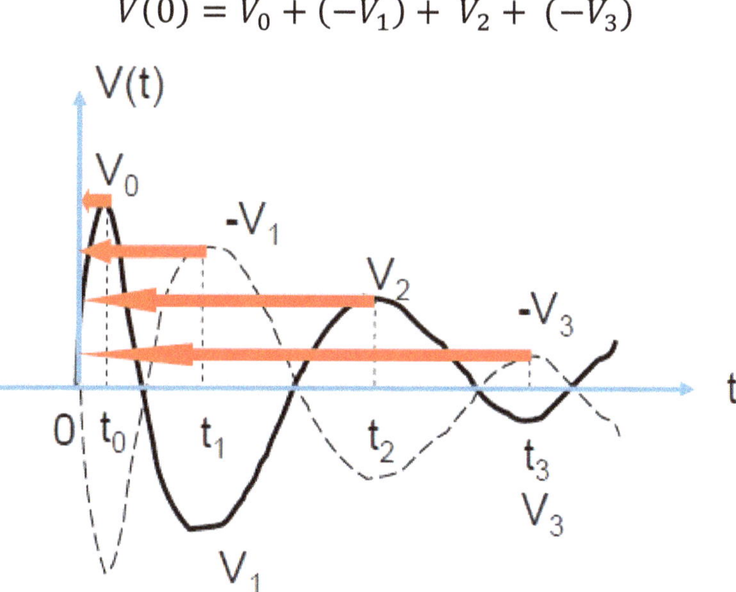

Figure 18.2: Shift of the Positive and Negative System Voltage Response Waves

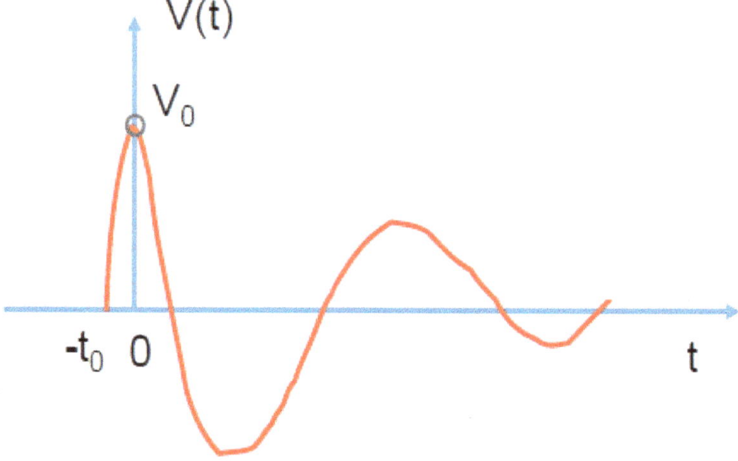

Figure 18.3: Shift of the System Voltage Response by ($-t_0$)

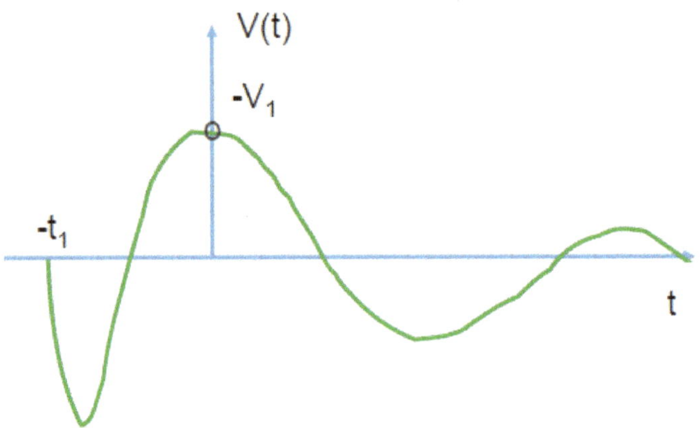

Figure 18.4: Shift of the Negative System Voltage Response by ($-t_1$)

In Equation 1, items V0 and V2 are maxima of V(t), (-V1) and (-V3) - maxima of -V(t) (minima of V(t)). In practice, Equation 1 will include as many maxima and minima of the system voltage response as it is necessary for the sake of accuracy. The minimum of the system voltage excursion can be achieved if we substitute in the current profile positive transitions by negative and negative by positive, respectively.

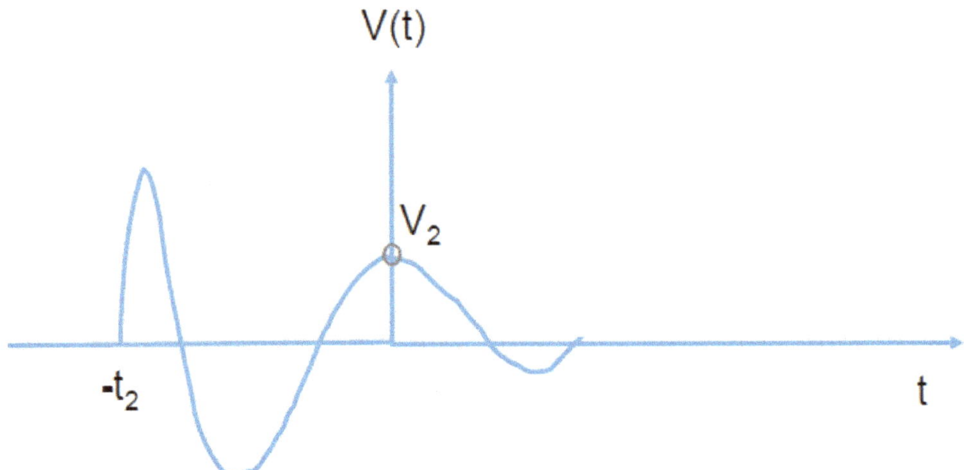

Figure 18.5: Shift of the System Voltage Response by ($-t_2$)

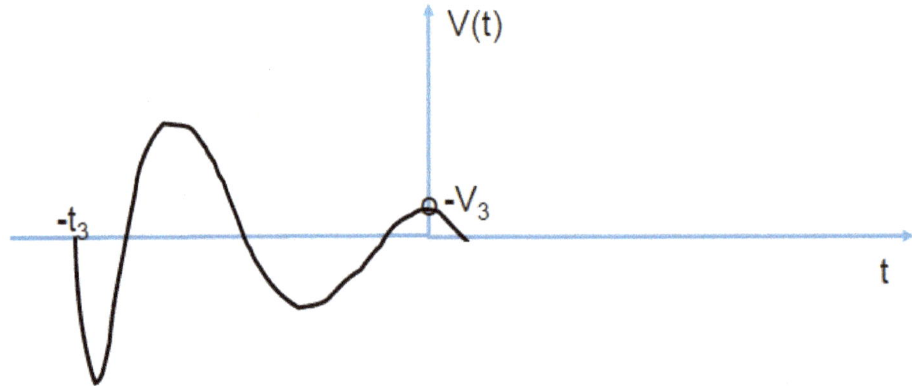

Figure 18.6: Shift of the Negative System Voltage Response by ($-t_3$)

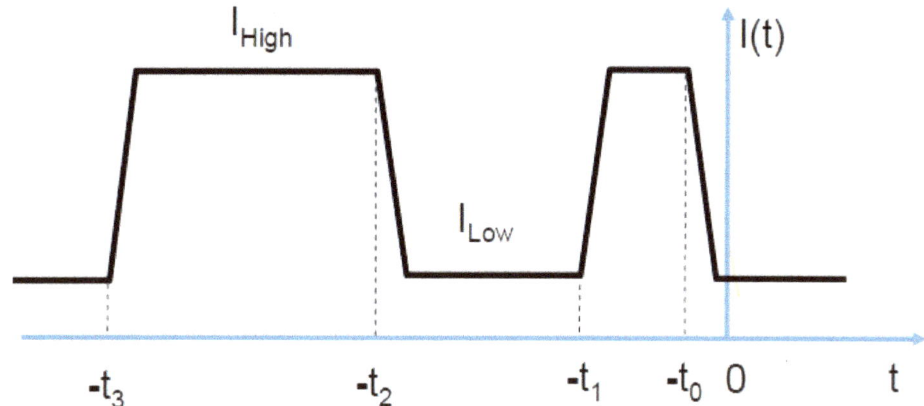

Figure 18.7: The Worst-Case Input Stimulus

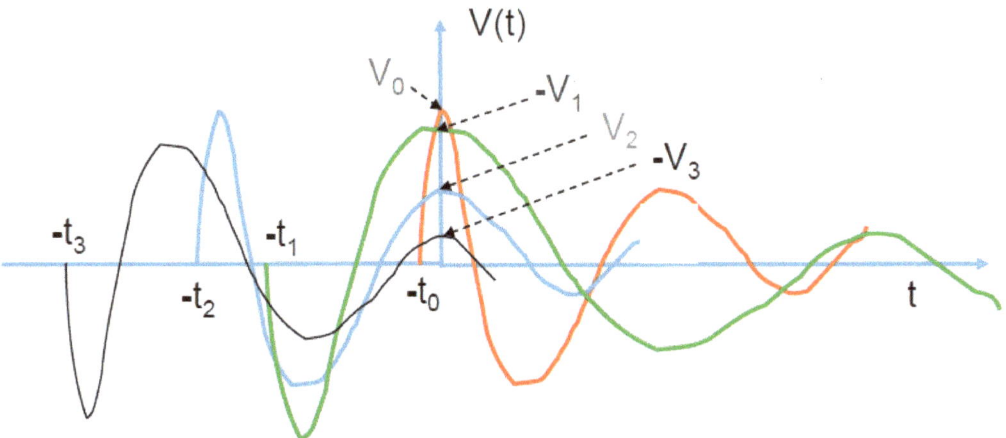

Figure 18.8: This Is How the Worst-Case System Response Is Built: $V(0) = V_0+(-V_1)+V_2+(-V_3)$

In general case $V(t) \neq -V(t)$ and $V(0) \neq 0$, also points of maxima of V(t) may not coincide with points of minima -V(t) and vice versa. This should be kept in mind while using the algorithm. We can denote minima and maxima of the system response V(t) like $V_{m1}, V_{m2}, \ldots, V_{mN}, \ldots$ and $V_{M1}, V_{M2}, \ldots, V_{MN}$ respectively.

We will use bars to denote minima and maxima of the negative system response, i.e. -V(t) in form $-V_{m1}$, $-V_{m2}, \ldots, -V_{mN}, \ldots$ and $-V_{M1}, -V_{M2}, \ldots, -V_{MN}$ respectively. We will also write Vdc to denote a DC voltage level.

While the algorithm presented is useful to create the input current wave form, the idea of summarizing voltage extreme values can be also used directly as follows:

$$V_{min} = V_{m1} + \bar{V}_{m1} + V_{m2} + \bar{V}_{m2} + \cdots + V_{mN} + \bar{V}_{mN} - (2N-1)V_{dc}$$

$$V_{max} = \bar{V}_{m1} + V_{m1} + \bar{V}_{m2} + V_{m2} + \cdots + \bar{V}_{mN} + V_{mN} - (2N-1)V_{dc}$$

...but Equation 2 and Equation 3 in general case should be considered as approximate ones. Nevertheless the developed approach is useful because when the system voltage responses, V(t) and -V(t), are measured on a real device the above equations can be used to acquire an experimental assessment of worst-case voltage noise.

The derived algorithm picks up all maxima of the system step responses to create the global maximum value of the system response. It also uses all minima of the system step responses to create the global minimum value of the system response. This makes the algorithm credible. In the next section of this work we will provide a strict proof of the algorithm correctness for the linear time-invariant systems.

The proposed algorithm builds the input current as a sequence of the current transitions in the reverse order to the sequence of standard voltage response extreme values (see Figure 18.9). This is why the algorithm has been named the reverse pulse technique.

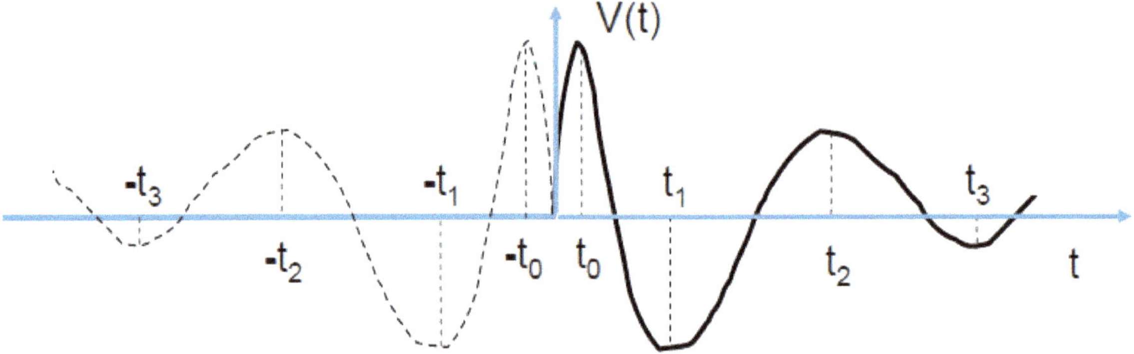

Figure 18.9: System Voltage Response and its Inversion

Reverse Pulse Techniques—Proof

It is not traditional to furnish a technical paper with a mathematical proof of the proposed algorithm, but we believe that this proof is important here. In the first stage of the algorithm development, the stimulus function consisted of quasi delta functions (narrow pulses of the limited magnitude). For the next stage, step functions substituted the quasi delta functions improving the results: system voltage response of greater magnitude was found.

There was a temptation to continue searching for algorithm improvements causing bigger system voltage responses within the given conditions. The proof ends such searches and shows that the published result is final and cannot be improved within the assumptions of the theorem.

Here we will prove that for the linear time-invariant systems, the algorithm named reverse pulse technique gives an absolute maximum voltage and absolute minimum voltage values at the system when the current varies within the given limits.

In the theorem below, 1(t) denotes the unit step function.

$$1(t) = \begin{cases} 0, & t < 0 \\ 1, & t \geq 0 \end{cases}$$

V(t) is a step response of the system. Usually V(t) is a solution of some ordinary differential equation with constant coefficients.

J(t) is a normalized input current:

$$J(t) = \frac{(I(t) - Imin)}{(Imax - Imin)}, 0 \leq J(t) \leq 1 \; at \; -\infty < t < \infty$$

E(t, J) is a voltage response of the system when the input current is J(t). $\hat{J}(t)$ is the normalized input current, maximizing the voltage response of the system at time t = 0, $\hat{j}(t)$ is the normalized input current, minimizing the voltage response of the system at time t = 0.

Theorem

Let:

a. J(t) be a normalized piecewise continuous function (with a finite number of discontinuities of the first kind on any finite interval), $0 \leq J(t) \leq 1$, $-\infty < t < \infty$;

b. E(t, J) be a voltage response of some linear time-invariant system to the exciting current J(t) and E(t, 1(t)) = V(t);

c. the function V(t) is equal to zero when t < 0 is a continuous function at $0 \leq t < \infty$ but can have a discontinuity of the first kind at t = 0, has a continuous derivative at 0 <t < ∞, and has a finite set of points of extreme values on any finite interval. The set {t0, t1, t2, …, tn , …} is a set of points of extreme values of V(t), where t0 < t1< t2 <…< tn …;

d. Emax is a maximum of the functional E(0, J) on the set of functions according to the condition a. with the maximizing function $\hat{J}(t)$, or $E(0, \hat{J})$ = Emax;

e. Emin is a minimum of the functional E(0, J) on the set of functions according to the condition a. with the minimizing function $\hat{j}(t)$, or $E(0, \hat{j})$ = Emin, then:

$$Emax = \sum_{i}^{\infty} Vmax_i - \sum_{i}^{\infty} Vmin_i$$

$\hat{J}(t) = \hat{J}1(t)$, if t0 is a point of maximum of the function V(t), $\hat{J}(t) = 1 - \hat{J}1(t)$, if t0 is a point of minimum of the function V(t), where:

$$\sum_{i}^{\infty} Vmax_i \text{ is the sum of all maxima of the function } V(t) at \; 0 \leq t \leq \infty,$$

$$\sum_{i}^{\infty} Vmin_i \text{ is the sum of all minima of the function } V(t) at \; 0 \leq t \leq \infty,$$

$$\hat{J}_1(t) = 1 + (t + t_0) + \sum_{i=1}^{\infty} [1(t + t_{2i}) - 1(t + t_{2i-1})],$$

$$\text{and } Emin = -Emax, \text{and } \hat{j}(t) = 1 - \hat{J}(t)$$

Proof

Let us consider the set $\{t_0, t_1, t_2, \ldots, t_n, \ldots\}$. Maxima and minima points of any function alternate. There is exactly one minimum between two adjacent maxima and exactly one maximum between two adjacent minima. The point t_0 can be a point of maximum or a point of minimum. Let us assume that t_0 is a point of maximum. Then $t_2, t_4, \ldots, t_{2i}, \ldots$ are also points of maximum, $t_1, t_3, \ldots, t_{2i-1}, \ldots$ are points of minimum.

For this definite case we denote:

$$t_0 = tmax_0,$$
$$t_1 = tmin_1, t_2 = tmax_1$$
$$\ldots$$
$$t_{2i-1} = tmin_i, t_{2i} = tmax_i$$
$$\ldots$$

We define the function $\hat{J}_1(t)$ as:

$$\hat{J}_1(t) = 1(t_0 + t) + \sum_{i=1}^{\infty} [1(t_{2i} + t) - 1(t_{2i-1} + t)]$$

For the case where t_0 is a point of maximum:

$$\hat{J}_1(t) = 1(tmax_0 + t) + \sum_{i=1}^{\infty} 1[(tmax_i + t) - 1(tmin_i + t)]$$

Let us note that each expression $1[(tmax_i + t) - 1(tmin_i + t)]$ is equal to 1 only when $-tmax_i \leq t \leq -tmin_i$

The voltage response E(t, J) of the system to the current J(t) can be represented in form of the convolution integral:

$$E(t) = \int_{-\infty}^{\infty} V'(\tau) J(t - \tau) d\tau$$

The derivative of the step response V′(t) is the impulse response.

We denote:

$$Em = \int_{-\infty}^{\infty} V'(\tau) \hat{J}_1(t - \tau) d\tau \mid_{t=0}$$

The substitution of the $\hat{J}_1(t)$ equation into the Em equation gives:

$$Em = \int_{-\infty}^{\infty} V'(\tau)\left\{1(tmax_0 - \tau + \sum_{i=1}^{\infty}[1(tmax_i - \tau) - 1(tmin_i - \tau)]\right\} d\tau$$

The expression in braces is the expression for $\hat{J}_1(-\tau)$. We can see that $\hat{J}_1(-\tau)= 1$, when:

$$\begin{gathered} -\infty < \tau \le tmax_0 \\ tmin_1 < \tau \le tmax_1 \\ tmin_2 < \tau \le tmax_2 \\ \dots \\ tmin_{2i} < \tau \le tmax_i \\ \dots \end{gathered}$$

...otherwise $\hat{J}_1(-\tau) = 0$. Now we can find the Em integral. For this we will represent it as a sum of integrals on the intervals given by the $\hat{J}_1(-\tau)= 1$ inequalities.

$$Em = \int_{-\infty}^{tmax0} V'(\tau)1(tmax_0 - \tau)d\tau$$

$$+ \sum_{i=1}^{\infty}\left(\int_{t\min i}^{t\max i} V'(\tau)\,[1(tmax_i - \tau) - 1(tmin_i - \tau)]\,d\tau\right)$$

$$= V(tmax_0) - V(-\infty) + \sum_{i=1}^{\infty}[V(tmax_i) - V(tmin_i)] = \sum_{i}^{\infty} Vmax_i - V\min_i$$

Here we use that $V(-\infty) = 0$. Notice that outside of the set of intervals of the $\hat{J}_1(-\tau)= 1$ limits, the expression in braces in the Em equation is equal to zero.

Now let J(t) be any function according to the condition a. We can represent the function J(t) in the form:

$$J(t) = \hat{J}_1(t) + I(t)$$

where:

$$\begin{gathered} -1 \le i(t) \le 0, \text{when}\, \hat{J}_1(t) = 1, \\ 0 \le i(t) \le 1, \text{when}\, \hat{J}_1(t) = 0 \end{gathered}$$

Substitute the $J(t)$ equation into the convolution integral with t=0.

$$E(0,J) = \int_{-\infty}^{\infty} V'(\tau)\hat{J}_1(t-\tau)d\tau|_{t=0} + \int_{-\infty}^{\infty} V'(\tau)i(t-\tau)d\tau|_{t=0}$$

The first integral in the $E(0,J)$ equation is coincident with (7) and is equal to Em.

The second integral can be represented as a sum of integrals on the intervals where V(τ) is a monotonic function: ($tmax_0$, $tmin_1$), ($tmin_1$, $tmax_2$), …, (tmaxi-1, tmini), (tmini, tmaxi), … On each interval (tmaxi-1, tmini) the derivative V′(τ) is negative and the function i(t) is positive or equal to zero according to the $-1 \leq i(t) \leq 0, \text{when } \hat{J}_1(t) = 1,$ limit (because the function Ĵ1(-τ) is equal to zero on these intervals).

So each integral of V′(τ)i(τ) on the interval (tmaxi-1, tmini) is negative or equal to zero. As well each integral of V′(τ)i(-τ) on the interval (tmini, tmaxi) is negative or equal to zero because the derivative V′(τ) is positive, while the function i(t) is less or equal to zero according to the $0 \leq i(t) \leq 1, \text{when } \hat{J}_1(t) = 0$ limit (the function Ĵ1(-τ) is equal to 1 on these intervals). So the second integral in the $E(0,J)$ equation is negative or equal to zero and equality takes place only when i(-τ) = 0 and therefore Emax = Em.

Hence, for each normalized excitation function J(t): E(0,J) ≤ Em, if t_0 is a point of maximum of the function V(t). We now need to consider the case, when t_0 is a point of minimum of the function V(t). In this case the points t_2, t_4, …, t_{2i}, … are also points of minima of the function V(t), while t_1, t_3, …, t_{2i-1}, … are points of maxima.

We will define the function Ĵ2(t) as:

$$\hat{J}_2(t) = 1 - \hat{J}_1(t)$$

It is easy to see that Ĵ2(t) = 1 when Ĵ1(t) = 0 and vice versa. Let us substitute Ĵ2(t) instead of Ĵ1(t) to the Em equation. One can also see that the function V′(τ) is positive on exactly the same intervals, where the function Ĵ2(−τ) is equal to 1.

We will find for this case that:

$$Em = -V(t_0) + V(t_1) - V(t_2) + \cdots + V(t_{2i-1}) - V(t_{2i}) + \cdots$$

Recalling that in the case $V(t_0)$, $V(t_2)$, $V(t_{2i})$ are minima and $V(t_1)$, … $V(t_{2i-1})$ are maxima of the function V(t), we see that (15) is equivalent to the equation:

$$Em = \sum_{i}^{\infty} Vmax_i - \sum_{i}^{\infty} Vmin_i$$

Similar to the previous case, the Em equation provides maximum of E(t, J(t)) on the set of functions J(t) according to the condition a.

The proof follows from representation of any function J(t) as the sum:

$$J(t) = \hat{J}_2(t) + i(t)$$

...where:

$$-1 \leq i(t) \leq 0, \text{when } \hat{J}_2(t) = 1,$$

$$0 \le i(t) \le 1, when\,\hat{J}_2(t) = 0$$

...with substitution of the J(t) equation to the Em convolution integral.

$$E(0,J) = \int_{-\infty}^{\infty} '(\tau)\hat{J}_2(t-\tau)d\tau \mid_{t=0} + \int_{-\infty}^{\infty} V'(\tau)i(t-\tau)d\tau\mid_{t=0}$$

...and evaluation of the second integral above, which is less or equal to zero, because the integrand is non-positive unless the function i(t) = 0.

The maximum of the functional E(0, J) is equal to the sum of all maxima minus the sum of all minima of V(t), according to the Em equation.

The maximizing function is J(t) = Ĵ1(t), if the first extreme value of the function V(t) is maximum, otherwise the maximizing function is J(t) = Ĵ2(t). Using the property of linearity of the system, we conclude, that Emin= -Emax, with the minimizing function j(t) = Ĵ2(t), if the first extreme value of the function V(t) is maximum, or j(t) = Ĵ1(t), if the first extreme value of the function V(t) is minimum.

Hence, the theorem is proved.

Conjecture

The method described here also finds the worst-case system current excitements at the set of the normalized current excitements with a limited slope. Instead of the unit step function, it is necessary to use a function:

$$Z(t) = \begin{cases} 0, \text{when } t < 0 \\ \frac{t}{T}, \text{when } 0 \le t \le T = \frac{1}{T}\int_{0}^{t}[1(t) - 1(t-T)]dt \\ 1, \text{when } t > T \end{cases}$$

...while finding the system step response and creation of the worst case exciting current profile.

Development

Modeling has confirmed that the current profile created by the reverse pulse technique provides the largest values of positive and negative extreme values encountered by the authors. Figures 18.10, 18.11, and 18.12 show illustrations of a microprocessor supply system. Extreme values provided by resonant frequency modeling are consistently lower, often much lower in the case of multiresonant systems.

This inspired us to name the phenomenon "aperiodic resonance," and this is analogous to classic resonance where there is a voltage increase in response to some periodic excitement. Aperiodic resonance can be inherent in multi-resonant circuits, both linear and nonlinear.

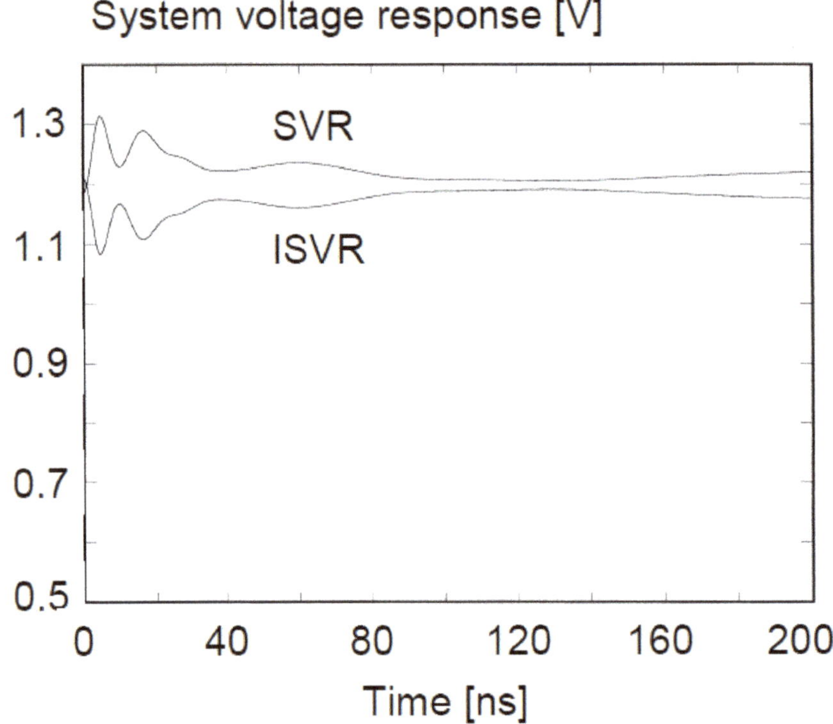

Figure 18.10: Microprocessor Supply System Voltage Response Waves (SPICE Modeling)

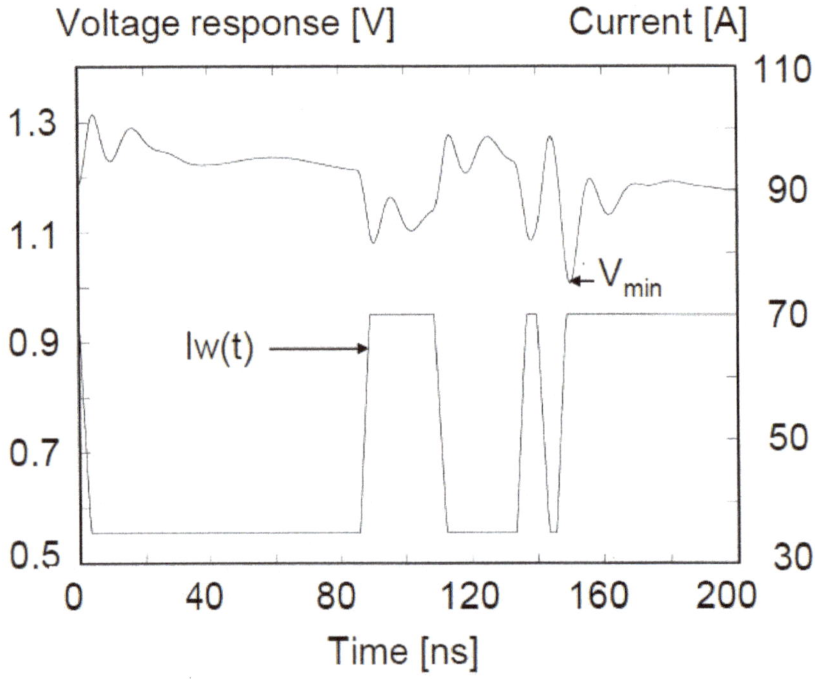

Figure 18.11: Worst-Case of the Microprocessor Supply System Minimum Voltage and Stimulus Current (SPICE Modeling)

For the single resonance systems the worst-case excitation is close to the excitation at the resonant frequency. For systems with multiple resonant frequencies, the reverse pulse technique provides the absolute extreme values of the voltage response, which cannot be achieved by other methods.

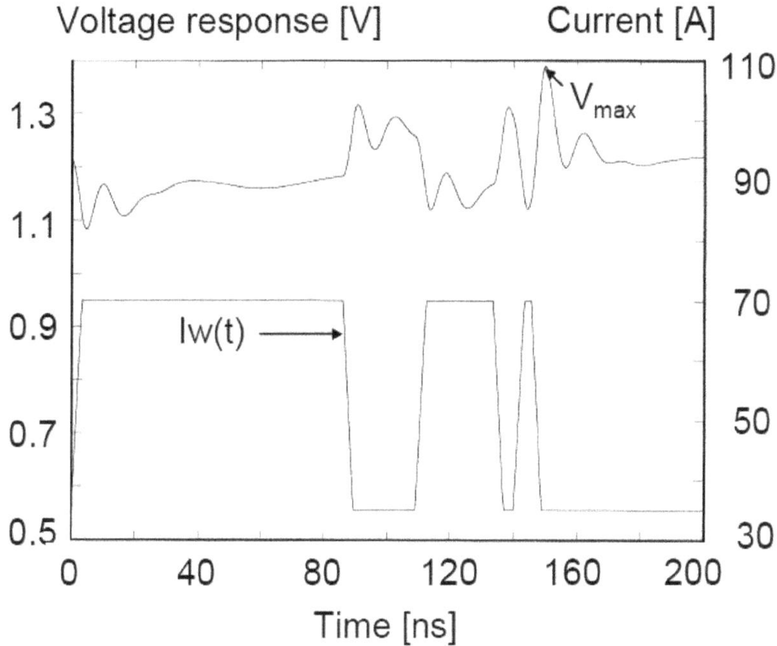

Figure 18.12: Worst-Case of the Microprocessor Supply System Maximum Voltage and Stimulus Current (SPICE Modeling)

Conclusion

This paper provides a brief introduction to a methodology that uses a circuit model of a microprocessor power distribution system and architectural specifications on the time-domain variation of the supply current to determine the worst-case voltage excursions. The reverse pulse technique can also be extended to more complicated current waveforms and, with some modifications, it can handle the slight nonlinearities that occur in the realistic modeling of microprocessor power distribution systems.

References

[1] M. Tsuk, D. Dvorscak, R. Dame, C. Houghton, and J. St. Laurent, *Modeling and Measurement of the Alpha 21364 Package*, Proc. IEEE 10th Topical Meeting on Electrical Performance of Electronic Packaging, 2001, pp. 283-286.

[2] L. Smith, *Packaging and Power Distribution Design Considerations for a Sun Microsystems Desktop Workstation*, Proc. IEEE 6th Topical Meeting on Electrical Performance of Electronic Packaging, 1997, pp. 19-22.

[3] V. S. Cyr, I. Novak, N. Biunno, J. Howard, *ARIES: Using Annular Ring Embedded Resistors to Set Capacitor ESR in Power Distribution Networks*, Proc. IEEE 10th Topical Meeting on Electrical Performance of Electronic Packaging, 2001, pp. 269-272.

[4] B. Young, *Digital Signal Integrity: Modeling and Simulation with Interconnects and Packages*, Prentice Hall, 2000.

Chapter 19— Modeling Noise on Printed Circuit Board Power Planes

John Grebenkemper, Ph.D., Development Engineering Manager, Hewlett-Packard

This paper was presented at DesignCon 2004, Santa Clara, California, January 29-February 1, 2004.

NOISE GENERATED BY di/dt switching on printed circuit board (PCB) power planes is becoming an increasingly important issue in computer design. Both switching rates and switching current amplitudes are increasing with each new generation of chip design.

This noise is one source of intermittent failures in digital logic. Conventional techniques to reduce the noise, including the use of multiple bypass capacitors, are becoming less effective at higher frequencies. Progress in improving designs is also hampered by the lack of good noise prediction tools. The prediction tool described in this chapter uses a frequency domain analysis solution. A frequency domain solution was selected since it can incorporate many subtle effects that are not easily implemented in time-domain tools. These effects include losses in the copper power/ground planes due to skin effect and losses in the laminate used to separate PCB ground and power planes. These effects are significant in today's higher-frequency system designs, and ignoring them results in substantial errors in noise prediction. This prediction tool has been used to test a number of PCB topologies. The tool shows that some strategies are much more effective than others for reducing noise. Comparisons between the predictions of the power plane noise modeling tool and actual test boards will be shown in the paper.

Introduction

Modern PCBs are constructed of alternating layers of conducting material, usually copper, and an insulating laminate. Some of the conducting layers are formed into traces to create signal layers.

Others are left nearly intact to create the power and ground layers that are used to form a reference plane for the signal layers and distribute power to the active devices on the PCB. When the active devices change their current demand, they create noise transients between the power and ground planes. This paper addresses the prediction of noise on a power plane relative to its adjacent ground plane.

Predicting the noise on a power plane in a PCB is a difficult problem to analyze for the following main reasons:

- The transmission line formed between a power and ground plane is two-dimensional. The impedance that a wavefront encounters decreases as it radially propagates away from a disturbance.
- The edge of the power plane is approximately equivalent to an open circuit, which will reflect most of the energy back into the PCB.
- The current transients generated by an active device that feed into the power and ground planes are not well known. No manufacturer can provide a good model of the waveform, and most cannot even tell you the di/dt of the device.
- The impedance load of the power, along with ground connections of the active devices to the power plane, is not specified. This will impact the noise that propagates on the power plane.
- The passive devices connected across the power and ground planes do not have well-specified characteristics. The manufacturer may provide a simple LCR model, but this rarely provides the true impedance of the device over a wide frequency range. Even if you measure a sample of the passive devices, the manufacturer may change their process, which can change some of the parameters that are important to noise on power planes. The losses in the dielectric material are generally not well specified. The best you can usually get is the loss tangent, which does not always characterize the material over the frequency range of interest.
- The surface roughness of the copper used in the power and ground planes impacts the skin effect resistance in these conductors. This characteristic is not specified and will impact the noise on the power planes.

All of these effects will impact the measured noise on a PCB power plane. The noise could be precisely predicted if these characteristics are all known. However, we have almost no information about some of them. The only way that we can make any prediction about the power plane noise is to make a number of simplifying assumptions.

Phases of Power Plane Noise

Let us suppose that an active device is connected between a power and ground plane through an ideal zero inductance connection.

This active device than instantaneously increases its current demand by ΔI. What then happens to the voltage between the power and ground plane?

Phase 1: Dielectric

Let us first look at the voltage ~100 ps after the initial current transient. There is a voltage wavefront radially propagating away from the point of the current transient. At the active device, the extra current can only be provided by the energy stored in the electric field between the power and ground planes. The voltage at the active device initially decreased, and it is now increasing back toward its original value.

In what ways can we decrease this initial noise step? The noise step can only be affected by the design of the power and ground planes on the PCB. If we halve the thickness of the laminate separating the power and ground planes, the electric field between them will double, and this will halve the size of the noise step.

There is also some belief that increasing the dielectric constant will decrease this noise. This is not true for Phase 1 noise, but it can be helpful for the Phase 4 noise. If we double the dielectric constant, the electric field strength will double. However, the propagation speed in the dielectric will decrease by the square root of two.

Thus, for a given time, the area that can supply charge will halve.

The net effect is that an increase in the dielectric constant will not affect the size of the initial noise step. In some cases, an increase in the dielectric constant may be harmful, since it will increase the time that it takes to propagate the noise step to the bypass capacitors.

The only effective way to control noise on a power plane in its earliest phases is to reduce the thickness of the laminate that separates the power and ground planes. If multiple power and ground planes are available, they may be placed in parallel to also reduce this noise. As we will see later, the reduction in laminate thickness is the most effective means of noise control available to the designer.

Phase 2: Bypass Capacitors

As the time increases to the hundreds of picoseconds, the transient noise event generated by the step increase in current will start to reach the first bypass capacitors. These bypass capacitors will begin to provide additional charge beyond what is available in the electric field to reduce the voltage transient. However, their series impedance limits the amount of charge that they can supply. The series impedance may be calculated from the equivalent series resistance (ESR) and the equivalent series inductance (ESL). In Phase 2, it is always desirable to minimize the ESR and ESL of the bypass capacitors.

The ESR is primarily determined by the capacitor construction; good 0.1 μF 0603 bypass capacitors should have an ESR less than 20 milliOhms. The ESL is determined by both the capacitor design and how it is mounted to the PCB. It should be possible to mount a single 0.1 μF 0603 capacitor on a PCB with an ESL of less than 1 nH. There are some specialized eight-pin capacitors available that achieve an ESL in the hundreds of picohenries. For this phase, the actual values of the capacitors are not that important as long as they are large enough to supply the required charge to reduce the transient. The ESR, ESL, and the number of capacitors are the most important factors. Ten capacitors with an ESL of 1 nH will work as well as one capacitor with an ESL of 100 pH.

Phase 3: Edge Reflections

The radial transient waveform will continue to travel until it reaches the edge of the board. A small amount may radiate as EMI, but the vast majority of the energy is reflected back into the board.

The radial waveform will reflect from the edge of the board at differing times, creating a multitude of interference patterns after the reflections. It is like throwing a rock into a quiet swimming pool and watching the waveforms reflect from the edges.

Series connected capacitors and resistors may be placed around the edge of the board to reduce the reflected energy. They are particularly effective at the board edges since the voltage transients double when reflecting from the edge of the board [1, 2].

Phase 4: Transient Damping

The transient noise will continue to bounce off the edges of the board for multiple times. I have observed a single transient reflect more than a dozen times before it was finally damped out. The voltage waveform at a single point looks chaotic as the multiple reflections pass through this point. There are a number of mechanisms that dissipate this energy until it finally decays back to its quiescent value.

The resistance in the copper dissipates the energy, particularly at higher frequencies as skin effect becomes more dominant, and decreasing the thickness of the laminate enhances this effect. If the laminate thickness is halved, then the electric field will double. This will cause the current flowing in the copper to double, and the I^2R loss will increase by a factor of four. A similar effect will occur if the dielectric constant of the laminate is increased. The loss in the dielectric also damps out the energy in the noise transients.

Increasing the loss tangent of the dielectric material will enhance this effect.

The components connected between the power and ground planes will have a resistive component in their impedance. This resistance will dissipate energy in the noise transient due to I^2R loss in the resistance. This effect applies both to the active and passive components. However, we have some control over this loss in the passive components. Laboratory work has demonstrated that a capacitor ESR of a few ohms is particularly effective at reducing Phase 4 noise.

Tuning capacitor values can also reduce the Phase 4 noise.

However, we have found that these methods work better and, once they are employed, the reduction in noise from capacitor value tuning is minimal.

The best ways to control Phase 4 noise are to use thin laminates between the power and ground planes to increase the dielectric constant and the loss tangent of the laminate and use high ESR-area capacitors distributed across the board.

Methods to Predict Noise

There are several ways to predict noise on a power plane. A time-domain method works best to predict the peak noise, but it requires exact knowledge about the waveforms of all of the noise sources on the board. It also needs to use a solution method that takes into account all of the frequency dependent effects, which become much more important as frequency approaches the Gigahertz range.

Finally, it may take a large solution time to find the peak noise, since it is dependent on the phase of all of the noise sources and the interactions between the reflections from the board edges.

There is a special case that can be easily solved for the self-induced peak noise from each source. In Phase 1, the peak noise is only dependent on the current switching waveform and the local physical characteristics of the power and ground planes. The equations for this can be directly solved from the mathematics of a transverse electromagnetic (TEM) wave emanating from the current disturbance.

In principle, this can be extended to Phase 2 for nearby bypass capacitors. This is most easily solved in the time domain by meshing the power and ground plane into its equivalent electrical circuit and including an electrical model for the bypass capacitors.

The special case can compute whether the local design can handle the initial transient from the switching current of a local active device. It cannot determine if the overall board design is sufficiently robust to handle many switching devices. It does not include the effects of standing waves that are created from the signal reflections off the board edges; these waves can create large voltage swings, especially when one of the board resonant frequencies is excited by a switching current waveform that contains significant energy at the same frequency.

The method described in this paper uses a frequency domain solution. This allows all of the frequency domain effects to be easily included in the solution. The main disadvantage is that it does not allow the computation of the peak noise, but instead computes a root mean square (RMS) noise. The peak noise is certainly several times larger than the RMS noise, but methods that reduce this noise should also reduce the peak noise.

This method could be used to predict the peak noise by preserving the phase information in the frequency domain and taking the Fourier transform to convert to the time domain. This will require preserving the spectral content of the switching current source.

The phasing of multiple current sources on a board must be preserved if the interference between these current sources is to be accurately modeled.

Noise Prediction Method

A PCB consists of alternating parallel layers of a conducting material separated by layers of insulating material, see Figure 19.1.

For most PCBs designed today, the conducting material is copper and the insulating material is a glass-and-epoxy resin generically referred to as FR-4. On some of the conducting layers, some of the copper is removed to leave numerous lines that are used to interconnect the various integrated circuits and other parts. This layer is referred to as a signal layer, since it contains the various signals that are used to interconnect the parts in the design. Other conducting layers are left nearly intact to distribute the ground and power to the active devices mounted on the PCB and to serve as a reference planes for the signal layers. These are called ground or power planes.

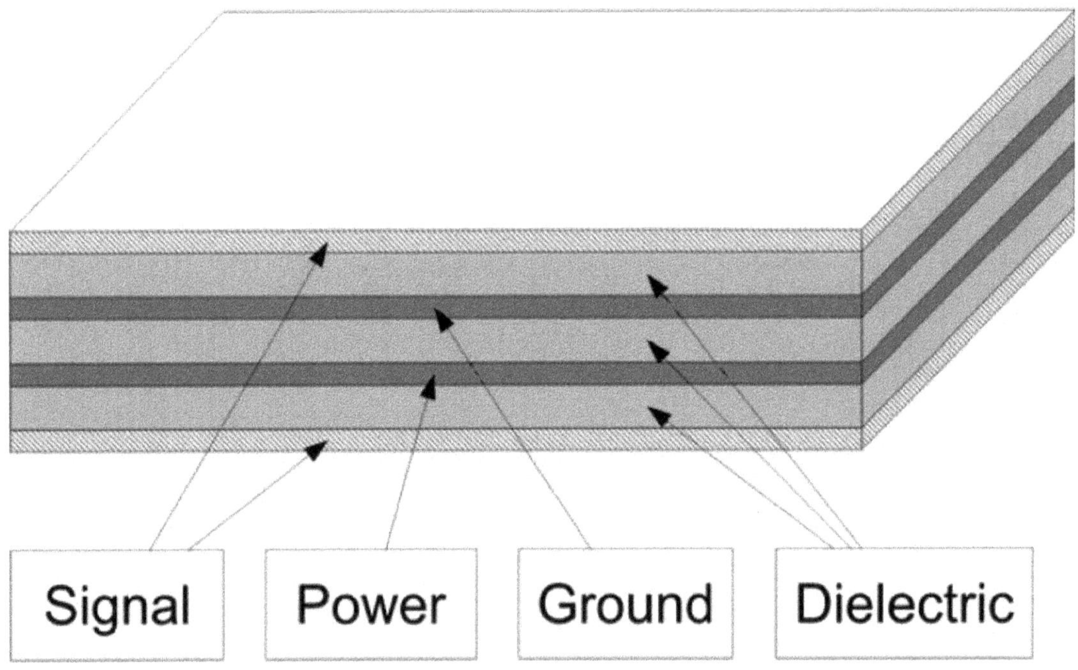

Figure 19.1: Stackup of a Four-Layer PCB

A low-cost board for a low-end PC may contain as few as four conducting layers with a signal layer on the top surface, a ground layer, a power layer, and a signal layer on the bottom surface. A high-end server board may contain more than 30 layers with multiple ground and power planes. The power and ground planes may be adjacent to each other or may have one or more signal layers between them.

Bypass capacitors are connected between the power and ground planes to help reduce the noise on the power planes. The noise reduction capacity of bypass capacitors decreases as the frequency increases, and they are becoming less effective for newer designs.

This means that a higher-frequency computer design must use more bypass capacitors or accept higher noise levels if no other changes are made to the design.

A power and ground plane pair may be modeled by meshing the planes and dielectric into rectangular sections in the (x, y) coordinate space [3–5]. The (z) coordinate represents the spacing between the planes. An equivalent electrical circuit as shown in Figure 19.2 represents these rectangular sections. Each connection point creates a node in the circuit.

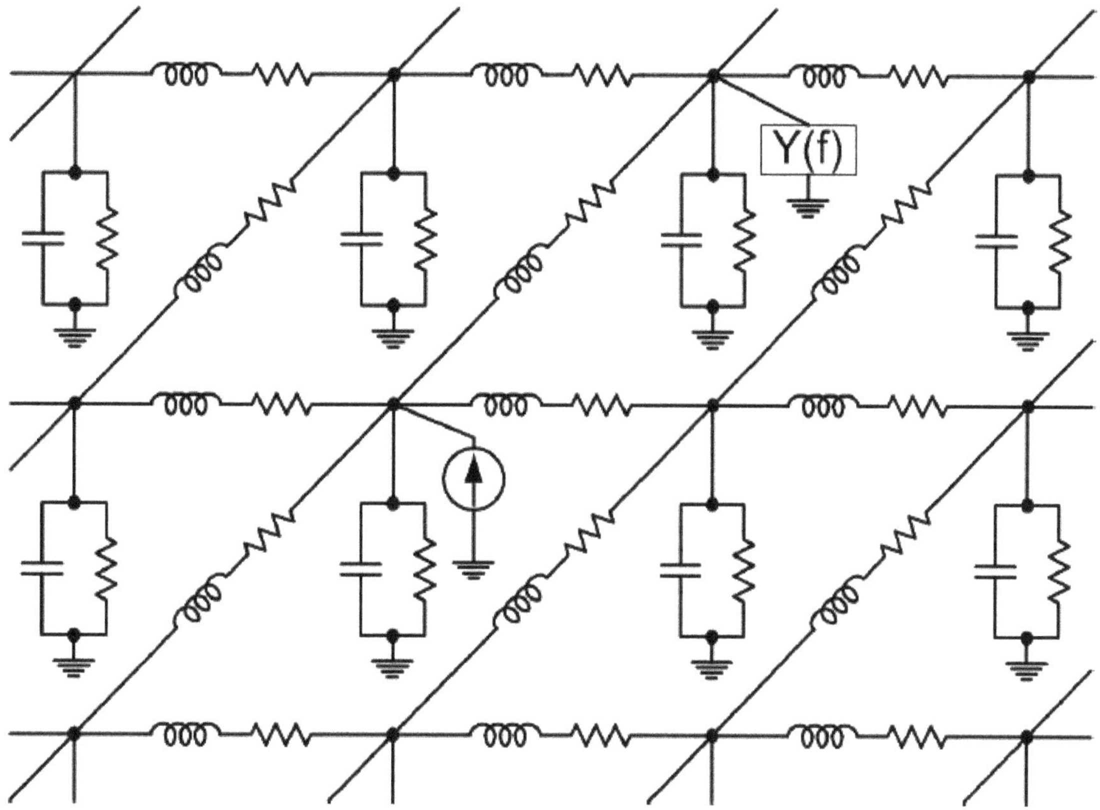

Figure 19.2: Equivalent Electrical Circuit of a Power and Ground Plane Pair. The Current Source Represents the Switching Current from Active Devices and the Y(f) Box Represents the Frequency Dependent Admittance of a Bypass Capacitor

In the simplest representation, the sections along the plane in the (x, y) directions are represented by an inductance in series with a frequency dependent resistance. In the (z) direction, the equivalent circuit is a capacitance in parallel with a frequency-dependent conductance. The varying current demands of the active devices are represented by a switched current source connected between the ground and power planes. The bypass capacitors and other circuit elements connected between the power and ground planes are represented by a frequency dependent admittance, Y(f). The L, R, G, and C terms of the equivalent circuit are given in the following equations.

$$L_x = \mu\frac{\Delta_x d}{\Delta_y};\ L_y = \mu\frac{\Delta_y d}{\Delta_x}$$

$$R_x = 2\frac{\Delta_x}{\Delta_y}\sqrt{\frac{\pi f\mu}{\sigma}}\left(\frac{sinh\left[2t\sqrt{\pi f\mu\sigma}\right] + sin\left[2t\sqrt{\pi f\mu\sigma}\right]}{cosh\left[2t\sqrt{\pi f\mu\sigma}\right] - cos\left[2t\sqrt{\pi f\mu\sigma}\right]}\right)$$

$$R_y = 2\frac{\Delta_y}{\Delta_x}\sqrt{\frac{\pi f \mu}{\sigma}}\left(\frac{sinh\left[2t\sqrt{\pi f \mu \sigma}\right] + sin\left[2t\sqrt{\pi f \mu \sigma}\right]}{cosh\left[2t\sqrt{\pi f \mu \sigma}\right] - cos\left[2t\sqrt{\pi f \mu \sigma}\right]}\right)$$

$$C_z = \varepsilon\frac{\Delta_x \Delta_y}{d}$$

$$G_z = 2\pi f \varepsilon \, tan[\delta]\frac{\Delta_x \Delta_y}{d}$$

Here, (Δx, Δy) are the mesh spacings in the x and y coordinates, d is the dielectric thickness separating the two planes, t is the thickness of the power and ground planes, σ is the conductivity of the power and ground planes, ε is the permittivity of the dielectric, μ is the permeability, and f is the frequency.

This equivalent circuit may then be computed as a set of admittances between the adjacent points in (x, y, z) space. These admittance equations include both the frequency-dependent skin effect loss and frequency-dependent dielectric loss.

$$Y_x(f) = \frac{\Delta_y}{2\Delta_x(\sqrt{\frac{\pi f \mu}{\sigma}}\frac{sinh\left[2t\sqrt{\pi f \mu \sigma}\right] + sin\left[2t\sqrt{\pi f \mu \sigma}\right]}{cosh\left[2t\sqrt{\pi f \mu \sigma}\right] - cos\left[2t\sqrt{\pi f \mu \sigma}\right]} + j\pi f \mu d}$$

$$Y_y(f) = \frac{\Delta_x}{2\Delta_y(\sqrt{\frac{\pi f \mu}{\sigma}}\frac{sinh\left[2t\sqrt{\pi f \mu \sigma}\right] + sin\left[2t\sqrt{\pi f \mu \sigma}\right]}{cosh\left[2t\sqrt{\pi f \mu \sigma}\right] - cos\left[2t\sqrt{\pi f \mu \sigma}\right]} + j\pi f \mu d}$$

$$Y_z(f) = 2\pi f \varepsilon\frac{\Delta_x \Delta_y}{d}(tan[\delta] + j)$$

To reduce noise, numerous bypass capacitors are connected between the power and ground planes. A bypass capacitor may be modeled as a capacitance with resistance and inductance in series with it. This is commonly referred to as ESR and ESL.

In addition, the vias that connect the capacitor to the power and ground planes will add additional series inductance and resistance. In the simplest form, the admittance of the bypass capacitors may be modeled as:

$$Y_c(f) = \frac{1}{\left(R_c + j\left(2\pi f L_c - \frac{1}{2\pi f C_c}\right)\right)}$$

...where R_c is the total ESR, L_c is the total ESL, and C_c is the capacitor value.

In real capacitors, values of the capacitance, ESL, and ESR all vary with frequency. To get more accurate results, the capacitor may be measured in an impedance analyzer and an estimated inductance and resistance added to these values to account for the vias. In the electrical model, the bypass capacitors are connected between the nearest node and ground.

The integrated circuits mounted on the PCB also have connections between the power and ground planes. These devices may provide both a capacitive and resistive load across the plane, which may have a beneficial effect on noise. Each active device may be represented by a frequency-dependent admittance connected between the nearest node and ground.

Once all of the admittance terms are calculated or measured, they are added to an admittance matrix that shows the admittance relationship between each of the nodes of the meshed planes.

This is a sparse matrix, since most of the nodes have no direct connection between them. In general, a particular node will only have a connection to its four adjacent neighbors and the ground plane node.

The current between each node and ground may be computed by multiplying the admittance matrix by the column matrix that represents the voltages between each node and ground.

$$|I(f)| = |Y(f)| \; * \; |V(f)|$$

This is not particularly useful, since the voltage distribution across the power plane is the unknown. However, if we take the inverse of the admittance matrix, we can rewrite the equation as follows.

$$|V(f)| = |Y^{-1}(f)| \; * \; |I(f)|$$

The |I(f)| column vector represents the switching current from active devices between each node and ground. Most of the terms in this vector will be zero, since most nodes have no active device at that location.

The |V(f)| column vector is the voltage between the power and ground planes at each node. The voltage distribution of noise across the power plane may be computed from the current switching in each active device that is attached between the power and ground planes. This computation is performed over the frequency range of interest.

The spectrum of each non-zero term in |I(f)| must be determined in order to do this calculation. We may compute the relative asymptotic approximation of this spectrum if we assume that the current transitions may be represented by a waveform that has equal rise and fall times t_r and a pulse width t_w.

$$I(f) = \begin{bmatrix} 1, & f \leq \dfrac{1}{\pi t_w} \\ \dfrac{1}{\pi t_w f}, & \dfrac{1}{\pi t_w} < f \leq \dfrac{1}{\pi t_r} \\ \dfrac{1}{\pi t_w t_r f^2}, & F > \dfrac{1}{\pi t_r} \end{bmatrix}$$

This spectrum is flat up to a frequency of 1/πtw, falls off at 20 dB/decade up to a frequency of 1/πtr, and then it falls off at 40 dB/decade above this frequency. For a pulse width of 10 ns and a rise time of 1 ns, the breakpoints occur at approximately 30 and 300 MHz as can be seen in Figure 19.3. If we used the exact response for a pulse as shown in Figure 19.3, the nulls created by the interference between the leading and falling edges would produce places in the spectrum to which we are not sensitive to noise.

In reality, the pulse widths and rise times are not accurately controlled and there will be energy in these null responses in a real system. The asymptotic response is a better way to estimate the effect of the noise on a design.

Multiple switching current sources on a board will create interference patterns between them. This will result in nulls and peaks on the board if we preserve the phase information. An alternative way to handle this problem is to assume that the sources are incoherent and sum the power instead of the voltage generated by an individual switching current source. This is the preferred way if there is no information about the timing of the switching current from the various active devices on a circuit board.

Most of the noise generated by the switching current transients tends to occur at lower frequencies. By the time we reach the gigahertz range, the noise is decreasing between 20 to 40 dB per decade. This implies that the noise reduction strategy needs to be more effective at lower frequencies. The higher frequencies will generate less overall noise and the package inductance will tend to filter this noise from the active silicon devices.

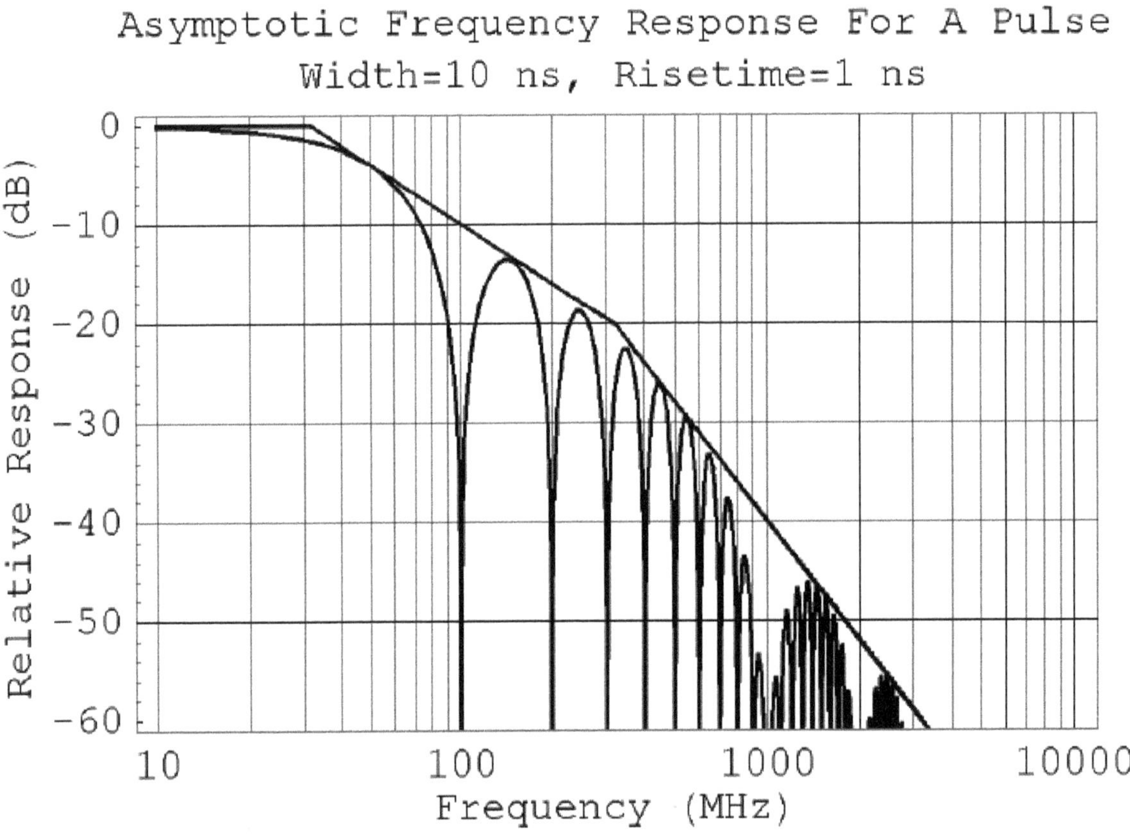

Figure 19.3: Relative Frequency Response of a 10 ns Pulse with a 1 ns Rise Time from a Switched Current Source. The Straight Lines Show the Asymptotic Approximation for the Frequency Response of the Pulse. The Other Curve Is the Exact Solution for a Pulse

The resulting column vector |V(f)| is the voltage as a function of frequency at each of the nodes in the rectangular mesh of the power planes. Each term in this column vector could be written as V(f, xi, yj), where (xi,yj) are the (x,y) coordinates of each point in the rectangular mesh.

We may compute the RMS noise power across the entire frequency range for each point in the mesh. We know, from Rayleigh's theorem, that the square of the voltage integrated over time equals the square of the voltage integrated over frequency [6].

$$\int_{-\infty}^{\infty} |v(t)|^2 dt = \int_{-\infty}^{\infty} |V(f)|^2 df$$

This means that the RMS voltage of the noise is the same whether we measure it in the time or frequency domain. Therefore, we may compute the RMS voltage at a given point on the PCB using the frequency domain spectral data.

$$V_{rms}(x_i, y_j) = \sqrt{\int_{-\infty}^{\infty} \left|V(f, x_i y_j)\right|^2 df}$$

Finally, we would also like to estimate the noise over the entire board. This is particularly useful to compare various strategies for noise reduction. The best way to do this is to compute the RMS average of the noise across the entire board.

$$V_{rms} = \sqrt{\frac{\sum_i \sum_j \left|V_{rms}(x_i, y_j)\right|^2}{N_x N_y}}$$

...where Nx and Ny are the number of points in the mesh in the (x) and (y) coordinates.

Maximum di/dt

The switching current from active devices is usually not available from most device vendors. Both the rate and the magnitude of the switching current of each active device is necessary to make an accurate prediction of the noise it will generate on a power plane.

While we may not be able to get this data, we can make a reasonable estimate of the rate of the current change. A changing current through an inductor gives the well-known equation:

$$V = L\frac{di}{dt}$$

This equation can be rewritten as:

$$\frac{di}{dt} = \frac{V}{L}$$

We can make reasonable estimates of the package inductance feeding the power plane and the voltage drop across this inductance. From this, we can then calculate a likely maximum di/dt of this device.

The on-chip silicon requires a reasonably steady supply voltage to function within its design parameters. If this voltage drops too much, the circuits in the silicon will fail. Typically, this will occur when the core supply voltage drops by 10 percent, but it could be different for various designs. From this line of reasoning, we can assume that the voltage across the supply inductance will not exceed 10 percent of the supply voltage.

The inductance of the supply voltage is determined by many power and ground pins in parallel. A single pair of power and ground pins might have a loop inductance of ~1 nH for a good BGA package to ~10 nH for a wire-bonded package. This loop inductance includes the effects of the vias connecting to the power and ground planes on the circuit board. Each additional pair of power and ground pins in parallel will reduce the loop inductance of the power distribution.

For example, assume that we have a 1.5 volt supply for the core voltage and it could operate without error down to 1.35 volts. This implies that we could have a maximum voltage drop across the inductance of 150 millivolts. Furthermore, we will assume that the loop inductance for the core voltage supply is 100 pH. Using the above equation we find that the maximum di/dt for this device is 1.5 A/ns. This type of calculation can be applied to each supply on an active device and to all of the active devices on a board. It provides a reasonable starting point for noise estimates given the lack of vendor data in this area.

Noise Control Measures

The previous section showed the mathematical formulas to compute the noise on the power plane of a PCB. This section will show the predicted results when the design parameters of a PCB are varied. The test results assume that the power plane and ground plane pair is 3" x 6" (7.6 cm x 15.2 cm) with a separation of 4 mils (10-4 meters). The noise source was assumed to be a 2 ns pulse width with a 200 ps rise time.

The calculations were done for a specific board size and noise source pulse. Changing these parameters will change the absolute magnitude of the results, but the general trends are still applicable.

The thickness of the laminate that separates the power and ground planes plays an important role in reducing the noise on the power plane. A thinner laminate will create more capacitance, which can provide additional charge when the active devices change their current demands. In addition, the decreased spacing between the power and ground planes will reduce the series inductance, which results in a lower impedance of the power distribution network (PDN). Finally, a thinner laminate also increases the dissipation of noise on the power plane. This occurs because a thinner layer generates a higher electric field between the power and ground planes, which causes more current to flow in the copper conductors and increases the I^2R loss in the conductor.

Figure 19.4 shows the predicted total RMS noise on the test board as the spacing between the power and ground plane is varied.

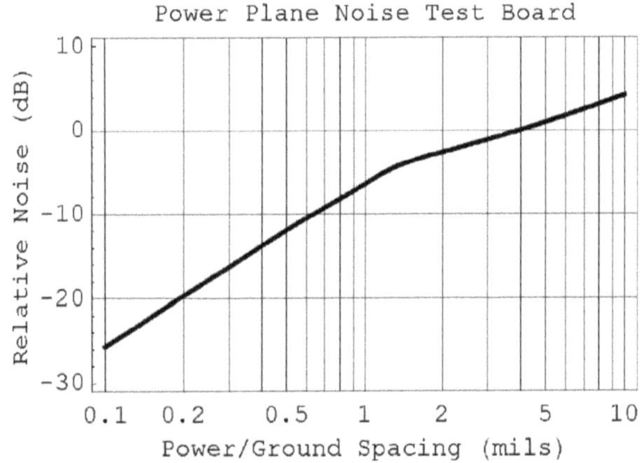

Figure 19.4: Total RMS Noise on a PCB Power Plane as a Function of the Power to Ground Plane Spacing. A Decrease in the Spacing Reduces the Noise

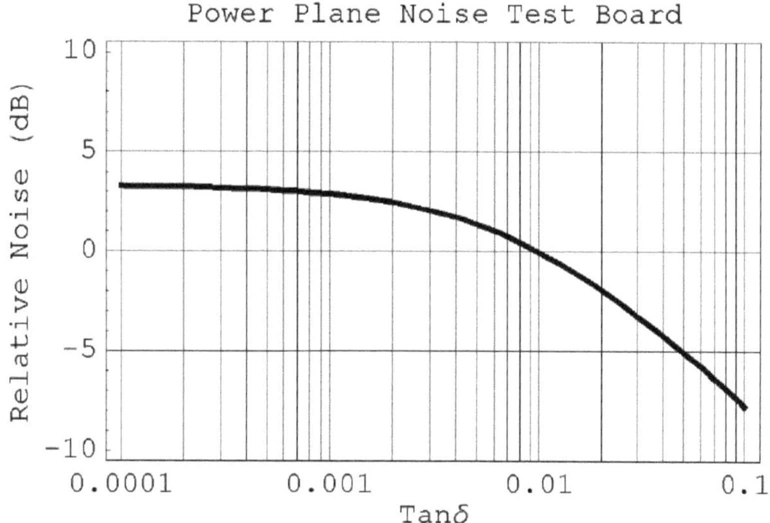

Figure 19.5: Total RMS Noise on a PCB Power Plane as a Function of the Dielectric Loss Tangent. An Increase in the Loss Tangent Reduces the Noise

The total noise falls by about 10 dB/decade for spacings greater than 1.5 mils and 20 dB/decade for smaller spacings. There is a significant benefit to make the spacing less than 1 mil. Reducing the dielectric thickness from 4 mils to 0.5 mils will reduce the total RMS noise by 11 dB.

The loss tangent of the laminate material also plays a significant role in reducing the total RMS noise as shown in Figure 19.5. The FR-4 dielectric typically has a loss tangent between 0.01 to 0.02, but some of the newer dielectric materials have even a smaller loss tangent. A much higher loss tangent is desirable for the dielectric separating the power and ground planes. Increasing the loss tangent from 0.01 to 0.1 will decrease the total RMS noise by 8 dB.

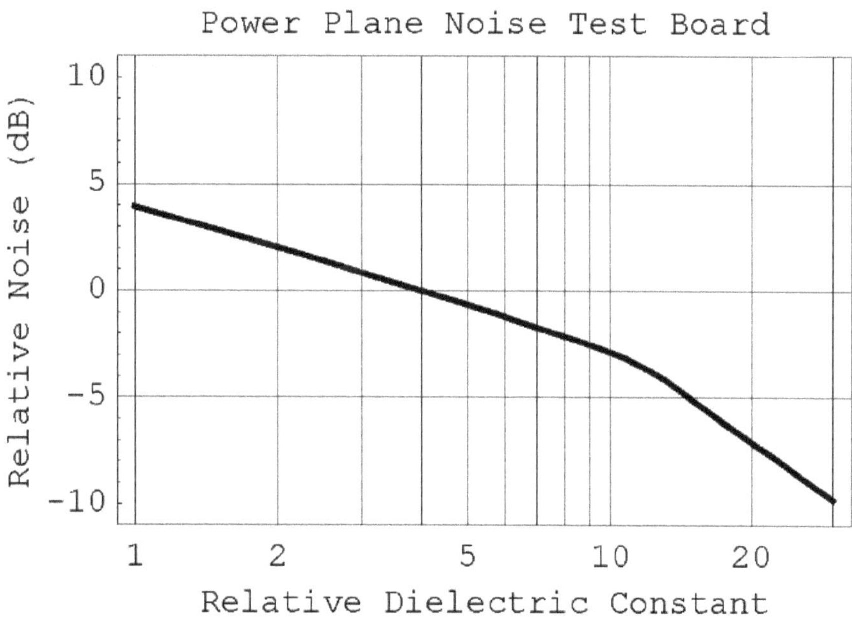

Figure 19.6: Total RMS Noise on a PCB Power Plane as a Function of the Dielectric Constant. An Increase in the Dielectric Constant Reduces the Noise

The relative dielectric constant of the laminate material will affect the total RMS noise on the power planes as shown in Figure 19.6. A higher relative dielectric constant will increase the capacitance between the power and ground planes and provide additional charge for the current switching transients of the active devices. Increasing the relative dielectric constant from 4 to 20 will decrease the total RMS noise by 7 dB.

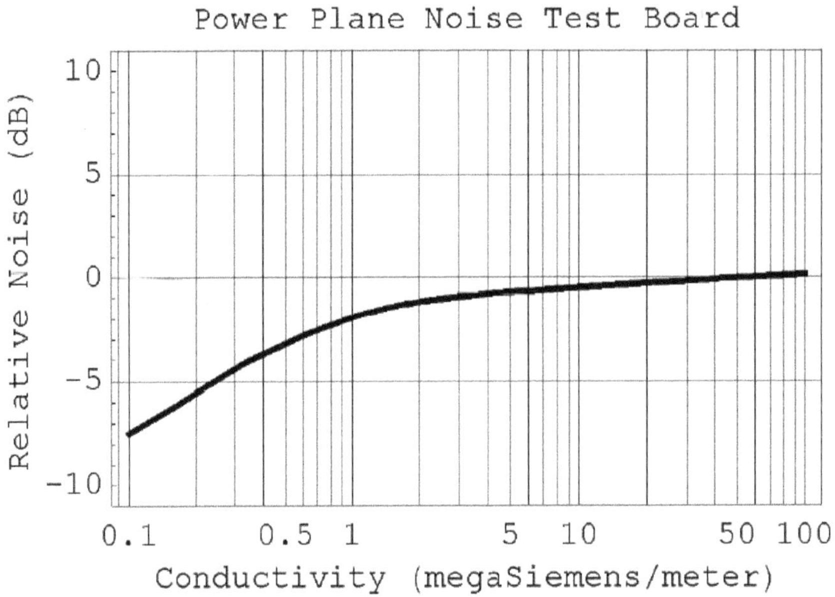

Figure 19.7: Total RMS Noise on a PCB Power Plane as a Function of the Conductor Conductivity. A Decrease in Conductivity Reduces the Noise

The conductivity of the conducting material in the power and ground planes will affect the total RMS noise as shown in Figure 19.7. The conductivity of the copper used in conventional printed circuit boards is typically about 58*106 Siemens/meter. Reducing the conductivity to 105 Siemens/meter would decrease the total RMS noise by 7 dB. This level of conductivity is available in a material called Ohmega-Ply®, which is used to manufacture buried resistors. This material could be coated on a copper plane to provide a low direct current (DC) resistance path for the power while providing a high loss path for the high-frequency components.

Bypass Capacitor ESR

Bypass capacitors are connected between the power and ground planes as one means to reduce the noise on the power plane.

Conventional wisdom for many years has been that the bypass capacitor should have a minimal ESR and ESL. Capacitor vendors have tried to minimize both of these variables. In particular, the ESR has been reduced from about 100 milliOhms a decade ago to around 10 milliOhms today.

The modeling software was used to check out the effect of ESR on noise. A total of 18 0.1 µF capacitors were placed on 1" spacings across a 3" x 6" test board. The initial noise calculation was done assuming that each capacitor had an ESR of 15 milliOhms. The ESR was then varied from 1 milliOhm to 10 Ohms. The ESL was assumed to be 1.5 nH. The calculation showed that the total RMS noise had a significant minimum when the ESR was in the range of 2 to 3 Ohms. As shown in Figure 19.8, the reduction in total RMS noise was 11 dB. This is a considerable improvement over the low-ESR capacitors, and it was quite unexpected that a high ESR would make a large reduction in the power plane noise.

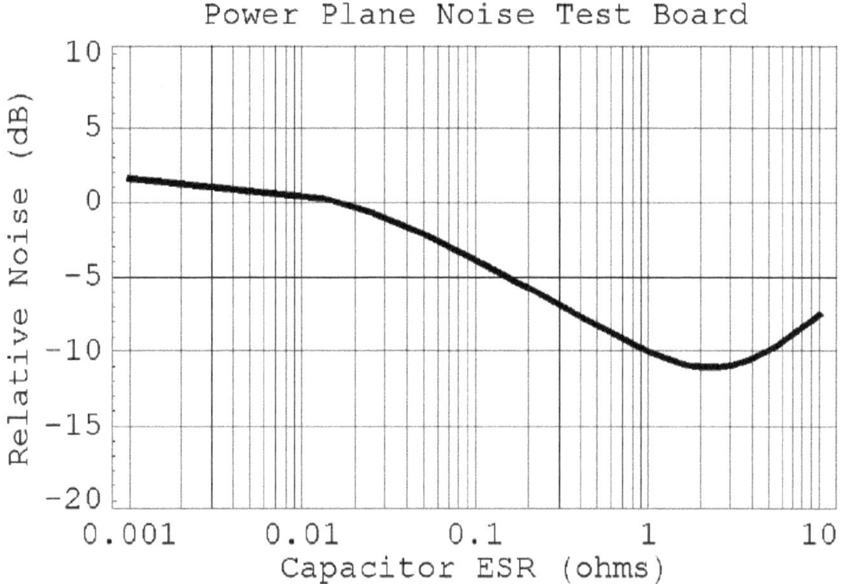

Figure 19.8: Total RMS Noise as Function of Bypass Capacitor ESR

The distribution of the noise across the PCB power plane is shown in Figure 19.9 and 19.10. Figure 19.9 shows the noise with low-ESR capacitors. The noise shows an increase at the board edges, particularly in the x-direction. This increase is due to the reflections from the open circuit at the board edge. The locations of the bypass capacitors can be seen as local minima in the response; the capacitor at (2.5, 1.5) is a good example.

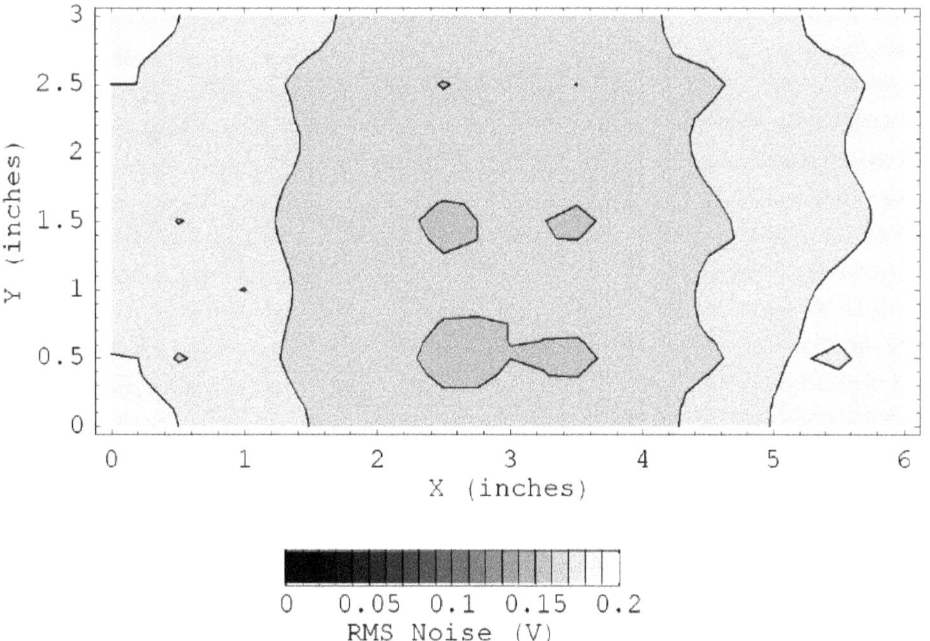

Figure 19.9: Noise Distribution across a PCB with Low-ESR Capacitors. The Board Is Driven in the Lower Left Corner 1 Inch from Each Edge. The Bypass Capacitors Are Located on 1-Inch Centers Starting ½ Inch from the Edges for a Total of 18 Bypass Capacitors. The Location of the Bypass Capacitors Can Be Seen as the Areas of Reduced Noise

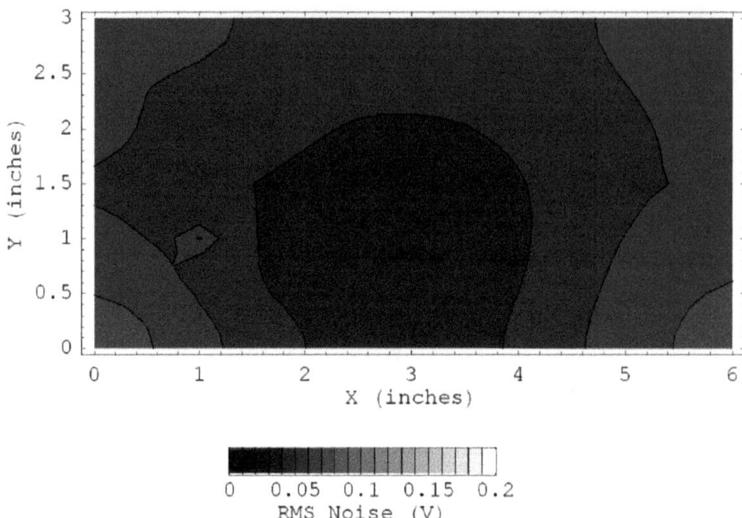

Figure 19.10: Noise Distribution across a PCB with High–ESR (3 Ohm) Bypass Capacitors. The Board Is Driven the Same as in Figure 19.9.

Figure 19.10 shows the noise distribution with high-ESR capacitors. The noise distribution is similar to Figure 19.9 with the noise maxima near the board edges, but the overall noise level is significantly reduced.

With both of these boards, the noise is lower near the center of the board. Noise-sensitive parts would see less noise if placed near the center of the board, and placement near the edge should be avoided.

It is clear from the earlier plots that 18 high-ESR capacitors are much more effective for noise reduction than 18 low-ESR capacitors. In some cases it would be desirable to use fewer capacitors to save product cost and board space.

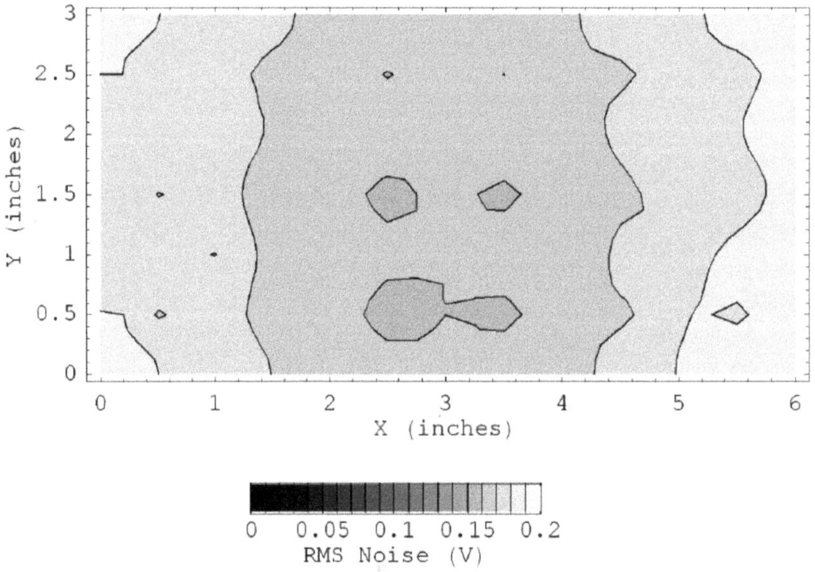

Figure 19.9: Noise Distribution across a PCB with Low-ESR Capacitors. The Board Is Driven in the Lower Left Corner 1 Inch from Each Edge. The Bypass Capacitors Are Located on 1-Inch Centers Starting ½ Inch from the Edges for a Total of 18 Bypass Capacitors. The Location of the Bypass Capacitors Can Be Seen as the Areas of Reduced Noise

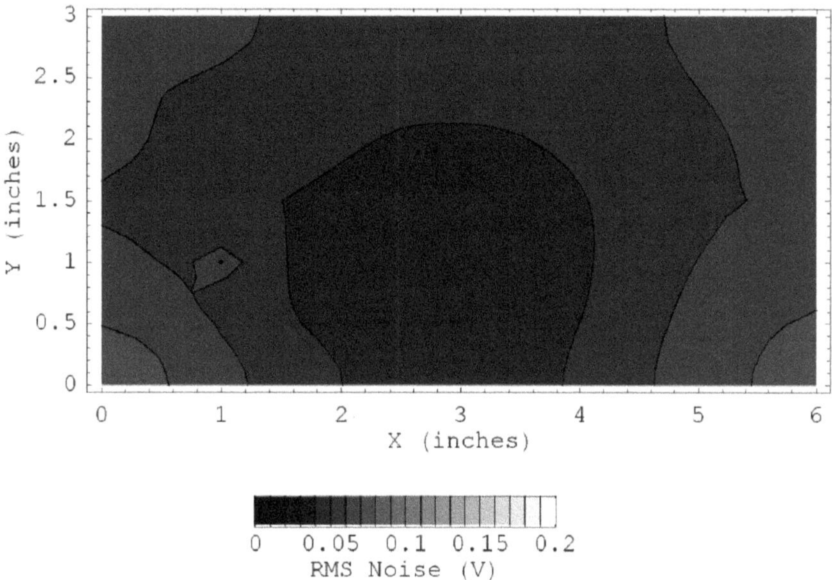

Figure 19.10: Noise Distribution across a PCB with High–ESR (3 Ohm) Bypass Capacitors. The Board Is Driven the Same as in Figure 19.9.

Figure 19.10 shows the noise distribution with high-ESR capacitors.

The noise distribution is similar to Figure 19.9 with the noise maxima near the board edges, but the overall noise level is significantly reduced. With both of these boards, the noise is lower near the center of the board. Noise-sensitive parts would see less noise if placed near the center of the board, and placement near the edge should be avoided.

It is clear from the earlier plots that 18 high-ESR capacitors are much more effective for noise reduction than 18 low-ESR capacitors. In some cases it would be desirable to use fewer capacitors to save product cost and board space.

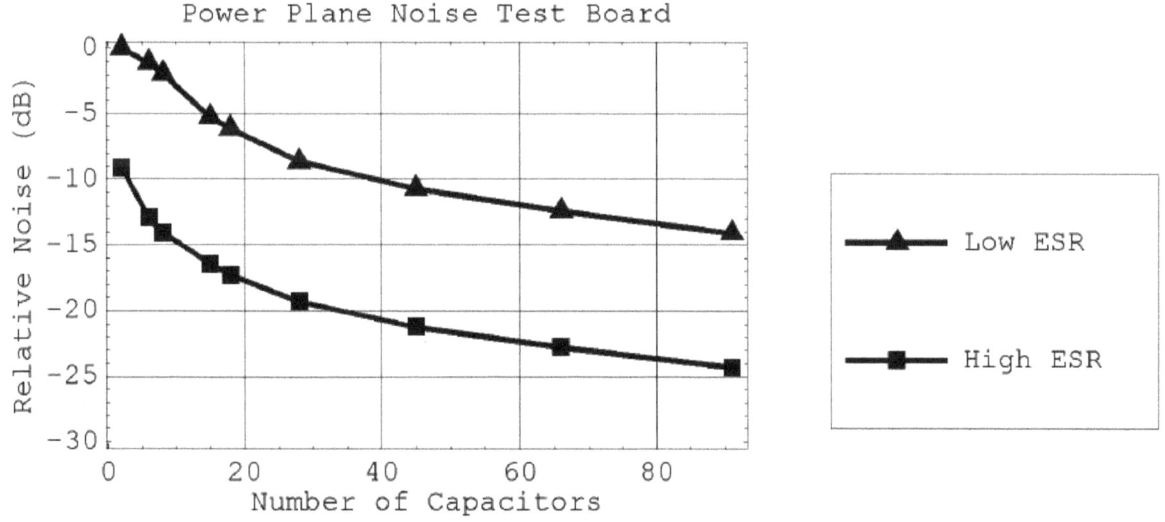

Figure 19.11: Effectiveness of High–ESR Capacitors Compared to Low-ESR Capacitors

Figure 19.11 shows the noise-reduction effectiveness of high- and low-ESR capacitors. The low-ESR capacitors have an ESR of 15 milliOhms, and the high-ESR capacitors have an ESR of 3 Ohms.

The 0 dB level was set using one low-ESR capacitor. It can be seen from the plot that six high-ESR capacitors reduce the noise level by 13 dB. It takes 80 low-ESR capacitors to reduce the power plane noise by an equivalent amount. Alternatively, using the same number of high-ESR capacitors would reduce the total RMS noise by more than 10 dB.

Measurement of Bypass Capacitors

The theoretical results show that a capacitor with a high ESR is much more effective for noise reduction than one with an ESR in the milliOhm range. This is a very surprising result, since it goes against conventional wisdom, which has tried to minimize the ESR in bypass capacitors as the best strategy to reduce power plane noise.

The size of the power plane noise test board was 3" x 6" (7.6 cm by 15.2 cm) and had a total thickness of 62 mils. The power and ground planes are located in the center of the board and separated by 4 mils. It was constructed using standard FR-4 glass-epoxy, which has a relative dielectric constant of about 4. The board contains 18 capacitor sites spaced 1" apart located on each side of the board. Eighteen sites on one side of the board can support a single surface mount capacitor. The 18 sites on the other side of the board can support a surface mount capacitor in series with a surface mount resistor. The test board also contains sites for 12 miniature SSMB-style (SSMB) connector locations with 1" spacings. This provides numerous points from which the board can be excited or measured. The ESL of the high ESR capacitors will be higher than if we had a true high-ESR capacitor.

A frequency domain S21 measurement was done between the diagonal corners of the power plane noise test board, as shown in Figure 19.12. This measurement directly shows the noise attenuation of the board and its bypass capacitors. The measurement was first done with only 36 low-ESR 0.1 μF capacitors as shown in the shaded line. The measurement was repeated by adding 3 Ohm resistors in series with 18 of the low-ESR capacitors as shown in the solid line. The measurements show that the first resonant peak at 400 MHz was reduced by more than 10 dB. These resonant peaks are the primary contributors to the noise on the power planes.

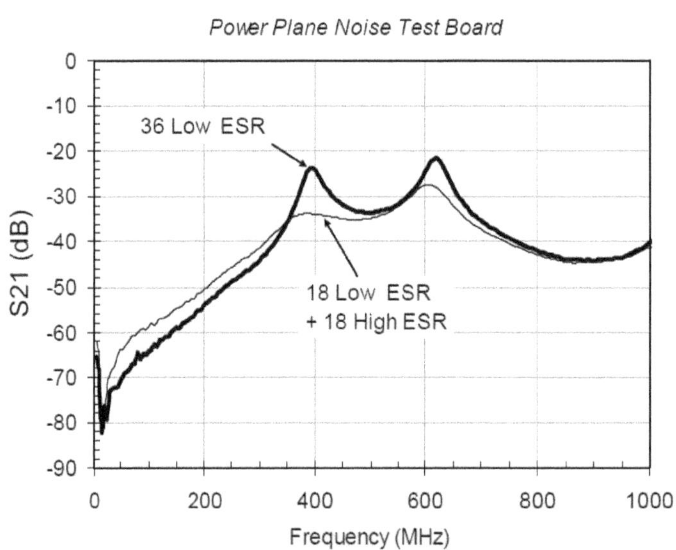

Figure 19.12: Frequency Domain Measurement of the Power Plane Noise Test Board with High-ESR (Thin Line) and Low-ESR (Thick Line) Capacitors

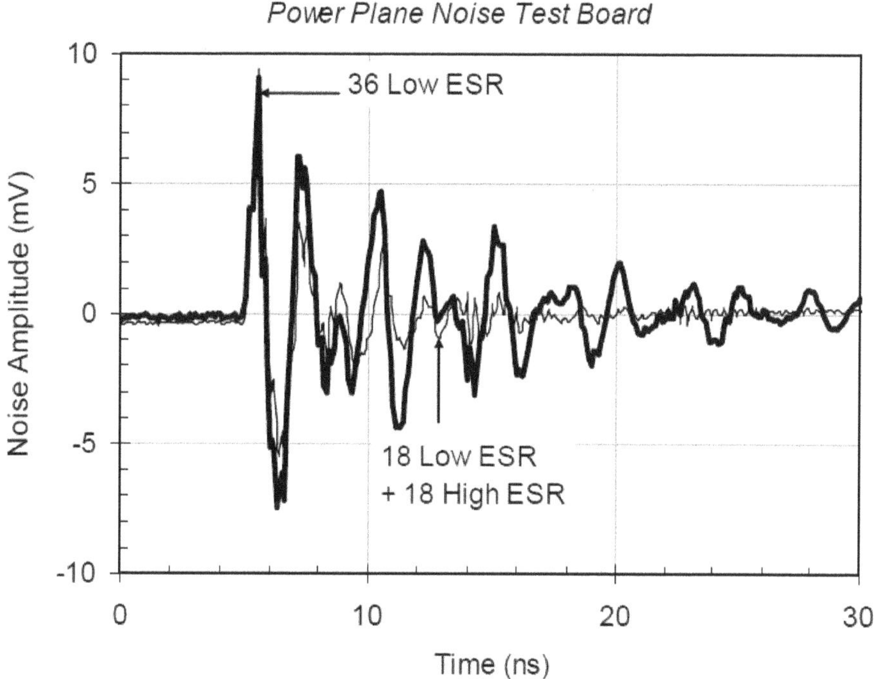

Figure 19.13: Comparison of Power Plane Noise as Measured with High-ESR (Thin Line) and Low-ESR (Thick Line) Capacitors

Figure 19.13 shows the results of time-domain measurement of the noise on the same set of test boards. The board was driven with a step generator at one corner of the board, and the output was measured at the opposite corner. The generator drove the board with a 200 ps rise time pulse that had a pulse width of 100 ns.

For the 36 low-ESR capacitors, the noise had decayed to 10 percent of its peak value 21 ns after the start of the pulse. When 18 of these capacitors were converted to high ESR by adding the 3 Ohm series resistors, the noise decayed to 10 percent of its peak value in 8 ns. This is a significant reduction in the ringing time.

Both of these tests show that the high-ESR capacitors do effectively reduce the noise on a power plane.

Conclusion

This paper has shown a method to predict the noise between the power and ground planes of a PCB. This prediction method may be used to compare various strategies to reduce the noise on the power planes. The following strategies work best for reducing this noise in the construction of a PCB:

- Decrease the thickness of the laminate between the power and ground planes
- Increase the dielectric loss in the laminate between the power and ground planes
- Increase the dielectric constant of the laminate between the power and ground planes
- Decrease the surface conductivity of the conductors that form the power and ground planes

The placement of noise sensitive parts on a PCB can impact the performance of these parts. The reflections from the edge of a power plane tend to increase the noise in this area.

- Avoid placing noise-sensitive parts near the edge of a PCB power plane; the lowest-noise areas tend to occur nearer the center of the power plane

The paper has also shown that the bypass capacitors that connect between the power and ground planes can greatly affect the noise levels. The calculations of the RMS noise show that it may be significantly reduced using high-ESR capacitors. However, these calculations do not show the peak noise that can be generated by a transient change in current demand. PCBs will still need low-ESR capacitors placed near active devices to supply charge when the current demands of the active devices change.

The best strategy to reduce overall board noise using bypass capacitors should incorporate the following elements:

- Place low-ESR capacitors near active devices that have significant changes in their current requirements
- Spread high-ESR capacitors over the entire board to dissipate the noise generated by current transients

This combination of both low- and high-ESR capacitors will produce a solution that requires fewer bypass capacitors to achieve the same noise levels, or it can achieve lower noise levels with the same number of bypass capacitors.

References

[1] Terrel L. Morris, *AC Coupled Termination of a Printed Circuit Board Power Plane in Its Characteristic Impedance*, U.S. Patent 5,708,400, January 13, 1998.
[2] István Novák, Reducing *Simultaneous Switching Noise and EMI On Ground/Power Planes By Dissipative Edge Termination*, IEEE Transactions On Advanced Packaging, Vol. 22, pp. 274-283, August 1999.
[3] Wiren Becker, Jim Eckhardt, Roland W. Frech, George A. Katopis, Erick Klink, Michael McAllister, Timothy G. McNamara, Paul Muench, Stephen R. Richter, and Howard H. Smith, *Modeling, Simulation, and Measurement of Mid-Frequency Simultaneous Switching Noise in Computer Systems*, IEEE Transactions on Components, Packaging, and Manufacturing Technology—Part B, Vol. 21, pp. 157-163, May 1998.
[4] Henry Hungjen Wu, Jeffrey W. Meyer, Keunmyung Lee, and Alan Barber, *Accurate Power Supply and Ground Plane Pair Models*, IEEE Transactions on Advanced Packaging, Vol. 22, pp. 259-265, August 1999.
[5] Larry Smith, Tanmoy Roy, and Raymond Anderson, *Power Plane Spice Models for Frequency and Time Domains*, Proc. EPEP Conference, October 2000.
[6] Ronald N. Bracewell, *The Fourier Transform and its Applications*, McGraw-Hill, Third Edition, 2000.

Chapter 20—Toward Developing a Standard for Data Input/Output Format for PDN Modeling and Simulation Tools

Ravi Kaw, Senior Staff Member, Agilent Technologies, Inc.
István Novák, Senior Staff Member, Sun Microsystems, Inc.
Madhawan Swaminathan, Professor, Georgia Institute of Technology

THIS CHAPTER EXPLORES the design and verification environment of power distribution networks (PDNs) in an attempt to point out areas for improving the tools and the methods. It also points out potential interfaces between PDN tools that can be standardized so that the models can be ported between various available tools.

Introduction

PDN design is a system level problem that includes elements of the system board, integrated circuit (IC) packages, and the ICs housed within those packages. In general, there are three basic requirements for PDNs that designers need to consider, either by simulations or by measurements:

- PDNs have to deliver sufficiently clean supply to the ICs
- PDNs have to provide low-noise reference path for signals
- PDNs should not radiate excessively

In order to make sure that the PDN delivers clean supply to active devices, one can and should simulate the time-varying voltage across the supply rail, which requires the knowledge of the currents and the (frequency-dependent) impedance profile of the PDN at all N points of interest. For two ports:

$$V_1(t) = invFFT\{Z_{11}(f)I_1(f)\} + invFFT\{Z_M(f)I_2(f)\}$$
$$V_1(t) = invFFT\{Z_M(f)I_1(f)\} + invFFT\{Z_{22}(f)I_2(f)\}$$

...where I(f) is the excitation current, V(t) is the resulting noise voltage and Z(f) is the 2x2 impedance matrix of the PDN.

An illustration of a PDN is shown in Figure 20.1.

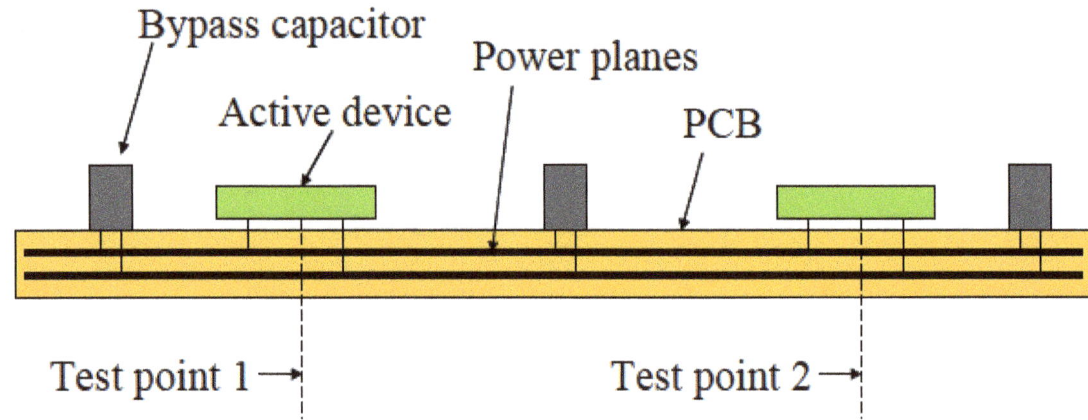

Figure 20.1: Sketch of a PDN with Two Test Points, PCB, Bypass Capacitors, and Active Devices

One aspect of the PDN simulation is to determine the i(t) current signature entering and/or exiting the PDN. Silicon designers may have the necessary information about the current signature, but for many silicon users, the current signature information is unavailable.

This leaves the only possibility for silicon users to measure and/or simulate the impedance profile [1] of the PDN and complete the equation with assumed i(t) equaling some percentage of watts/V [2], or stop short of calculating v(t) altogether and just stay with the impedance profile. The user needs the impedance matrix (or any other network matrix) as a function of frequency, for a number of ports, representing the noise sources and other points of interest: bypass capacitors, test points.

The reference path for signals is not an inherent function of the PDN but, in many designs, the PDN acts also as a reference to one or more signals. The PDN may be of significant size in terms of wavelength of the highest frequency of interest, and therefore a full-wave solution may be necessary to obtain the noise. Here the user may want to simulate the impact of reference-layer transitions, reference-plane changes over split planes, simultaneous switching noise (SSN) due to shared vias, due to finite plane resistance and inductance, including plane perforations [3].

In regards to radiation, the user again may be interested in the impedance profile of the PDN to avoid resonances that may get excited by the signals or noise sources [4]. There are some differences between cases (a) and (c): common-mode currents that may not create SI problems may create excessive radiation. Also, point-of-load PDN structures tend to have a progressively band-limited filtering as we move away from the active device through the package and onto the board, so high-frequency noise appearing on the board may not find its way back to the silicon, but it can create too much radiation from the board.

This design space includes an ever-growing mix of modeling tools, simulation tools, methodologies of design, evaluation metrics, and formats for transferring data across interfaces. So, the following discussion begins with a description of general elements of PDNs and metrics used to evaluate them. This is followed by a list of improvements required for modeling and simulation tools. A brief discussion is presented about the design methodologies used. This points to the lack of standards at the boundaries of silicon package and package board. The paper concludes with suggestions for such standards that can allow seamless exchange of design information across functional boundaries.

PDN Classification

PDNs can be classified into two categories: core PDNs and input/output (I/O) PDNs. The physical contents of core PDNs extend from the core switching networks and power/ground (P/G) grids that reside inside an IC, package P/G for core, core P/G network on printed circuit board (PCB), and voltage regulation module (VRM). Sockets and bypass capacitors on chip, on package, and on PCB also form parts of this network.

Cores of large application-specific ICs (ASICs) or central processing units (CPUs) may have a dedicated core PDN, feeding only one core called point-of-load circuits. In some designs, core P/G planes also act as reference planes for some of the signals, and therefore the return-path function of these PDNs must also be considered.

On the other hand, the I/O PDN generally includes signal delivery nets (SDNs). Its physical contents extend from the chip I/Os (including their P/G), on-chip bypass capacitors for I/Os, interconnections and redistributions, package and related PDN/SDN on the personal computer (PC) board, and VRM.

Sockets, connectors, and bypass caps at all stages are also included. Although it is not typical in today's designs, I/O PDNs feeding only the ICs I/O sections, without serving as signal reference, can also be constructed.

Finally, not only both classifications described above can be either pure PDNs or PDNs+SDNs, but there are PDNs that combine all of these functions: the same PDN may feed core(s) and I/O(s) and may also serve as signal reference path.

Core PDNs

Core PDNs consist of various elements: on-chip switching circuits with details that can be acquired from register transfer language (RTL) information and test vectors provided by a user; location(s) or potential locations of on-chip bypass capacitors and their value if already designed in; P/G grid (often provided in GDS format); package P/G structure and location(s) of bypass capacitors and their value if already designed; PCB P/G structures supporting the core P/G, including bypassing schemes (locations and values); and VRM(s). Test points may also be included as separate nodes in each structure.

An illustration of a possible physical realization is shown in Figure 20.2.

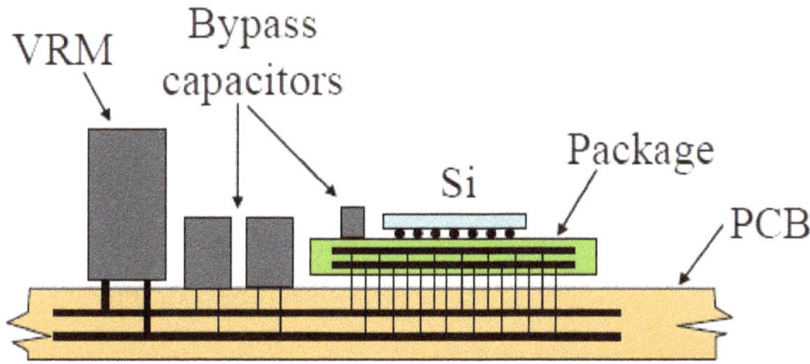

Figure 20.2: Illustration of a Core Point-of-Load PDN

On-chip switching activity depends on the circuit type and the logic vectors used. This core "current signature" is modeled in several ways. This data is provided by the silicon designer. Some tools use the core test vectors as a starting point. Due to the complexity of modeling the current signature, it is often measured under known conditions [5]. Core loading is best described as net-by-net resistance. This can be condensed for the entire core or divided by circuit blocks. Some vendors use statistical models as well [6].

Others have proposed the use of a Gaussian current pulse [7].

Bypass capacitor values and location should also be supplied by the chip designer. This should include non-switching gates that act as native bypass caps, which is switching pattern–sensitive. On-chip P/G grids can be modeled as RC circuits, RLC circuits, or more comprehensive element manager (EM)-based broadband circuits (tool development effort is required for some of this) [8]. Three commercial tools offer modeling capability of this structure with varying degrees of sophistication [9-11].

Package P/G nets can be modeled with several commercial tools available now. Most modeling tools use a single frequency for extraction. Modeling tools need some upgrades, and these have been listed later in this document. At least two modeling tools can create wideband models of the package and the PCB, either separately or together. One preferred PDN design methodology suggests use of separate models for each package and the PC board that can be put together in the simulator [12]. At the systems level, the individual package models should be simplified into a simple circuit with all bumps shorted and all balls shorted. Sometimes it is necessary to expand the model to provide finer granularity.

Figure 20.3 shows a simplified lumped equivalent circuit of the core PDN as seen from the silicon. The IC core transient current is represented by the current source on the right. The bandwidth of both the model and the transient noise is the widest at this point, extending way into the GHz region. The IC distribution with the interfacing package impedance creates the first major filtering, where the bandwidth usually drops below 1 GHz. Through large packages, the series distribution and the attached package capacitors further limit the bandwidth to the low MHz to few times ten MHz range. The board horizontal impedance with the bulk and mid-frequency capacitors create the next filtering step, reducing the bandwidth to the kHz range, which bandwidth eventually has to be handled by the VRM.

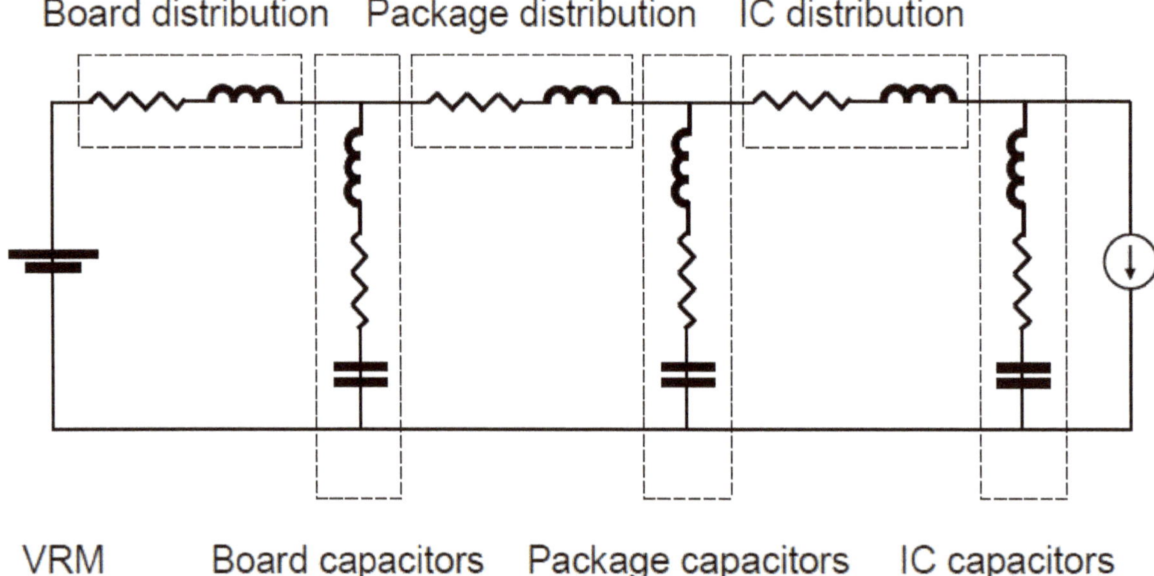

Figure 20.3: Simplified Lumped Equivalent Circuit of a Point-of-Load Core PDN

The output data of impedance versus frequency can be based on measurements also, although this can capture only a limited condition of logic activity. Still this is a good starting point. The output should be compatible with popular simulators such as SPICE, although there is a growing need for faster simulators [13].

In the case of large data file, macro-models may be used [14]. Because the PDN physically encompasses a big part of the system, noise appearing on the PDN may create not only signal-integrity issues, but also electromagnetic compatibility (EMC) problems.

Capturing the near-field and/or far-field radiation from a PDN with complex geometry is a very challenging task. As a first step in preventing EMC issues, the key requirement is properly capturing potential structural resonances.

Metrics for Evaluating Core PDNs

The generally accepted metric for the design of core power is impedance versus frequency. Alternately, one can use voltage ripple and current signature [2].

Figure 20.4 shows illustrative core–PDN impedance profiles for a high-power CPU. Trace "a" plots the self-impedance magnitude at the board-package interface, which is the parallel of the board impedance and package + silicon impedance. The peak at 100 MHz comes from the die-package resonance [15], and it is characteristic to most large package applications. The smaller peak at three decades lower frequency may be the result of the PCB-to-package inductance resonating with the on-package capacitance.

Trace "b" shows the same PDN at the silicon-package interface.

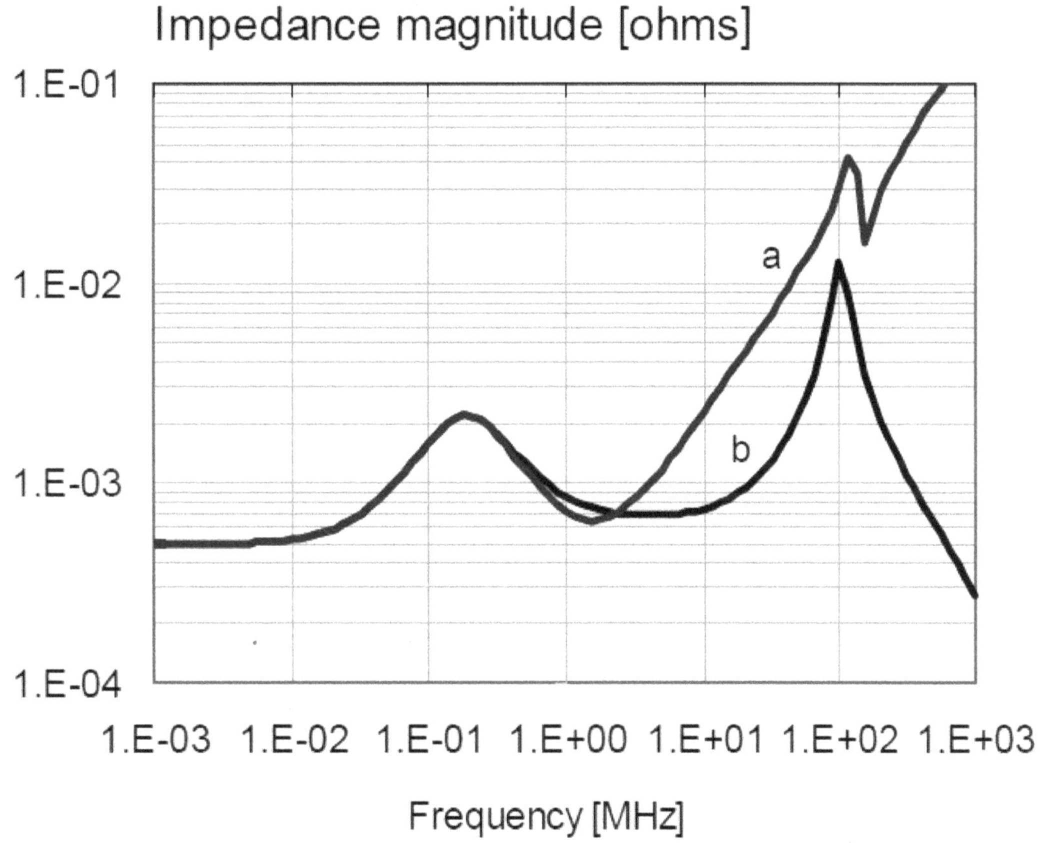

Figure 20.4: Illustrative Impedance Profiles of a High-Power CPU Core. Trace a: Impedance of Core PDN at the Board-Package Interface. Trace b: Core-PDN Impedance of the Same Network at the Silicon

Let us note that besides self-impedance profiles, transfer impedance curves are also useful for electromagnetic interference (EMI) purposes in determining the amount of core-clock leakage from the IC to the PCB.

I/O PDN + SDN

Power supply noise is dependent on the return currents on the planes. Therefore, the I/O PDN discussion also includes SDNs.

The elements of I/O PDNs involve the following:

- On-chip switching circuits (e.g., I/Os)—this can be in the form of driver models or an extracted net that captures the related I/O structures as well. These files can become voluminous. One fix is to create macro-models that can support the various nuances of the drivers, including variable drive strengths. A key item in these extracted models is the pre-emphasis. That should be included as well. Some of this sophisticated macro-modeling is still being developed at university level [16] and has yet to be commercialized. Changes have been proposed to the IBIS models to capture some of the complexities of modern I/Os [17].
- On-chip bypass, both add-on and native (supporting the I/Os), extracted by the user. The native capacitance is pattern-sensitive.
- On-chip P/G grid (supporting the I/Os).
- Package P/G and I/O nets, plus bypass caps.
- PCB P/G structures, signal nets, and bypass schemes used.
- Connectors for signal nets on the PCB and for P/G (if used).
- Far-end package I/O structures and input circuitry of the receiver circuit. This should include the terminations, the ESD network, etc. Again, a macro-model may be a useful substitute for a large extracted net.
- VRM(s).

Most comments listed under core PDN apply to I/O PDN as well.

The I/O power and signals interact nonlinearly via the drivers and should preferably be modeled and simulated together. Both frequency-domain and time-domain simulations are required. It is also preferable to have broadband models for both PDN and SDN structures.

The return currents on the planes generate the coupling between the PDN and the signal lines. Hence, signal referencing is a very important constituent of I/O design in the package and board.

Providing appropriate path for the return currents translates into the assignment for power and ground at the package-board interface. Since most ICs today support thousands of signal I/Os with comparable power and ground I/Os, thousands of interconnects need to be analyzed in the package to assess the impact of power supply noise on eye diagrams. To complicate matters further, the PDN consists of multiple plane layers containing thousands of vias and loaded using hundreds of capacitors. Although the complexity of the problem is enormous, approximations based on the understanding of the return currents can greatly simplify the problem to be solved. Macro-modeling is one of many methods that can be used to simplify the problem.

An example is shown in Figure 20.5, where a multilayered PDN is first modeled using an electromagnetic solver and the frequency response at specific points represented using a macro-model in SPICE. Power supply noise can be simulated by using transmission line models of interconnects referenced to the macro-model. Since the distributed nature of the PDN is captured in the macromodel, this method does not degrade the accuracy of the results.

Moreover, nonlinear macro-models can be connected to the signal lines to improve the accuracy of the simulations.

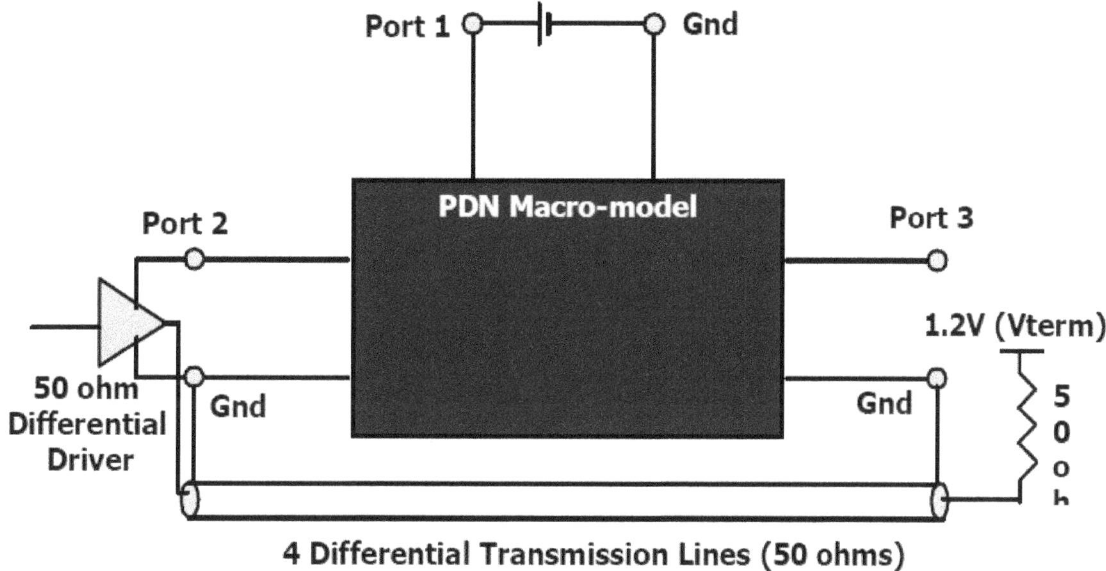

Figure 20.5: PDN and SDN Modeling Using Macro-Models

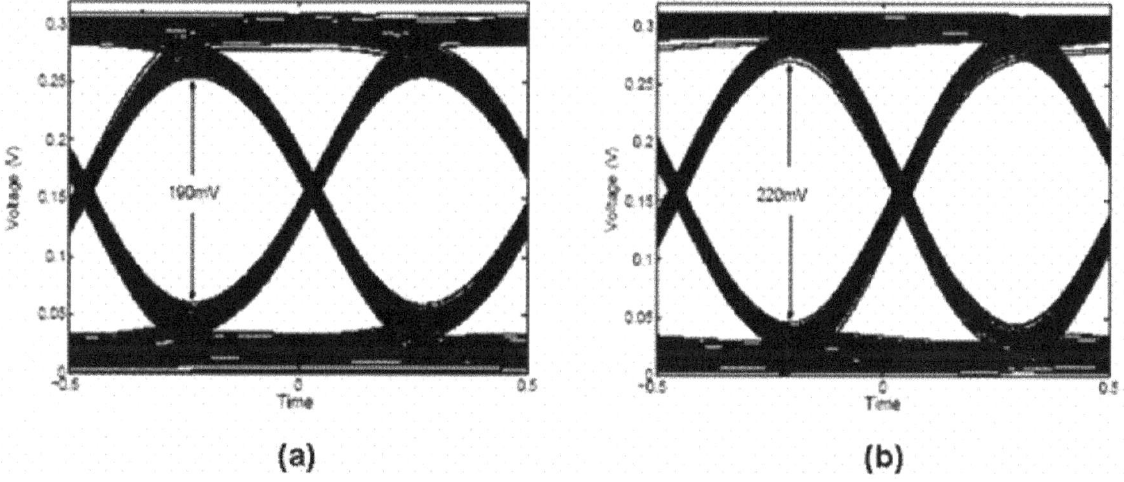

Figure 20.6: Impact of Power Supply Noise on Eye Diagrams (a) before and (b) after Causality Enforcement

Though macro-modeling provides a method for coupling the PDN and SDN models, it has limitations. Most macro-modeling methods are limited to a finite number of ports. Even if this limitation is overcome, the spice netlist for a PDN macro-model with many ports can become unmanageable.

In addition, since macro-models are generated using band-limited frequency data, these models can violate causality.

Macro-models violating causality can lead to the artificial closure of the eye, as shown in Figure 20.6 where a 30 mV voltage reduction is seen when causality is violated using macro-models [18].

Metrics for Evaluating I/O PDNs

- Eye diagram—the standard for this is best described by the user, depending on their requirements. Since the power supply noise on the PDN couples into the signal lines, the eye diagrams are a good metric for evaluating the impact of noise on signal propagation.
- Logic failure—this has been suggested by some, but is not commonly used. This is a more difficult problem to simulate, since logic failure mechanisms vary between systems.
- Delay—this has been suggested by some but is not commonly used. A more important parameter could be jitter.
- Impedance versus frequency—similar to the core PDN, impedance could be defined for the I/O power delivery network. However, since the return currents dictate the noise on the power planes, impedance at multiple points need to be computed.

The starting point for PDN design is always a direct current (DC) design, which requires a DC resistance calculator and has terminals that should be defined by the user. This may simply condense all the sources (solder-bumps or wire-bonds) and all the sinks (balls or pins) into a two-terminal net for each power and ground net, or break them into groups specific to some circuit blocks or simply group them into geographical regions.

It should be noted that DC resistance is not modeled by every modeling tool. Some tools try to mimic DC conditions by choosing a frequency at which the skin depth is equal to half the trace thickness, but this is not representative of the many via sizes or plane thickness, which may be different from the trace thickness.

Proposed Requirements for Modeling Tools

A very important requirement for modeling the PDN is the ability to analyze these structures at multiple levels and pass information between levels. An example is shown in Figure 20.7 consisting of the behavioral level, transistor level, and physical level.

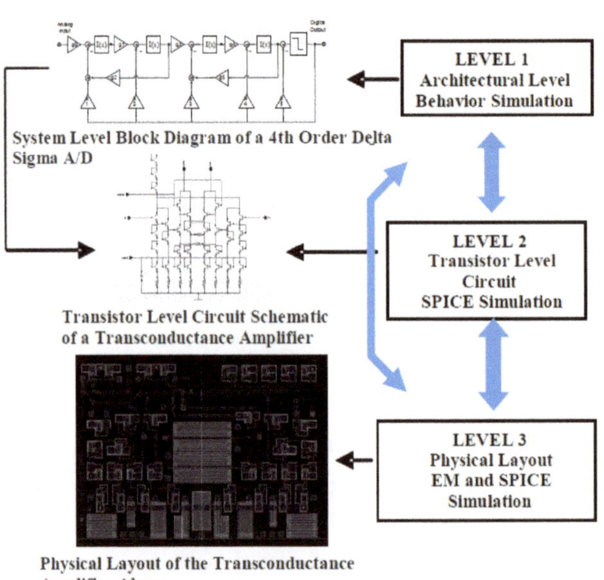

Figure 20.7: Hierarchy in PDN Analysis

The behavioral level captures the architectural details of a microprocessor such as current drawn, frequency, and power with compact PDN models. The transistor level consists of spice simulations with nonlinear circuits and circuit models of the PDN.

The physical level contains the three-dimensional structures of the PDN with detailed analysis using electromagnetic simulators. One method for passing information between levels is through both linear and nonlinear macro-modeling. Although system definitions and designs tend to follow these three options, currently, none of the commercial tools support these three levels. Most tools are confined to just one level.

The following is a wish list of upgrades that could be incorporated:

- Ability to download stackup info from various design tools and include a reasonable library of material properties.
- Support rapid extraction of frequency-dependent impedance of multiple P/G plane stackup for quick PDN PDN+SDN analysis.
- Support "condensation/merging" of P/G pins and/or vias when needed to simplify models.
- Support de-embedding of reference nodes inside a structure that does not need to be connected to the global ground; e.g., bypass cap nodes. This is especially true of tools that yield S-parameters.
- Accept incorporation of electrical models for decoupling caps and other passive elements when modeled together.
- Generate frequency-dependent impedance for all structures (IC, package, and PCB) to include PDN+SDN for IO power.
- Be output-compatible with network simulation tools.
- Have ability to convert between S, Y, Z, and ABCD formats, and also from frequency-dependent model parameters to equivalent time-domain sub-circuits without causality problems.
- Model PCBs, packages, on-chip P/G grid, sockets, connectors, VRMs (could be several tools).
- Capable of multiport macro-model generation to reduce complexity.
- Calculate bandwidth of validity for the user and provide methods to enhance it.
- Must be flexible enough to model portions of a system (PKG by PKG, or portions of PCB, or PKG + small portion of PCB around it) so that models of either can be used elsewhere.
- Should deliver output that can be used in a simulation engine elsewhere (outside the tool's framework) in a standard format.
- Have sensitivity analysis based on tolerances of geometry and material properties, which would be helpful in a combined modeling/simulation tool.
- Incorporate effect of trace edge shape; otherwise error in Zo and crosstalk of up to 10 percent or more can result [19].
- Use actual shape of via structures.
- Include all couplings (e.g., between traces, between vias).
- Include effective and accurate ways to account for "holy" planes, splits, irregular plane outlines, and large internal cutouts.

An EM-based numerical method is required to model the linear network consisting of planes, decoupling capacitors, vias, and other power distribution interconnect structures. Since the target impedance is a measure of the performance, the output is impedance parameters varying as a function of frequency.

This could be 1-port, 2-port, or n-port parameters, but it is advantageous to limit the number of ports. Modeling tools have evolved over time; however, most lack one capability or another.

Proposed Requirements for Simulation Tools

Most designers use simulation tools such as HSPICE or ADS. New simulation tools have also emerged ([9], [10]). The following are some requirements that are not fully incorporated into any of these tools:

- Fast and reasonably accurate simulators to handle large buses along with their P/G network (PDN+SDN).
- Certain design methodologies require these simulators to handle a very large number of nodes.
- Accept lumped as well as behavioral (macro-models) models compatible with IBIS-4. Pre-emphasis would be helpful and often quite necessary.
- Support time-domain and frequency-domain analysis with built-in software for conversion of waveforms between them.
- Should be able to handle required bandwidth. Allow user to test for passivity and bandwidth.
- With the shrinking noise margins in presence of structures with geometrical tolerances that are rather large (5 to 10 percent), statistical simulation methods other than Monte Carlo type are needed to get faster results [20].

Methodology

Modeling, simulation, and analysis of PDN is done in several ways, all depending on the size and type of system. Some tools include all these functions and are therefore used for small systems with one or two packaged ICs [9]. Others model packaged parts and PC boards separately and then place them all together in the simulator [12].

In other words, methodology is usually user-defined. The format of interface data used also varies by user: some use impedance matrices, others use transmission matrices based on ABCD parameters [21]. As a result, conversion between Z, Y, ABCD, etc., is necessary for all modeling tools. PDN design can get too complicated in a hurry due to the enormous number of nodes involved. This requires judicious choice of reduction of the problem based on an understanding of the macro-space (board-level) requirements versus micro-space (chip-package-level) requirements.

The IC design community learned by default about the importance of standards for data transfer, which involves data format as well as rules of interpretation of that data across tool boundaries. Unfortunately, it is so complicated that every design house must have a "methodology team." Hopefully that will not happen to the PDN design community if we adopt standards early.

Standards

One of the biggest stumbling blocks at the moment is the lack of a standardized interface definition at the boundaries (silicon package and package board) [22].

If we had a standard way to exchange simulation information at these boundaries, the "owners" of each section could go out and develop better simulation tools without the risk of being incompatible with tools for the other sections. It is instructive to think about the silicon-package-board-package-silicon path as a high-speed serial link, where we have several good examples of having interface standards. All of the serial-link standards have compliance points and specifications that people have to maintain at those points, together with the definitions of the interfaces. This allows a cable manufacturer to develop a compliant cable without knowing the board and silicon details. The PDN is physically more complex because it is a multi-node network.

The challenge here is to find the proper level of abstraction to describe the interfaces so that they capture all of the important aspects (such as a wide-area array package-board or package-silicon connection) and to maintain a simple and generic form. For each interface we may want different complexity levels; the package-board interface can be rightfully modeled with a single node when only the silicon is simulated, and similarly the silicon-package interface can be rightfully modeled as a single lumped node when only the board is simulated. Specific suggestions have been made for specifying broadband target impedance in the frequency domain at each interface. Similar suggestions have been made for specifying current signature and ripple voltage in the time domain.

Conclusion

Core and I/O PDN designs require complex tools, which should be capable of doing co-analysis of silicon, package, and board. Standardized interface definitions and data input/output formats are required to enable further developments of these tools.

Acknowledgments

The authors thank all IEEE TC-12 team members and all authors of [22], as well as Jiayuan Fang (Sigrity) and Ching-Chao Huang (Optimal), for their valuable contributions. For more information, visit the IEEE TC-12 Web site: www.ewh.ieee.org/soc/cpmt/tc12.

References

[1] A. Waizman et al., *Integrated Power Supply Frequency Domain Impedance Meter (IFDIM)*, Proceedings of the 13th EPEP, October 25-27, 2004, Portland, Oregon.

[2] L. Smith, et al., *Power Distribution System Design Methodology and Capacitor Selection for Modern CMOS Technology*, IEEE Tr. AdvP, August 1999, pp. 284-291.

[3] J. Miller, et al., *Modeling the Impact of Power/Ground Via Arrays on Power Delivery*, Proceedings of the 13th EPEP, October 25-27, 2004, Portland, Oregon.

[4] I. Novák et al., *Impedance and EMC Characterization Data of Embedded Capacitance Materials*, Proceedings of APEX2001, January 16-18, 2001, San Diego, California.

[5] R. Mandrekar et al., *Extraction of Current Signatures for Simulation of Simultaneous Switching Noise in High Speed Digital Systems*, Proceedings of the 12th EPEP, October 27-29, 2003, Princeton, New Jersey.

[6] Shen Lin et al., *Full-chip Vectorless Dynamic Power Integrity Analysis and Verification Against 100uV/100ps- Resolution Measurement*, Proceedings of CICC 2004, Oct. 3-6, 2004, Orlando, Florida.

[7] B. Garben et al., *Frequency Dependencies of Power Noise*, IEEE Tr. on AdvP, May 2002, pp. 166-173.

[8] Jae-Yong Ihm et al., *Comprehensive Models for the Investigation of On-Chip Switching Noise Generation and Coupling*, Proceedings IEEE EMC05 Symposium, Chicago, Illinois.

[9] www.sigrity.com.

[10] www.apache-da.com.

[11] www.ansoft.com.

[12] Om P. Mandhana and Jin Zhao, *Power Delivery System Performance Optimization of a Printed Circuit Board with Multiple Microprocessors*, Proceedings of ECTC-2004.

[13] Yuzhe Chen et al., *A New Approach to Signal Integrity Analysis of High-Speed Packaging*, Proceedings

of EPEP 1995, 2-4 October 1995, pp. 235-238.

[14] S. H. Min and M. Swaminathan, *Construction of Broadband Passive Macromodels from Frequency Data for Distributed Interconnect Networks*, IEEE Tr. On EMC, Vol. 46, No. 4, pp. 544-558, November 2004.

[15] L. Smith et al., *Chip-Package Resonance in Core Power Supply Structures for a High-Power Microprocessor*, Proceedings of IPACK'01, July 8-13, 2001, Kauai, Hawaii.

[16] M. Swaminathan, *Macro-Modeling of Non-Linear I/O Drivers using Spline Functions and Finite Time Difference Approximation*, Proc. of the 12th EPEP, October 27-29, 2003, Princeton, New Jersey.

[17] S. B. Huq et al., *BIRD95: Power Integrity Validation Using HSPICE*, IBIS Summit, January 31, 2005, Santa Clara, California.

[18] R. Mandrekar and M. Swaminathan, *Causality Enforcement in Transient Simulation of Networks through Delay Extraction*, to be presented at the Signal Propagation of Interconnects Workshop, May 11-13, 2005, Germany.

[19] Ravi Kaw et al., *The Effect of Metal Edge Profiles on the Accuracy of Electrical Modeling of Advanced Packages*, submitted to IEEE Tr. on AdvP.

[20] E. Matoglu, et al., *Statistical Signal Integrity Analysis and Diagnosis Methodology for High Speed Systems*, Tr. On AdvP, Vol. 27, No. 4, pp. 611-629, November 2004.

[21] J. H. Kim, M. Swaminathan, *Modeling of Irregular Shaped Power Distribution Planes using the Transmission Matrix Method*, IEEE Trans. on Components, Packaging and Manufacturing Technology—Advanced Packaging, pp. 334-346, Vol. 24, No. 3, August 2001.

[22] R. Kaw, S. Camerlo, A. Waizman, J. P. Miller, J. Fan, I. Novák, J. Drewniak, *Board and Package Level PDN Simulations*, panel discussion at DesignCon 2004, Feb 2-5, 2004, San Diego, California.

Chapter 21—Overview of Frequency Domain Power Distribution Measurements

István Novák, Senior Signal Integrity Staff Engineer, Sun Microsystems

This chapter is a shortened and revised version of the paper presented at DesignCon 2003 East, Boston, Massachusetts, June 23-25, 2003.

IN TODAY'S ADVANCED digital systems, the power distribution networks (PDNs) often have to deliver hundreds of watts at low voltages, and the required low impedance of the power distribution must be maintained over a wide frequency band. Traditional high-frequency measuring instrumentation has been tailored to handle impedances close to 50 Ohms. The very-low-impedance values in the PDNs create measurement and calibration challenges.

This chapter explains the benefits of the frequency-domain method and impedance measurements with two-port virtual network analyzer (VNA) setups. Extraction of component parameters, calibration options for different instruments and frequency ranges, probe connections and constructions are explained. Various compensation methods are described, and examples are given to measure PDN components from single elements to full working systems. The concept of attached impedance/inductance of bypass capacitors is introduced, and we show data supporting the claim that for the high-frequency inductance of bypass capacitors, the cover thickness of the part matters and the total height of the part does not (see the original full-length paper [1]).

Why Frequency Domain?

The PDN of digital circuits has to feed the chips with direct current (DC) power. The switching circuitry creates current transients, which generates voltage fluctuations across the PDN impedance. The voltage fluctuations (transient noise) must be kept below a predefined limit so that it does not interfere with the signaling.

For each signal line, a time window can be identified within which any supply-rail noise will reduce the noise margin. In synchronous systems, this is the window defined by the setup and hold requirements. Outside the sampling window, the noise will not harm the signal, except extra-large noise levels may drive the attached devices into nonlinear regions or even cause breakdown in low-voltage devices.

However, as noise propagation from multiple noise sources may be too complex to analyze in detail, a conservative approach assumes that the specified supply-rail noise limit should be kept at all times.

Because digital signaling is defined by its voltage or current levels in the time domain, it may seem obvious to characterize the behavior of the PDN also in the time domain. While the characterization itself is doable in the time domain, validation in a real system in the time domain is very hard.

Here is why.

Having multiple active devices connected to the same PDN rail, the noise voltage at each location becomes the sum of the products of the appropriate self- and transfer impedances and noise currents.

Let us assume we have a circuit where the PDN consists of a printed circuit board (PCB), one power-ground plane pair, three bypass capacitors, and two active devices. For sake of simplicity, let us assume that we are interested in the noise voltages at the active devices only, and the size of the devices and the highest frequency of interest allow us to use only one test point for each active device. We can further assume that we know from simulations or from measurements the impedance matrix of the PDN for the two test points. With very few exceptions, the PDN components are electrically reciprocal; therefore, $Z_{12} = Z_{21} = Z_M$. Electrical symmetry, however, cannot be assumed in a generic case; therefore Z_{11} and Z_{22} are, in general, different.

The noise voltages at test points 1 and 2, generated by the noise currents of $I_1(t)$ and $I_2(t)$ of the two active devices, can be expressed as:

$$V_1(t) = invFFT\{Z_{11}(f)I_1(f)\} + invFFT\{Z_M(f)I_2(f)\}$$
$$V_1(t) = invFFT\{Z_M(f)I_1(f)\} + invFFT\{Z_{22}(f)I_2(f)\}$$

If we want to close the PDN design cycle with validation in the time domain, we have to know both the impedance matrix and the transient noise vector. The PDN is built of passive components (except the voltage regulator modules); therefore, the PDN's impedance is less subject to statistical variations due to component tolerances. With well-behaving components, it is certainly not much time varying. In contrast, the transient noise in a complex system comes from many packages of active devices, and the sources have their own timing and activity schedule. On a large board, there may be hundreds of packages with thousands of active cells switching. Predicting or simulating the worst-case transient noise in the entire system is a daunting task. It is daunting, because transfer impedances and propagation delays among the noise sources and test points are usually not negligible, so the worst-case maximum transient noise will not necessarily occur when and if all the sources switch simultaneously at the same time.

If instead we wanted to find the transient current by measurements, the difficulty is very practical: measuring current in a small-pitch PCB is not easy, as it would require a shunt element in series to each current path, or a current-measuring loop around each current-carrying conductor to be measured. And, even if we could measure the current, it does not remove the problem stemming from the highly statistical nature of the noise current. Recently several publications described ways to indirectly measure the impedance profiles and/or transient currents (e.g., [2] and [3]), which assume that the active devices can be exercised in a controlled manner.

Although eventually we may want to measure and validate the noise in the time domain, it is more straightforward to segment the task and do the design and validation separately for the impedance matrix, followed by time-domain measurements/validations if necessary.

A suggested new methodology starts with the step transient responses corresponding to the test points and calculates the absolute worst-case transient noise magnitude [4]. In this case, obtaining the impedance matrix by measurements or simulations is sufficient because the step responses can be obtained from the impedance matrix through straightforward calculations. This way, we can eventually get the worst-case estimated noise without doing extensive time-domain measurements.

Finally, yet another reason suggesting frequency-domain instead of time-domain measurements is the fact that any random noise from the environment can be suppressed more readily in frequency domain. In fact, VNAs operate with narrow bandwidth in synchronous mode, so averaging will bring out the useful signal and suppress random noise. When we try to measure the worstcase transient noise instead, probably with an oscilloscope in infinite persistence mode, any random external noise getting into the measurement setup will corrupt our data.

How to Measure PDN Impedance

With increasing power levels and constantly dropping supply voltages, the PDN impedance of high-power systems must be low, sometimes in the milliOhm range. The low impedance values themselves create unique challenges in the selection of instruments, setups, and connections.

To measure the impedance of an unknown device over a wide frequency range, we can use inductor-resistor-capacitor (LRC) meters, impedance bridges, or VNAs. With just one port (two connection points), any of these choices will be limited to impedance values of a few hundred milliOhms or higher and inductances of a few hundred pH or higher.

The reason for this limitation is the practical difficulty of creating connections from the measuring instrument to the unknown impedance. One millimeter (about 40 mils) of wire can represent a $Z_{connection}$ impedance of a milliOhm or more resistance at low frequencies and hundreds of pH inductance at high frequencies. We can attempt to remove this extra impedance by calibration, but we will face a difficult task in defining the geometry and connection points with such a high accuracy that the calibration/de-embedding could be effective (for instance by using calibrated wafer probes) at high frequencies and low impedance values.

Figure 21.1 shows that with one-port impedance measurements we eventually always measure $Z_{DUT} + Z_{connection}$.

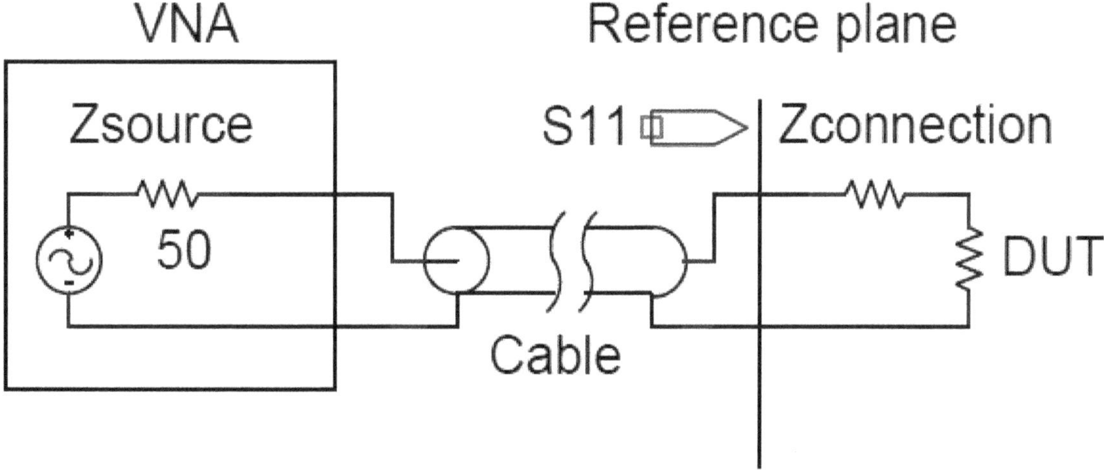

Figure 21.1: Setup with VNA, for One-Port Impedance Measurements. One-Port Impedance Measurements Are Limited to a Few Hundred MilliOhms Mid-Frequency Error

Why Two Ports?

On the other hand, VNAs offer a convenient way of measuring low impedances, similar to the four-wire DC resistance measurement setup.

As shown in Figure 21.2, two-port VNA connections use Port 1 to launch a known signal current into the unknown impedance, and Port 2 is used to measure the voltage drop. The same extra $Z_{connection}$ impedance that appears directly in series to the unknown impedance in one-port VNA measurements is now transformed into the two loops of VNA ports, each having a nominally 50 Ohm impedance. Although we now have two connections to care for instead of one, the connections to the device under test (DUT) will introduce very little error up to several GHz frequencies. For the error analysis on self- and transfer impedances, see [4].

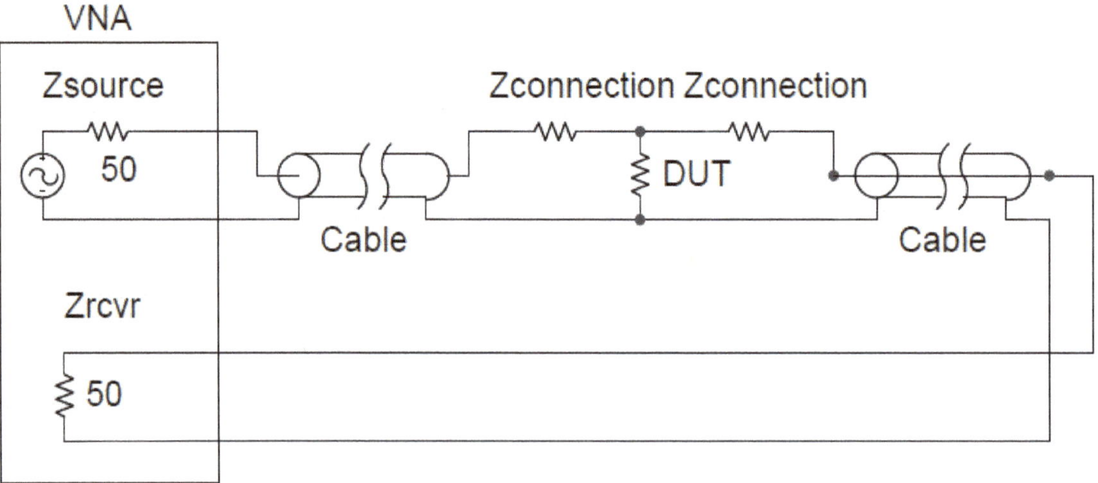

Figure 21.2: Two-Port Shunt-Through Setup with VNA. This Setup Greatly Reduces the Error Due to the Series Connection Impedance by Moving Them into the 50-Ohm Loops of VNA Ports

Connection between Self and Transfer Impedances

By using the two-port connection of VNAs for impedance measurements, we always literally measure transfer impedance between the connections from Port 1 to Port 2. If, however, the two ports are connected to the same (or almost the same) location on the PDN, the transfer impedance becomes self-impedance. This creates a single and unified way to measure both self and transfer impedances with the same instrumentation and setup.

The top graphs in Figure 21.3 illustrate the gradual change of transfer impedances into self-impedance over a bare pair of square planes. The figure assumes a pair of 10" x 10" square bare plane pair with 2-mil plane separation and dielectric constant of 4. To show the effects clearly, dielectric and copper losses are ignored.

The impedance curves are shown with one port always in the middle, while the other port moves away along the diagonal by the following distances along the x and y coordinates from the center: 0.25", 0.5", 1", 2", 3", 5".

Note that at low frequencies, all curves follow the impedance of the static plane capacitance. At higher frequencies some of the curves hit a sharp minimum and other curves go through a shallow bottom. Because of the loss-less assumption, all curves go through the same modal resonance peak values. When losses are not negligible, the modal resonance peaks depend slightly on the location on the planes. On the right, the same set of curves is shown on zoomed horizontal and vertical scales, with labels identifying the individual curves. Label 0" indicates self-impedance at the center; label 5" shows the transfer impedance between the center and corner.

The labels show port separation along the x and y coordinates; the line-of-sight probe spacing is $\sqrt{2}$ times larger.

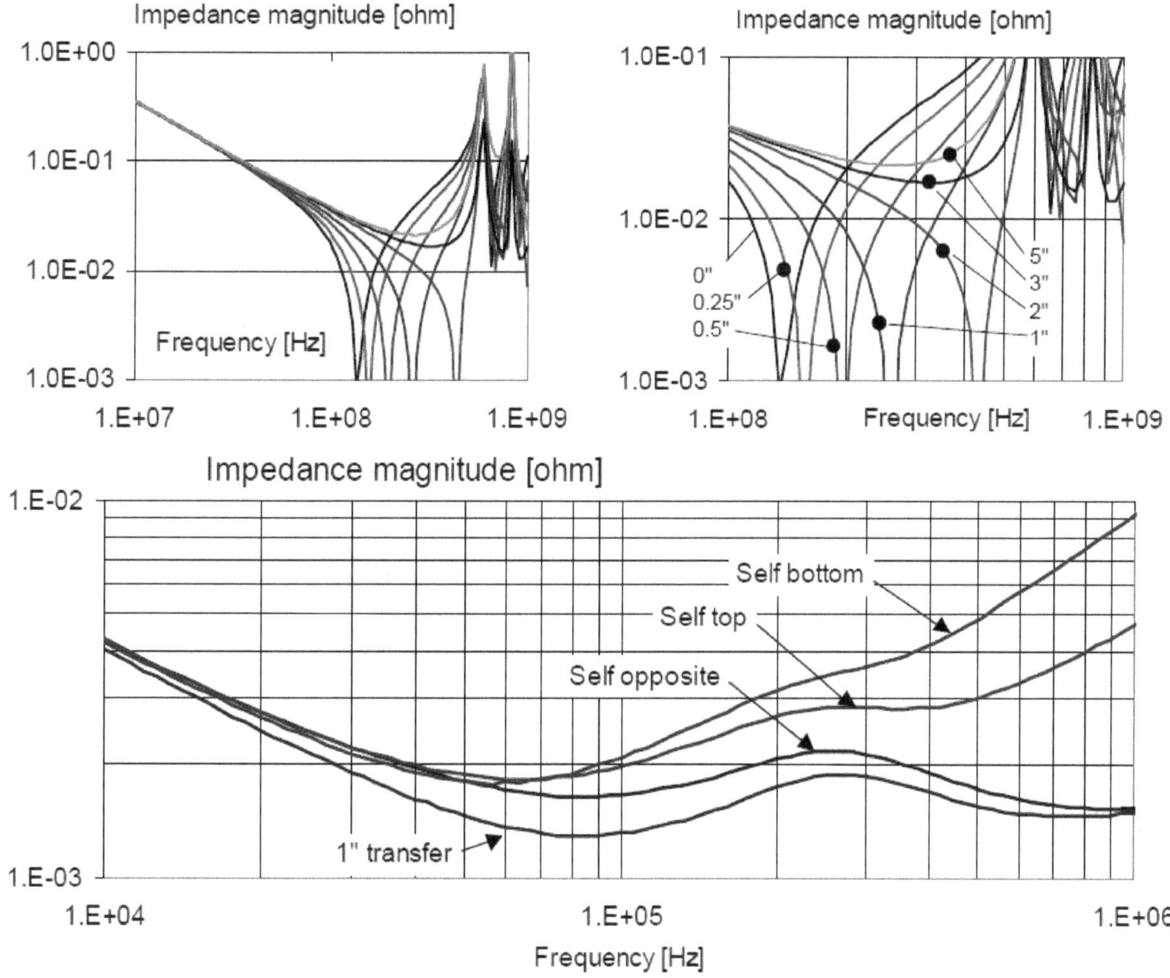

Figure 21.3: Top Graph: Simulated Self and Transfer Impedances on a Bare Pair of Power Planes. Left Graph: Overall Picture. Right Graph: Zoomed Horizontal and Vertical Scales, with Port Connection Spacing Labeled. Bottom Graph: Measured Self and Transfer Impedances at Low Frequencies

At lower frequencies, the usual assumption could be that self and transfer impedances may not differ significantly, as long as the spatial dimensions are much smaller than the shortest wavelength of interest. At very low impedance values, however, the series losses of the distribution network together with the bypass capacitors may create different impedances at and between various points.

This is illustrated on the measured low-frequency plots on the bottom of Figure 21.3. There were several large bulk capacitors on the plane, creating the shallow minimum around 100 kHz. The power/ground plane pair was close to the top in the stackup. The DC-DC converter was not powered up; hence we see the capacitive slope between 10 and 100 kHz. There was one pair of test vias, where the VNA ports were connected.

The curve labeled "self bottom" was measured with both VNA ports connected to the bottom pads on the test via pair, thus having the longer part of the via loop in series to the PDN impedance. At 1 MHz, the equivalent inductance is approximately 1.3 nH. The curve labeled "self top" was measured with both VNA ports connected to the top pads of the test via pair, thus going through the shorter part of the via loop in series to the PDN impedance. The equivalent inductance at 1 MHz was 680 pH. The curve labeled "self opposite" was measured with the two VNA ports connected to the opposite sides of the test via pair. The equivalent inductance at 1 MHz is 100 pH. The curve labeled "1" transfer" was measured with one VNA port connected to the test via pair, the other VNA port connected to a via pair 1 inch away.

Note that at 100 kHz, where the wavelength is 200 meters, a 1" (0.0254 meter) distance results in an impedance drop from 1.6 to 1.3 milliOhms. This happens because the low ESR of the bulk capacitors forms an attenuator with the series resistance of the planes.

Measuring Very Low Impedance Values

As long as the unknown impedance magnitude is much lower than 25 Ohms, the complex voltage divider of Figure 21.2 between the two 50 Ohm VNA connections and the unknown Z_{DUT} impedance can be simplified, and the magnitude of the unknown impedance can be approximated by:

$$Z_{DUT} = 25\ x\ S_{21}, S_{21} = 10^{\frac{S_{21}^{[db]}}{20}}$$

If we use the Bode-plot style output from the VNA, the dB values first have to be converted to ratios, as shown by the second expression in Equation 3.

If we do not need high accuracy or the phase of the unknown impedance, there is hardly any need for calibration with certain VNA types at low frequencies. We just have to establish the nominal full-scale reading as 25 Ohms.

Measuring Arbitrary Self-Impedances

The impedance of an inductor or capacitor varies linearly or inversely with frequency, so chances are that sooner or later the impedance magnitude will not be much smaller than 25 Ohms. In such cases we have to solve the voltage divider expression for the unknown impedance.

With Z_{VNA} (usually 50 Ohms) connecting impedance to both VNA ports, the unknown Z_x impedance can be expressed as:

$$Z_X = \frac{Z_{VNA}}{2} \frac{S_{21}}{1 - S_{21}}$$

To get the correct answer, we cannot avoid calibration, because we also need the phase information to solve the complex equation. For medium accuracy, a simple through calibration is usually sufficient. To obtain the highest possible accuracy, a full two-port calibration should be done.

Due to the robustness of the two-port impedance measurement scheme, small discontinuities in the connections to the DUT can be neglected up to several GHz frequencies [3].

By rearranging the Z_X equation, the real and imaginary parts of the unknown impedance can be expressed as:

$$Re(Z_{DUT}) = 25\ x\ \frac{Re(S_{21})\ x\ (1 - Re(S_{21}) - Im(S_{21})^2}{(1 - Re(S_{21}))^2 + Im(S_{21})^2}$$

$$Im(Z_{DUT}) = 25\ x\ \frac{Im(S_{21})}{(1 - Re(S_{21}))^2 + Im(S_{21})^2}$$

Note that these two equations can be easily programmed in a spreadsheet that takes the measured VNA data.

The two-port shunt-through connection is useful to measure low impedances. If and when the unknown impedance is high, it can be placed in series to the instrument ports, creating a two-port series through connection scheme. The unknown impedance can then be expressed as:

$$Z_x = 2Z_{VNA}\frac{1 - S_{21}}{S_{21}}$$

The graphs of Figure 21.4 show the extracted impedance magnitude and phase measured on a small-size PCB with three paralleled plane pairs with and without complex solution of the voltage divider.

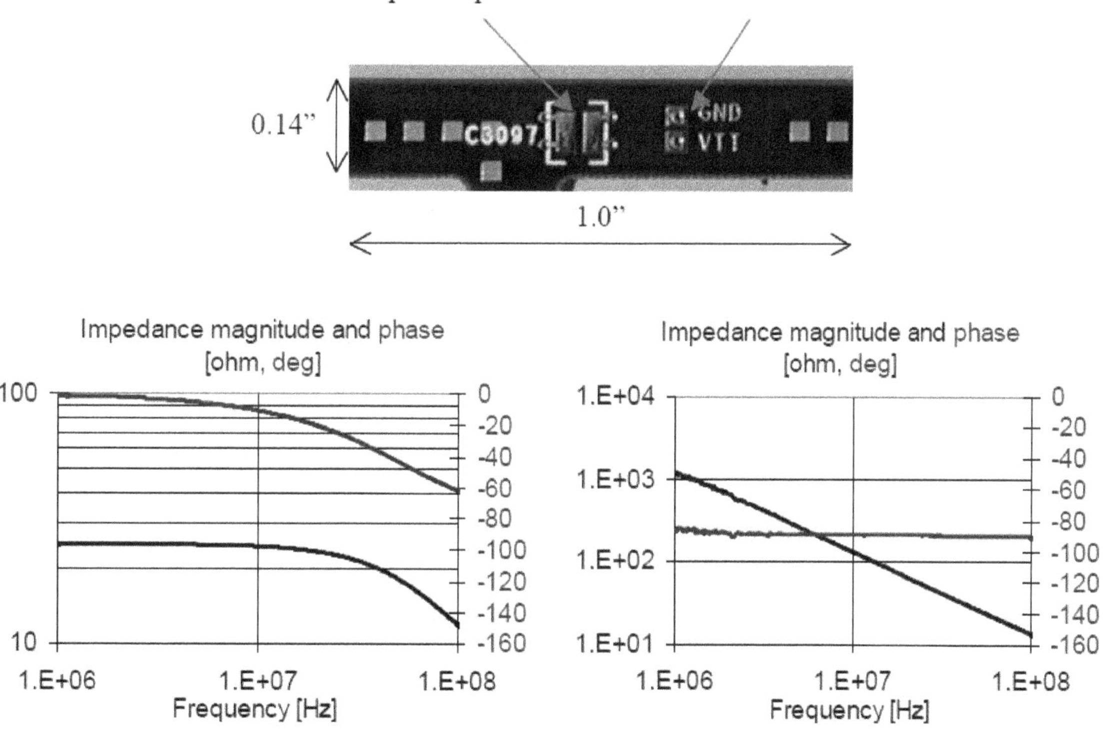

Figure 21.4: Measured Impedance Magnitude and Phase of Three Paralleled Bare Plane Pairs of 1" x 0.14" Size. Top: Picture of the Board with Dimensions and Connections. Lower Left: Impedance Obtained without Complex Inversion of Voltage Divider Formula, Using (3). Lower Right: Same Measured Data Using Complex Inversion with (5). On Both Graphs, Left Axis Shows Logarithmic Impedance Magnitude and Right

Axis Shows Linear Phase.

Note that without solving the complex attenuator equation, the extracted impedance magnitude at low frequencies saturates at 25 Ohms. Similarly, instead of the -90 degree phase angle that we expect from a capacitance, at low frequencies, the uncorrected phase angle approaches zero degrees.

Extracting Component Values

Often times the measured PDN component can be approximated with a very simple equivalent circuit in a given frequency range. As shown in Figure 21.5, the most common equivalent circuits for PDN components are series C-R-L, or parallel C-R-L circuits.

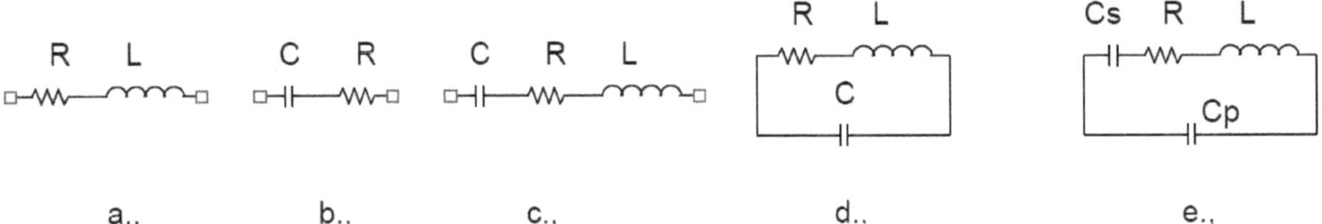

Figure 21.5: Simple Equivalent Circuits of PDN Components

Series C-R-L equivalent circuits can be applied to bare plane pairs (from very low frequencies up to close to the first modal parallel resonance), and to many bypass capacitors. Parallel C-R-L equivalent circuits can be applied to shorted plane pairs and capacitors mounted on plane pairs (around their parallel resonance). The parameter extractions can be built into spreadsheet calculations that capture the measured data. When a frequency independent R-L or C-R model is sufficient, the extraction is straightforward and accurate. When series or parallel C-R-L models have to be used and/or when the components show non-negligible frequency dependency, an iterative solution is necessary to obtain the values for all of the equivalent-circuit elements.

Extracting Capacitance

Assume we have to measure a single piece of capacitor or a one-dimensional transmission line (PCB trace) or two-dimensional transmission line (PCB parallel planes). Below their first series resonance frequencies, the measured impedance will follow the $|Zc| = 1/(2\pi fC)$ trend. We can then reverse-calculate the capacitance from the imaginary part of impedance. As we need the phase information of the measured impedance, a minimum of through calibration is necessary before taking data (full two-port calibration is necessary for best accuracy). The capacitance estimate becomes:

$$C = -\frac{1}{2\pi f Im\{Z_{DUT}\}}$$

Note that for low-loss capacitors, where the phase angle of the impedance is very close to -90 degrees, $Im\{Z_{DUT}\}$ can be replaced with magnitude$\{Z_{DUT}\}$, and in that case there may be no need for calibration as long as $Z_{DUT} << 25$ Ohms.

Figure 21.6 is an example of a low-frequency bulk capacitor, showing its impedance magnitude and phase as well as the extracted capacitance. Because, in the given frequency range the impedance stays <<25 Ohms, there is no noticeable difference whether we plot the impedance from the Z_{DUT} equation or the $Re(Z_{DUT})$ *and* $Im((Z_{DUT})$ equations.

However, in either case, the extracted capacitance has a sharp increase as frequency approaches the series resonance frequency.

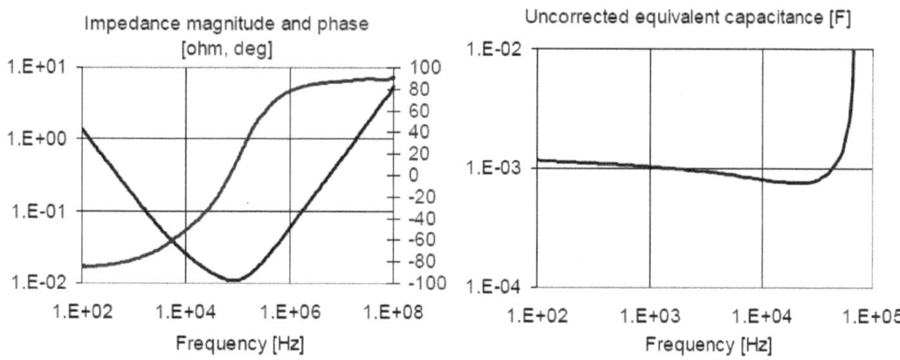

Figure 21.6: Left: Measured Impedance Magnitude and Phase of a 1200 μF Polymer Electrolytic Capacitor. Impedance Magnitude Axis Is on the Left; Phase Axis Is on the Right. Right: The Extracted Capacitance versus Frequency Using the $\boldsymbol{Re(Z_{DUT})}$, $\boldsymbol{Im((Z_{DUT})}$ and C Equations

Compensating for Series Inductance

As shown on the right-hand graph of Figure 21.6, the capacitance value given by Equation 7 will show a gradually increasing error as we approach the series resonance frequency of the device. Note the sharp rise of capacitance just before 100 kHz. As long as the inductance in the equivalent circuit can be estimated with reasonable accuracy, a correction can be applied to Equation 7.

With an inductance estimate of L, the corrected capacitance can be expressed as:

$$C = -\frac{1}{2\pi f(Im\{Z_{DUT}\} - 2\pi f L)}$$

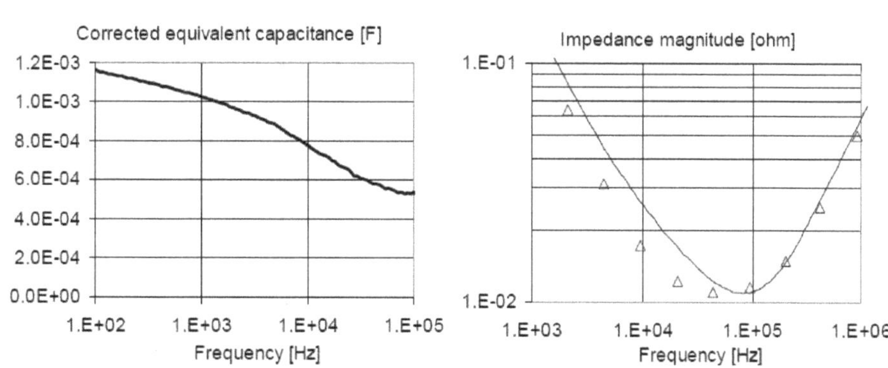

Figure 21.7: Left: Extracted Capacitance of the 1200 μF Polymer Electrolytic Capacitor, Using the C Equation with an Estimated Inductance of L = 9.4 nH. Right: Impedance Magnitude on Zoomed Scale: Continuous Line Is Measured Impedance, Triangles Show the Approximation with a Series RL-C Circuit, with Frequency Independent Values of C = 1200 μF, R = 0.01 Ohm, and L = 9.4 nH

Note in Figure 21.7 the frequency dependence of the extracted capacitance and the large error between the measured impedance plot and simulated impedance when we assume a frequency independent capacitance value.

Extracting Equivalent Series Resistance

Any one-port complex linear time-invariant circuit can be approximated with a single impedance or admittance at a given fixed frequency. In circuit theory, equivalent series resistance (ESR) is the real part of the equivalent impedance. The equivalent impedance and its elements are, in general, frequency dependent.

Assuming dominant series losses in the component, it equals the real part of impedance at the series resonance frequency. If the real part of impedance varies with frequency, ESR will also depend on the connection inductance, as the loop inductance together with the capacitance determines the series resonance frequency.

ESR values many times are much smaller than 25 Ohms, so in simple measurements there is no need to solve for the complex voltage divider. Dependent on the component and connection geometries, some parts will exhibit significant variation of equivalent series losses, because at higher frequencies the internal current path may change significantly [5]. Figure 21.8 shows the impedance magnitude and real part as a function of frequency for a 1200 µF polymer electrolytic capacitor.

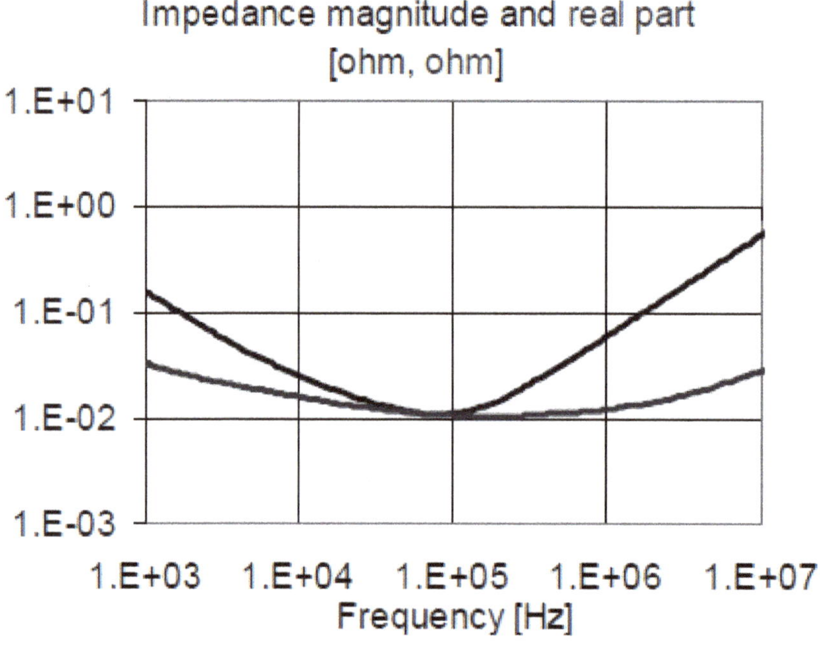

Figure 21.8: Impedance Magnitude (Upper Trace) and Real Part of Impedance (Lower Trace) of a 1,200 µF Polymer Electrolytic Capacitor

Extracting Inductance

Shorted traces and power/ground planes, as well as capacitors above their series resonance frequencies, become inductive. The measured impedance will follow more or less the $|ZL| = 2\pi fL$ trend. We can then reverse-calculate the inductance from the imaginary part of impedance.

As we need the phase information of the measured impedance, a minimum of through calibration is necessary before taking data. For higher accuracy, full two-port calibration is preferred.

The inductance estimate becomes:

$$L = \frac{Im\{Z_{DUT}\}}{2\pi f}$$

...where f is frequency and Z_{DUT} is the measured impedance of device.

Note that for low-loss inductors where the phase angle of the impedance is close to 90 degrees, $Im\{Z_{DUT}\}$ can be replaced with $Magnitude\{Z_{DUT}\}$, and in that case there may be no need for calibration as long as $Z_{DUT} << 25$ Ohms.

Compensating for Series Capacitance

Figure 21.9 shows the extracted equivalent inductance of a 1,200 μF polymer electrolytic capacitor, shown in Figure 21.6 through 21.8.

Similar to the extracted capacitance, the extracted inductance value will also show a false frequency dependency, an increasing negative error, as frequency approaches the series resonance frequency. As shown on the left-hand graph of Figure 21.9, the inductance value given by Equation 9 will show a sharp decrease around 100 kHz.

As long as the capacitance at the series resonance frequency in the equivalent circuit can be estimated with reasonable accuracy, a correction can be applied to Equation 9.

With a series-capacitance estimate of C_S, the corrected inductance can be expressed as:

$$L = \frac{Im\{Z_{DUT}\} + \frac{1}{2\pi C_s}}{2\pi f}$$

The graph on the right of Figure 21.9 was calculated with the L equation on the same measured data, using $C_S = 550$ μF. Note that instead of the nominal 1200 μF capacitance, the $C_S = 550$ μF capacitance value at the series resonance frequency was used. This simple correction extends the valid frequency range of the extracted inductance by at least half a decade.

Note that the equivalent inductance after correcting for the series capacitance still shows a frequency dependency, a negative slope, but this is now real: it tells us that as frequency increases from 100 kHz to 100 MHz, the current loop changes such that inductance drops from 9.5 nH to 8.5 nH.

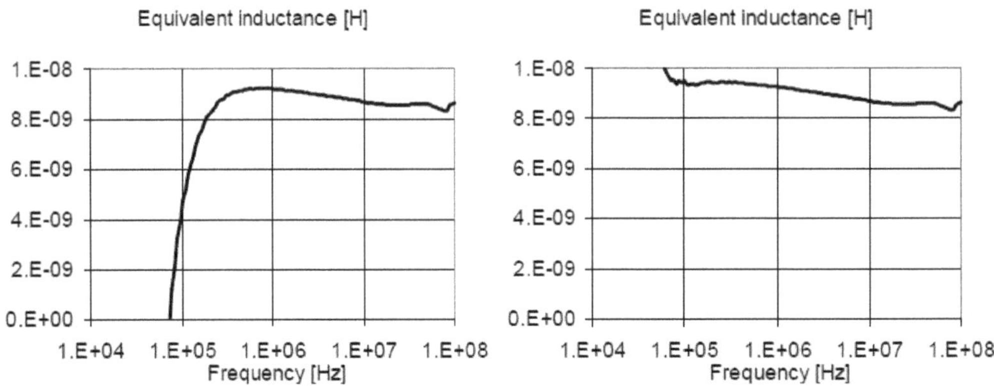

Figure 21.9: Left: Extracted Equivalent Inductance of a 1,200 μF Polymer Electrolytic Capacitor, Using Equation 9. Right: Corrected Inductance on the Same Set of Data, Using Equation 10 with $C_S = 550$ μF. Note that for the Particular Capacitor, the Capacitance was found to be a Strong Function of Frequency: Equation 10 Requires the Capacitance Value around the Series Resonance Frequency

Compensating for Parallel Capacitance

In case of shorted planes, the inductance of short is in parallel to the static plane capacitance, which creates a parallel C-R-L circuit, similar to that of Figure 21.5d. The extracted inductance using the inductance estimation equation again will show an increasing error as frequency approaches the parallel resonance frequency. Having an estimate on the C_P shunt capacitance, the extracted inductance can be corrected to take into account the susceptance of C_P:

$$L = \frac{Im\{Z_{DUT}\}}{2\pi f(1 + 2\pi f C_p Im(Z_{DUT}))}$$

Figure 21.10 shows the measured impedance of a 1.0"x0.144" plane pair with a metal short soldered over the capacitor pads.

Note that we have a similar equivalent circuit when a capacitor is connected to PDN planes, and the capacitor's inductance creates a parallel L-C circuit with the static plane capacitance.

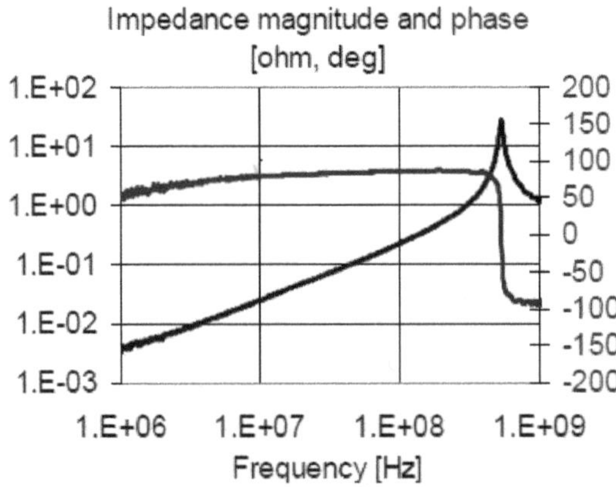

Figure 21.10: The Impedance Plot Shows the Impedance Magnitude and Phase of the Small Test Board Shown in Figure 21.5, Measured at the Test Points, with the Capacitor Pads Shorted

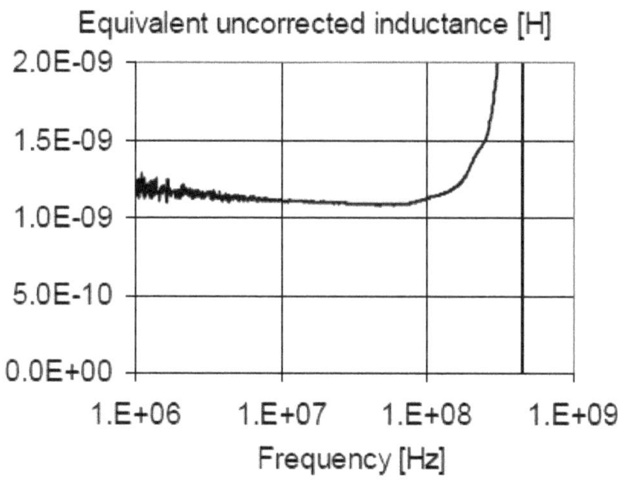

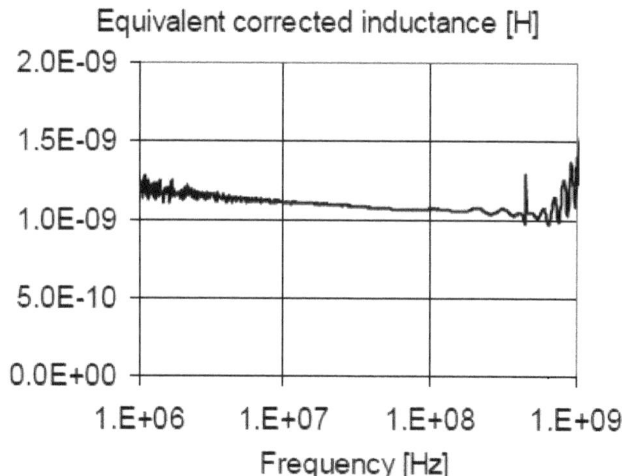

Figure 21.11: Extracted Inductance of the 1.0"x0.14" Plane Pairs (Shown in Figure 21.5) with Short on the Capacitor Site. Left: Calculated from Equation 9 with No Correction for Parallel Plane Capacitance. Right: Corrected for Plane Capacitance, Calculated from the Inductance Equation.

Calibration and Probes—No Calibration

At low and mid-frequencies, where the source and receiver flatness of the VNA is good, quick measurements can be done with no calibration at all. DUT impedance magnitude can be obtained with reasonable accuracy as long as its value is much less than 25 Ohms.

At 1 MHz, the wavelength in cables is 150 meters or more (depending on the dielectrics in the cable), so cables in a lab environment are usually much shorter than the wavelength. This means that mismatches in the cables and VNA-port impedances will result in a small constant error in the result, as opposed to the frequency-dependent ripple error when cable length is longer than the wavelength.

We still have to watch for cable-braid ground-loop errors (explained later).

Figure 21.12 shows the measured impedance profile of the same 1,200 μF polymer capacitor that was shown earlier. The data was taken with the 4395 VNA with no calibration, isolation transformer, or isolation amplifier.

The measured data may appear to be correct, but a reference measurement with the same setup on a short (shown in Figure 21.12 on the right) reveals that the ground-loop error at low frequencies makes the measured data questionable.

Note also that the phase information is totally irrelevant due to the lack of calibration, so capacitance and inductance could be extracted only from the impedance magnitude, further limiting the accuracy of data.

Note, however, that measuring a smaller-value (<=100 μF) and higher-ESR (say aluminum electrolytic) capacitor would be sufficiently accurate in this frequency range without any calibration.

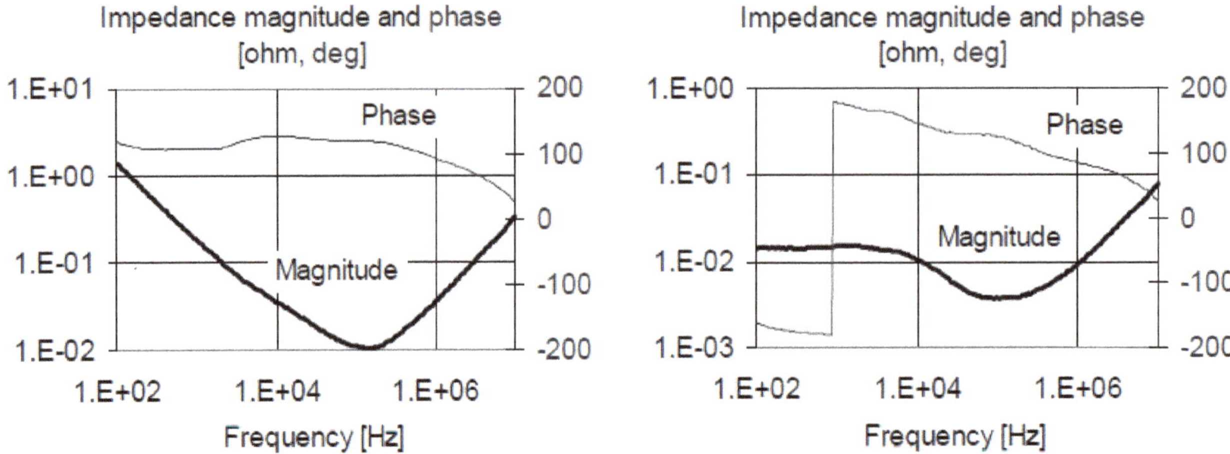

Figure 21.12: Left: Impedance Profile of a 1200 µF Polymer Capacitor, Taken with No Calibration. Right: Impedance Reading with the Same Uncalibrated Setup, with the Capacitor Location Shorted (Note the Different Vertical Scales). Around 10 kHz, the Impedance Reading of the Capacitor Is Only Three Times Bigger than the Impedance Reading of the Shorted Fixture, Creating a Visible "Dent" in the Capacitor's Impedance Profile at around 10 kHz. For Such Large Capacitance and Low-ESR Parts, the Cable-Braid Loop Error Has to Be Removed.

For low-frequency measurements, there is little concern about probe size. Often times, the probes can be replaced with short pigtail soldered coaxial-cable connections. For reproducible measurements, a small fixture is recommended.

Calibration and Probes—Through Calibration

If cable and probe reflections can be neglected, errors of source flatness and cable/probe attenuation can be partly removed just by doing a simple through calibration. The through calibration establishes the full-scale impedance value and the phase reference.

Inversion of the complex voltage divider becomes possible, and expressions (5) can be used to obtain the real and imaginary parts of the unknown impedance without the <<25 Ohms restriction.

Depending on the frequency range and quality of cables/probes we use, the still uncalibrated impedance mismatches may show up at higher frequencies in form of an increasing ripple on the measured data.

The left-hand graph of Figure 21.19 is an illustration of the ripple error we can expect at high frequencies with having only through calibration. The right-hand graph shows the same DUT measured at high frequencies with probes after full two-port calibration.

Note the different horizontal scales: the graph on the left spans 100 kHz to 1 GHz; the series resonance frequency of the capacitor is seen at 20 MHz.

The graph on the right spans 100 MHz to 10 GHz, in which frequency range we see a 600 MHz parallel resonance of capacitor's inductance and plane capacitance, followed by three plane resonance peaks.

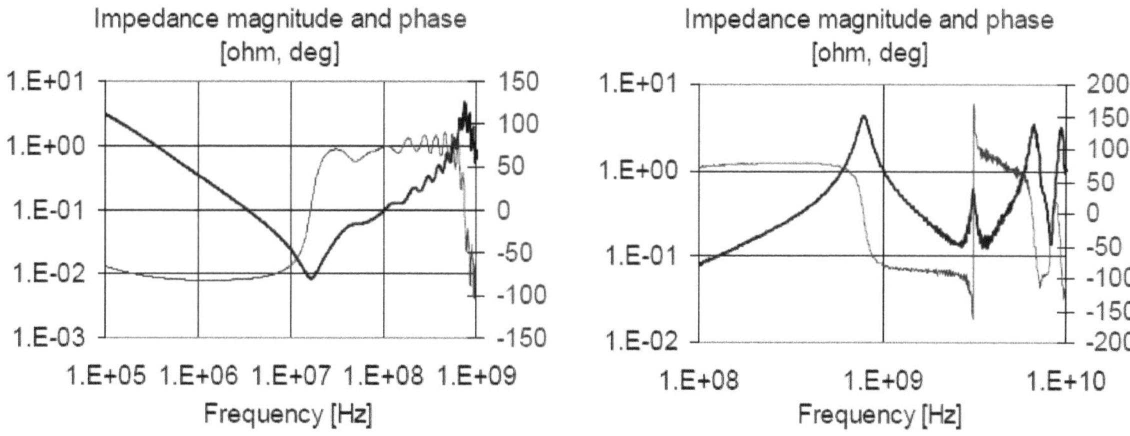

Figure 21.13: Measured Impedance Magnitude and Phase of a Ceramic Capacitor Soldered on a Small Test Board. Low-Frequency Response with Only Through-Calibration Is Shown on the Left; High-Frequency Response of the Same DUT with Full Two-Port Calibration Is on the Right

Calibration and Probes—Full Two-Port Calibrations

To probe PCB through-hole test points, short semi-rigid probes can be used to connect the VNA to the DUT. Figure 21.14 shows a simple homemade probe with 50-mil pin spacing made of standard 0.082" diameter copper-sleeve semi-rigid coaxial cable. For maximum high-frequency isolation, there is a ferrite sleeve on the coaxial cable. The center wire of the coaxial cable serves as the hot pin. A short wire, soldered to the copper sleeve, is the ground pin.

The pin-to-pin spacing is 1.25 mm (50 mils).

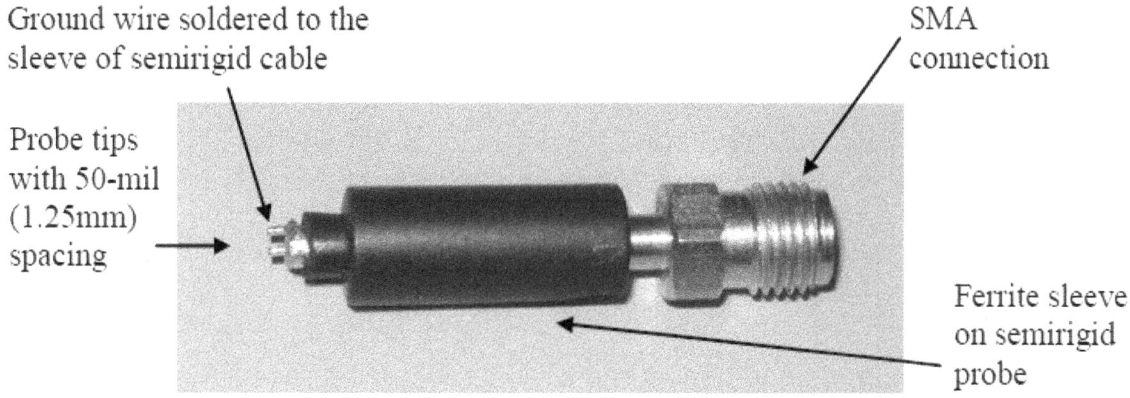

Figure 21.14: Short Semirigid Coaxial Probe with SMA Connection

With semirigid coaxial probes, full two-port calibration can be done up to the end of the two coaxial cables leading to the probes. Standard SMA open, short, and load elements can be used for the reflection calibrations. For the through-calibration part, a semirigid SMA-SMA cable is used with a total length equaling the two semirigid probes. The only uncalibrated error is due to the impedance mismatch in the short semirigid probes, but mismatch and attenuation of the connecting cables are all removed.

With miniature two- or three-pin probes such as PicoProbes from GGB Industries, full two-port calibration can be done with the calibration kit available on ceramic substrate. Figure 21.15 shows probes with 450-micrometer pin pitch. The probes are positioned over the shorting pads of the calibration substrate.

With the calibration substrate, calibration can be performed up to the tips of the probes.

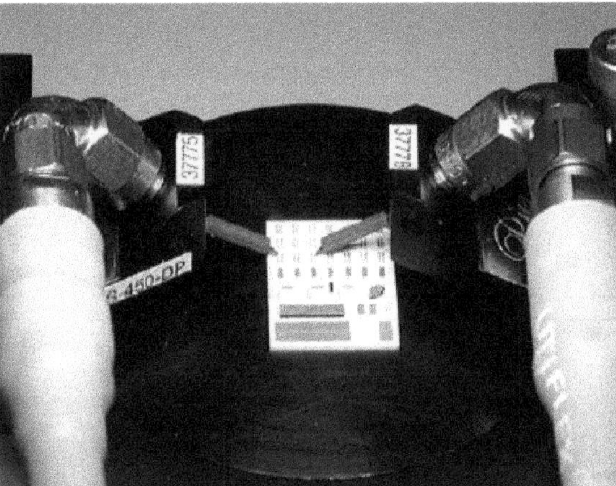

Figure 21.15: Left: S-G 40 mil (0.45 mm) PicoProbe. Right: Two Probes on Ceramic Calibration Substrate

Making the Proper Connections—Eliminating Cable-Braid Loop at Low Frequencies

The ground returns of Port 1 and Port 2 in the VNA are connected together inside the instruments. When we measure a low-impedance DUT, for instance an active DC-DC converter output, or a metal short, the current of Port 1 will create a voltage drop across the parallel equivalent of the two cables' braid resistances.

The measured low-frequency impedance value, instead of Z_{DUT}, becomes:

$$Z_{measured} = Z_{DUT} + R_{b1} x\, R_{b2}$$

The cable-braid ground loop is not a limit at high frequencies, because the cable inductance creates a -20dB-per-decade roll off this residual reading above the corner frequency determined by the braid resistance and inductance. The inductance of the cable can be increased, and the corner frequency beyond which the error rolls off can be reduced by placing ferrite clamps or ferrite beads around the cables.

The right-hand graph of Figure 21.16 shows the impedance magnitude measured with different cable configurations, with short across the test points. All cables for this illustration were 24" long, type RG174 flexible coax and 0.082" semi-rigid coax. The plots show the residual error with and without ferrite clamps around the cables.

The measurements were done with an HP4395 VNA, between RF Out and Input B.

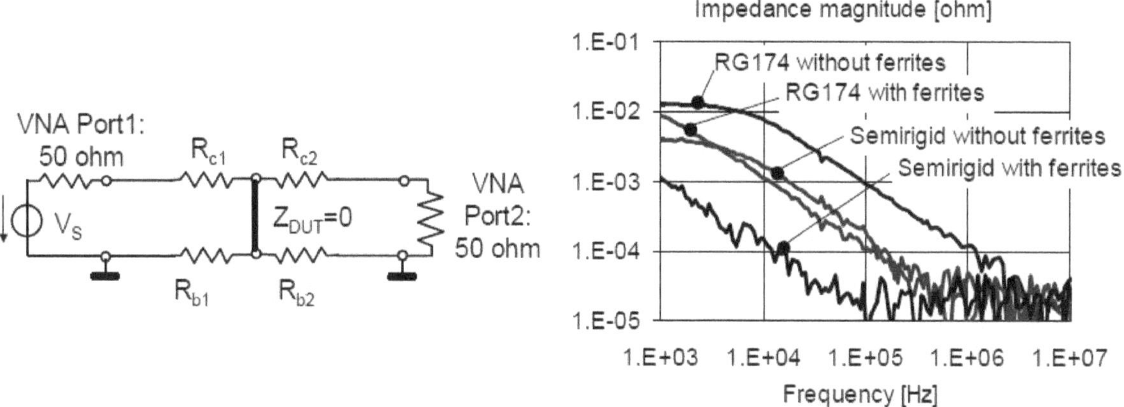

Figure 21.16: Left: Equivalent Circuit of Cable-Braid Loop. Right: Measured Impedance Magnitudes of Residual Error on Shorts with Different Cables, with and without Ferrite Sleeves.

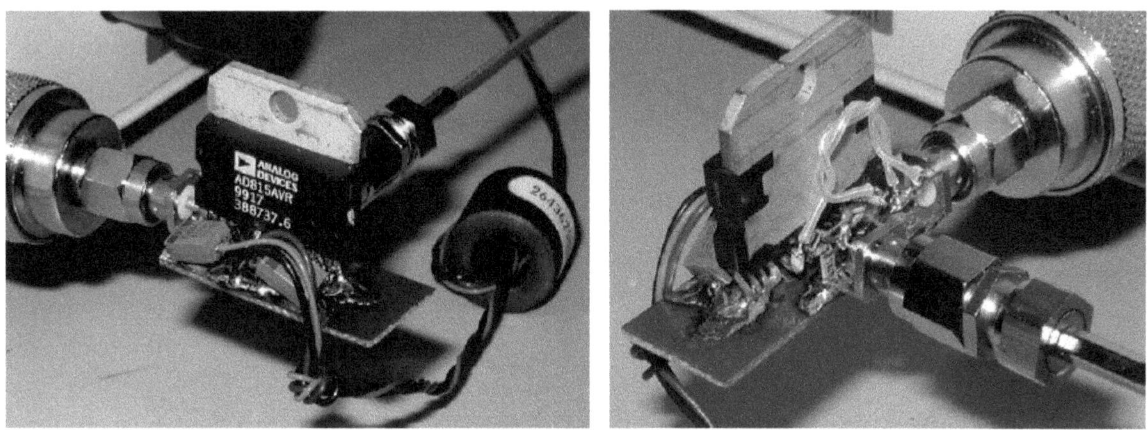

Figure 21.17: Differential-Input Single-Ended Output Amplifier for the DC-10 MHz Frequency Range. Left: Circuit Schematic. Front and Back Photo of the Amplifier, Connected to the RF Output of the 4395A VNA

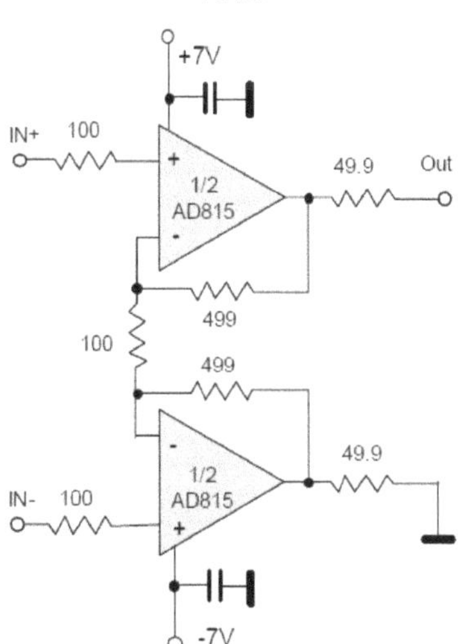

Figure 21.18: Circuit Schematic of an Isolation Amplifier

The only drawback of the isolation amplifier is the need for external power supplies. The supply voltages must be selected such that the connected VNA input should survive the maximum saturated output voltage in case transients should occur. The lowest value of supply voltage is limited by the specified minimum rail voltage of the operational amplifier.

With the circuit shown in Figure 21.18, proper operation was achieved at and above ±4V supply voltage. For the HP4395 VNA model, the maximum input DC voltage is 7V; this suggests a ±7V maximum supply voltage for the differential amplifier. The output impedance of the amplifier is very close to 50 Ohms in the entire 0-10 MHz frequency range, and active DUTs (such as DC-DC converters and voltage regulator modules [VRMs]) can be connected to its output without any problem.

In this simple homemade setup, both the supply-voltage sensitivity and the absolute frequency response limit their application to frequencies up to 10 MHz.

Measuring VRMs

VRMs are DC-DC converters that convert the incoming (e.g., 48V or 12V) DC to the final regulated voltages. VRMs should provide the specified DC current, and at low frequencies, we also expect them to provide low impedance. The bandwidth in which the low impedance can be maintained primarily depends on the switching speed. As a second-order effect, the loop transient response and bandwidth also depend on the style of the control loop and range of allowable load impedances. Currently the mainstream switching frequency is a few hundred kHz per phase, achieving way over a MHz effective switching frequency with multiphase converters.

The output impedance in the fast point-of-load converters can be as low as a fraction of a milliOhm up to at least 10 kHz. VRM response, on the other hand, can be heavily nonlinear for large load-current steps. Guaranteeing the stability of control loop could also be a challenge for widely varying load currents and bypass capacitor impedances.

For PDN-design purposes, the most convenient is if the VRM stays in the linear mode over the entire specified line and load range. In this case, the simulated or measured output impedance is maintained under all operating conditions, there is no need to separately consider the large-signal conditions. One possible way to achieve this condition is to use voltage positioning with current-mode control loops [6].

Measuring the output impedance of VRM requires considerations similar to that of measuring low-ESR bulk capacitors. Because VRM control loops tend to have high gain at DC, the typical output impedance response is "inductive-like," achieving very low impedances at low frequencies. Therefore the cable-braid ground loop must be opened by either transformer isolation or by differential-input amplifiers; ferrite-loaded cables are usually not sufficient if we want to measure the VRM impedance below a few kHz.

If we use the differential-input single-ended output isolation amplifiers, no further precaution is necessary. If we have the amplifier in front of the VNA input, the input impedance of the amplifier is high enough that it will not interfere with the VRM output, and the differential-to-single-ended conversion will suppress the DC output voltage of the VRM so that the VNA input is not stressed with DC voltage. If the amplifier is placed to the Port 1 output, in front of the DUT, the series 50 Ohm resistor protects the amplifier's output from the DC voltage of the regulator. The proper DC load current of the VRM can be set either with a resistive load or by using an active electronic load device.

If the VRM is not powered up, its output impedance can be measured similar to any low-frequency bulk capacitor. Figure 21.19 shows the setup to measure the output impedance of a small-size socketed VRM. The DC input power was connected through wires on the left. The output voltage appeared at multiple pins on the socket. To maintain the low output impedance, a small double-sided thin PCB was soldered to the appropriate pins: all positive output pins were soldered to the top copper plane, all negative output pins were soldered to the bottom copper plane.

The solid copper planes across a thin dielectric separation ensure that the VRM output pins are connected to the measurement points through a low-resistance and low-inductance path. The heavy isolated wire on the right shows the connection to the electronic load. The VNA probes were soldered in the middle of the two-sided plane connection. The remote sensing wires were soldered to the same location.

Figure 21.19: Setup and Connections to Measure VRM

The VRM can be measured without input power (inactive loop) or powered (active loop). The left graph of Figure 21.20 shows the VRM's output impedance with no input power. It follows the impedance of its filter capacitors at the output. Note that if the loop stability requires or assumes external capacitors, those should also be connected to the VRM output while we measure its output impedance.

On the right of Figure 21.20, two versions of the same VRM are shown, with input power applied. Note the different vertical and horizontal scales for the two graphs. All data was taken with no additional bypass capacitors. The unpowered output impedance clearly follows the impedance profile of the on-board filter capacitors. The two versions of active output impedances illustrate the potential problem of instability. One of the curves maintains less than one milliOhm up to 20 kHz frequencies, but at 85 kHz this version had a sharp impedance peak. While the magnitude of the impedance peak itself may not be a problem (it is around 15 milliOhms), the sharp peak indicates a very low phase margin, and it indicates the risk of self-oscillation under slightly different tolerance constellation. Separate analysis of the phase margin of the control loop indicated, in fact, low phase margin. This impedance peak can be viewed as the classic intercapacitor and resonance peaking, when a bypass capacitor's inductive impedance creates a peak with another bypass capacitor's capacitance. In this case, the inductive behavior comes from the decreasing loop gain in the VRM with increasing frequencies.

The second plot on the right-hand graph exhibits no sharp peaking. However, as the relative positions of the two plots show, unless the switching frequency is increased or the control-loop topology is changed, modifying an existing loop to increase phase margin almost always comes with an increase of output impedance below the peak frequency. In this particular case, the output impedance at 10 kHz increased from 0.4 milliOhms to 3 milliOhms. Finally, above the peak, at about 200 kHz and above, both versions' output impedances follow the impedance curve of the output filter capacitors.

Note that the output impedance profile, while not providing any quantitative data about the loop stability, gives an easy and quick way to qualitatively check stability. Note also that measurement of loop stability usually requires opening up the control loop, while measuring the output impedance can be done without any modifications to the VRM.

In fact, with the methodology described above, the VRM output impedance can be easily measured in a fully assembled working system.

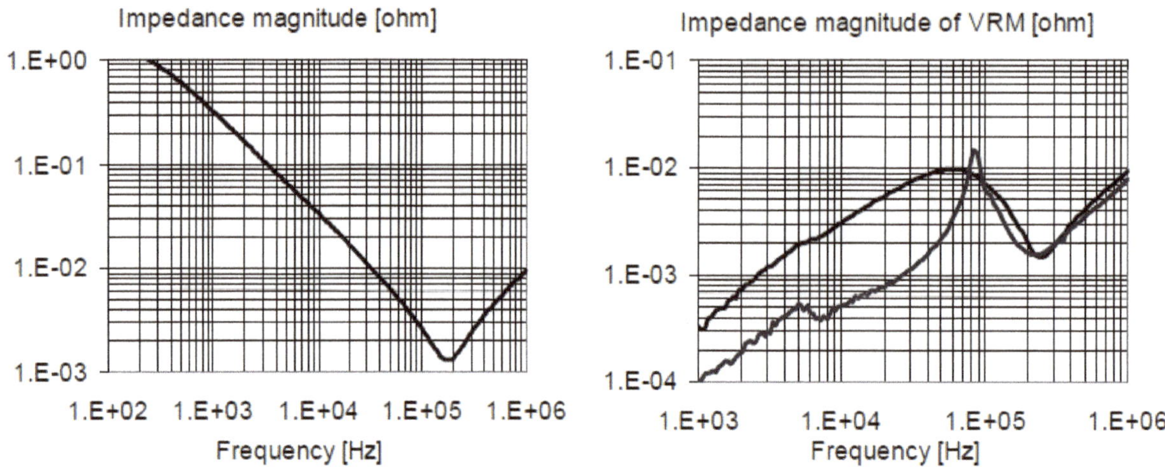

Figure 21.20: Measured Low Frequency Output Impedance of a VRM with No Input Power (Left) and with Input Power for Two Loop-Compensations (Right)

Measuring Low Impedances at High Frequencies

Cable performance limits the two-port VNA measurements at low frequencies due to the cable-braid ground loop.

At high frequencies, besides the noise floor of the instrument itself, the finite surface transfer impedance and resonances in the cable braids set the error floor. As PDN impedances are very seldom close to the 50 Ohm instrument impedance, there is usually a big mismatch/reflection at the connection to the DUT. Through the finite surface transfer impedance of the cable braids, the reflections and leakage show up as erroneous peaks in the response.

The left graph of Figure 21.21 shows the equivalent impedance reading from two RG178 coaxial cables connected to a VNA port with their ends shorted with SMA caps. The two cables were a couple of inches apart. The same cables in the same position were also measured with a different number of small ferrite clamps put around them. The graph also shows a trace with just a few ferrite clamps around the cable: the erroneous peaks are shifted toward lower frequencies and the peak values are reduced.

To achieve the lowest possible error floor, a flexible coaxial cable requires ferrite clamps or sleeves all along its lengths.

The graph on the right in Figure 21.21 was measured across a plane short in a medium-size PCB.

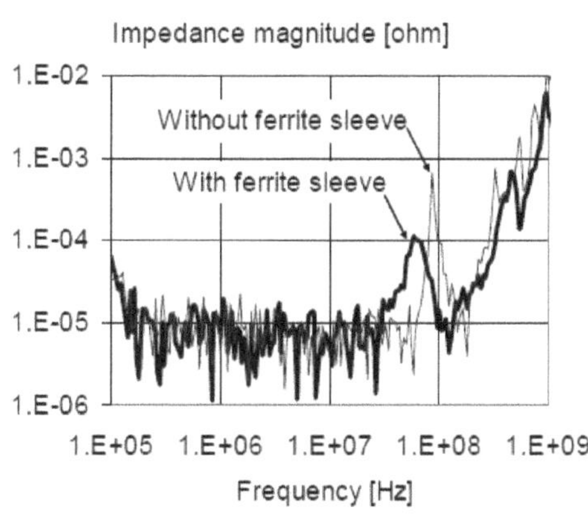

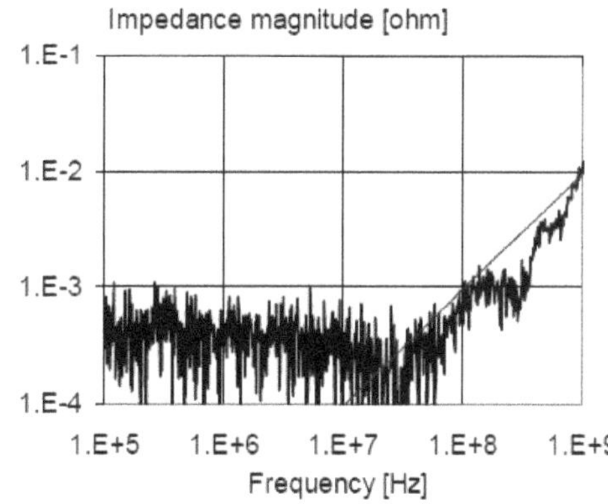

Figure 21.21: Illustration of Cable Performance at High Frequencies. Left: Equivalent Impedance Magnitude with Two Port VNA Measurements, the Two Cables Are Shorted at Their Ends, Two Inches Apart, with and without Ferrite Sleeves. Right: Residual Impedance Magnitude Measured on Plane Short with 50-mil Semi-rigid Coax Probes and Ferrite Covered Coaxial Cables. The Thin Straight Slope Shows the Impedance Magnitude of a 1.5 pH Inductance.

The few pH equivalent inductance of the error floor is achievable only if we not only cover the entire length of the cable braid with ferrite absorber sleeves, but also place ferrite sleeves around the semi-rigid probe and isolate the probe-holder metal parts from the cable.

There are coaxial cables available with good surface transfer impedances, which would reduce this error. However, cables with better braids and shields tend to be not only more expensive, but also more stiff. When we have to reach with two probes inside a bigger system or connect to a pre-mounted fixture, flexible cables are essential.

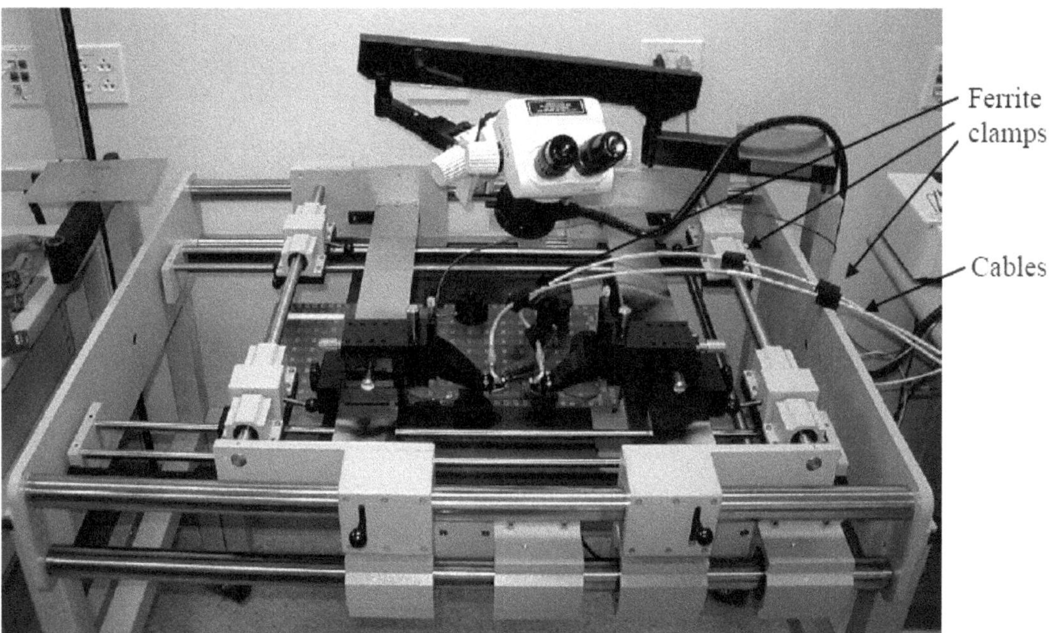

Figure 21.22: Large Probe Station with High-Performance Coaxial Cables with a Few Ferrite Clamps. Note the Ferrite Clamps at All Locations Where the Cables Otherwise Would Touch Either Each Other or the Metal Frame of the Probe Station.

Measuring the Full PDN

Figure 21.23 shows the setup for measuring the PDN of a powered board. The semi-rigid coaxial probes provide reliable measurements up to about 1-2 GHz. The coaxial probes are mounted on small probe positioners. For self-impedance measurements, the probes are inserted into plated-through holes from the opposite sides. Measuring the self-impedance from the opposite sides will ensure that the residual coupling through the probe-pin loops will not limit the measurement.

Furthermore, by measuring from the opposite sides, we will measure the impedance as it appears across the PDN planes. If we measure the impedance at a via pair from the same side, what we get is the PDN impedance at the planes in series to the impedance of connecting vias. For transfer impedance measurements the probes can be inserted into the through holes from the same or from the opposite sides. Note the ferrite sleeves on the coaxial cables.

For low-frequency measurements with active VRMs, the coaxial cables are connected to a low-frequency VNA (in this case an HP4395) through an isolating transformer and an extra length of RG178 coaxial cable (to increase the DC resistance), or an isolation amplifier. For mid-frequency measurements, the same coaxial cables are connected to another VNA (in this case HP4396).

Figure 21.23: Probe Connections for Full PDN Self-Impedance Measurements

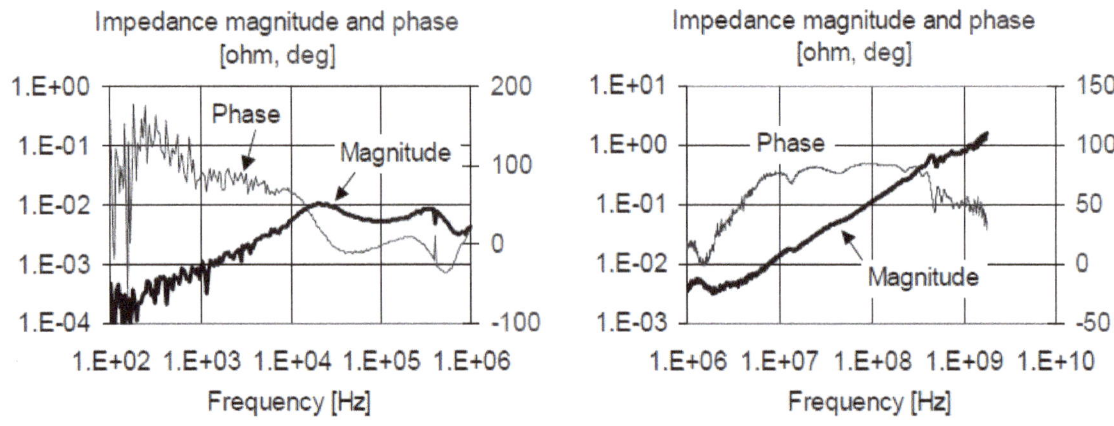

Figure 21.24: Measured Impedance Profile on One of the Supply Rails of a Powered Board. Left: Low-Frequency Impedance, Measured with an HP4395 VNA and Isolation Transformer. Right: High-Frequency Self-Impedance, Measured with an HP4396 VNA at One Pair of Test Vias

Calibrations can be performed with both VNAs, and measurement can then be performed over the entire frequency range only by switching cables at the VNA inputs; the probes can stay connected to the DUT.

The two graphs in Figure 21.24 show the measured self-impedance curves on one supply rail on a powered-up board. The two graphs together cover the 100 Hz to 1.8 GHz frequency range.

Repeatability of Data

In making measurements on low-impedance structures, the repeatability of the connections has to be looked at. There is a valid concern that the finite contact resistance between the probe tips and DUT connection points may alter the data. To illustrate the robustness of the two-port measurement setup, a low-ESR bulk capacitor was measured repeatedly in the same small PCB fixture without soldering. The sample was inserted into the through holes of the fixture and was held in place by slightly twisting the capacitor body so that the leads make connections to the through-hole walls.

Figure 21.25 shows the setup.

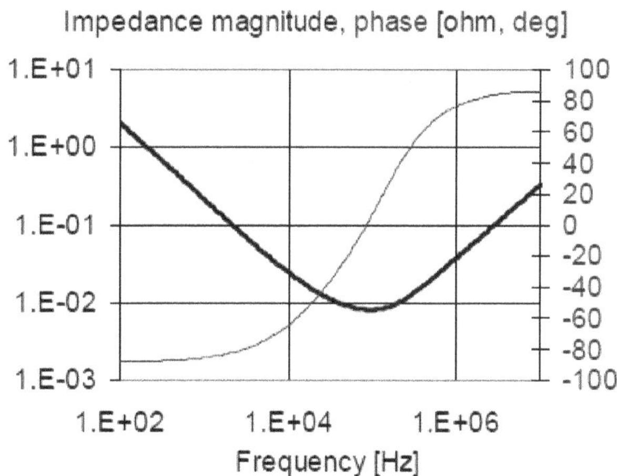

Figure 21.25: Illustration of Repeatability of Data at Low Frequencies in Fixtures without Soldering. Left: Photo of the Fixture with One of the Samples Inserted. Right: Impedance Magnitude and Phase versus Frequency

Figure 21.26 shows the result. Note that the extracted impedance minimum (ESR) shows a variability of about 1 milliOhm, but the inductance reading is very consistent. This suggest that to measure low-ESR bulk capacitors and/or active VRMs, soldered connections may be needed. However, at higher frequencies, where inductance dominates the path, the repeatability is sufficient without soldered connections.

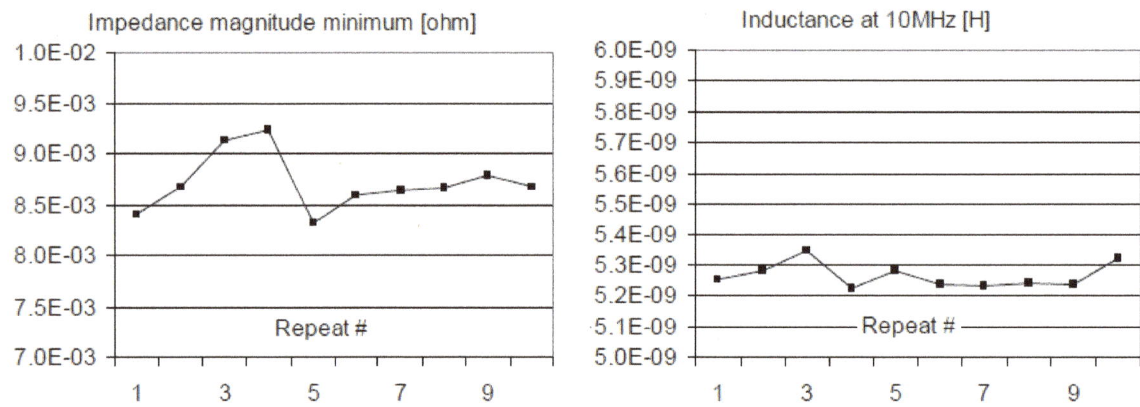

Figure 21.26: Left: Minimum of Impedance Magnitude. Right: Extracted Inductance at 10 MHz. The Same Sample Piece Was Measured 10 Times in the Same Fixture, Without Soldering

Acknowledgments

The author wishes to thank Sreemala Pannala, Leesa Noujeim, and Jason R. Miller for their contributions and Eric Blomberg, Mark Kanda, and Kevin Hinckley for the useful discussions.

References

[1] István Novák, *Overview of Frequency-Domain Power-Distribution Measurements*, DesignCon 2003 East, Boston, Massachusetts, June 23-25, 2003.

[2] Greg Taylor, Craig Deutschle, Tawfik Arabi, and Brian Owens, *An Approach to Measuring Power Supply Impedance of Microprocessors*, Proceedings of the 10th Topical Meeting on Electrical Performance of Electronic Packaging, October 29-31, 2001, Cambridge, Massachusetts, pp. 211-214.

[3] Isaac Kantorovich, Chris Houghton, Steve Root, and Jim St. Laurent, *Measurement of MilliOhms of Impedance at Hundred MHz on Chip Power Supply Loop*, Proceedings of the 11th Topical Meeting on Electrical Performance of Electronic Packaging, October 27-29, 2002, Monterey, California.

[4] István Novák, *Measuring Milliohms and PicoHenrys in Power Distribution Networks*, DesignCon 2000, Santa Clara, California, February 1-4, 2000.

[5] Larry D. Smith and David Hockanson, *Distributed SPICE Circuit Model for Ceramic Capacitors*, Proceedings of the 51st Electronic Components and Technology Conference, May 29-June 1, 2001, Orlando, Florida.

[6] Richard Redl, Brian P. Erisman, and Zoltan Zansky, *Optimizing the Load Transient Response of the Buck Converter*, Proceedings of APEC98.

Chapter 22—Simple Transmission Line Causal Model for Multilayer Ceramic Capacitors

István Novák, Senior Staff Engineer, Sun Microsystems
Gustavo Blando, Staff Engineer, Sun Microsystems
Jason R. Miller, Senior Staff Engineer, Sun Microsystems

This chapter is a revised version of the paper *Slow-wave Causal Model for Multi-layer Ceramic Capacitors*, DesignCon 2006, Santa Clara, California, February 6-9, 2006.

THERE IS AN ongoing interest in refining the simulation models for passive components in electronic circuits. For simple analyses, bypass capacitors are modeled by a series C-R-L equivalent network.

To capture the frequency dependency of the circuit parameters, more complex equivalent circuits can be used: ladder L-R networks to model the frequency-dependent inductance and resistance and/or C-R networks to model the frequency-dependent capacitance. These equivalent circuits can be compatible with both time-domain and frequency-domain SPICE simulations, but the optimum topology of the equivalent circuit may depend on the type and construction of capacitor.

This chapter first describes the actual current distributions inside multilayer ceramic capacitor (MLCC) parts simulated with a bedspring model, and we make some counterintuitive observations about the frequency dependency of inductance and resistance in tall MLCC parts.

Based on those observations, we then derive a causal model, which represents the capacitor with a periodically loaded lossy transmission line. It is shown that the load circuit in the unit cell corresponds to the waveguide formed by two adjacent capacitor plates. The unit cells are further simplified and lumped together into one lossy, open-ended transmission line with a series R-L circuit capturing the impedance of the cover layer of the capacitor. The parameters of the single lossy transmission line are derived from the geometry and material properties. It is shown that the important features are captured in the model by an effective capacitance and dielectric loss tangent, which combine the dielectric loss of the ceramic material and the resistive losses of the capacitor plates.

The model is shown to capture the important characteristics of measured data, and it is simple enough to be used in multiple copies in circuit simulators.

Introduction: Present Modeling Options

When considering the parasitics of bypass capacitors, a widely used simple model is a series C-R-L network, where C is the capacitance of the part, R is the equivalent series resistance (ESR), and L is the equivalent series inductance (ESL), as shown in Figure 22.1. In its simple form, C, ESR, and ESL are assumed to be frequency-independent constants. However, measured data indicates [1] that all three of these parameters are eventually frequency-dependent and furthermore may be interrelated through the application geometry.

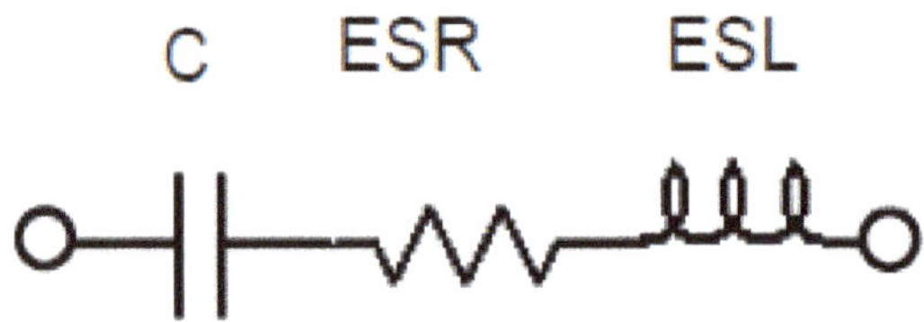

Figure 22.1: Simple RLC Equivalent Circuit of a Capacitor

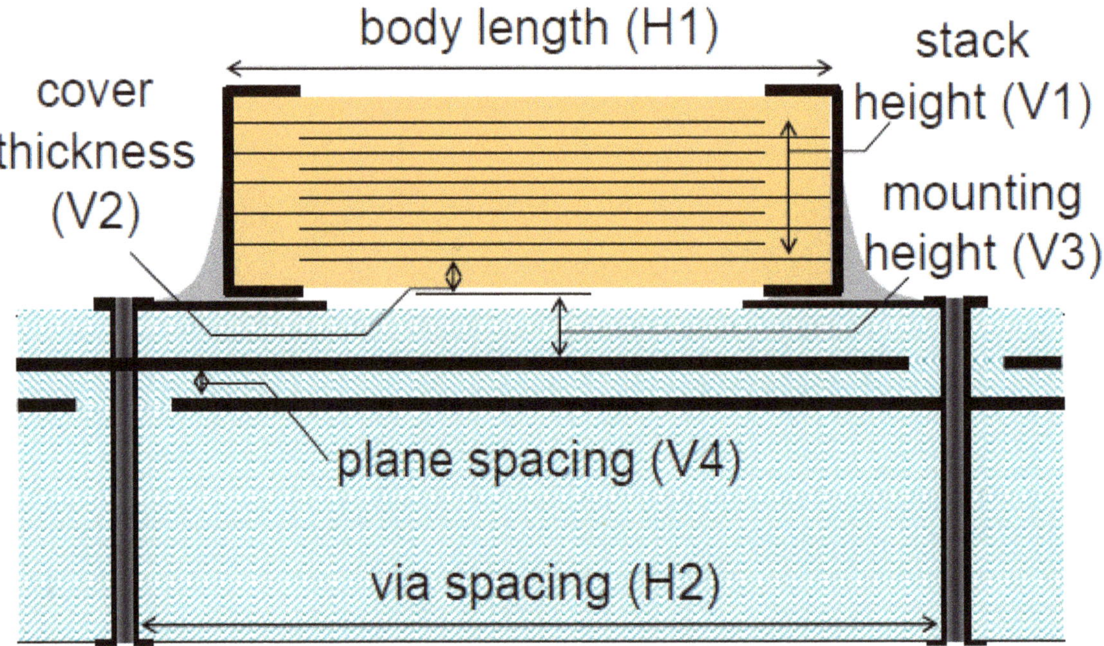

Figure 22.2: Vertical Cross-Section of an MLCC Mounted to PCB Planes

The capacitance may be frequency-dependent, primarily because of dielectric losses [2].

ESR is the result of transforming the parallel dielectric losses and series-conductive losses into a single series resistance value.

As long as the tangent delta of the dielectric material varies little with frequency, the parallel loss resistance drops inversely with frequency. The series resistance of the part comes from the terminals and conductive layers on the dielectric sheet(s).

Besides bulk capacitor constructions such as tantalum brick capacitors, the capacitor plates in high–CV MLCCs are thin enough that in the frequency range of interest, their thickness is less than the skin depth, and therefore the alternate current (AC) resistance contributions of the plates do not vary much with frequency.

Overall, ESR still varies at high frequencies because of non-uniform current distribution in the plates [4].

Inductance depends both on the internal construction of the part and the external geometry forming the closed current loop. As illustrated in Figure 22.2 for the case of a reverse-geometry MLCC attached to a pair of planes on a PCB, the measured impedance of the part exhibits strong frequency dependency on all three of the parameters.

When extracting capacitor parameters from measured data, we face a further complication: ESR is simply the real part of the measured impedance (after the proper calibration and/or de-embedding process), but the capacitive and inductive reactances show up in a superimposed way in the imaginary part of the measured impedance.

If capacitance and inductance were frequency independent, extracting them from the imaginary part of measured impedance would be easy. Because the capacitive and inductive reactances change with frequency in the opposite way, we know that at frequencies much below the series resonance frequency (SRF) the inductive reactance is negligible and from the measured reactance we can calculate the capacitance. Similarly, we could calculate the inductance from a measured reactance value at a frequency much above SRF, which is assumed for instance in [3] and [4].

With relatively strong frequency dependency of capacitance and/or inductance, which is the case of tall capacitor stacks with lossy dielectrics and aggressive mounting, using a low-frequency capacitance and a high-frequency inductance cannot resolve the frequency-dependent capacitance and inductance values around SRF.

One step further is to iteratively approximate the capacitance and inductance close to SRF and use those (constant) values to extract the frequency dependent capacitance and inductance curves [5]. This improves the accuracy and range of validity for the extracted capacitance and inductance curves, but unless we have further data points or constraints, we still cannot uniquely resolve the two unknowns, C(f) and ESL(f), from one data point of $\mathrm{Im}\{Z(f)\}$.

Instead of trying to blindly extract the parameters from measured data, more sophisticated equivalent circuits can also be used to fit the measured data on the model. Equivalent circuits composed of frequency-independent resistors, capacitors, and inductors unconditionally guarantee causality and easy compatibility to circuit simulators. To describe frequency-dependent capacitance and ESR of bulk capacitors, [6] suggests a composite RLC network. To capture the frequency-dependent ESR and ESL of MLCCs, [7] uses a resonant ladder network, while [8] proposes a transmission-line model.

For MLCC parts, it is customary to assume that the capacitive and inductive currents balance at SRF and, thus, at that frequency the current uniformly penetrates all plates and for this reason the lowest ESR value occurs at SRF. Similarly, it is usually assumed that inductance monotonically drops from its low-frequency value toward the high-frequency loop inductance.

In contrast to usual expectations, data on Figure 22.3 indicates that the minimum of the impedance real part is not at SRF: at 600 kHz ESR is 2.8 milliOhms; whereas, at the 2.1 MHz SRF, the ESR is at 3.6 milliOhms. Moreover, the ESL(f) value extracted according to the procedure in [5] results in 600 pH at SRF, but the inductance first increases with frequency, instead of decreasing, reaching a 660 pH peak at 4.2 MHz before it starts going down.

Is this due to measurement errors or a deficiency in the extraction procedure, or do ESR and ESL really behave contrary to common assumptions?

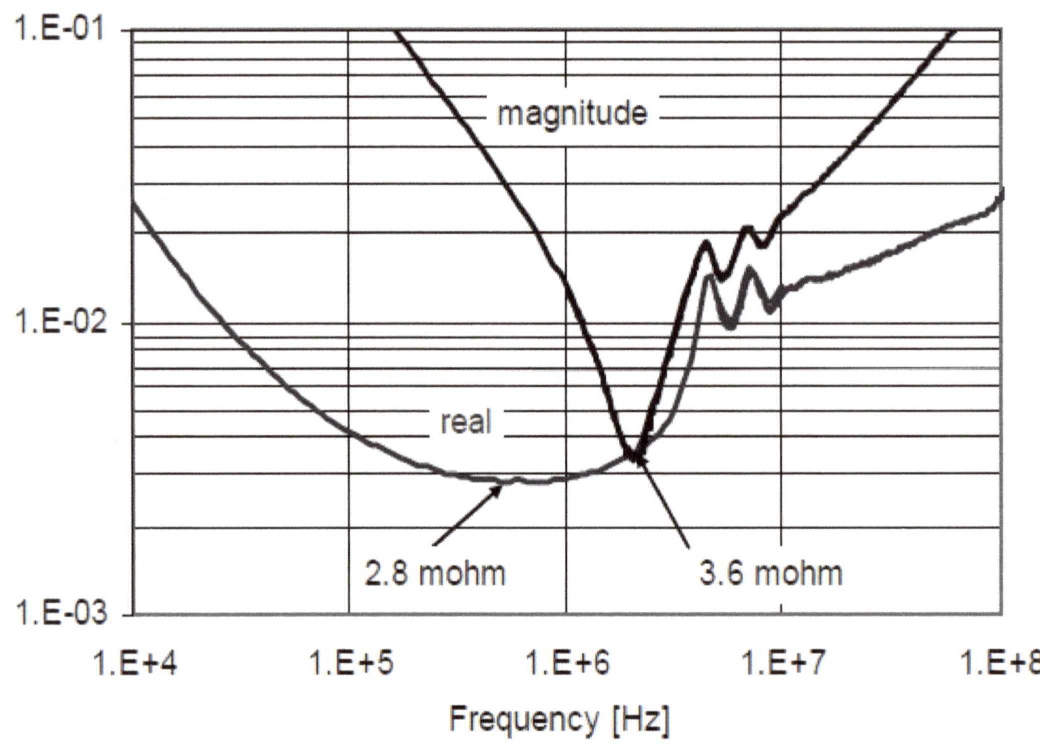

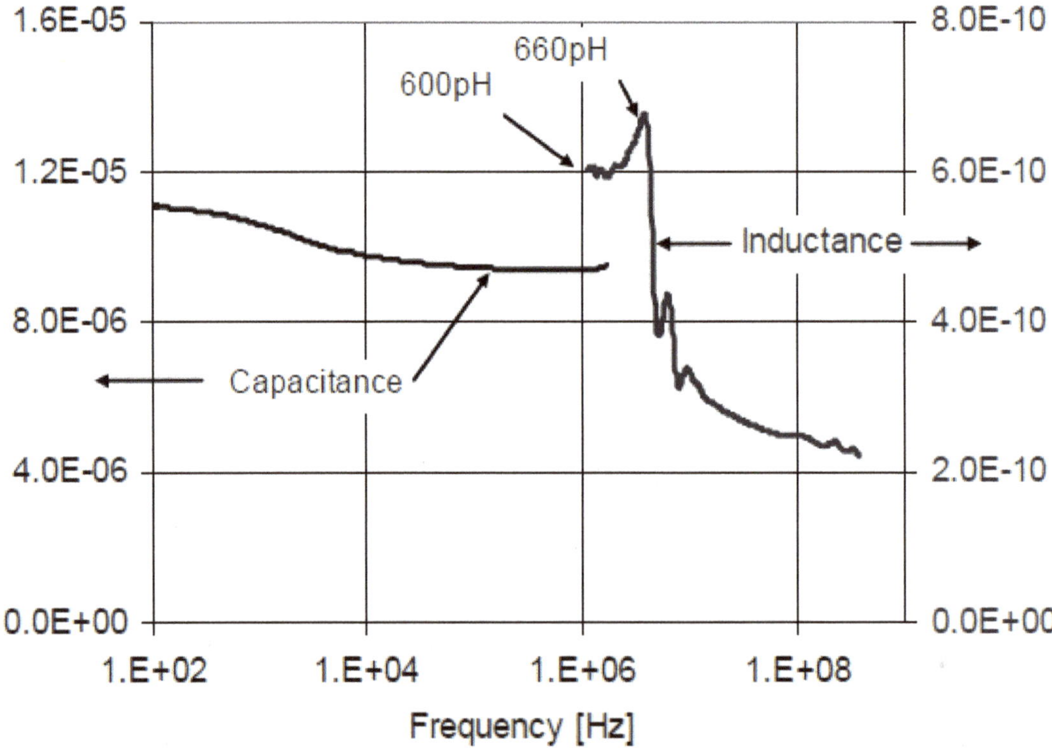

Figure 22.3: Measured Impedance Magnitude of a 10 μF 0508 MLCC with the Real Part of the Impedance (Top), and Extracted Capacitance and Inductance versus Frequency (Bottom)

As it was shown in detail in [9], this counter-intuitive behavior comes from the vertical resonances (and at higher frequencies, to a lesser degree, from horizontal resonances) along the capacitor body. These features can be captured by using a two-dimensional transmission-line or RLGC bedspring array. Figure 22.4 shows the partial schematics of the bedspring model, where the capacitor plates are grouped into 10 horizontal layers, each layer broken down into 10 segments. Note that the grid size of 10 by 10 is somewhat arbitrary but proved to be sufficient to capture the major features.

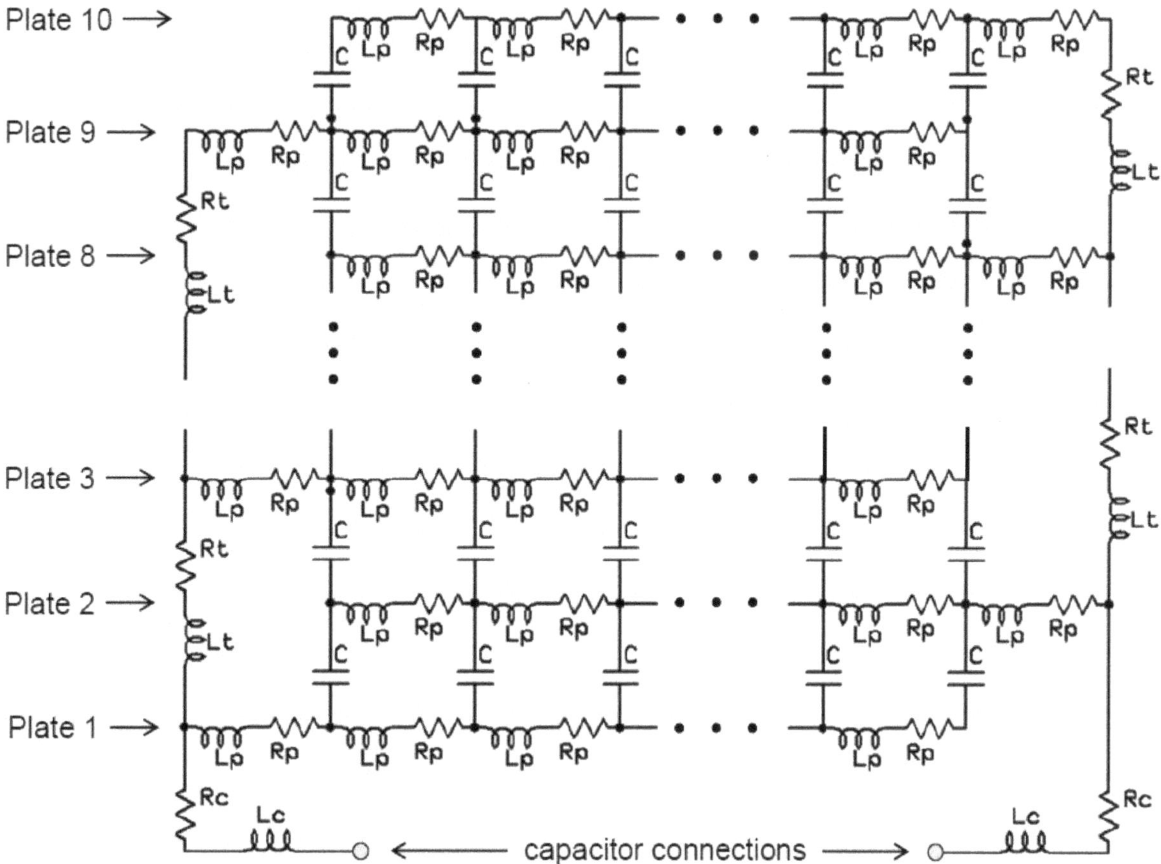

Figure 22.4: Partial Schematics of the Bedspring Capacitor Model. The Model Consists of 10 Capacitor Plates, 1 through 10. The Lowest Capacitor Plate, Connecting to the PCB, is Plate 1. Plates 1, 3, 5, 7, and 9 Are Connected to the Left Terminal. Plates 2, 4, 6, 8, and 10 Are Connected to the Right Terminal. Each Capacitor Plate Is Divided into 10 Equal Segments (Plus an End Piece), Represented by Series RL Networks. At Each Internal Plate Node, a Capacitor Represents a Piece of the Dielectric Material.

Figure 22.5 shows the impedance magnitude and real part of impedance simulated at the connection terminals. The schematics entries used for the simulation are listed below the chart. Note that the bedspring model captures all of the important features that we want to study: it shows that impedance real part starts to increase below SRF, and there is a set of dampened, but pronounced secondary resonances. Since we assume no dielectric loss, the impedance real part at low frequencies does not rise.

The current distribution plots in [9] also showed the reason why the inductance starts to rise around SRF before it eventually goes down. For a fully animated illustration of currents in the capacitor plates and in the dielectrics, see [10]. Here, we reproduce the current distribution plots only at SRF. In Figure 22.6, we see the current distribution along the capacitor plates, whereas the bottom plot shows the current distribution in the dielectric layers. The nonlinear current distribution is a clear indication that both ESR and ESL are increased.

While the bedspring model is unconditionally causal, and it is useful to study the vertical and horizontal resonances in MLCC parts, the model is clearly too complicated to be used in large PDN simulations, where we may need dozens, or hundreds, of such capacitor models in the same network.

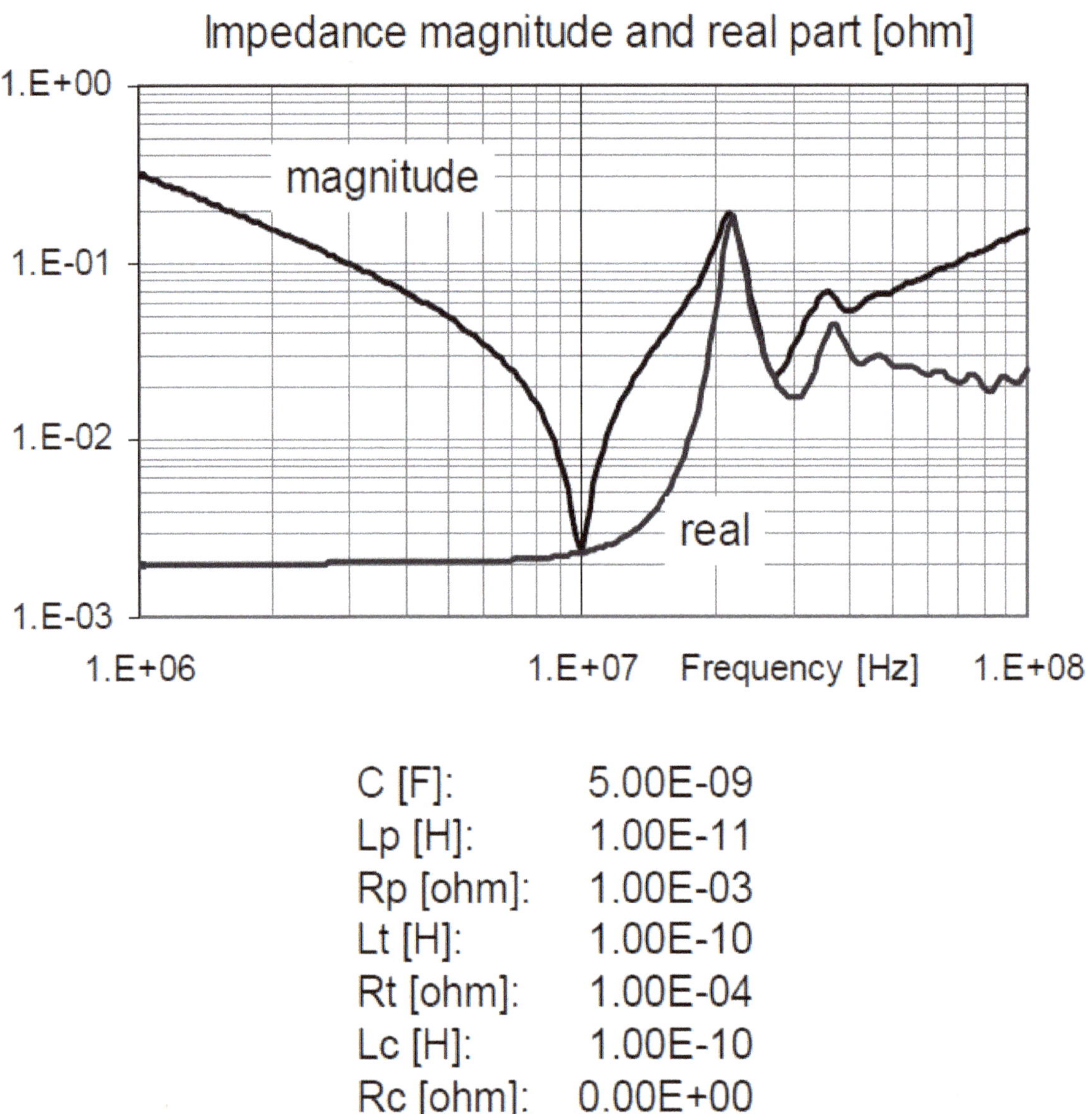

Figure 22.5: Simulated Impedance Magnitude and Impedance Real Part at the Capacitor Connections Terminals of the Circuit Shown in Figure 22.4. Circuit Parameter Values Are Listed below the Graph.

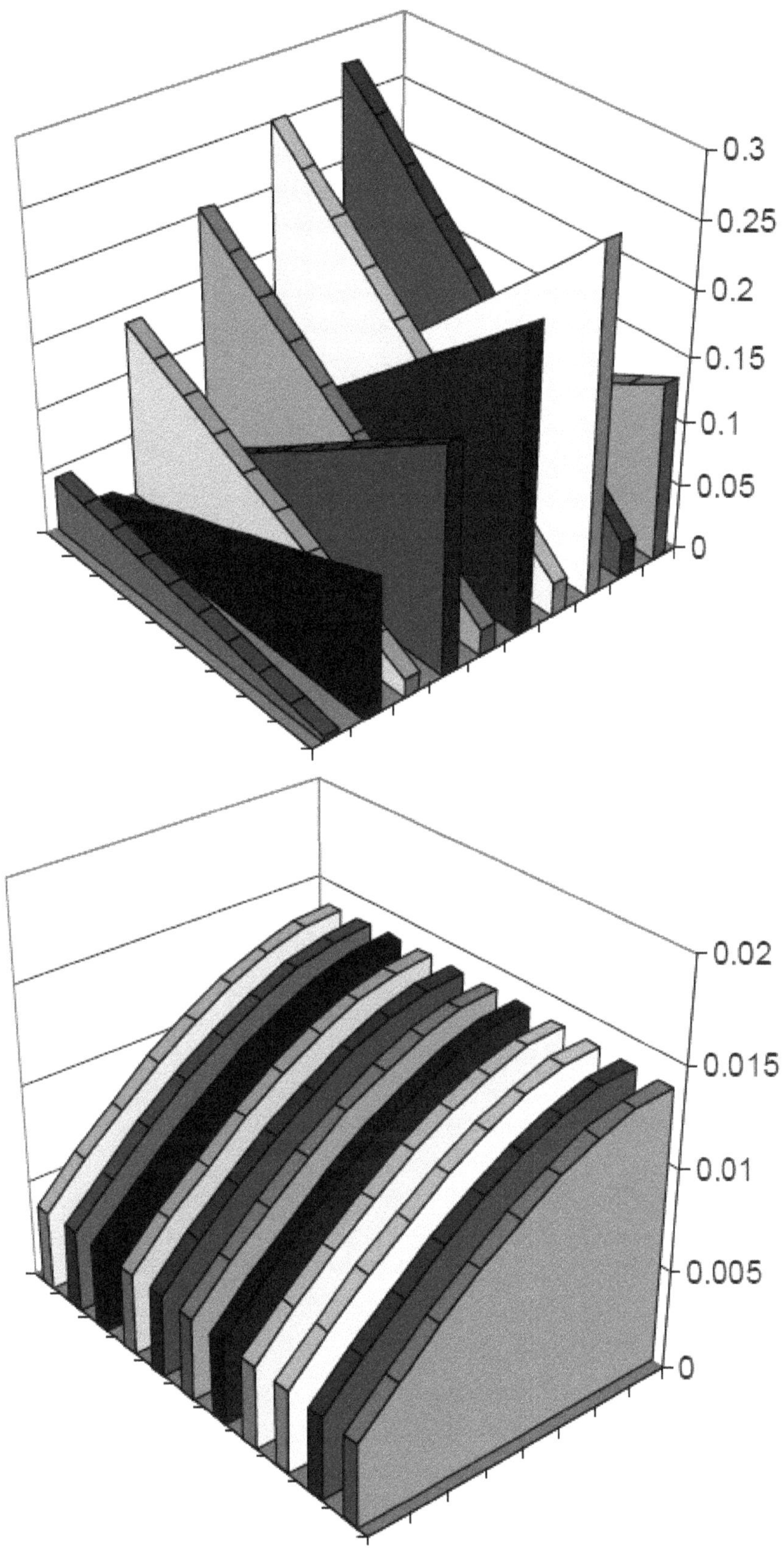

Figure 22.6: Current Distribution along the Capacitor Plates (Top) and Inside the Dielectrics (Bottom) at the 10 MHz SRF. Vertical Axis Unit on Both Plots Is Ampere

The black-box behavioral model introduced in [9] offers a unified model for bypass capacitors, simple enough to use many of them in frequency-domain PDN simulations. The model uses only three components, a series C-R-L circuit, but all three elements are frequency dependent. The frequency dependency of each element is captured and described by seven parameters, resulting in a total of 21 parameters for one model. The expressions capturing the frequency dependencies are based on the behavior of measured capacitor pieces. The parameters can be obtained either by manual or by semi-automatic curve fitting. However, the black-box model does not guarantee causal behavior, and because the three components are frequency dependent, it is not well suited for time domain simulations.

Periodically Loaded Causal Transmission-Line Model—The Unit-Cell Model

The slow-wave causal model is built on the realization that a multilayer ceramic capacitor is a periodically loaded lossy transmission line.

The unloaded transmission line is formed by the two vertical terminals of the capacitor by removing the capacitor plates but leaving the dielectric material in place. This vertically oriented unloaded transmission line in itself is already lossy: the terminals have finite resistance, and the dielectric material has a finite dielectric loss tangent. It is also known that causality dictates capacitance to change with the log of frequency in proportion to the dielectric loss tangent.

In an MLCC part, the large capacitance is achieved by inter-digitated capacitor plates attached alternating to the opposite terminals. These capacitor plates form a set of periodically arranged lossy transmission lines, attached orthogonally to the capacitor terminals. As will be shown, the multitude of capacitor plates will not only increase the total capacitance of the part, but also behave like a dielectric material with increased loss tangent and additional frequency dependency of capacitance. The effective loss tangent is a mix of the loss tangent of the original dielectric material and the resistive loss of the capacitor plates. The expectation is that if we properly assign the dimensions and material constants, or if we do a blind optimization of these parameters to match the measured behavior of a capacitor, all of the major features will be simultaneously captured without the need to change and optimize independently capacitance, resistance, and inductance values (which was the case in [9]). In addition, the model is based on the physical properties of the structure, and it guarantees that the model will be causal.

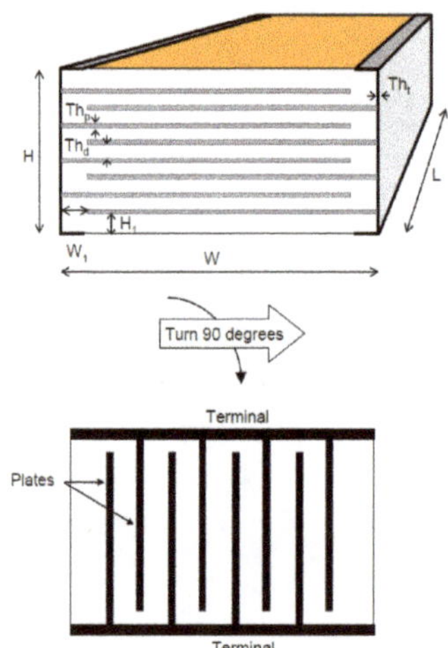

Figure 22.7: Top: Side View of MLCC with the Important Dimensions. Bottom: Representation of the Same MLCC by turning it 90 Degrees

The sketch on the top of Figure 22.7 defines the major dimensions of an MLCC, relevant to our calculations. We assume that the rectangular capacitor body is W wide, L long and H high. The illustration shows a reverse-geometry capacitor, because for the correlation we will use the measured data on the previous 10 μF 0508 MLCC example. The calculations and methodology, however, do not mandate a reverse-geometry capacitor. For other geometries, such as regular or inter-digitated capacitors, L, W, and H can be changed appropriately as needed.

To simplify the calculations, the capacitor cross-section is assumed to be symmetrical, both horizontally and vertically. We assume the same H_l cover thickness both on top and bottom. A similar horizontal symmetry assumes that capacitor plates are stopping W_l distance from the unconnected terminals at both sides. This symmetry is assumed here only for the sake of convenience; the procedure can be easily extended to different top and bottom cover thicknesses and/or for different end gaps at the left and right terminals.

The capacitor plate thickness is Th_p, and each dielectric layer between the plates is Th_d thick. The vertical capacitor terminals are assumed to be Th_t in thickness. We further assume that the dielectric material has ε_d dielectric constant and tan_δ loss tangent at a given working frequency, and their frequency dependent values are inter-related through the causal requirement [2]. The capacitor plates have a conductivity of σ_p, and the terminal material has a conductivity of σ_t.

The periodically loaded transmission-line model becomes apparent when we turn the capacitor sideways and distort the aspect ratio.

As shown on the bottom of Figure 22.7, the height (H) of the original capacitor body becomes the length of the transmission line, and the capacitor plates will represent a periodical loading along the transmission line.

The capacitance of the unloaded transmission line equals the capacitance of the capacitor body between the vertical terminals, without the capacitor plates:

$$C_0 = \varepsilon_0 \varepsilon_d H \frac{L}{W}$$

The propagation delay of the unloaded transmission line equals the propagation delay along the empty vertical capacitor body, with the capacitor plates removed but dielectric material in place:

$$t_{pd0} = \frac{H}{v} = H\sqrt{\varepsilon_0 \varepsilon_d \mu_0}$$

With C_0 and t_{pd0} known, we can calculate the characteristic impedance and the inductance of the unloaded vertical transmission line:

$$L_0 = \frac{t_{pd0}^2}{C_0} = \mu_0 H \frac{W}{L} \text{ and } Z_{00} = \sqrt{\frac{L_0}{C_0}} = \frac{120\pi}{\sqrt{\varepsilon_0}} \frac{W}{L}$$

The resistance of each terminal along its entire vertical length is:

$$R_t = \frac{1}{\sigma_t} \frac{1}{Th_t} \frac{H}{L}$$

From (1) through (4), we can calculate the parameters of the two end pieces simply by scaling the unloaded transmission line length by the ratio of H_1/H for each end piece:

$$t_{pd\_end} = t_{pd0}\frac{H_1}{H}$$

The end-piece transmission lines are denoted by suffix _end.

As shown on the top of Figure 22.8, adjacent capacitor plates along the terminal will create the periodical loading. The impedance of the entire capacitor is observed at the left end of the transmission line; the right end of the structure is open.

The loading is created by the lossy, open-ended transmission lines formed by adjacent plates. By following (1) through (4), we can calculate the parameters of a transmission line formed by adjacent capacitor plates. This transmission line is denoted by suffix -p.

$$C_p = \varepsilon_0\varepsilon_d L\frac{W - 2W_1}{Th_d}$$

$$t_{pdp} = \frac{W - 2W_1}{v} = (W - 2W_{1)}\sqrt{\varepsilon_0\varepsilon_d\mu_0}$$

$$L_p = \frac{t_{pdp}^2}{C_p} = \mu_0\frac{Th_d(W - 2W_1)}{L}$$

$$Z_{0p} = \sqrt{\frac{L_p}{C_p}} = \frac{120\pi}{\sqrt{\varepsilon_0}}\frac{Th_d}{L}$$

$$R_p = \frac{1}{\sigma_p}\frac{1}{Th_p}\frac{W - W_1}{L}$$

From N capacitor plates, we get N-1 pairs to create the periodical loading. For N>>1, we can approximate the number of cells in the periodically loaded structure with N.

The number of plates, the dielectric and plate thicknesses and the top/bottom cover thicknesses are inter-related through the following formula:

$$N = \frac{H - 2H_1}{Th_p + Th_d}$$

The conductive and dielectric losses of the transmission lines are, in theory, frequency dependent.

The δ skin depth in conductors is defined as:

$$\delta = \sqrt{\frac{1}{\pi f \sigma \mu}}$$

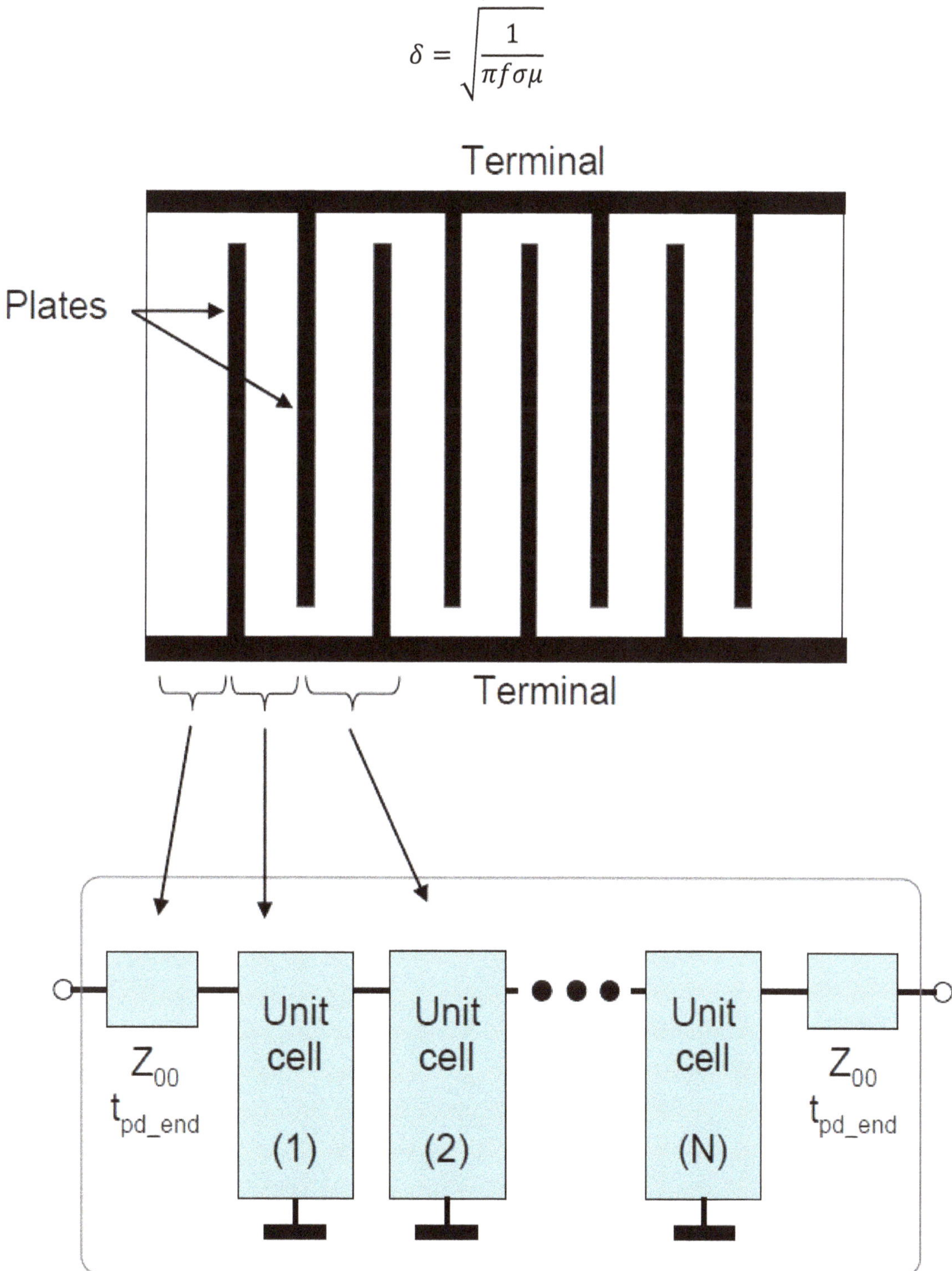

Figure 22.8: Slow-Wave Periodically Loaded Model of MLCC. The Periodically Loaded Transmission Line Is Broken Down into Symmetrical Unit Cells, Each Cell Representing a Length Equal to the Capacitor-Plate Pitch. The End Pieces, Corresponding to the Bottom and Top Dielectric Covers, Are Represented by Unloaded Transmission-Line Sections.

The skin depth in copper reaches 2 µm at 1 GHz frequency. To meet the requirements of the high ceramic firing temperature, the capacitor plates today use nickel, which has lower conductivity than that of copper but is ferromagnetic. The lower conductivity increases, the higher permeability decreases the skin depth. The resistance of individual capacitor plates is skin-depth-limited, and it is relatively frequency independent up to tens of MHz frequencies.

The terminals are usually much thicker than the capacitor plates and, therefore, their resistance and inductance may show somewhat more frequency dependency.

The dielectric losses are represented by a parallel conductance G in the transmission-line model:

$$G = 2\pi\, f\, C\, tan\_\delta$$

In the parallel conductance G equation, we can substitute the appropriate capacitance and loss tangent values for the unloaded transmission line of terminals or the lossy transmission line of capacitor plates.

This generic model links the geometry and material properties of an MLCC to a causal electrical model, which can be directly used to calculate the impedance of the capacitor. Although this model is still too complex to include in an actual circuit simulator in multiples of copies, it is very suitable for correlation purposes. We can use any of the computer math packages to obtain the input impedance of the periodically loaded transmission-line circuit, which represents the impedance of the capacitor.

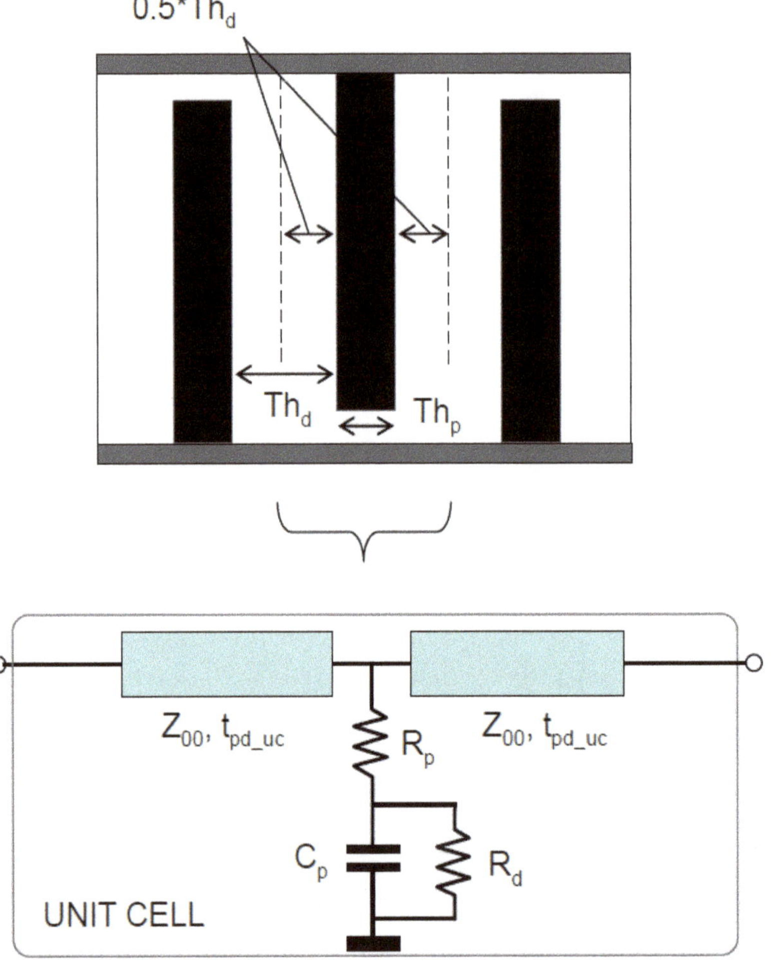

Figure 22.9: Generating Unit-Cell Parameters from the Geometry

The Lossy Transmission-Line Model

The model derived in Figure 22.7 through Figure 22.9 is generic, and as such, it is valid over a wide range of parameters. When we look at the actual geometry and resulting model numbers for a typical MLCC, we can achieve substantial simplifications without a major loss of accuracy.

Since we are interested in the input impedance of the structure with open at its end (on the top), the second end piece of unloaded transmission line can be totally neglected. Since it is open terminated on the right, only the static capacitance of the end piece matters. Not having capacitor plates in the end piece, for large N, its static capacitance is orders of magnitudes lower than the total capacitance, and therefore it can be rightfully ignored.

The end piece on the left is in series to the external connections, and therefore cannot be completely ignored. We can still, however, simplify the left end piece by using the earlier arguments, and neglect its parallel capacitance and conductance. This leaves us with its series inductance and resistance. These values can be obtained from (3) and (4), by substituting H_1 for H.

$$L_{0\_end} = \mu_0 H_1 \frac{W}{L}$$

$$R_{t\_end} = \frac{1}{\sigma_t} \frac{1}{L} \frac{H_1}{Th_t}$$

The unit cells can be simplified in a similar way. In fact, the capacitances and conductances of the series unloaded transmission line pieces can be ignored, leaving only a series L-R term. For one unit cell, using the suffix uc, and substituting Th_p + Th_d for H_1, we get:

$$L_{0\_uc} = \mu_0 W \frac{(Th_p + Th_d)}{L}$$

$$R_{t\_uc} = \frac{1}{\sigma_t} \frac{1}{L} \frac{Th_p + Th_d}{Th_d}$$

The open-ended loading transmission lines formed by adjacent capacitor plates can be simplified by neglecting the inductance. The C_p capacitance and R_p series resistance of the transmission line were already given in the C_p and R_p equations.

There is one more element that should not be ignored: the parallel conductance of the loading open-ended transmission line. It can be calculated from the G equation, by substituting the values for one plate pair:

$$G_p = 2\pi\, f\, C_p \tan\_\delta$$

These simplifications lead to the equivalent circuit of the unit cell shown in Figure 22.10. The shunt capacitance has its own G_p conductive loss term originated from the dielectric loss tangent and an R_p series resistive loss term originated from the resistance of the adjacent capacitor plates. At any given frequency, the series and parallel loss terms can be combined into a single term. The schematics on the right of Figure 22.10 shows a parallel equivalent, where G'_p represents a combination of R_p and G_p.

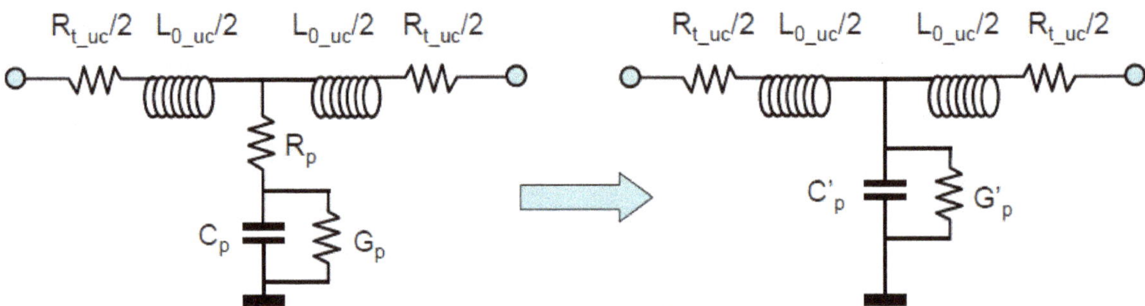

Figure 22.10: Equivalent Schematics of the Simplified Unit Cell. We Get the Circuit on the Left by Neglecting the Capacitance and Conductance of the Series Transmission Line and by Neglecting the Inductance of the Parallel Transmission Line. We Get the Circuit on the Right by Combining the Series and Parallel Loss Terms around the Shunt Capacitance

Let us note that during the transformation, in a general case, both the capacitance and conductance will change. Assuming that tan_δ is small, we get:

$$C'_p = \frac{C_p}{1 + (\frac{\omega}{\omega_p})^2} \text{ and } G'_p = G_p + \omega C'_p \frac{\omega}{\omega_p}$$

Where $\omega_p = 1/\tau_p$, and $\tau_p = R_p C_p$.

If $\tau_p << 1$, the formulas can be further simplified to:

$$C'_p \approx C_p \text{ and } G'_p \approx G_p + \omega C_p \frac{\omega}{\omega_p}$$

By combining the G_p and C'_p equations, we get:

$$G'_p = \omega C'_p \tan\_\delta + \omega C'_p \omega \tau_p$$

From C'_p and G equations, we can calculate an effective loss tangent for the lossy transmission line:

$$\tan\_\delta' = \frac{\tan\_\delta + \frac{\omega}{\omega_p}}{1 + (\frac{\omega}{\omega_p})^2}$$

In the final step, we can realize that the cascaded unit cells represent the ladder equivalent of a uniform lossy transmission line.

To obtain the per-unit-length transmission-line equivalent parameters, we multiply the unit-cell parameters by N. The inductance and the resistance are then simply the inductance and resistance of the (H-2*H_1) section of the vertical terminals. The capacitance will become approximately the full capacitance of the part itself, though as indicated by the C_p' and G_p' equations, it drops sharply above the ω_p corner frequency. Moreover, to obey causality, the capacitance also drops slightly with frequency, due to the finite loss tangent.

The parallel conductance can be calculated from the full capacitance and the virtual loss tangent. This eventually leads us to the simplified equivalent circuit of Figure 22.11. One should note that this simple circuit is causal, and it works on many time-domain and frequency-domain simulators.

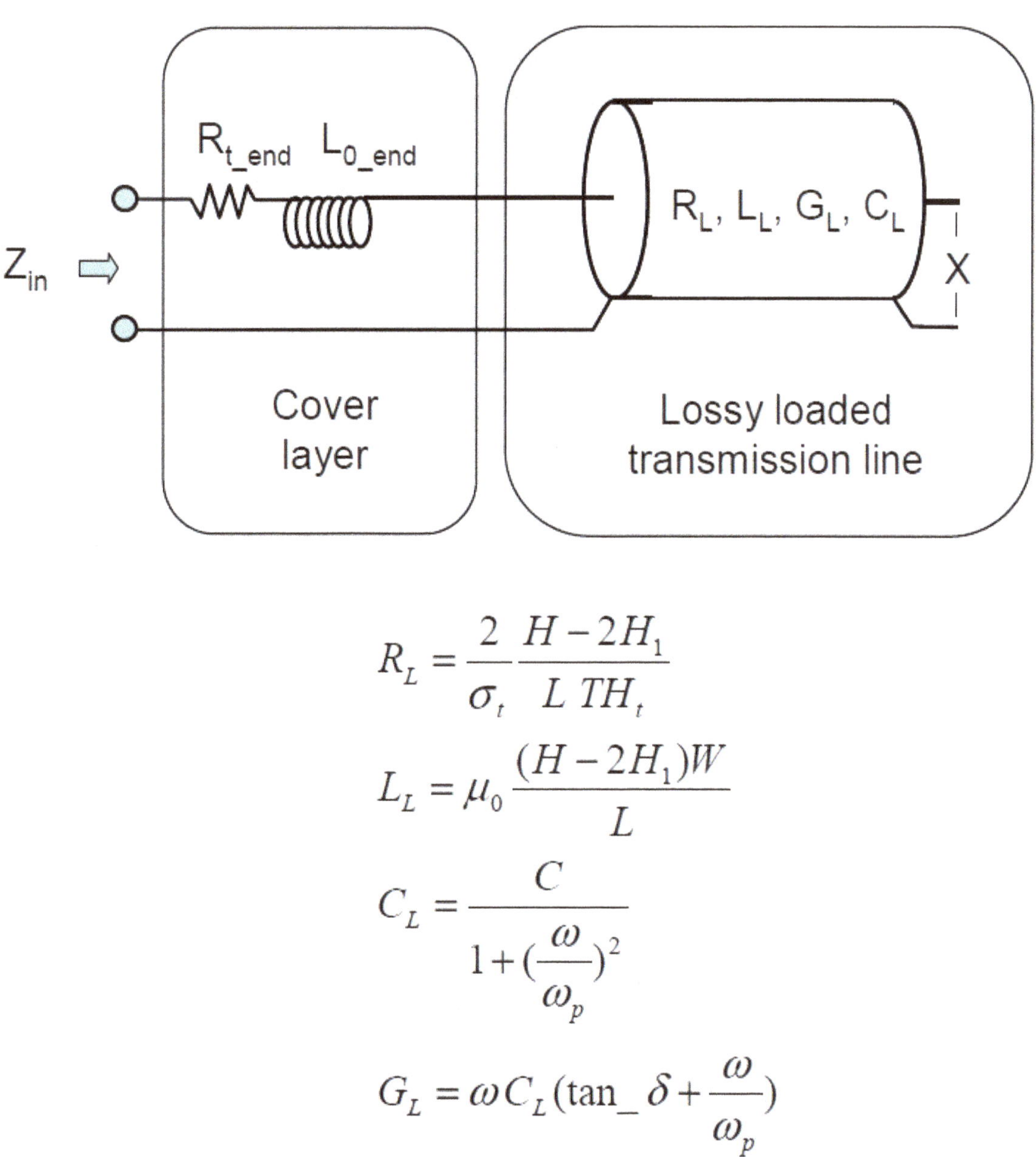

Figure 22.11: Simplified Causal Equivalent Circuit of a Multilayer Ceramic Capacitor

Let us note that the important aspect of this simplified model is not the lossy transmission line itself, but rather the unique way how it captures the convoluted effect of conductive and dielectric losses in the frequency-dependent capacitance and conductance per unit length.

Correlations

Now, we return to the example shown in Figure 22.3. The 10 μF 0508 MLCC part was measured in a small fixture, shown in Figure 22.12. The fixture has 22 layers. Capacitor sites are on top and bottom, connecting to 400x600 mil rectangular plane shapes further down in the stack. The horizontal layout of the site used for this test is shown on the top of Figure 22.12. The capacitor pads for the 0508 site are on the bottom side (layer 22), connecting to power planes on layers 20 and 21 with a set of blind vias.

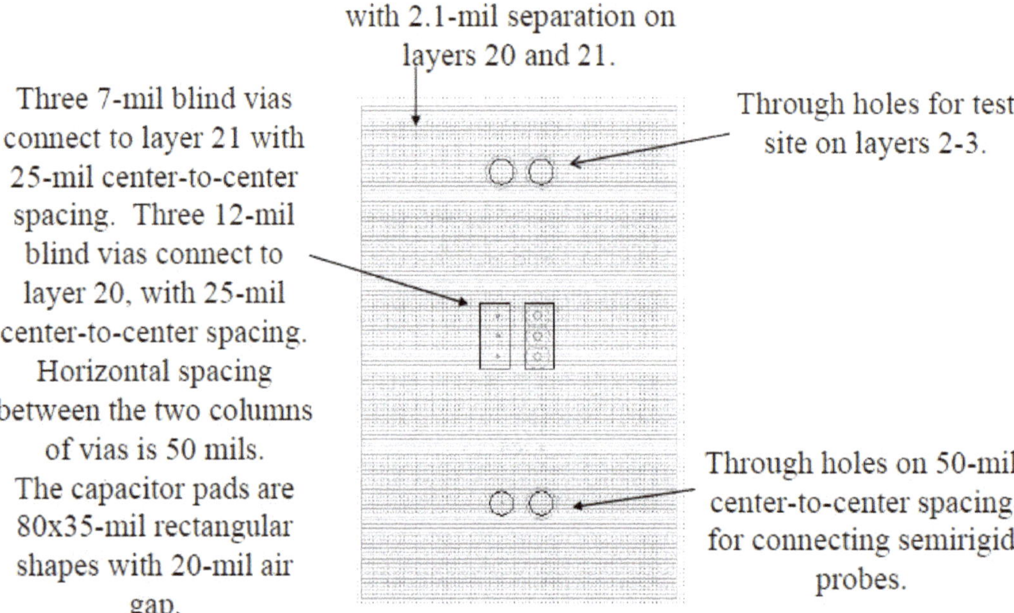
400x600 mil plane shapes with 2.1-mil separation on layers 20 and 21.
Three 7-mil blind vias connect to layer 21 with 25-mil center-to-center spacing. Three 12-mil blind vias connect to layer 20, with 25-mil center-to-center spacing. Horizontal spacing between the two columns of vias is 50 mils. The capacitor pads are 80x35-mil rectangular shapes with 20-mil air gap.
Through holes for test site on layers 2-3.
Through holes on 50-mil center-to-center spacing for connecting semirigid probes.

Figure 22.12: Photo of the Test Fixture with Capacitor and Probes (Top) and Geometry of Test Fixture for Measuring the 10uF 0508 MLCC Sample (Bottom)

First the test site was characterized with the capacitor pads open and shorted with two vector network analyzers (VNAs): 4395A in the 100 Hz to 10 MHz frequency range and 4396A in the 100 kHz to 1,800 MHz frequency range. The measured data was correlated to the lossy transmission-line model. Figure 22.13 shows the correlation after a brief manual optimization of parameters.

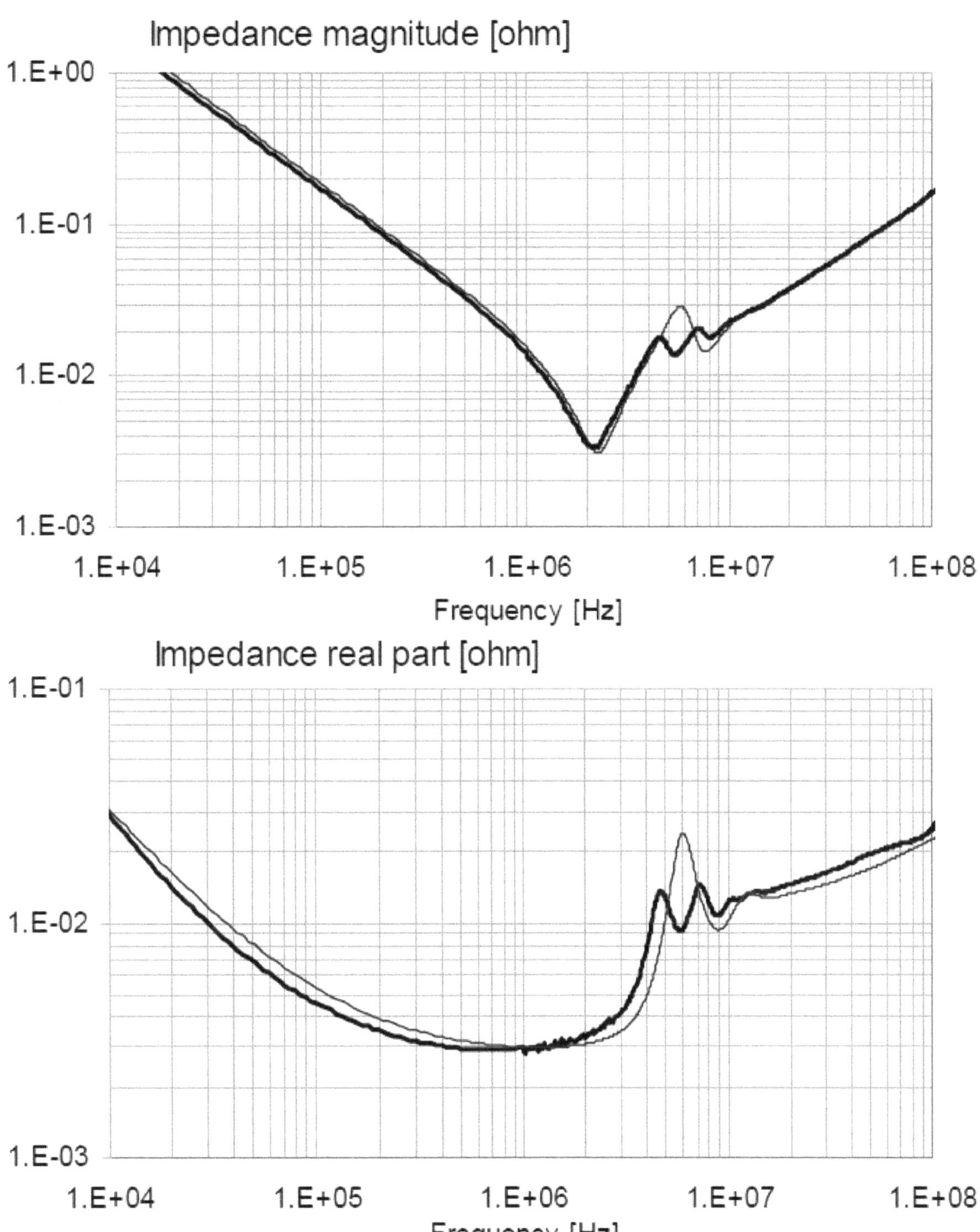

Figure 22.13: Correlation with Lossy Transmission-Line Model. Top: Impedance Magnitude; Bottom: Impedance Real Part

Conclusion

It was shown that a causal model can be constructed for MLCC, which can capture the primary and secondary resonances of the part. The model is based on physical parameters of the capacitors, but the exact knowledge of these parameters is not a must: the parameters can be obtained by fitting the model to measured data.

The parameters of cascaded unit cells can be combined into a single lossy, frequency- and parameter-dependent transmission-line equivalent circuit. The simplified model is suitable many frequency and time-domain simulators, and simple enough to be used in multiple copies in PDN simulations.

References

[1] István Novák and Jason R. Miller, *Frequency-Dependent Characterization of Bulk and Ceramic Bypass Capacitors, Proceedings of EPEP,* October 27-29, 2003, Princeton, New Jersey.

[2] Antonije R. Djordjevic et al., *Wideband Frequency-Domain Characterization of FR-4 and Time-Domain Causality*, IEEE. Transactions on ElectromagneticCompatibility, November 2001, p. 662.

[3] Michael J. Hill and Leigh Wojewoda, *Capacitor Parameter Extraction—Techniques and Challenges*, Intel Technology Symposium, Fall 2003.

[4] Larry Smith, *MLC Capacitor Parameters for Accurate Simulation Model*, in TF7 "Inductance of Bypass Capacitors; How to Define, How to Measure, How to Simulate," Proceedings of DesignCon 2005, January 31-February 3, 2005, Santa Clara, California.

[5] István Novák, *Frequency-Domain Power-Distribution Measurements—An Overview*, HP-TF1 TecForum, DesignCon East, June 23, 2003, Boston, Massachusetts.

[6] Hideki Ishida, *Measurement Method of ESL in JEITA and Equivalent Circuit of Polymer Tantalum Capacitors,* in TF7 *Inductance of Bypass Capacitors; How to Define, How to Measure, How to Simulate*, Proceedings of DesignCon 2005, January 31-February 3, 2005, Santa Clara, California.

[7] L. D. Smith, D. Hockanson, and K. Kothari, *A Transmission-Line Model for Ceramic Capacitors for CAD Tools Based on Measured Parameters*, Proceedings 52nd Electronic Components & Technology Conference, San Diego, California, May 2002, pp. 331-336.

[8] Charles R. Sullivan and Yuqin Sun, *Physically-Based Distributed Models for Multi-Layer Ceramic Capacitors*, Proceedings of EPEP2003, October 2003.

[9] István Novák, *A Black-Box Frequency Dependent Model of Capacitors for Frequency Domain Simulations*, DesignCon East 2005, September 19-22, 2005, Worcester, Massachusetts.

[10] http//home.att.net/istvan.novak/papers/Grid_sweep.zip

Authors

István Novák
https://www.linkedin.com/in/istvan-novak-865792/

Mark Alexander
https://www.linkedin.com/in/mark-alexander-san-francisco/

Bruce Archambeault
https://www.linkedin.com/in/bruce-archambeault-62414b45/

Dale Becker
https://www.linkedin.com/in/dale-becker-7262831/

Nick Biunno
https://www.linkedin.com/in/nick-biunno-303a7b4/

Gustavo Blando
https://www.linkedin.com/in/gustavo-blando-624488/

Chris Burket
https://www.linkedin.com/in/chris-burket-5871011a/

Chee-Yee Chung
https://www.linkedin.com/in/chee-chung-4054704/

Samuel Connor
https://www.linkedin.com/in/sam-connor-43025/

Valerie St. Cyr
https://www.linkedin.com/in/valerie-st-cyr-815b1911/

Tom Dagostino
https://www.linkedin.com/in/tom-dagostino-a2936114/

Erik S. Daniel
https://www.linkedin.com/in/erik-daniel-44188038/

Victor Drabkin
Mr. Drabkin received his B.S. and M.S. degrees in computers from the Moscow Institute of Electronic Machinery in Russia.

James L. Drewniak
https://www.linkedin.com/in/james-drewniak-04584011/

Brian P. Erisman
https://www.linkedin.com/in/brian-erisman-119a904/

Jun Fan
https://www.linkedin.com/in/jun-fan-49a2042/

Jonathan L. Fasig
https://www.linkedin.com/in/jonathan-fasig-b11b8527/

June Feng
Ms. Feng received her M.S. from University of California at Davis and her B.S. from Beijing University in China.

Barry K. Gilbert
https://engineering.purdue.edu/ECE/Alums/OECE/2005/gilbert.html

John Grebenkemper
https://www.linkedin.com/in/john-grebenkemper-8769355/

Christopher Houghton
https://www.linkedin.com/in/christopher-houghton-a45a525/

Hideki Ishida
Mr. Ishida was a leader of the working group of the measurement method of ESR and ESL in JEITA. The specification was published by JEITA in 2006 ("JEITA RC – 2002: Low ESR measurement method on surface mount capacitors for use in electrical and electronic equipment").

Isaac Kantorovich
https://www.linkedin.com/in/isaac-kantorovich-aaa57ab1/

Ravi Kaw
https://www.linkedin.com/in/ravikaw/

Joong-Ho Kim
https://www.linkedin.com/in/joong-ho-kim-732a248/

Woopoung Kim
https://www.linkedin.com/in/woopk/

James L. Knighten
https://www.linkedin.com/in/jim-knighten-85ab3911/

George Korony
https://www.linkedin.com/in/george-korony-348ba345/

Hai Lan
https://www.linkedin.com/in/hai-lan-7468939/

Chris Madden
https://www.linkedin.com/in/christopher-madden-4aa7071a9/

Scott McMorrow
https://www.linkedin.com/in/scott-mcmorrow-b784b21/

Jason R. Miller
He received his Ph.D. and M.S. in electrical engineering from Columbia University.

Al Neves
https://www.linkedin.com/in/alfred-p-neves-b7563a/

Leesa Noujeim
https://www.linkedin.com/in/leesa-noujeim-8802972/

Dan Oh
https://www.linkedin.com/in/kyung-suk-dan-oh-08b5b05/

Atul Patel
He received his degree in automobile engineering from Spartan Institute in India in 1984.

Richard Redl
https://www.linkedin.com/in/richard-redl-30580012/

Andrew P. Ritter
He has a B.A. in geology from Franklin and Marshall College, awarded in 1977.

Ralf Schmitt
https://www.linkedin.com/in/ralf-schmitt-062a009/

Giuseppe Selli
https://www.linkedin.com/in/giuseppeselli/

Larry D. Smith
https://www.linkedin.com/in/larry-smith-power-integrity/

Madhavan Swaminathan
https://www.linkedin.com/in/madhavan-swaminathan-446702/

Masaaki Togashi
Mr. Togashi has been awarded 33 U.S. patents for his work in a vast array of MLCC products. He has cowritten numerous technical papers and has presented at various technical forums in Japan.

Michael Tsuk
https://www.linkedin.com/in/michael-tsuk-a94268131/

Brian Vicich
https://www.linkedin.com/in/brian-vicich-2654116/

Alex Waizman
https://www.linkedin.com/in/alex-waizman-4657786/

Steve Weir
https://www.linkedin.com/in/steve-weir-5015155/

K. Barry A. Williams
He holds a bachelor's degree from Northeastern University.

Xingchao (Chuck) Yuan
https://www.linkedin.com/in/chuck-yuan-6840b57/

www.ingramcontent.com/pod-product-compliance
Lightning Source LLC
Chambersburg PA
CBHW051014240326
41458CB00145B/6456
* 9 7 8 1 9 4 9 2 6 7 6 7 9 *